METHODS OF WORKING COAL AND METAL MINES

VOLUME 1

Theory and Application of Rock Mechanics
to
Roof Control and Support Problems

METHODS OF WORKING COAL AND METAL MINES

VOLUME 1

Theory and Application of Rock Mechanics
to
Roof Control and Support Problems

BY

SETH D. WOODRUFF

Min. Eng., E.M.
Member American Institute of Mining, Metallurgical and Petroleum Engineers
Member American Society of Civil Engineers

PERGAMON PRESS
OXFORD · LONDON · EDINBURGH · NEW YORK
TORONTO · PARIS · FRANKFURT

Pergamon Press Ltd., Headington Hill Hall, Oxford
4 & 5 Fitzroy Square, London W.1
Pergamon Press (Scotland) Ltd., 2 & 3 Teviot Place, Edinburgh 1
Pergamon Press Inc., 44-01 21st Street, Long Island City, New York 11101
Pergamon of Canada, Ltd., 6 Adelaide Street East, Toronto, Ontario
Pergamon Press S.A.R.L., 24 rue des Écoles, Paris 5^{e}
Pergamon Press GmbH, Kaiserstrasse 75, Frankfurt-am-Main

First edition 1966

Library of Congress Catalog Card No. 65-28058

1727/66

CONTENTS

PREFACE

THIS book deals with the structural behavior of soils and rocks around mine workings and similar underground openings and with methods for designing such openings, and for stabilizing the ground around them.

The book has been written in order to fulfil the need for a work which condenses the knowledge and the experience in this field and presents it logically arranged so that it can be put to use by the student in the earth sciences and by the engineer who works in the field with soils and rocks.

Since the end of World War II technical advances in the mining industry have been rapid. Knowledge regarding the structural behavior of rocks around mine workings has been vastly increased through numerous theoretical investigations and through extensive investigations in the field. New devices and techniques have facilitated the measurement of pre-existing stresses in rocks and soils, and the measurement of stresses induced in these materials by the excavation of openings.

Techniques and equipment are being constantly improved and the writer believes that it will some day be commonplace to install rock strain measuring devices in mine working places to indicate just how safe the rock "structure" is at any given time, and to warn of any impending structural failure.

The material presented herein has been gathered from the experience of the writer and from the technical literature. Several thousand technical publications were perused in the course of writing the book.

This book describes the structural behavior of ground around sub-surface openings, explains the conditions which produce such behavior, and presents mathematical bases for predicting the nature of behavior under a given set of conditions.

The writer has attempted to present the basic principles in such a manner that the fundamental principles of natural structures in earth material can be visualized and grasped by the practical man in the field, and has also presented the basic mathematical relationships from which the student and the engineer can make predictions as to the stability of subsurface openings in various types of earth materials.

Many of the illustrations used herein have been borrowed from the periodical literature and are used by permission of the publishers of the said literature. A large number of the illustrations are taken from publications of the U.S. Bureau of Mines and are used by permission of that organization.

The sources of illustrations used herein are acknowledged by means of number references to a bibliography at the end of each chapter.

The author wishes to express his thanks to all those who have assisted him in obtaining material for this book.

SETH D. WOODRUFF

INTRODUCTION

A BRIEF HISTORY OF THE DEVELOPMENT OF THE SCIENCES OF SOILS MECHANICS AND ROCK MECHANICS

THIS volume is concerned primarily with the phases of soil mechanics and rock mechanics which treat with the structural behavior of these naturally occurring materials around mine workings and similar underground openings. The following brief account of the history of development of soil and rock mechanics may overlook some of the important contributions to these fields simply because these contributions do not apply to the behavior soils or rocks around underground openings.

DEVELOPMENT OF THE SCIENCE OF SOIL MECHANICS

The science dealing with the structural behavior of soils (soil mechanics) began a period of rapid development in the middle nineteen-twenties. Impetus was given to this development by the publication of Terzaghi's *Erdbaumechanik* in 1925. Prior to this time a number of individuals had contributed to various phases of the science, but Terzaghi's book was the first integrated work and was the most important single contribution to the science of soil mechanics.

At the end of this section is appended a bibliography listing in chronological order many works which touch upon the application of soil mechanics in determining the stability of sub-surface openings.

DEVELOPMENT OF THE SCIENCE OF ROCK MECHANICS

The science dealing with the structural behavior of rocks has developed much more slowly than the science of soil mechanics because of the difficulty involved in measuring the displacements and stresses in rock masses *in situ*. During the past century hundreds of thousands of miles of mine workings and hundreds of miles of civil works tunnels have been driven, yet until recently no satisfactory apparatus had been devised to measure displacements in the solid rock outside the boundaries of the accessible openings. Methods of measuring stresses in elastic transparent materials by photographic means which were developed in the late nineteen-twenties and the early nineteen-thirties also aided in analyzing stresses around simulated mine openings.

Availability of these tools, together with the necessity for increasing mining productivity and reducing costs, has resulted in accelerated research on the structural behavior of rocks. Such research began in earnest in the nineteen-thirties, was dropped during the war, but was resumed at a greatly accelerated pace after 1946.

Early Investigators

Early investigators in the field of rock mechanics were concerned chiefly with the surface subsidence accompanying mining operations. In 1825 a commission, appointed to investigate surface cracks in the vicinity of coal mines near Liège, concluded that there was no danger from subsidence over workings deeper than 300 ft. In 1838 Gounot proposed his theory of the law of the normal; that is, he suggested that planes normal to the dip of a bed limit the fracturing from an excavation. Investigations relating to surface subsidence carried on in France, Belgium, Germany, and Austria.

Probably the best known of the nineteenth-century investigators is Fayol, who conducted extensive laboratory tests as well as mine observations on mining subsidence and in 1885 published a summary of his findings and of the contradictory opinions which had *been advanced up until that time.*

Developments in the Nineteen-thirties

Not until the middle nineteen-thirties did the science of rock mechanics begin to crystallize. Among the most important of the recorded observations were those of Faulkner and Phillips, published in 1935, who recognized the fact that many of the fractures occurring in the rock in coal mines were not natural fractures but were the result of stresses induced by extraction operations. These fractures were termed "induced cleavage", and their occurrence was explained mathematically as the results of shear stresses induced by abutment pressures.

Probably the most important single theoretical contribution to the science of rock mechanics was the treatise by the German mining engineer Rudolf Fenner which was published in 1938. This remarkable mathematical work demonstrated the relationship between soil mechanics and rock mechanics; that is, it explained the structural behavior of a mass of ground of which a portion is granular material and a portion is unfractured elastic material.

Fenner's paper was published in 1938, just prior to the outbreak of the war, and did not become generally available in English language translations until some years after the war. Most of the theory dealing with rock mechanics has been developed in Europe and is based upon the principles outlined by Fenner.

Another important contribution was the paper by Evans, published in 1941. He demonstrated experimentally, and explained quantitatively, the "voussoir beam" or "voussoir arch" effect which enables a fractured rock to support itself around an opening.

Developments since 1945

Since the end of the World War II increasing amount of effort has been devoted to research in the field of rock mechanics. A number of private and governmental agencies support such research. These include governmental agencies in the field of mining such as the U.S. Bureau of Mines in America, the National Coal Board in Great Britain, and similar governmental bureaus in other coal-producing countries. Also interested in rock mechanics are those governmental agencies which are concerned with the construction of civil works projects, such as dams and tunnels.

The work is also carried on by colleges and universities, and research staffs are supported by some of the larger mining companies. With so many men working in this field, it is not possible to list here all those who have made important contributions to the advancement of the field, therefore the writer lists only those contributions which he considers to be "highlights".

References to numerous other important papers will be found in the bibliography at the end of each chapter of the text.

Among the important contributions to a knowledge of stress distribution are two by Duvall, published in 1948, which deal with stress analysis applied to underground mining problems. These deal with stresses around openings in elastic material. Other papers dealing more thoroughly with the same subject were published by Panek in 1951, Terzaghi and Richart in 1952, and Caudle and Clark in 1955.

Among the most important contributions to the theory of rock behavior under stress are those by Labasse, which were published during the years 1949–52. These deal with the mechanics of those rocks commonly encountered in the coalfields of Europe, and include a comprehensive treatment of the relationship between natural rock stresses and the pressure on mine-support systems.

Successful research in the field of rock mechanics requires the coordinated effort of a number of individuals and the coordination and analysis of a great many measurements. Such a program was carried out at the Neumuhl Colliery in Germany in connection with the extraction of a panel of coal from underneath critical surface structures. The results were published in a series of papers in 1956 and have been summarized by Jacobi.

Similar campaigns of measurements were carried out by Tincelin and Sinou in the iron mines of Lorraine, France, and the results have been reported at intervals in various technical publications, with the conclusions published in 1958.

Research carried on in the deep gold mines of the Witwatersrand, South Africa, has led to the discovery of a method for de-stressing the rock to reduce the frequency and severity of rock bursts. The results of the research were reported by Roux, Leeman, and Denkhaus in 1957.

In Sweden a method of measuring the stresses in rock *in situ* has been perfected by Hast. The method employs magnetostrictive gages which are installed in drilled holes. Deformation of the elements after the gages are cored out of the rock changes the magnetic properties and allows deformations to be accurately measured. Results of measurements in some Swedish mines were published in 1957.

Lane reported, in 1957, the results of an extensive series of tests and measurements carried out in test sections of the Garrison Dam Tunnels in North Dakota. These measurements gave important information as to pressures and deformation of steel supports, concrete lining, and movement in the surrounding rock.

Roberts, Emery, Chakravarty, Hawkes, and Williams published in 1962 a progress report outlining the development of photoelastic coating techniques for determining strains and stresses in rocks and supporting structures. This is a relatively simple technique and promises to make the measurement of rock strains a routine mine procedure.

The foregoing has presented what the author considers to be some of the highlights in the development of the mechanics of soils and rocks. There are, of course, a larger number of other important contributions, most of which will be found in bibliographies at the end of each chapter.

Two bibliographies are presented hereinafter. Each is arranged chronologically. The first lists some of the important contributions to the field dealing with soils and subsurface structures. The second lists those which deal with rocks and their relationships to subsurface structures.

HISTORICAL BIBLIOGRAPHY

SOIL PRESSURES ON SUBSURFACE STRUCTURES

1. Culmann, *Die graphische Statik*, Zürich, 1866.
2. Ritter, *Die Statik der Tunnelgewolbe*, Berlin, 1879.
3. Engesser, Über dem Erddruck gegen innere Stutzwande, *Deut BauZtg*. 1882, p. 92.
4. Janssen, Versuche über Getreidedruck in Silozellen, *Zentr. Bauv*. 1896, p. 446.
5. Praute, Messungen des Getreidedrucks gegen Silowandungen, *Z. Ver. deut. Ing*., 1896, p. 1122.
6. Leon, A. and Wilheim, F., Über die Zerstorungen in tunnelartig gelochten Gesteinen, *Oesterr. Wochenschr. Offent*., Baudienst, 1910 and 1912.
7. Wilmann, V., *Über einige Gebirgsdruckerscheinungen und ihren Beziehungen zum Tunnelbau*, Leipzig, 1911.
8. Heim, Zur Frage des Gebirges — und Gesteinfestigkeit, *S.B.Z.*, 24 February 1912.
9. Bierbaumer, *Die Dimensionierung des Tunnelmauerwerks*, Engelmann, Berlin, 1913.
10. Terzaghi, K., *Erdbaumechanik*, 1925, pp. 211 and 215.
11. Schmid, H., *Statische Probleme des Tunnel- und Druckstollenbaues und ihre gegenseitige Beziehung*, Springer, 1926.
12. Ros, M. and Eischinger, A., Versuche zur Klarung der Frage der Bruchgefahr, *Diskussionbericht* No. 28, June, 1928.
13. Yamaguti, N., On the stresses around a horizontal circular hole in gravitating elastic solid, *J. Civil Eng., Japan*, No. 40, 1929.
14. Rendulic, Spannungszustand in der Umgebung eines Hohlraumes, *Wasserwirtschaft*, Nos. 18 and 19, 1934.
15. Vollmy, A. Die Bruchsicherheit einzebetteter Rohre, *Diskussionsbericht* Nr. 35, E.T.H., 1937.
16. Fenner, R., Untersuchungen zur Erkenntnis des Gebirgsdrucks, *Glückauf*, Nos. 32 and 33, 1938.
17. Mindlin, R. D., Stress distribution around a tunnel, *Proc. Am. Soc. Civil Engrs*. Vol. 65, pp. 619–642, 1939.
18. Bussman, K. H. and Stocke, K., Modelversuche zur Klarung der Spannungsverteilung in der Umgebung von Strecken im Gebirge, *Ber. Staatl. Materialpruf*. Berlin-Dahlem. Wiss. Abt. dtsch. Materialprufung-Anst., 1. Folge, Heft 3, 1939.
19. Kommerel, O.: *Statischeberechnung von Tunnelmauerwerk*, Ernst und Sohn, Berlin, 1940.
20. Wiedemann, K.: *Neuere Anwendung der Unterfangungsbauweise im Tunnel und Stollenbau*, Ernst und Sohn, Berlin, 1940.
21. Earth pressure and shearing resistance of plastic clay — a Symposium, *Proc. Am. Soc. Civil Engrs*., Vol. 108, 1943, contains the following contributions:
    Terzaghi, K., Liner-plate tunnels in the Chicago (Ill.) Subway, pp. 862–899.
    Peck, R. B., Earth pressure movements in open cuts. Chicago (Ill)., Subway, pp. 900–928.
    Housel, W. S.: Earth pressure on tunnels, pp. 929–950.
22. Bendel, *Ingénieur Géologie*, Springer, Vienna, 1943.
23. Terzaghi, K., *Theoretical Soil Mechanics*, John Wiley, New York, 1943.
24. Krynine, D. P., *Soil Mechanics*, 2nd ed., McGraw-Hill, New York, 1947.
25. Caquot, A. and Kerisel, J., *Traité de mécanique des sols*, Gauthier-Villars, Paris, 1956.

ROCK MECHANICS

26. Fayol, M., Sur les mouvements de terrain provoqués par l'exploitation des mines, *Bull. de la soc. de l'Industrie Min*., 2nd series, Vol. 14, 1885.
27. Faulkner, R. and Phillips, D. W., Cleavage induced by mining, *Trans. Institution of Mining Engineers*, vol. 89, 1934–1935, pp. 264–288.

28. FENNER, R., Untersuchungen zur Erkenntnis des Gebirgsdrucks, *Glückauf*, Vol. 74, No. 32, August 13, 1938, pp. 681-695; Vol. 74, No. 33, August 20, 1938, pp. 705–715.
29. VAN ITERSON, F. K. TH., La pression du toit sur le charbon près du front dans les exploitations par tailles chaussantes, *Revue Universelle des Mines*, September, 1941, pp. 241–253.
30. EVANS, W. H., Strength of undermined strata, *Trans. Institution of Mining and Metallurgy*, vol. 50, 1941, pp. 475–532.
31. DUVALL, W. I., Stress analysis applied to underground mining problems. Part 1. Single openings, *U.S. Bureau of Mines R.I.* 4192, March, 1948.
32. DUVALL, W. I., Stress analysis applied to underground mining problems. Part 2. Stress analysis applied to multiple openings and pillars, *U.S. Bureau of Mines R.I.* 4387, November, 1948.
33. LABASSE, H., Les pressions de terrains dans les mines de houilles. *Revue Universelle des Mines*. Part I. Le terrain houiller, January, 1949, pp. 3-15; Part II. Les pressions de terrains autour des poits, March, 1949, pp. 78-88; Part III. Les pressions de terrains autour des travers-bancs horizontaux, January, 1950, pp. 3-18; Part IV. Les pressions de terrains autour des galeries en couche, February, 1950, pp. 41–44; Part V. Les pressions de terrains autour d'une taille chassante; la fissuration préalable, March, 1951, pp. 85–91; Part VI. Les pressions de terrains autour d'une taille chassante; l'influence des exploitations à distance, March, 1951, pp. 91–106; Part VII. La mécanique des bancs détendu, July, 1952, pp. 251-264.
34. PANEK, L. A., *Stresses about Mine Openings in a Homogeneous Rock Body*, New York, 1951.
35. TERZAGHI, K. and RICHART, F. E., Stresses in rock about cavities, *Geotechnique*, Vol. 3, June 1952, pp. 57–90. (Also Paper No. 41, Harvard Soil Mechanics Series.)
36. CAUDLE, R. D. and CLARK, G. B. Stresses around mine openings in some simple geologic structures, *University of Illinois Engineering Experiment Station Bulletin*, No. 430, Vol. 42, No. 69, May, 1952.
37. JACOBI, O., *The Pressures On Seam and Goaf*, paper delivered at International Strata Control Congress, Essen, October, 1956. Published by Steinkohlenbergbauverein, Essen, 1956. (*See also Glückauf*, 26, March 1960, pp. 409–418).
38. LANE, K. S., Garrison Dam — Evaluation of results from tunnel tests section, *A. S. C. E. Proceedings*, Vol. 83, No. SM 4, November 1957 (SM Paper No. 1439).
39. ROUX, A. J. A., LEEMAN, E. R. and DENKHAUS, H. G., De-stressing: a means of ameliorating rockburst conditions. Part I. The conception of de-stressing and the results obtained from its application.
40. HILL, F. G. and PLEWMAN, R. P., Part II. Implementing de-stressing with a Discussion on the results so far obtained.
Journal of the South African Institute of Mining and Metallurgy, October 1957, pp. 101–127.
41. HAST, N., Bergtrycksmatningar i gruvor, *Jernkontorets Annaler* (Sweden), Vol. 141, No 10, 1957, pp. 633–752.
42. TINCELIN, E. and SINOU, P., *Summary of Results Obtained from Eight Years Research in Strata Control*, Paper delivered at the International Strata Control Congress, Leipzig, October, 1958. Sponsored by Deutsche Akademie der Wissenschaften zu Berlin, Sektion für Bergbau.
43. TALOBRE, J., *La Mecanique des Roches Appliquée Aux Travaux Publics*, Dunod, Paris, 1957.
44. ISAACSON, E. DE ST. Q., *Rock Pressure in Mines*, Mining Publications Ltd., London, 1958.
45. OBERT, L., DUVALL, W. I., and MERRILL, R. H., Design of underground openings in competent rock, *U.S. Bureau of Mines Bulletin* 587, 1960.
46. Photoelastic Coating technique applied to research in rock mechanics, *Transactions of the Institution of Mining and Metallurgy*, Vol. 71, Part 10, 1961–62. *Also* Bulletin No. 668, July 1962., pp. 581–617.
ROBERTS, A., EMERY, C. L. and CHAKRAVARTY, P. K., Part 1. Laboratory Investigations.
ROBERTS, A., EMERY, C. L., HAWKES, I., CHAKRAVARTY, P. K. and WILLIAMS, F. T., Part 2. Field Investigations
47. ROBERTS, A. and HAWKES, I., The application of photoelastic devices for measuring strata pressure and support loads, *Mine and Quarry Engineering*, July 1963, pp. 298–308.
48. MULLER, L., *Der Felsbau* Vol. I, Ferdinand Enke Verlag, Stuttgart, Germany, 1963.

PART A

The Influence of Geological Conditions

CHAPTER 1

TYPES OF ROCK

ROCKS are divided into three broad classifications:

Sedimentary: Consolidated.
Unconsolidated.

Igneous.

Metamorphic.

SEDIMENTARY ROCKS

Consolidated Sediments

Sedimentary rocks include all those whose constituents were deposited from suspension or from solution in water. Shale, mudstone, sandstone, conglomerate, are composed respectively of clay; clay and silt; sand; sand, gravel, and boulders, transported by streams and generally deposited along the lower reaches of streams where the streams spread out on flood plains or where they discharged into large quiet bodies of water such as lakes or seas. The solid particles settled, forming layers on the bottom. As successively more layers were deposited the originals became buried and the pressure of overlying beds, plus heat, plus cementing materials deposited from percolating waters, consolidated the clay beds into shales, the silty clays into mudstones, the sands into sandstone, and the mixed sand, gravel, and boulders into conglomerate.

Degrees of hardness and strength vary greatly in these rocks. A shale may consist of clay beds pressed by only a few hundred feet of overburden so that they cohere in layers, or shale may have been buried under thousands of feet of rock and be almost as hard as slate. Similarly, sandstones from different geological formations, or even from different parts of the same formation, may vary from a loosely cemented crumbly sandstone to one which is tightly cemented and almost as hard as quartzite. Conglomerates may vary similarly, depending upon their age and geologic history.

Limestones are formed through the extraction of calcium carbonate from sea water by marine life, or by the precipitation of calcium carbonate from sea water. Limestone usually occurs as a thick-bedded homogeneous rock. Because of its strength it is one of the best rocks in which to excavate underground openings.

The largest part of the world's mineral production comes from the sedimentary rocks. Coal itself is a sedimentary rock, and coal mining comprises by far the largest volume of mineral production (excluding petroleum) which is extracted from sedimentary rocks.

The principal metalliferous deposits found in sedimentary rocks are the typical lead–zinc deposits in limestone and the uranium deposits in sandstone such as those in the Colorado Plateau, as well as some bedded iron ore deposits such as those in Alabama, and some iron ores in France.

Unconsolidated Sediments

These blanket the earth, covering the bedrock, usually with a few feet, or a few tens of feet, but in some cases to depths of thousands of feet. These sediments consist of organic material, clay, silt, sand, gravel, boulders. Most such unconsolidated material has been transported to its present location by flowing water. A small proportion, such as loess and dune sand, has been transported by the wind and some has migrated short distances by the process called "soil creep" – that is, movement down slope under the action of frost, rain, and the activity of vegetable and animal life. Some material only recently weathered from bedrock is *in situ* – that is, it still remains at the location where it formed.

Since unconsolidated materials blanket most of the earth it is necessary to pass through this blanket when it is desired to construct structures in the underlying "bedrock". Many civil works structures are constructed entirely within the unconsolidated material. Typical of these are the subway tunnels in some of the world's great cities. In Chicago and London these are located principally in clay beds. Since unconsolidated materials do not possess any appreciable structural strength, excavations within them usually require support with a tight lining. Concrete is almost invariably chosen for the interior permanent lining except that segmented cast iron linings are frequently used in sub-aqueous tunnels.

In selecting sites for underground factories or storage facilities, the unconsolidated sediments are the least satisfactory site choices. However, if a central location is of paramount importance (for instance bomb shelters in the heart of a city) and cost is secondary then it may be desirable to locate shelters in these materials in spite of the support problems involved.

IGNEOUS ROCKS

Igneous rocks are those which have formed by cooling of a molten magma. They include intrusives and extrusives.

Intrusives

Intrusive igneous rocks are those which have forced their way upward into previously consolidated rocks and have eventually cooled to form a solid granular mass. Most common of the intrusive rocks exposed on the Earth's surface are the granites and the related diorites. Other fairly common intrusives are monzonites, diabases, and porphyries.

Since intrusive rocks solidified under a thick cover which acted as an insulating blanket, they cooled slowly and each of the constituent minerals has solidified as a crystal of appreciable size. This is illustrated by the texture of a typical granite; each of the visible grains is a mineral crystal. The light colored grains are crystals of quartz or feldspar while the dark colored are biotite (mica) or one of the ferro-magnesian minerals. Because the crystals had limited space in which to grow as the melt cooled and interfered with each other, perfectly formed crystals are rare. Nearly always the boundaries of a mineral grain are rough and irregular, interlocking with its neighboring crystals.

Extrusives

Extrusive igneous rocks formed by the extrusion, flow, or ejection of the same molten magmas from the interior on to the Earth's surface. In this manner they cooled and solidified rapidly. As a result individual minerals composing the rock have not had an opportunity to form large crystals before solidifying and these rocks have a fine grained, sometimes glassy, appearance. Individual crystals can usually be seen only under a microscope. The most common extrusive igneous rock is basalt which is the extrusive equivalent of a diabase (intrusive). Other fairly common extrusives are rhyolite (extrusive equivalent of a granite), trachyte, obsidian. Other types, such as volcanic lavas, occur less widely distributed.

Igneous rocks are host rocks for certain types of metalliferous deposits, principally copper and gold. The copper deposits of Butte, Montana, which occur in a diorite, and the copper deposits of Bingham Canyon, Utah, which occur in an intrusive porphyry, are typical of deposits in igneous intrusives. Generally the extrusives do not make good host rocks although the copper deposits of Michigan and the nickel ores of Quebec occur partially in metamorphosed volcanic extrusives.

Igneous intrusives, particularly granites, are favorable locations for underground structures because of their strength and massive homogeneous structure. This is illustrated by experience in Sweden where underground hydroelectric plants, factories, shelters, and even garages are excavated from granite. Because of the strength of the rock and the fact that it has sustained little fracturing, very little support is necessary and construction costs are very low.

Igneous extrusives are not generally favorable locations for subsurface structures. They have usually been extruded in relatively thin layers (25 ft or less) and as each layer cools shrinkage produces contraction joints. These form planes of weakness in addition to those formed by the surface between successive flows. In addition, the extrusives, particularly the basalts, being high in ferro-magnesian minerals, are easily weathered. Percolating water causes weathering along joints which may extend to considerable depths.

METAMORPHIC ROCKS

Metamorphic rocks are formed by the alteration of pre-existing rocks through heat, pressure, or chemical action. They may be created by the alteration of any pre-existing sedimentary, igneous, or metamorphic rock. Some common metamorphic rocks and the rocks from which they were derived are listed below:

Metamorphic Rock	*Original Rock*
Slate	Shale
Quartzite	Sandstone
Argillite	Mudstone
Marble	Limestone
Gneiss	Granite
Schist	Slate
Amphibolite	Basalt

The largest proportion of the world's metalliferous ores are extracted from the metamorphic rocks. Quartzite is particularly favorable as a host rock, because of its chemical composition and because of its brittleness and tendency to shatter when fractured. The shattering leaves open channels through which ore bearing solutions can migrate. Lead, zinc, copper and gold are found in metamorphic rocks. Typical mining districts are the Coeur d'Alene in Idaho, and the Mother Lode in California.

As a class the metamorphic rocks are more unpredictable as to subsurface conditions and amenability to excavation of underground workings. This is due, partially, to the fact that the same processes which produced the metamorphism also have produced faults, shear zones, cleavage, brecciation and zones of hydrothermal alteration. Aside from this is the uncertainty as to the degree of metamorphism of a rock mass from place to place. For instance at one location a schist may contain a high proportion of mica and be so soft as to be difficult to support, while at another location, a short distance away, it may contain little mica and be relatively hard.

Of the metamorphic rocks quartzite is one of the strongest. Compressive strengths of 15,000–30,000 psi are typical of quartzites. All of the metamorphic rocks, having been subjected to heat and pressure, are quite stable chemically except for the schists and gneisses. In these mica is an important constituent and affects the structural strength as well as resistance to weathering. Most of the other metamorphic rocks offer favorable locations for location of underground facilities providing they have not been badly fractured by tectonic stresses.

CHAPTER 2

STRUCTURAL DEFECTS IN ROCK

FRACTURES

The principal structural weaknesses in rocks are those planes along which fracture has occurred or planes of weakness along which fracture can readily take place.

SHEAR PLANES

Shear planes are the most common of the defects in rock. These vary in magnitude from single planes along which only microscopic movement has taken place, to fault zones which are hundreds of feet in width and consist of an infinite number of shear planes. The aggregate movement on each side of such a fault may amount to several tens of miles. In such large zones the rock in the center of the zone has been

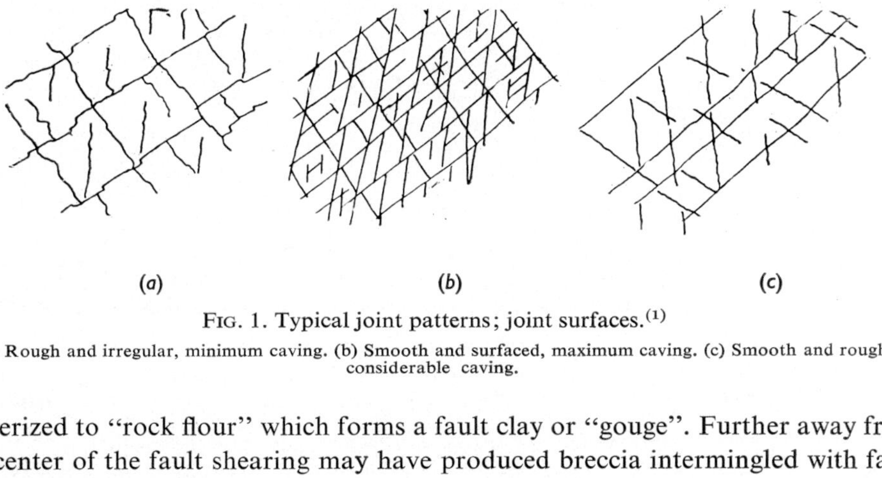

(a) (b) (c)

FIG. 1. Typical joint patterns; joint surfaces.[1]

a) Rough and irregular, minimum caving. (b) Smooth and surfaced, maximum caving. (c) Smooth and rough, considerable caving.

pulverized to "rock flour" which forms a fault clay or "gouge". Further away from the center of the fault shearing may have produced breccia intermingled with fault clay, and still further away the shear planes will be spaced further apart, and divide the rock into large "slices" or into angular blocks. The shear planes which mark the dividing line between intensive sheared rock (gouge) and relatively sound rock are called the walls of a fault. If a fault dips (is not vertical) then the underlying wall is the "foot wall" of the fault and the overlying wall is the "hanging wall". In some faults the dividing line between intense shearing and sound rock is sharply delineated and the walls are easily identified. In others the intensity of shearing grades outward from the center and no sharp dividing line exists. In such case the designation of locations of walls will be arbitrary. The whole zone of shearing is designated as a *fault zone.* In some cases rocks are sheared without any detectable concentration of the shear planes; that is, they are uniformly spaced a few inches apart. In such case the zone is designated as a "shear zone".

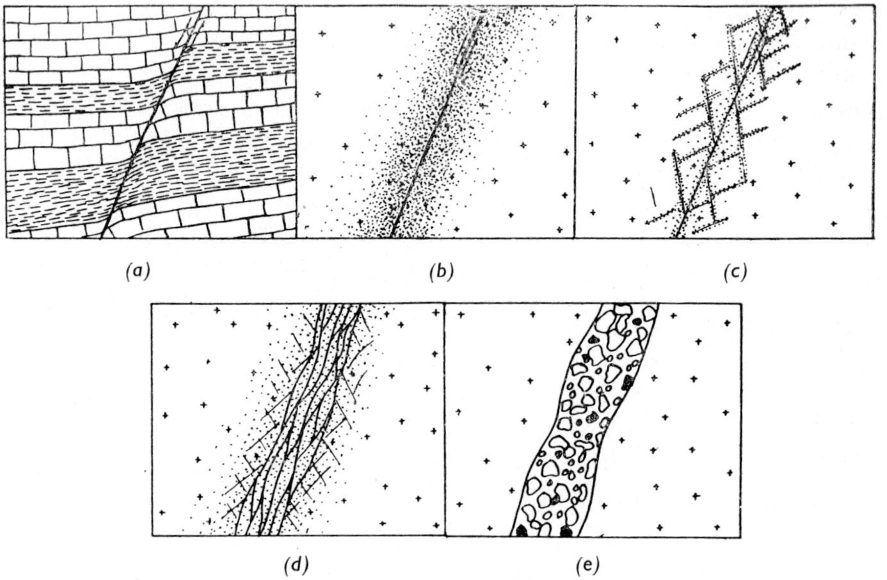

FIG. 2. Types of faults.[1]

(a) Clean sharp break, no alteration of wall rocks. (b) Sharply defined, no subsidiary fractures, pervasive alteration. (c) Sharply defined, with subsidiary fractures, localized alteration. (d) Braided, with subsidiary fractures, general alteration. (e) Open, porous, filled with breccia.

The nature of fault "filling" is determined by the type of rocks which are sheared. In quartzite, which is a brittle rock, shearing produces a mixture of angular sand and breccia fragments through which water can readily percolate. The result is similar in brittle sandstones. Igneous rocks usually produce a mixture of fault clay and angular fragments, while shales, mudstones, siltstones, argillaceous sandstones, and schists usually result in a fault zone filled principally with clay containing some angular fragments. This mixture ordinarily forms a barrier which is almost completely impervious to water. Types of faults are shown in Fig. 2.

FOLDING

If folding in sedimentary rocks is intense (if folds are sharp) considerable shearing movement takes place between individual beds and beds may be cut by cleavage, thus creating angular fragments. Generally the softer rocks such as shales, schists, and slates bend to form folds without breaking while the harder ones subjected to similar deformation shatter and crush. The strong, brittle rocks are known as "competent" rocks while the weak, plastic rocks are called "incompetent".

JOINTS

Joints may be formed as tension fractures during the cooling of an igneous rock or may be formed in other types of rocks as the result of bending or as the result of release of high compressive stresses. Most rocks are cut by three systems of joint planes, approximately at right angles to each other. In even the most massive and intact rocks, such as the granite selected for quarrying, joints are normally encountered about every 30–40 ft. However, in intrusive igneous rocks the joint planes tend to be rough and to interlock and most such rocks will not be seriously weakened by jointing alone, although they may be weakened by other factors such as shearing or decomposition. In sedimentary rocks joints normal to the bedding planes are often quite prominent in weathered surface outcrops. Generally it will be found that at a short distance below the surface the joints are difficult to find and appear to have no important effect on structural condition of the rock. Probably part of the accentuation of joints exposed on the surface is due to weathering and part is due to elastic expansion of the rock which took place as the surface load was removed by erosion. This probably explains the fact that joints are more noticeable in work at shallow depths than in deep work.

Typical joints are shown in Fig. 1.

BEDDING PLANES

Bedding planes are a feature common to all sedimentary rocks and to some metamorphic rocks which were derived from them. Bedding planes are particularly important in the relatively thin bedded rocks, that is, the shales, siltstones, mudstone, and sandstones. Bedding planes are the surfaces which divide the different layers of water-deposited sediments and are planes of structural weakness in the rock as the coherence between different beds is usually less than that within a bed.

These structural weaknesses are very important economically because the world's coal supply is produced from the sedimentary rocks and the mine roofs overlying the coal beds consist most often of shale or sandstone. A great deal of research has been devoted to the most economical methods for supporting these roofs.

In excavating for permanent structures in dipping sedimentary beds the long dimensions of a structure should be oriented at right angles to the strike of the bedding, as this avoids exposing wide expanses of strata parallel to the *planes of weakness*.

LAMINATIONS

Laminations such as occur in schists and sometimes in gneisses are similar in structural effects to bedding of sedimentary rocks, and the same precautions should be taken.

CHEMICALLY DECOMPOSED ROCK

Weathered Rock

Weathering of rock results from the action of rainwater as it percolates downward through cracks, joints, or through the interstices between the mineral grains composing the rock. The water absorbs small quantities of carbon dioxide from the air to form carbonic acid which reacts with constituents of the feldspars to form clay and soluble compounds. The latter are carried away in solution while the clay remains in place. At the same time the dark minerals in the rock (ferromagnesian minerals and micas) are attacked by oxygen and moisture to form iron oxides and hydrates. These form a crumbly mass without appreciable strength.

The depth to which weathering extends in a given type of rock depends principally upon the climate. In a dry climate rock may be weathered to a depth of only a few feet while in a warm, wet climate decay of the rock may extend to many tens of feet and even hundreds of feet in the vicinity of fractures which act as channels. Usually weathering takes place above the water table and rock below this will be unweathered. This is not always true, however, because the water table may have been lower at some time in the geologic past and weathered zones which were then formed may extend below present groundwater surface.

The rocks most susceptible to weathering are those containing large proportions of the micas or ferromagnesian minerals (dark colored igneous or metamorphic rocks) such as basalt, diabase, schists, gneisses, greenstones.

Hydrothermally Altered Rock

Hydrothermal alteration is effected by hot aqueous solutions rising from great depths. These can convert some of the original mineral constituents of a rock into entirely different minerals. The most serious, and common, detrimental effect from a structural standpoint, is the alteration of the hard feldspars into clay (kaolin) or to a fine grained mica (sericite). Thus an otherwise sound rock, such as granite, may contain zones having the structural characteristics of a stiff clay. If located at an appreciable depth the overburden pressure will cause the altered rock to squeeze into any opening and will make support very difficult. Such alteration is sometimes difficult to detect by surface examination as the altered zone may not extend to the surface. Most frequently hydrothermal alteration has taken place along fault and fracture zones. However, altered zones also occur which are not associated with any apparent fissuring.

Soluble Rock

Solution is most active in limestone which is the most common of the soluble rocks. Rainwater dissolves carbon dioxide from the atmosphere and forms a weak carbonic acid. This, in percolating down through fractures in the limestone, dissolves its main constituent (calcium carbonate) by converting it to the soluble calcium bicarbonate. Thus fractures in the limestone are progressively enlarged to form channels for flowing groundwater. Absence of surface drainage in limestone formations is a good indication that extensive sub-surface channels exist. However, channels in limestone are so erratic that it is practically impossible to locate all of them in advance of actual excavation.

STRUCTURAL CLASSIFICATION OF ROCK

Presented below is a classification (used by the writer) which gives an approximate indication of the structural competence of rocks which are fractured in various degrees. The general classifications are:

STRATIFIED OR BEDDED ROCK (Sedimentary Rocks)

Thin Bedded: Most of the beds are less than 12 in. thick, with a large proportion being less than 6 in. thick.

Medium to Thick Bedded: Most of the beds in a sequence are over 12 in. thick and some are at least 24 in. thick.

(It is assumed that there are no fractures across the bedding; if there are transverse fractures then the stratified rock must be placed in one of the fractured classifications used for the Non-Stratified rocks below)

NON-STRATIFIED ROCK (Igneous and Metamorphic Rocks)

Massive Rock: Fractures and/or joints are spaced several feet, or more, apart.

Moderately Fractured or Jointed: 1 Fractures or joints generally 2–4 ft apart and tightly closed.

Blocky and Seamy: Fractures generally 1–2 ft apart and tightly closed.

Brecciated Rock: Pieces are generally less than 1 ft and most pieces are less than 6 in. in the largest dimension.

No Clay on Fracture Planes.

Fault Clay, "Gouge", or Slickensides on Fracture Planes.

Granular, Pulverized or Disintegrated.

Containing No Clay.

Containing Clay.

Fault Clay or "Gouge" with Fine Breccia.

Sedimentary Clays.

(See Chapter 5, page 250, for method of calculating support pressure.)

USE OF THE STRUCTURAL CLASSIFICATION TO PREDICT SUPPORT PRESSURES

During the preliminary geological examination, and subsequent preliminary geological exploration, the rock can be classified under one of the classifications listed above.

The Section titled "Simplified Method for Predicting Pressures on Supports in Workings at Shallow or Moderate Depths" (in Chapter 5 of Part B) gives the approximate coefficient of internal friction for each of the rock types classified above. This is expressed as the angle of internal friction, Φ, which may be substituted into equation (16c) or (16d) of Chapter 5, part B, to compute the probable maximum pressure on supports for openings in the various types of rock.

In many cases the rock at any particular point may not fit one of the classifications given above. In such cases it is necessary to interpolate and to use a value for internal friction intermediate between those given.

GENERAL LOAD CONDITIONS

Supports installed in single openings in rock which is structurally classified as moderately fractured, or better, are seldom required to support any appreciable load if the depth below the surface is not more than 1000–1500 ft.

Relatively heavy loads may develop however, if supports are installed in one of the following types of soft or structurally defective rock: (1) Fault clay and/or breccia; (2) fault breccia; (3) soft shales or schists; (4) fine weathered material; or (5) hydrothermally altered zones in which alteration has produced a large proportion of clay or mica in the rock, or which contains a large proportion of serpentine.

In the unconsolidated sediments those containing a high proportion of saturated clay may exert great pressure on supports, even at shallow depths.

Supports which are installed in openings situated in those rocks not having the aforementioned structural deficiencies serve two purposes:

(1) To prevent the falling of individual fragments and slabs created by blasting and/or natural planes of weakness.

(2) To apply a slight pressure to the rock surface to hold the natural structural rock "blocks" in position and thus stabilize the natural "voussoir arch".

The principle that a fragmented cohesionless (clastic) material is essentially self-supporting is an important one and is treated in detail in Chapter 5 of Part B. This principle implies that in an opening with an arched roof, or back, the rock around the perimeter will form a structural arch, the individual fragments being under compressive stress, and only enough pressure need be exerted against the rock surface to prevent units from dropping out of the arch, or from shearing under the compressive stress.

Thus support in the average single opening in a rock no more than moderately fractured carries little, if any, actual load but merely acts to hold the elements of the self-supporting arch in place, and to counteract shear stresses in the arch.

A massive rock with few cross fractures, or a thick bedded rock may support itself by true natural beam action. However, rocks with cross fractures such as "moderately jointed" or "moderately blocky and seamy" rock depend upon the principle of the natural voussoir arch for holding the components in place. The arch is self-supporting when the individual fragments are irregularly fractured, rough of surface, and tend to interlock. If fragments are loose or separated by spaces, or by layers of fault clay then obviously they cannot furnish enough compressive strength and will be too weak in shear to form a natural voussoir arch. Probably the dividing line (as regards self-supporting ability) between rock classifications would be between "moderately blocky and seamy" and "very blocky and seamy". The first mentioned can usually support itself, without assistance, while the latter usually requires installation of support.

CONDITIONS REQUIRING HEAVY SUPPORT

When the following natural structural defects are present the material will usually exert moderate to heavy pressure on the temporary supports even when an opening is located at shallow depths.

Faults: contain much soft clay and brecciated rock fragments.

Brecciated zones: may consist of finely broken fragments.

Hydrothermally altered zones: large proportion of original rock converted to clay.

Weathered zones: large part of original rock to clay and fragments.

Soft shales: these consist of poorly consolidated clay.

Swelling rock: this condition is usually due to presence of montmorillonite, a clay which swells when it becomes moist.

Fault Zones

These contain fault clay, breccia, and sometimes considerable water. Under average conditions most such zones encountered will not exceed more than a few feet in width. Faults less than 5 or 6 ft wide seldom develop great pressure on supports because friction of fault filling with fault walls tends to support the filling material, particularly if the fault dips at an angle other than vertical. In fault zones wider than 5 or 6 ft, however, the clay and breccia fault filling settle on to the supports and develop pressures which may require heavy support. If the fault filling is wet it will probably squeeze in around the sides of opening and cause the bottoms of the ribs to "kick in", necessitating the installation of struts across the invert to hold the ribs against the side pressure.

Theoretically, in a wide fault zone which extends to the surface, the full pressure of semi-fluid fault clay could be developed on the support. Yet the author has seldom encountered conditions where pressure on supports of a single opening was equivalent to more than 15–20 ft of rock, even in fault zones up to 250 ft wide.

In some fault zones in the more brittle rocks, such as quartzite, little clay is found, and fault filling consists mainly of angular crushed fragments with the characteristics of a coarse sand. Such materials exert top pressure on supports similar to that of a pile of sand, and will exert some side pressure as well.

Fault zones in brittle rocks are likely waterways and reservoirs and, when first intersected, may discharge considerable quantities of water and rock debris such as sand breccia.

Brecciated Rock

Extensive zones of breccia are sometimes encountered in which a combination of orogenic shear and compressive forces has broken the rock into numerous irregular fragments. Fragments as small as a fist may form an arch if they are in tight contact and rough of surface. They may support themselves, requiring only light pressure from support to keep them in place. If such breccia is loose, and with spaces between the fragments, however, such fragments will not support themselves and exert heavy pressure on supports.

Hydrothermal Alteration

Although this type of chemical alteration can act to solidify fractured rock or to harden soft rock, it more frequently tends to soften the rock by converting the feldspathic constituents to clay, fine grained micas or serpentine. The resulting rock has a very low shear strength and, though it may stand well at shallow depths, when it is subjected to heavy overburden pressures it will tend to squeeze into the opening.

Weathered Rock

Normally, weathered rock is encountered at relatively shallow depths only. It usually consists of rusty-appearing angular grains and fragments intermixed with clay and frequently accompanied by large fragments or blocks of intact rock. The rusty appearance is imparted to weathered rock by the oxidation of iron minerals. Fine material resulting from weathering consists of angular grains of quartz, and feldspar, and its behavior is similar to that of sand. If the mixture contains clay it may stand well if a tunnel is driven through it in dry weather but during the rainy season may soften and begin to squeeze the support. If thoroughly saturated such clayey material may develop static pressure equal to the full depth pressure.

Soft Shales

These are commonly encountered in the younger sedimentary rocks, and may occur sporadically in formations of well consolidated shale, and occasionally in sandstone formations. Such soft and poorly consolidated shale strata and stiff clays are commonly encountered in coal-bearing strata.

At shallow depths most of the weak shales present no support problems. If the depth is more than a few hundred feet such rocks begin to yield plastically under the influence of depth pressure, and to "flow" toward and "squeeze" the support. A bed of soft shale squeezed between two beds of sandstone, for example, is "extruded" from between these beds and into the adjacent opening.

This phenomenon accounts for much of the downward buckling of soft roof strata observed over narrow openings as well as for "floor heave". The author has seen cases where the latter phenomenon has raised the floor of a 14-ft diameter tunnel by as much as 4 ft. Instances are on record where openings have been completely filled and closed by soft strata "extruded" into the opening. (See Chapter 5 of Part B.)

The initial static pressure exerted on rigid support by such plastic material may approach the value of the depth pressure. However, slight yielding of the support allows internal frictional forces in the soft material to extend over a larger area and the ground pressure decreases rapidly as the support yields.

Swelling Rock

Fortunately, swelling rocks are seldom encountered in underground workings. Most commonly, swelling is caused by the clay mineral montmorillonite which can absorb several times its volume of water and in swelling generates pressures on supports.

The mineral anhydrite (calcium sulphate) also swells when wet. It combines with water to form gypsum, and a volume increase results. Such difficulties with anhydrite strata were encountered in the Haunenstein Tunnel (1917) and in the Simplon Tunnel (1898–1905).[2]

Unconsolidated Sediments

These are encountered ordinarily at relatively shallow depths and for that reason the overburden pressure is not great. A soft clay at shallow depth may stand fairly well and exert little pressure on supports while the same clay with a hundred feet or more of cover would be difficult to hold.

Sand and gravel do not offer difficult support problems provided that the excavation is located above the water table, and provided that such sand or gravel has enough cohesion that it does not "run" in at the face. If such material is very loose it may be desirable to consolidate it by means of chemical injections, or "chemical grouting".

If an underground excavation is to be made below the water table in unconsolidated sediments then difficulties will occur due to inflows of water carrying in mud and sand and also heavy pressure on supports or lining may be expected.

ESTIMATION OF ACTUAL LOADS ON EXISTING SUPPORTS

In the absence of measuring devices installed on tunnel supports, the equivalent rock load may be roughly estimated by observing the maximum deflection of lagging which bridge across from set to set.

The formula for calculating maximum deflection of a simple beam is

$$y = \frac{5Wl^3}{384EI}$$

and taking the weight of rock (in place) as 170 pcf and $y = 0.1$ in.

$$\text{Depth of rock load} = \frac{0.54Ed^3}{l^4}$$

Taking the modulus of elasticity E as 1,500,000 psi (approximately that for wet Douglas fir) and substituting in the formula we find the following loads produce a deflection of 0.1 in.:

TABLE 1

Span	For 2-in. lagging Equivalent rock load	For 3-in. lagging Equivalent rock load
30-in. span	7.9 ft	26.7 ft
48-in. span	1.2 ft	4.0 ft
60-in. span	0.5 ft	1.7 ft

The figures above are the depths of uniform loads of solid rock which would be required to produce maximum deflection of 0.1 in. in the center of a lagging supported as a simple beam. The figures in Table 1 are computed for rough lagging (e.g. full dimension). If 2-in. finished lagging ($1^5/_8$ in. thick) is used, as in some tunnel operations, deflections will be 1.8 times as great as for 2 in. rough lagging.

The deflection is directly proportional to the load, therefore a deflection of 1 in. would indicate 10 times the load of a 0.1 in. deflection, etc.

The depths are indicated as loads of solid rock, actually the rock would be fractured and an equivalent weight of fractured rock would, of course, occupy more space than that indicated in the table for solid rock.

The deflections shown are computed from formula for wet Douglas fir. It would be preferable to make field check of the deflections, under various loads, with the type of lagging which is actually being installed for support.

It is probable also that the steel rib, being relatively rigid, will take a larger proportion of the rock load than an equivalent length of lagging; this would tend to be the case in the fractured rock but, in the case of soft and plastic rock, the rock will tend to flow away from the points of high stress and toward the zones of lower stress, tending to equalize the pressure distribution. The load thus tends to be more uniformly distributed in soft rocks and less uniformly in those consisting of hard fragments, where point pressure may develop on ribs or even along the span of the lagging.

Another method of load estimation, in the absence of load measuring cells, is observation of the amount of squeeze of the blocks and wedges which are inserted between lagging and ribs, or between lagging and rock. With Douglas fir, or similar timber, fairly heavy loads must develop before any squeeze on the blocking is apparent.

The deflection of the lagging is a much more sensitive indication of rock load than is the amount of squeeze of the blocking.

EXAMPLES OF ROCK LOADS OBSERVED IN A TUNNEL

The author has kept detailed geological records as well as detailed records of structural conditions for subsurface workings. One of these, a tunnel, 15,400 ft long, will be cited as a typical example here because of the variety of geological structures and rock conditions through which it passed.

This tunnel was excavated to a modified horseshoe cross-section with a nominal diameter of 8.3 ft. Temporary support was by 4-in. 13-1b. H-beam steel arch ribs normally spaced at 5-ft center to center and lagged with 2 in. lagging. Upon holing through the tunnel was lined with concrete to form a circular conduit 6 ft in inside diameter.

Average daily advance during driving was about 30 ft per 24 hr in each heading with the maximum being about 50 ft. Maximum depth of cover was about 800 ft; average depth in the rock section was about 350 ft; depth of cover in the alluvium ranged from 20 to 50 ft.

The tunnel alignment was nearly north and south. The strike of sedimentary bedding and contacts of other formations was generally between N. 50 °W. and N. 70 °W. Sedimentary bedding and contacts of other formations dipped northerly at 40–60°. Sedimentary formations were Upper Miocene.

The geological cross-section, proceeding from north to south, was:

Granite (including 70 ft of weathered material)	4760 ft
Conglomerate	1040 ft
Sandstone	490 ft
Basalt	1310 ft
Conglomerate	470 ft
Sandstone (tuffaceous and conglomeratic)	210 ft
Sandstone (tuffaceous)	620 ft
Shales, sandstones, and interbedded shales and sandstones	4770 ft
Alluvium	1730 ft

The geology was "logged" in detail foot by foot from end to end of the tunnel. In addition a graphical record was kept of "Structural Conditions in the Tunnel". The tunnel was inspected at intervals and on the graphical record were noted the sets subjected to pressure, the degree of pressure. Moderate pressure was classified as that which produced detectable deflection of more than one or two lagging in a set.

In summary all zones of significant ground pressure which developed in the tunnel were located in *weathered* or *faulted zones* in the granite, or in the softer beds of the thin-bedded shales.

About 600 ft of the 4760 ft of granite (about 13 per cent) could be classified as producing moderate to heavy pressures. About 800 ft of the 4770 ft of shales and sandstones (about 17 per cent) developed pressure. Total lengths of zones of moderate to heavy ground pressure in the 13,670 ft of rock tunnel was 1400 ft, or about 10 per cent. In the remainder of the rock tunnel no significant pressure on supports could be detected.

The portion of the tunnel through the alluvium was driven by spiling over steel sets. The water table was first lowered to a level below the tunnel invert by a series of wells sunk from the surface along the tunnel line. Pressures sufficient to deform the lagging developed at a few points where saturated sand, silt, or blue mud were encountered. However, the dewatering had stabilized the unconsolidated material so that not more than 10 per cent of it exerted moderate to heavy pressure on support.

GEOLOGICAL HAZARDS

The foregoing has listed some of the types of structurally defective rocks which are difficult to support. Listed below are the types of rock in which the various types of defects may be found. These are the principal rock defects which should be looked for when conducting a geological survey or a sub-surface exploration program in connection with proposed underground structures.

Faults occur in all types of rocks.

Brecciated zones occur in all types of rocks.

Hydrothermally altered zones occur in all types of rocks but the most damage occurs in the light colored igneous rocks, particularly granite.

Weathered zones occur in all types of rocks. The dark colored igneous rocks are the most susceptible to breakdown by weathering, and the open joints in igneous intrusives allow deep penetration of weathering.

Soft shales (and bedded clays) are found in the sedimentary rocks, particularly the younger in geologic age and usually at relatively shallow depths. Such strata are commonly encountered in coal mining operations.

Swelling rock most often encountered is in sedimentary rocks, particularly shales.

Nearly all those structural weaknesses or defects which result in heavy pressure on supports have one characteristic in common; that is the material in the weak zone contains a significant proportion of clay or "rock flour". When wet such material loses its shear strength and is subject to viscous flow. If intermixed with granular material such saturated clay may act as a lubricant to encourage movement of the whole mass.

PRECAUTIONS IN ROCKS CONTAINING CLAY

When tunneling through clay, whether it be fault clay and breccia, or residual clay weathered from rock, sedimentary clay, or soft shale, it will be found that at shallow to moderate depths dry clay stands very well and requires little support. In fact very fast progress can be made in working through such ground since the material is easily excavated, trims to a neat line, and requires only light support. If there is any possibility that seasonal rains, or groundwater will percolate down from the surface to saturate the clay then heavy support should be installed at the time of excavation.

In driving workings through such ground wide footblocks should be installed under the posts, and struts should be provided across the bottom to keep ribs from "kicking in". These footblocks and struts should be installed even if the clay is dry, if there is a possibility that drainage water will flow along the bottom.

If rainwater percolates down to saturate the clay it will begin to squeeze against the supports. Light support will ordinarily yield slowly by bending and eventually break, necessitating replacement.

In an instance familiar to the author an 8-ft diameter tunnel was driven for 150 ft through fault clay and breccia (fault zone in granite). Initial support was by 4-in. 13-lb. steel ribs at 5 ft centers with tight 2-in. lagging. The dry material exerted no pressure on the supports, although the depth of cover varied from 40 to 130 ft. The tunnel stood perfectly for about 8 months during the dry season. When the rainy season commenced, the clay became moist and began to squeeze. Within a month's time lagging was "bowed" inward to the breaking point and ribs were "toeing in" at the bottom. This necessitated removing the ribs, one at a time, enlarging the tunnel to its original dimensions and replacing support. The same ribs were placed at $2\frac{1}{2}$ ft centers, equipped with invert spreaders, and 2 in. lagging. This support was sufficient to hold the ground against further encroachment and, in fact, little additional evidence of pressure or movement could be detected.

At another location in the same tunnel, with cover about 700 ft deep, the tunnel passed through a zone about 240 ft wide in granite consisting of breccia with considerable clay between the rock fragments. The dry material stood well and exerted no pressure on supports which were initially 4-in. 13-lb. steel ribs at 5-ft centers lagged with 2-in. lagging. The material made for fast progress and heading was advanced about 40 ft every 24 hr. After it had been penetrated for about 100 ft, water began to appear dripping from seams throughout the section previously excavated. The ground began to squeeze and exerted enough pressure to break 2-in.

lagging, the ribs began to "kick in" and the bottom rose as much as 8 in. in places. Ribs were removed, one at a time, the tunnel enlarged to its original dimensions and the ribs replaced on $1\frac{1}{2}$–2 ft centers, and equipped with struts; 3-in. lagging was used.

This support was sufficient to hold the rock and it appeared that the maximum pressure had been dissipated with the first squeeze, as pressure on the reinstalled support was nowhere greater than that exerted by a few feet of loose rock. The tunnel was advanced through the remainder of the fault zone by spiling.

The water which appeared at this location could not be correlated with any seasonal rains. It appeared rather that the excavation of the tunnel had caused some realignment of groundwater channels, possibly due to relaxation of ground surrounding the tunnel and opening up of new fissures.

Where a tunnel is constructed through rock which contains a large proportion of unconsolidated, but dry, clay which may be subject to subsequent saturation it is usually more economical to install adequate support initially rather than to wait for pressure to develop before installing heavy support. In some situations it will be apparent that there is no possibility of water seepage entering the clay and in such case the dry "clayey" rock will require little support.

GROUNDWATER

Next to the structural condition of the soil or the rock the presence or absence of groundwater is the factor which has the greatest influence on excavation and support methods.

A soil which is only moist will stand quite well while the same soil located below the water table will flow into the excavation, and will exert heavy pressure on supports.

In rock those zones containing appreciable proportions of poorly consolidated clay will stand well if fairly dry, but will flow and squeeze the support if saturated.

In the harder rocks water may not greatly influence pressures on supports, but when it flows from fissures in heavy streams working conditions become very difficult. In the case of water-bearing fault zones large quantities of pulverized material may be washed from a fault by water trapped in the zone.

Another type of trap for groundwater is formed when a pervious formation, such as a basalt, or fractured igneous rock is in contact with an impervious rock such as a shale. The latter acts as a dam against which the water table in the pervious rock can build up. Although no sudden large inrush of water may be expected at such a contact when intersected by an underground working, still continuous flows from cracks and fissures may continue for a long period of time until the pervious rock is drained down to the tunnel level.

Preliminary Determination of Groundwater Conditions

Methods of determining groundwater conditions include the measurement of water levels in drill holes, the pumping of drill holes, and pressure testing of drill holes. In addition a thorough study of surface runoff conditions should be made prior to starting underground work. Such a study will give a clue as to the possible sources and total quantities of groundwater which may be encountered and will allow pumping facilities to be planned.

BIBLIOGRAPHY

1. Wahlstrom, E. E., Application of geology to tunneling problems, *Trans. Am. Soc. Civil Engrs.* October, 1947, pp. 1245–56.
2. Ott, J. C., Quelques aspects du problème de la poussée sur les tunnels, *Bull. Techn. Suisse Romande*, Vol. 71, No. 1, pp. 13–22; No. 4, pp. 41–50, 1945.

CHAPTER 3

PRELIMINARY EXPLORATION

METHODS for determining sub-surface conditions prior to planning the location of a sub-surface structure, or a mining operation, include the following:

(1) Geophysical methods.
(2) Geological mapping and study of the surface (with trenching or stripping of the surface when necessary to expose bedrock for geological inspection).
(3) Drilling or boring.
(4) Exploration tunnels or shafts.

GEOPHYSICAL METHODS

Geophysical methods have become very important in the search for new mineral deposits. Most of those deposits which outcropped on the surface have already been discovered and methods now used must be able to detect concealed deposits. Geophysical equipment and techniques are constantly being improved. The introduction of the airborne magnetometer immediately after World War II is a typical example. The first spectacular success with this instrument was the discovery, in Pennsylvania in 1948, of a large body of iron ore concealed under 1500 ft of barren overburden.

Those areas which show anomalous readings (that is, readings which deviate from normal) on geophysical instruments are investigated more closely by geologists on the ground and if still promising may then be explored by drilling and/or shaft sinking and tunneling.

Jakosky[(1)] classifies geophysical methods into those which measure a *natural field of force* and those which measure an *artifical field of force*. Under *natural field of force* he includes the following methods: geothermal, magnetic, gravitational, geochemical, radioactive, electrical self potential, slim-hole drilling with mud logging, and finally core-hole drilling with paleontological studies and core orientation. Under *artificial field of force* he lists: electrical (inductive), electrical (conductive), seismic (refraction), seismic (reflection), and slim-hole drilling with electrical logging. The foregoing items under each heading are listed in the generally recognized order of increasing resolving power; that is, in the order of their ability to diagnose subsurface conditions. Under certain specific circumstances a method of generally low resolving power may become a method of high resolving power, such as the magnetic method when applied to a near-surface magnetite deposit.

In the field of mineral exploration the airborne magnetometer, which was developed for submarine detection during World War II and introduced to civilian uses immediately after the war, greatly accelerated the rate of location of magnetic anomalies and, supplemented by other methods, has produced some important ore finds. However, it was soon discovered that magnetite is a very common and widespread constituent of Pre-Cambrian rocks.[5]

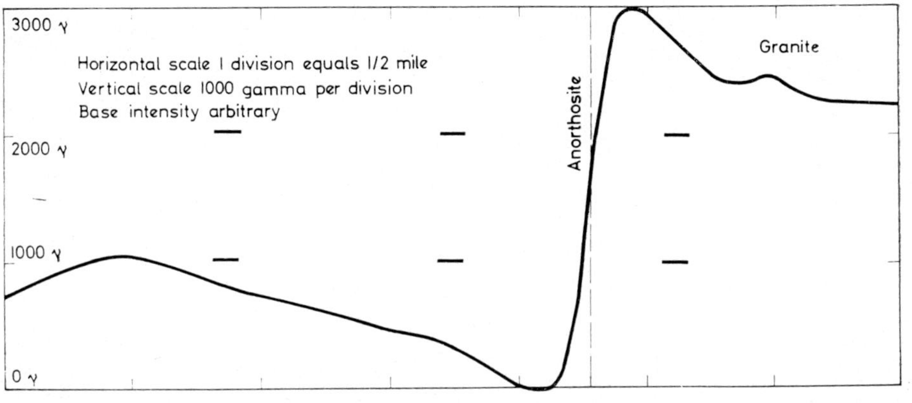

FIG. 1. Magnetometer reading shows a sharp rise at the contact between the granite and the anorthosite formations in Quebec.[2]

The development of airborne electromagnetic equipment then furnished a new tool for locating sub-surface conducting bodies. However, in addition to the sulphides which are valuable for their mineral content other conductors such as graphite, certain types of water-bearing shear zones, and barren sulphides exist. Electromagnetic exploration showed that these conductors were much more abundant and widespread than had been imagined prior to the development of airborne electromagnetic equipment.

The International Nickel Company of Canada[5] has perfected these geophysical techniques, and supplemented them by geological techniques to the point of being reasonably certain that nine out of ten anomalies drilled will be caused by sulphides. However, the records show that at least 100 barren sulphide bodies exist in most Pre-Cambrian areas of Canada for every deposit which contains economic values. These latter may or may not prove to be ore bodies.

The airborne magnetometer has proved a valuable tool for obtaining supplemental information on regional geology, and because of the rapidity of the method large areas of the country have been covered by magnetometer surveys. Records from such surveys permit deductions as to the depths to the crystalline basement rocks, and also serve to outline the boundaries of intrusive bodies.

A particular type of geophysical investigation may be most effective when applied for locating one particular type of mineral deposit. For instance, the "Overvoltage" method has been developed, by Newmont Mining Co.[3] since the war to the point where it is effective in locating and outlining deposits of disseminated metallic minerals.

With this method an interrupted current is applied to the earth and the voltages induced in the earth by current interruption are measured. Disseminated conducting particles act as individual condensers which delay decay of the impressed voltage.

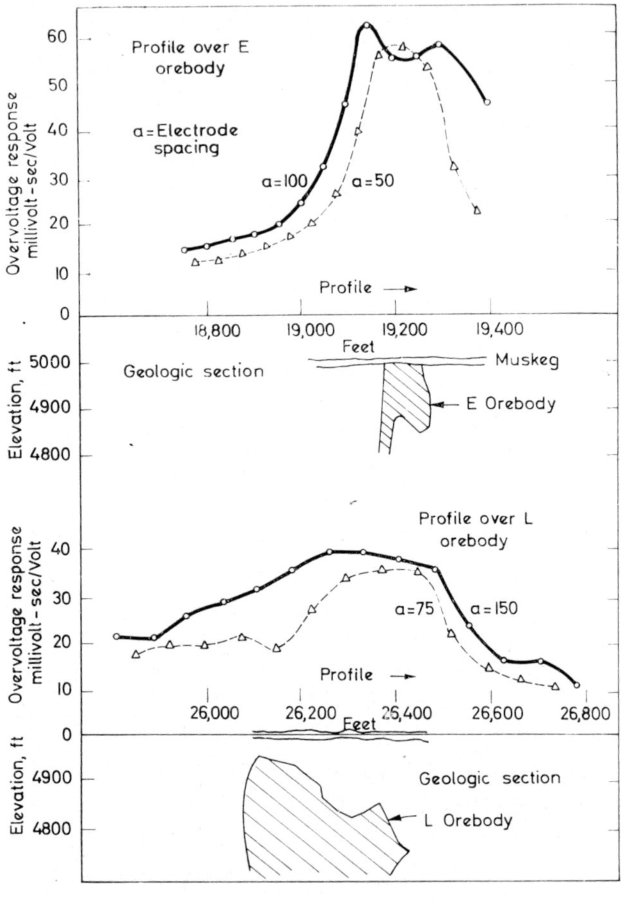

FIG. 2. Overvoltage profiles at Lynn Lake, Manitoba.[3]

Seismic methods are used most extensively for locating and outlining geologic structures which are favorable for the occurrence of petroleum. With the seismic method elastic waves (sound waves) are produced in the earth by explosives and the time of travel of these waves along various paths are measured. The velocity of a sound wave in any medium depends upon the elastic modulus and the density of that medium. Therefore, sound waves travel at different velocities in different types of soil and rock, and in addition are partially reflected or refracted at boundaries between different materials. These factors are utilized in locating and outlining geological structures.

The seismic method has been used recently for foundation exploration work in connection with construction of a railroad fill across Great Salt Lake and the method has also been used within the past few years on several highway location jobs to determine the depth of alluvial material covering bedrock along proposed highway routes.

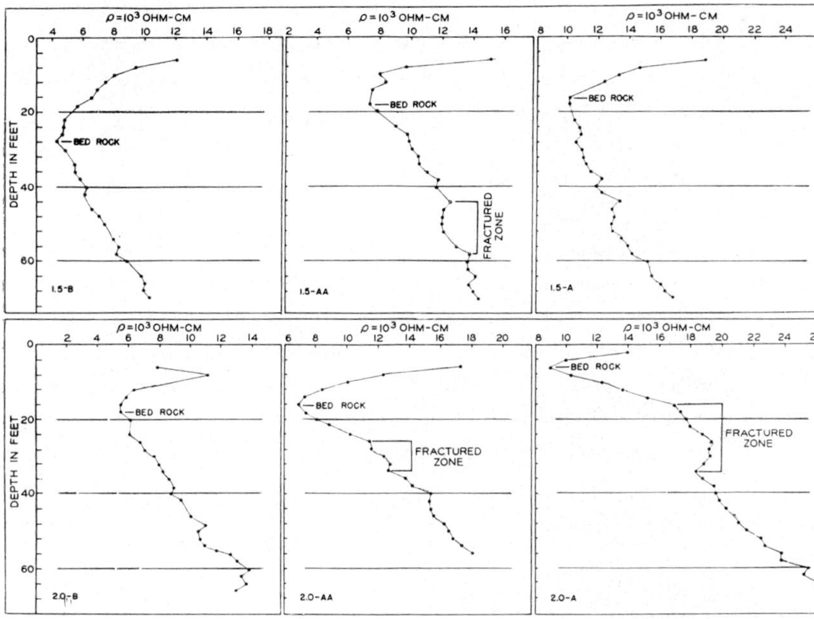

FIG. 3. Results of an electrical resistivity survey. Correlation chart showing method of delineating fractured zones in bedrock.[(4)]

The method can be similarly employed as a supplement to drilling, for determining depths of loose surface formations where it is desirable to locate subsurface structures sufficiently deep to be in solid rock.

The "electrical resistivity" method has recently been applied in diagnosing subsurface conditions in the portal areas of the Lehigh Tunnel on the new Northeastern Extension of the Pennsylvania Turnpike System.[(6)]

The studies were made for the purpose of indicating the approximate depths to bedrock. The results showed (a) bedrock much deeper than originally postulated; (b) the most desirable areas for the rock portals where bedrock approached the ground surface.

Test holes were subsequently drilled and it was found that the bedrock was indeed deeper than 10 ft and deeper even than was indicated by the resistivity survey.

It was found that preliminary geophysical interpretations at the South Portal area had been made on top of the weathered shale rather than on sound rock. Thus, with this particular rock the resistivity method was unable to delineate between weathered shale and sound shale since both have a low resistivity. This would not be the case with a limestone or sandstone bedrock.

In the case of the exploration of a mineral deposit an extensive program of geological study of the surface, supplemented by drilling, and sometimes by the sinking of exploration shafts, and driving of tunnels, is normally carried out to determine the grade and the size and shape of the deposit. This program furnishes sufficient information for planning the location of major permanent underground facilities. In the case of civil works facilities, however, such detailed advance information is not available and a special effort must be made to obtain it.

GEOLOGICAL STUDY OF THE SURFACE

For a surface geological study all exposed rocks and soils are examined and pertinent information relative to each is recorded. The following items of information are of interest in a geological survey: the type and description of each rock together with notation of any unusual characteristics; the presence of bedding planes, schistocity, or gneissic banding as well as the strike and the dip of such features; the presence of joints, shear planes, gouge seams, or folds together with the strike and dip of such fractures; the presence of springs, seeps, and watercourses.

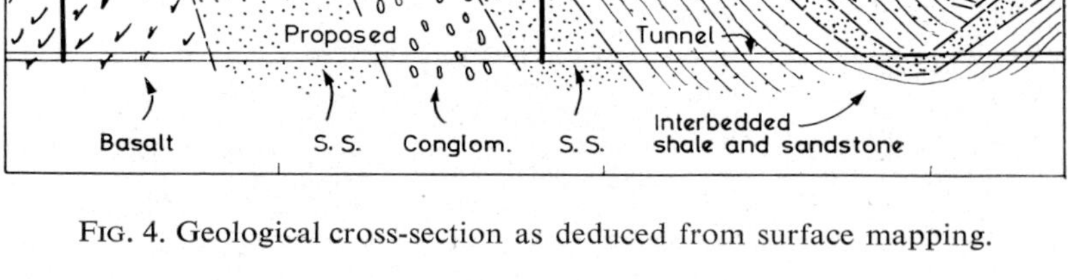

FIG. 4. Geological cross-section as deduced from surface mapping.

If the exposed rock is relatively uniform over a large area then fewer data need to be noted. Generally, notations should be made at any location where the rock type, character, strike, or dip change markedly.

Hand specimens may be collected for examination and identification of constituent minerals in the laboratory, as well as for tests of the physical characteris-

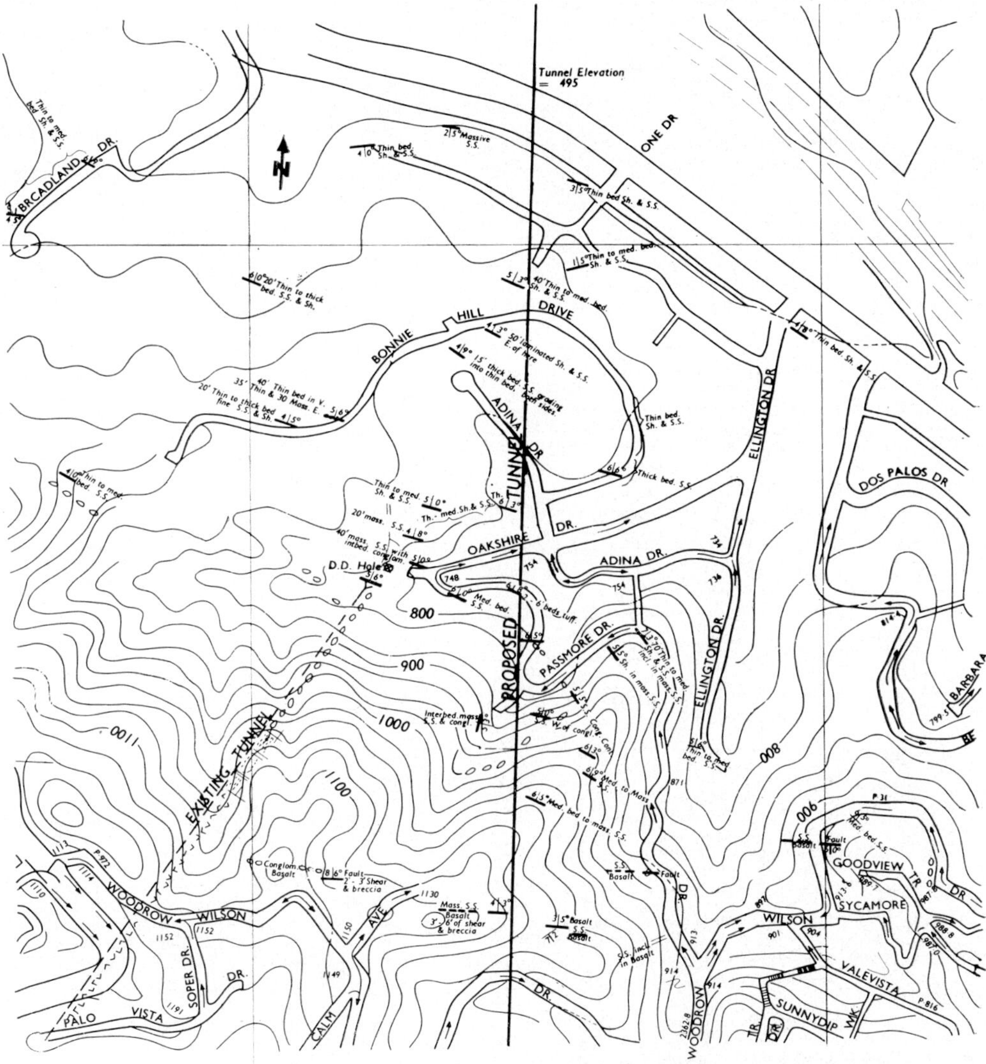

FIG. 5. Geological map of rocks exposed on the surface.

tics. Generally, the rock in surface exposures will be somewhat weathered and therefore weaker than a similar type located below the zone of surface weathering.

The location of each outcrop examined should be marked on a map together with its strike and dip, and identified by a number. More detailed description, when required, may then be recorded in a notebook and keyed by number to its location on the map.

Generally, the area examined should extend for a considerable distance on each side of the proposed subsurface structure. Along the proposed line for a deep tunnel it may be desirable to examine and map the principal geological features for 1000–2000 ft on each side of the tunnel line in order to detect any major geological structures which may dip at angle sufficiently flat to intersect the proposed tunnel line at depth. In some instances where the overburden is heavy and outcrops are few it may be desirable to range further afield in order to find enough exposed rock to obtain a reasonably good overall picture of sub-surface conditions.

When the information from a surface geological survey is all assembled cross-sections may be drawn and those geologic structures which are exposed on the surface may be projected to depth. It will then become apparent as to where preliminary subsurface exploration is most needed. Such additional exploration may be needed to determine the presence or absence of suspected fault zones, to more accurately determine the location and extent of known fault zones, to locate the contacts between different formations, to determine the groundwater conditions at depth, to determine the depth to which surface weathering has progressed, or to locate sound rock in which to excavate an opening.

GEOLOGICAL MAPPING TECHNIQUES

Base maps on which to record geological information may be made from the standard U.S. Geological Survey topographic maps. These maps as available are on a scale of 1 in. to 2000 ft and, at nominal cost, may be "blown up" to a larger scale by photographic means.

A scale of 1 in. to 200 ft or 1 in. to 400 ft is convenient for use in the field. The larger scale gives sufficient space for recording more geological information directly on the map. If the geology is very complex then it may be desirable to employ a scale even larger than 1 in. to 200 ft, in order that all pertinent strikes, dips, and related information may be properly placed on the map in the field.

Maps which show topography are preferred for work in the field since the physical features aid the geologist in locating himself on the ground as well as aiding him in visualizing the relationships of geologic structures in three dimensions. It is desirable also that the map show additional natural features such as streams, wood, peaks, swamps, etc., and also show accurately such man-made improvements as streets, roads, buildings, railroads, quarries, etc. All these features aid the geologist in accurately locating surface geologic features.

DRILLING OR BORING

The surface geological examination will indicate locations where drilling or boring is desirable to further resolve subsurface conditions. ("Drilling" as used herein implies the boring of small holes – usually 3 in. in diameter – in rock, while "boring" implies the boring of larger diameter and shallower holes for testing soil conditions.)

For exploration of alluvial material truck-mounted augers are available which can rapidly put down large diameter holes to depths up to 75 or 100 ft. In loose material, such holes will require casing. In order to obtain undisturbed samples of the material in place it may be necessary to use a "drive sampler" in the bottom of the hole, at intervals. Such large diameter holes offer good opportunities for pump tests, and can also serve as dewatering wells during construction.

For delineating an ore body, or for exploring rock prior to location of a subsurface structure, diamond drills are usually employed. The size most commonly used for moderately deep exploration uses an NX diamond bit which cuts a hole about 3 in. in diameter and takes a core about $2\frac{1}{8}$ in. in diameter. Truck-mounted rigs are available which have a rated capacity of 1500 ft of hole.

Some of the factors which should be observed and recorded while drilling are given by Loofbourow[(7)].

(1) Record where drill rods "float" as a very simple and definite indication of tight ground.

(2) If severe water problems are anticipated, it may be desirable to record water levels and changes in them, to check recovery of drill water, and to make porosity, permeability, and possibly pump tests in the manner familiar to petroleum engineers and dam designers.

(3) Record all core loss and, as far as possible, determine its cause, obtaining information from sludge or core particles, drill operation, and opinion of drillers. A gouge seen along vein walls may be of great importance, but difficult to core. Its presence may be indicated by relief of rod pressure, mudding of the bit, jump in motor speed, or slickensides and parallel breaks in adjacent core.

(4) Record drilling progress in various rocks with notes of average footage per shift or average time required to run out 1 or 2 ft. Mechanical recorders, like those used in oil drilling, may be employed on hydraulic or "rotary" type drills.

(5) Note and record bit footages in various types of ground.

(6) Illustrate structural and physical information on models.

(7) Study minutely and record all data on the physical character of rocks and ore. If all core is sent for assay, logging will be particularly vital. If core is to be split, it ought to be split in such a way that physical features are not destroyed, say, by diamond saw.

(8) Laboratory study of various kinds may be appropriate. Laboratory study of rock properties should become of increasing usefulness. However, experience,

judgment, and a keen awareness of structural conditions will be needed to take samples properly and interpret the results. Results of physical testing at Rifle, Colorado have been generally confirmed by underground work.

(9) Drillability with rotating bits often does not correspond to drillability by percussion. However, if rotary drilling is carefully observed, and if physical character of rocks is taken into account, and analogies studied, the drillability by percussion may be approximated closely.

(10) It may be vital that core rejects, whether from samples sent for analysis or split core or both, be saved for metallurgical study.

(11) Other lines of investigation appropriate to each case should be given weight as to importance.

Other methods of exploration sometimes used in rock include churn drilling and percussion drilling. Both of these methods destroy the structure of the rock and therefore are not suitable when detailed geological information is required. They are used principally in exploring mineral deposits, and drill cuttings are collected at intervals for assaying.

Drill cores will show the condition of rock at the location and elevation of a proposed structure. From drill cores the type and the condition of rock can be determined; amount and nature of fracturing, rock porosity, dip of bedding, laminations, faults, etc., can all be determined from drill cores.

Groundwater

It is highly desirable also to be able to predict groundwater conditions which will be encountered during underground construction operations. The most satisfactory method for testing water conditions is to drill a hole 8 in. or more in diameter and to employ a deep well pump to determine the water yield and "drawdown curve" of the hole. These are indicators of water conditions to be expected during construction operations.

Although the pump method is the most satisfactory for determining sub-surface water conditions it is not practicable, in most locations, to drill holes as large as 8 in. in diameter. This applies in hard rock where exploration drilling is normally done with a diamond drill equipped with an NX bit. The NX bit cuts a hole about 3 in. in diameter and takes a $2\frac{1}{8}$-in. core. With a heavy drill rig holes 1000–1500 ft deep can be put down. It is not possible, however, to install a pump in a hole as small as this.

Water is sometimes lifted from holes by means of compressed air blown down the hole. The string of drill rods is lowered down to within a few feet of the hole bottom, the top of the drill string is connected to an air compressor, and the air turned on. Air passes down the hollow drill string and up the annular space between drill rod and hole wall, with sufficient velocity to lift out any water in the hole. This can be caught in a sump for measuring purposes. This method presents difficulties when water stands more than 230 ft deep in the hole because the hydrostatic pressure at the bottom of the hole equals or exceeds the pressure of the compressed air (usually about 100 psi). Under this condition air cannot pass out the lower end of the drill string. Even if head of water is somewhat less than 230 ft the unbalanced air pressure may not be sufficient to force sufficient air volume out of the end of drill string to lift water out of a deep hole.

Another method of checking subsurface water conditions is to make periodic checks on level of water standing in test holes by lowering a float tied to the end of a string, down the hole. A measurement should be made immediately after pulling the drill string at the end of a work day and another measurement the following morning before drilling starts again. These will detect any change in level of water surface. After drilling is completed it may be desirable to leave the holes open for several weeks, or months, and to make periodic checks of water levels. From these a good picture of groundwater levels and seasonal fluctuations can be obtained.

An indirect method of testing porosity and permeability of formations at elevations of proposed structures is by means of pressure testing. With the method a "packer" is attached to the end of the drill string and the string is run into the hole so that the packer is located at the top of the portion to be pressure tested. The packer is then expanded to block off the hole and water is pumped through the drill string at constant pressure, the quantity being measured. The water passes through the drill string and into the open hole below the packer. If rock is tight and unfractured no water can be pumped. The quantity which can be pumped, at any given pressure, is indicative of the degree of fracturing and permeability of the rock below the elevation of the packer. The quantity is also a function of the groundwater surface elevation. This can be determined by measuring water surface in the hole before pressure testing and the net effective pressure on inside surface of the hole can then be calculated so that the test can be correlated with others.

By measuring the water level in the hole the actual groundwater elevation can be determined and the height of groundwater above the elevation of the proposed structure computed. Designating this figure by h and remembering that h ft of groundwater counterbalance h ft of the water column inside the drill string, then pressure at the water pump should be such as to create a pressure equal to h ft of water in the drill string at groundwater level.

In other words, if the hole were empty, water would flow into it under a head of h ft. Now if the hole is allowed to fill with water to groundwater level and a pressure equal to h is imposed on this surface then water should flow out of the hole at approximately the same rate it originally flowed in. With proper procedures it should be possible to duplicate a pumping test, in reverse.

Betonite clay slurry is frequently used as a circulating medium during core drilling operations. Such drilling muds lubricate a hole as well as sealing the hole walls and supporting soft or broken rock to prevent it from falling into the hole. While drilling mud facilitates drilling it also seals off cracks and seams in the rock and impedes groundwater circulation. For this reason drilling mud should not be used if reliable tests for sub-surface water conditions are desired. Instead, holes in soft, crumbly, or broken ground should be cased with steel drill casing, which may be perforated to admit water. Although casing a hole is a much more expensive process than supporting hole walls with drilling mud, it is worth the extra expense if a true record of water conditions is required.

BIBLIOGRAPHY

1. JAKOSKY, J. J., *Exploration Geophysics*, Trija Publ. Co., Newport Beach, Calif., 1957.
2. How aeroplanes speed mining progress, *Eng. Mining J.*, Vol. 159, No. 1, Jan. 1958, pp. 77–82.
3. BALDWIN, R. W., A decade of development in overvoltage surveying, *Mining Eng.*, March, 1959, pp. 307–314.
4. SCHARON, L., UHLEY, R., and TSVI, M., Subsurface investigations of a plant site, *Mining Eng.*, Feb. 1959, pp. 215–218.
5. Exploration procedure at Inco, *Mining Congr. J.* April, 1959, pp. 107–109.
6. SCHARON, H. L., and CLEAVES, A. B., Symposium on "Geological Factors in Tunnel Construction". Geophysical investigations for the Lehigh tunnel, *Proc. Am. Soc. Civil Engrs. J. Soil Mech. Found. Div.*, Vol. 84, No. SM 2, May, 1958 (Paper 1960).
7. LOOFBOUROW, R. L., Why not predesign underground mines? *Eng. Mining J.*, Vol. 153, No. 6, June, 1952, pp. 88–90.

PART B

The Structural Behavior of Soils and Rocks

INTRODUCTION

For the purpose of indicating its general structural characteristics a rock may be classified either as "competent" or as "incompetent".

OPENINGS IN "COMPETENT" ROCK

"Competent" is the term applied to a rock which will stand without support when an opening is excavated in it. Such rock may be presumed to act as an elastic solid. Rock which contains pre-existing fractures of such nature that they do not affect the stability of an opening may be considered as competent.

OPENINGS IN "INCOMPETENT" ROCK

Rocks containing pre-existing fractures or weaknesses of such nature that they will not stand without support may be designated as "incompetent". Also included would be those rocks which are intact but which lack strength to resist the stress concentrations induced by the creation of openings.

The terms "competent" and "incompetent" are relative terms. For example, a small tunnel driven through a medium soft shale at a depth of 100 ft below the ground surface will probably be self-supporting and the rock would be classed as "competent", whereas an opening the same size driven in the same rock at a depth of 3000 ft would probably require closely spaced supports to keep it open and the rock would be considered "incompetent".

DESIGN OF UNSUPPORTED OPENINGS IN COMPETENT ROCK

Underground openings in competent rock may, for design purposes, be classified as follows:

Openings in Massive Rock

(a) Single openings in massive rock.

(b) Multiple openings in massive rock.

Design of these openings involves the selection of a shape and spacing for the openings which will meet practical needs for usage and at the same time which will avoid destructive concentrations of stress in the rock around the openings.

Openings in Bedded Rock

(a) Single openings in bedded rock.
(b) Multiple openings in bedded rock.

In designing single openings in bedded rock the roof strata immediately above an opening are presumed to act as beams, and calculations are made to determine the maximum lengths of beams which will safely support themselves across the opening.

For the design of multiple openings in bedded rock it is necessary to investigate both the safe span for the roof strata, which are presumed to act as beams, and to investigate also the compressive stresses which are created in the pillars by the overburden load which is supported by each pillar.

Investigations for the design of openings in bedded rock may include theoretical investigations as well as excavation of experimental openings which are instrumented to record roof sag, surface, strain in the rock, strain measurements within the rock, and pillar compression.

The Elastic Boundary

Material around the boundary of an underground opening which requires no support may be considered to be in the elastic state. No pressure need be exerted against the boundary to prevent the material from fracturing and falling into the opening.

Stress Distribution around a Circular Opening

To illustrate stress distribution around an elastic boundary the simplest case will be cited. This consists of a circular hole in a plate of elastic material. The plate is shown in Fig. 1 with pressure applied to the edges, prior to the creation of the hole. The pressure on any plane such as AB, which is normal to p, is equal to p. A circular hole is now cut through the plate and, since stress from block $CDFE$ cannot be transmitted across the hole as a compressive stress it must be carried around the hole. Thus the stress at the sides of the hole is greatly increased and is maximum at the ends of the horizontal diameter EF, but decreases away from the boundary of the hole until it decreases almost to the original stress value, p, at a distance of about one diameter from the edge.

At the boundary of the hole the radial stress, commonly designated by σ_r must be zero. Therefore, the only stress acting will be the stress tangential to the boundary of the hole. Progressing away from the boundary the radial stress increases and the tangential stress decreases.

The stress illustrated in the diagram of Fig. 1 is the tangential stress, which will be designated in this book as σ_t. It is maximum at the ends of the horizontal axis

EF and is minimum at the ends of the vertical axis where its value under the conditions shown is $(-p)$, that is, a tensile stress equal in magnitude to p.

It should be noted that the sum of the areas *IHL* and *JKG* under the curves *IL* and *JK* represent additional load on plane *AB* and must be approximately equal to the original load represented by the cross-hatched area *HGFE*.

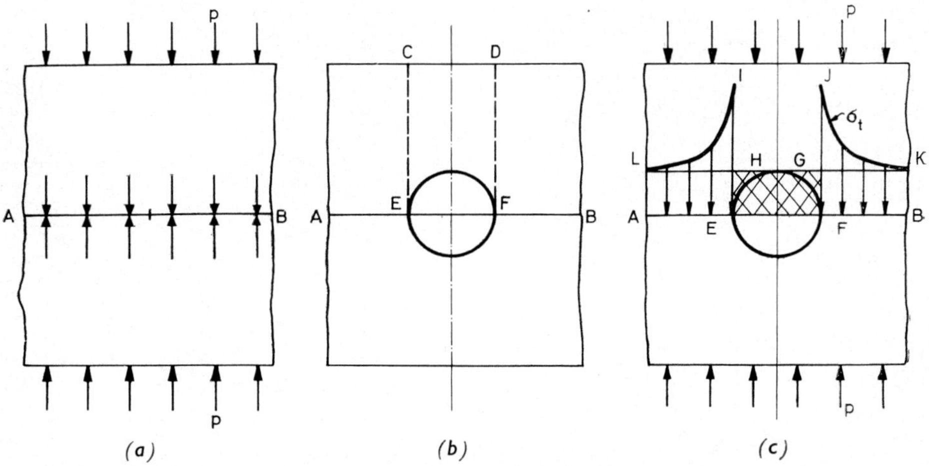

FIG. 1. (a) Elastic material subjected to unit pressure "p". (b) Hole created in elastic material. Pressure on block *CDFE* must be transmitted around the hole. (c) Stress distribution around a hole in elastic material.

The design of openings in competent rock is treated in Chapter 1, Part B. Design procedures are based on the assumptions that the rocks are unfractured and are perfectly elastic materials; it is further assumed that the physical properties of any given bed of a stratified rock are uniform throughout that bed. It is also assumed that non-stratified rocks are homogeneous.

DESIGN OF OPENINGS OF INCOMPETENT ROCKS

When an opening is to be excavated in a rock which will not support itself then the problem becomes one of designing support for the rock around the opening as well as designing the opening itself. The following design information is needed: (1) degree of fracturing of the rock; (2) orientation of fractures, or other planes of weakness, with respect to the opening boundary; (3) coefficient of internal friction of the rock; (4) rock compressive strength; (5) magnitudes of the natural rock stresses.

Information concerning some of these items may be unavailable, or difficult to obtain, and it may be necessary to use estimates based on past experience.

Design of an opening involves comparison of stresses which will be created around openings of various cross-sectional shapes, and selection of a shape which is the best compromise between working requirements and support requirements. Design of support involves calculation, or estimation, of the pressure which will be exerted on the support by the ground, selection of the type of support to be used and determination of the required size and spacing of said support.

The structural behavior of competent rock may be analyzed by relatively simple means but the behavior of incompetent rock can only be predicted, approximately, by making a number of assumptions to idealize the problem and then applying one of a number of rather complicated formulas.

There is no single formula which may be used for predicting the structural behavior of rocks over a wide range; rather, there are several formulas, each of which may be applied within a narrow range of rock conditions and for certain types of rock structures. The actual range of application, of any formula depends upon the structural condition of the rock, the configuration of the excavation, and the magnitudes of natural rock stresses.

Natural Structures

The excavation of any opening beneath the Earth's surface results in a redistribution of the natural (pre-existing) earth stresses. These may be visualized as realigning themselves around the perimeter of the opening and concentrating at certain critical points on the boundary.

If the soil or rock surrounding the opening is strong enough to resist these induced stresses then no significant fracturing occurs and the opening requires no lining or support. This is the condition encountered in most subsurface openings situated in rock at shallow to moderate depths beneath the surface.

However, at greater depths, or in weak rocks or soils, the over-stressed material at the boundary fractures. In such cases one of two conditions may occur.

(1) If the fractured material takes the form of interlocking fragments, or slabs, or if it retains sufficient cohesion, then the opening may stabilize itself and become self supporting after adjustment by fracturing has penetrated a certain distance from the opening boundary. This is the condition encountered in many openings located at shallow depths in soils or weak rocks, and also of openings located at great depths in strong rocks.

(2) If fractured material is "pulverulent", then fragments fall into the opening; or if it is plastic then material squeezes into the opening. Eventually the opening is filled and stability of the surrounding ground is restored.

These conditions are encountered in soils and weak rocks at moderate depths and in many of the stronger rocks at great depths. Obviously, such openings will not remain serviceable without lining or supports.

The behavior of soil or rock around an opening depends upon a number of factors of which the following are the most important:

(1) Magnitudes and directions of natural earth stresses.

(2) Size and shape of opening.

(3) Strength of the ground in which the opening is excavated.

When a sub-surface opening stands without support it is evident that natural structural principles are in action.

The two structural members which support themselves across openings are:

(1) The beam.

(2) The arch.

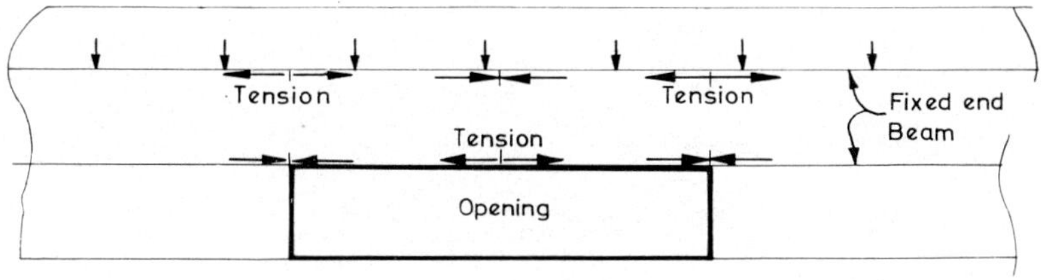

FIG. 2. General distribution of stress in a "fixed-end" beam over an opening.

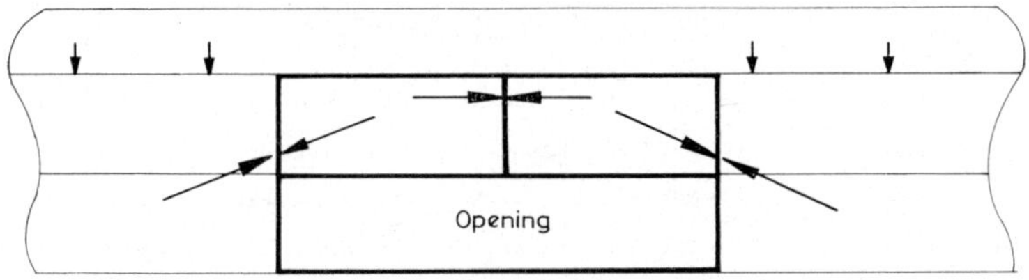

FIG. 3. A self-supporting "linear arch" formed by cutting the "fixed-end" beam. Stresses are all compressive.

The beam depends for its strength upon its internal resistance to bending and this resistance requires that the material composing the beam possess strength both in compression and in tension. In a "fixed-end" beam the forces are approximately as shown in Fig. 2.

The arch is a means of spanning an opening by resolving vertical pressure into horizontal or diagonal thrust.

If we take the beam of Fig. 2 and cut it as shown in Fig. 3 it will still support itself across the opening. However, the nature of the stresses has changed and stresses are compressive. These stresses at the points of concentration are much larger than in the case of a beam, but it is a characteristic of soils and rocks that compressive strengths are much greater than the tensile strengths, and are infinitely greater if the soil or rock is fractured.

The form of arch shown is the simplest possible form and illustrates the basic principle of "arching action".

Arching Action

When the term "arch" is mentioned it usually conveys a picture of vaulted opening, and the term "arching action" seems to signify the process by which the vaulted opening is formed. As used herein, however, the term *arching action* does not signify the shape of the opening but refers rather to the natural process by which a fractured, a pulverulent, or a plastic material acquires a certain amount of ability to support itself partially through the resolution of the vertical component of its weight into diagonal thrust.

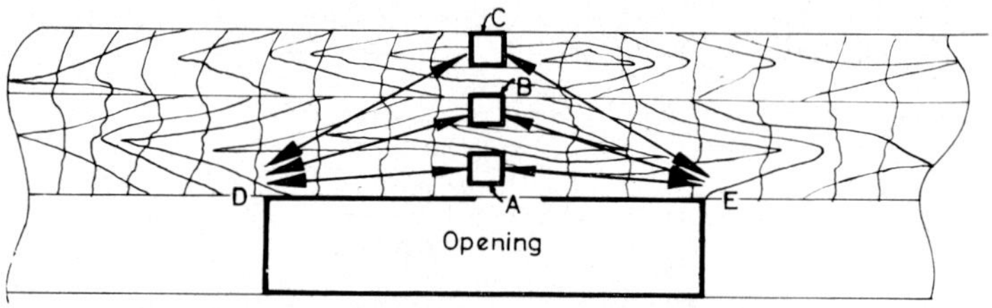

FIG. 4. Stress conditions and "arching action" in a pulverulent material.

To illustrate, if we now cut the "linear arch" of Fig. 3 with numerous random fracture planes, the material approaches a "pulverulent" condition as shown in Fig. 4. It is evident that, without cohesion, material will fall into the opening. Although the material cannot support itself it still retains a great deal of inherent strength as evidenced by the fact that in soils or fractured rocks a relatively small amount of installed support is sufficient to stabilize the opening. The strength of such support is only a fraction of that which would be needed to support the full weight of the ground overlying the opening.

Thus the material derives its self supporting tendency from the aforementioned "arching action", that is, the vertical component of its weight is resolved into diagonal thrust.

As shown in Fig. 4, the lower element *A* has little tendency to support itself because, as it tends to move downward, it creates little deformation along the reaction lines *AD* and *AE*. Consequently, little stress is created along these lines and there is almost no vertical component resisting downward movement of this element.

When element *C* attempts to move downward it creates a great deal of strain (shortening) along reaction lines *CD* and *CE*, and therefore a great deal of stress. The vertical component of this stress may be sufficient to prevent further downward movement of this element. Similar reasoning applies to all other elements above the opening.

It is evident that a relatively small force applied upward on the plane *DE* will stabilize the ground. This force need be only a fraction of that which would be required to support the dead weight of the column of ground overlying the opening.

The supporting force need only be sufficient to prevent failure (by shearing) of the ground under the compressive stresses along the resultants such as *AD* and *AE*, *BD* and *BE*, *CD* and *CE*.

If no support were installed then a high, vaulted arch would probably form. If support is properly installed it strengthens the ground structurally and enables it to support itself but no sharp dividing line can be drawn between supported ground and self supporting ground. The tendency of the ground to support itself increases with distance from the boundary of the opening.

Chapter 1, which follows, deals with the stresses which are induced in competent rocks around underground openings and with the design of such openings.

Chapter 2 deals with the natural (primitive or pre-existing) stresses which exist in undisturbed soils and rocks.

Subsequent chapters deal with the behavior of incompetent rocks and with methods for design of openings and their supporting structures in incompetent rocks.

CHAPTER 1

DESIGN OF UNDERGROUND OPENINGS IN COMPETENT ROCK

THE following discussion of stress distribution around underground openings is based on the following assumptions:[1]

(1) Massive rock obeys Hooke's law and is isotropic and homogeneous with respect to its elastic properties.

(2) The opening is in an infinite medium. This condition is satisfied if the distance from the opening to an adjacent boundary is greater than three times the dimension of the opening in the direction of the boundary.

(3) The opening is long compared with its cross-section, and the long axis of the opening is horizontal.

(4) The cross-section of the opening can be represented by a simple geometric shape, such as a circle, ellipse, ovaloid, or by a rectangle with rounded corners.

(5) The axes of the cross-sections of openings are vertical and horizontal unless otherwise specified.

(6) The stress distribution along the length of the opening is uniform and independent of length. For this condition the problem of determining the stress distribution around openings reduces to one of plane strain and may be solved by considering a hole in a wide plate subject to a uniform two-directional stress field in the plane of the plate.

(7) The vertical stress on a unit horizontal section of the rock is equal to the weight of the rock above the section; thus

$$\sigma_z = \gamma z$$

where σ_z is vertical compressive stress, γ is the density of the rock, and z is the vertical distance below the surface.

(8) The horizontal stress on a unit vertical section of the rock is given by

$$\sigma_x = M\sigma_z$$

where σ_x is horizontal compressive stress, σ_z is vertical compressive stress, and M is a constant depending on the type of stress field.

TYPES OF STRESS FIELDS

There are four general types of stress fields which might be encountered in practice. Three of these are represented graphically in Fig. 1 and given algebraically by the equations under the diagrams.

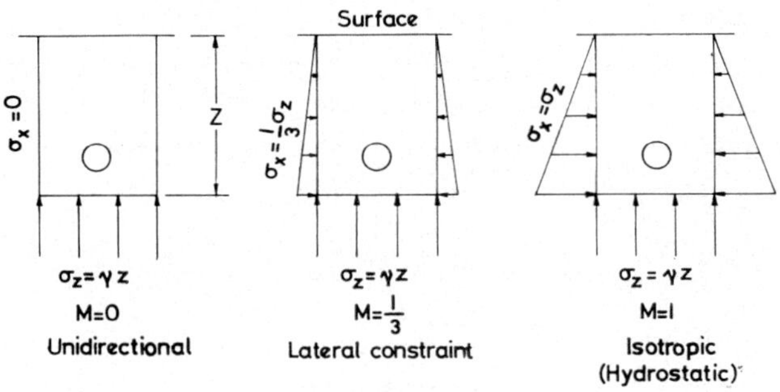

FIG. 1. Three types of stress fields assumed.[1]

The fourth type of stress field is that in which the horizontal field stress exceeds the vertical field stress. This stress condition is frequently encountered in the field and is probably much more common than is the condition of no lateral constraint. Stress concentrations around a circular opening for such a condition may be visualized by rotating Fig. 2 through 90°.

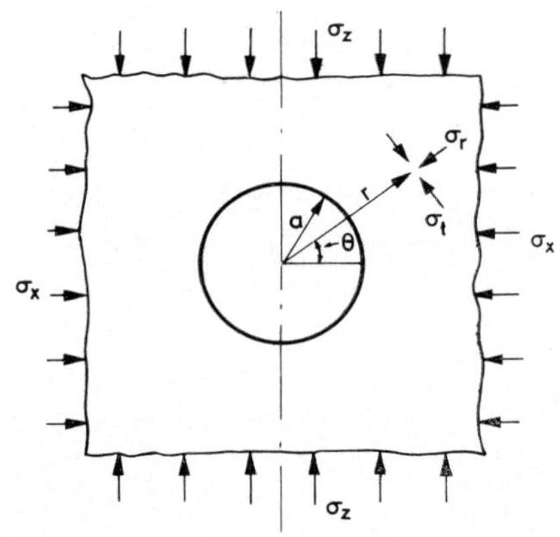

FIG. 2. Stress equilibrium around a circular opening in elastic material.

The following diagrams showing stress distribution around openings have been prepared on the assumption that $M = 0$; $M = \frac{1}{3}$; and $M = 1$.

The first condition assumes no lateral constraint.

The second assumes that lateral stresses at depth are the resultants of depth pressure, γz, and constraint which prevents lateral expansion of the rock. Under such conditions the lateral stress would be

$$\sigma_x = \frac{\mu}{1-\mu}\,\sigma_z\,.$$

Taking Poisson's ratio, μ, as $\frac{1}{4}$, the value of M would be $\frac{1}{3}$.

The third condition assumed is that of "hydrostatic" or "isotropic" stress; that is, that the vertical and lateral stresses are equal. The author prefers to use the term "isotropic" for this stress condition in intact rock since it avoids the implication that such stress conditions exist because the rock is in a plastic or fluid state, when such is not the case.

STRESSES AROUND A CIRCULAR OPENING

Equations (1), (2), and (3) show that stresses around the opening are concentrated at, or near, the boundary of the opening, and are independent of the elastic constants of the material and the radius of the hole. The radius a appears in these equations only in the dimensionless ratio a/r, which specifies the distance from the boundary of the hole.

For convenience the magnitude of the stress near an opening is expressed as a ratio of the stress at a certain point to the average applied stress outside the zone of disturbance. This ratio is referred to as the stress concentration. (A positive stress concentration means that the stress at a certain point has the same sign as the applied stress. A negative stress concentration means that the stress at a certain point has the opposite sign of the applied stress.)

Magnitudes of Boundary Stresses

Fig. 3 shows the tangential stress distribution at the boundary of a circular opening for stress fields of the three types shown in Fig. 1. For $M = 0$, the maximum boundary stress concentration at $\theta = 0°$ is 3 and at $\theta = 90°$ is (-1), with the minus indicating tensile stress.

For $M = \frac{1}{3}$ the critical stress concentration is 2.67 at $\theta = 0°$ and there is no stress at $\theta = 90°$. For $M = 1$ the critical stress concentration is 2 and occurs at all points on the boundary of the circle.

Maximum Compressive and Tensile Stresses

Stress concentrations along the axes from a circular opening in a unidirectional stress field ($M = 0$), have been calculated from equations (1), (2) and (3) and are shown in Fig. 4. This figure shows that tangential stresses are a maximum at the boundary of the opening, whereas radial stresses are zero. The maximum shear stress also occurs at the boundary and equals one-half the maximum tangential stress.

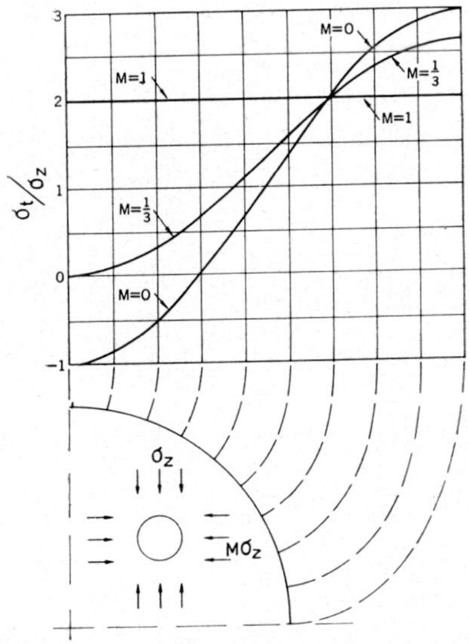

FIG. 3. Boundary stress concentrations for circular openings.[1]

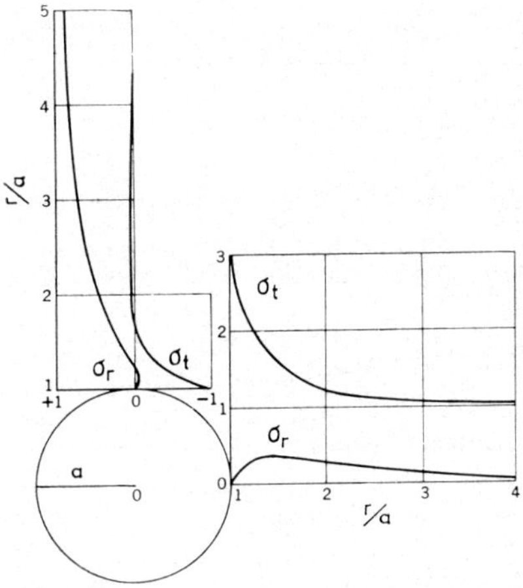

FIG. 4. Stress concentrations along axes of symmetry for circular openings; unidirectional stress field.[1]

General Equations for Stresses around Circular Openings in Elastic Material[1,2]

(See Fig. 2)

$$\sigma_r = \left(\frac{\sigma_x+\sigma_z}{2}\right)\left(1-\frac{a^2}{r^2}\right)+\left(\frac{\sigma_x-\sigma_z}{2}\right)\left(1-\frac{4a^2}{r^2}+\frac{3a^4}{r^4}\right)\cos 2\theta \qquad (1)$$

$$\sigma_t = \left(\frac{\sigma_x+\sigma_z}{2}\right)\left(1+\frac{a^2}{r^2}\right)-\left(\frac{\sigma_x-\sigma_z}{2}\right)\left(1+\frac{3a^4}{r^4}\right)\cos 2\theta \qquad (2)$$

$$\tau_{rt} = \left(\frac{\sigma_z-\sigma_x}{2}\right)\left(1+\frac{2a^2}{r^2}-\frac{3a^4}{r^4}\right)\sin 2\theta \qquad (3)$$

Where:

σ_x = horizontal applied stress;
σ_z = vertical applied stress;
σ_r = radial stress;
σ_t = tangential stress;
τ_{rt} = shear stress;
a = hole radius;
r = radial distance from the center of the hole;
θ = polar coordinate; horizontal axis represents $\theta = 0°$.

STRESSES AROUND ELLIPTICAL OPENINGS

General Equations

The general equation for tangential stress at the boundary of an elliptical opening in an infinitely wide plate subjected to a uniform two-dimensional stress field (see Fig. 5).

$$\sigma_t = \frac{[\sigma_x - \sigma_z][(p+q)^2 \sin^2\beta - q^2] + 2pq\sigma_z}{(p^2 - q^2)(\sin^2\beta + q^2)} \tag{4}$$

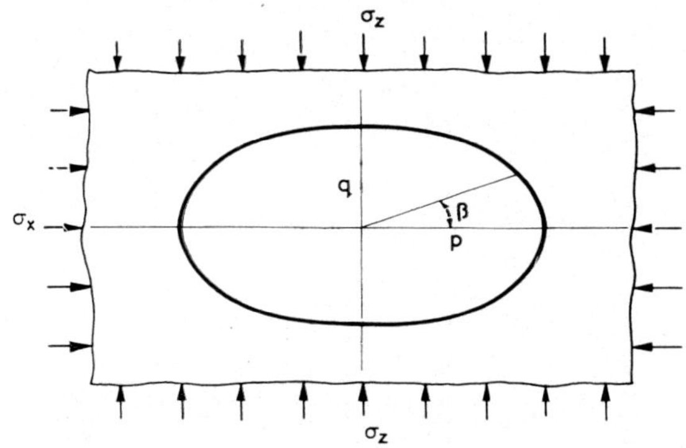

FIG. 5. Stress equilibrium around an elliptical opening in elastic material.

For a unidirectional stress field: maximum and minimum stresses at the boundary occur at the ends ot the axes, that is, when $\beta = 90°$ and $\beta = 0°$. Assuming that $\sigma_x = 0$ then;

$$\sigma_{t(\beta=0°)} = \sigma_z \left(1 + 2\frac{q}{p}\right) \tag{5}$$

$$\sigma_{t(\beta=90°)} = -\sigma_x \tag{6}$$

For an isotropic stress field the basic equation is

$$\sigma_t = 2\sigma \left[1 - (p-q)\frac{p \sin^2\beta - q \cos^2\beta}{p^2 \sin^2\beta + q^2 \cos^2\beta}\right] \tag{7}$$

Maximum and minimum stresses occur at the ends of the axes

$$\sigma_{t(\beta=0°)} = 2\sigma \frac{p}{q} \tag{8}$$

$$\sigma_{t(\beta=90°)} = 2\sigma \frac{q}{p} \tag{9}$$

(Additional equations for boundary stresses in bi-directional stress fields are given in Chapter 5.)

Magnitudes of Compressive and Tensile Boundary Stresses

Boundary stresses for ellipses with four different width/height ratios have been calculated and are shown graphically in Fig. 6.

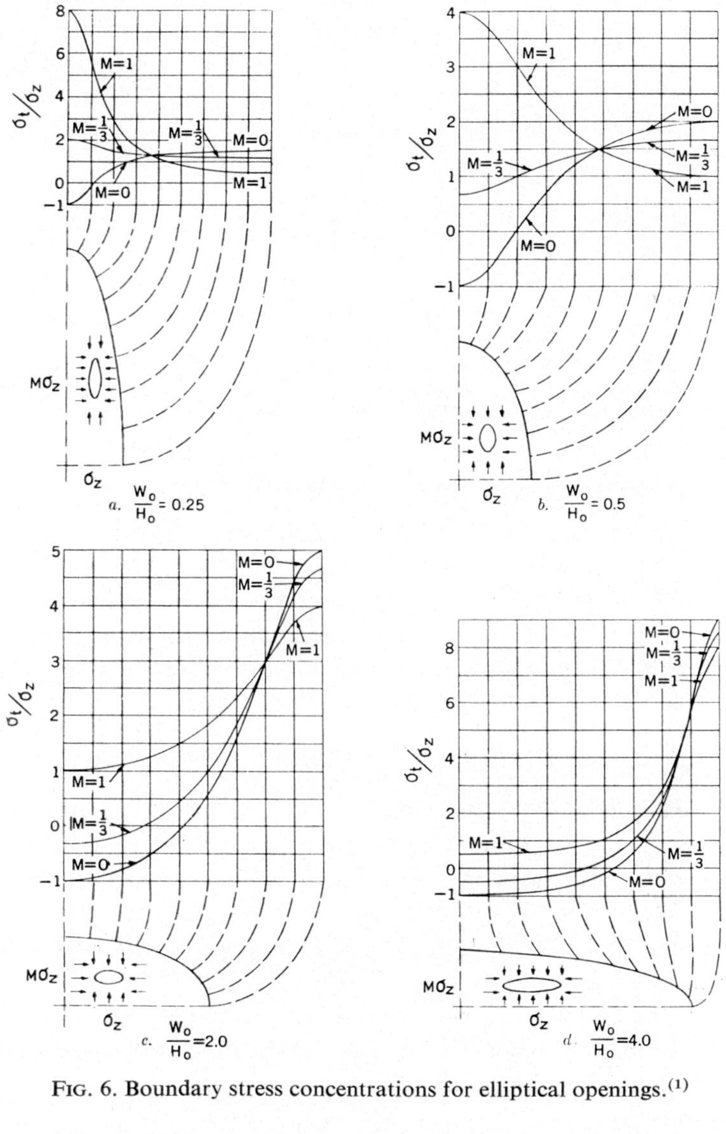

FIG. 6. Boundary stress concentrations for elliptical openings.[1]

For the unidirectional stress field $M = 0$, the maximum stress concentration at the end of the horizontal axis increases as the width/height ratio increases, whereas the stress concentration at the top and bottom of the opening remains constant at a value of minus one, signifying tension.

For the two-directional stress field, $M = \frac{1}{3}$, the boundary stress concentration at the end of the horizontal axis increases with the width/height ratio, and the stress concentration at the end of the vertical axis changes from large positive values to small negative values.

The isotropic stress field produces maximum stresses on the horizontal axis for width/height ratios greater than 1 and on the vertical axis for width/height ratios less than 1.

STRESSES AROUND OVALOIDAL OPENINGS

Ovaloids can be approximated by the following parametric equations:

$$x = p \cos \beta + r \cos 3\beta \tag{10}$$

$$y = q \sin \beta - r \sin 3\beta \tag{11}$$

where p, q and r are parameters of ovaloid, β is the angle, and x and y are rectangular coordinates.

Magnitudes of Compressive and Tensile Boundary Stresses

Boundary stresses for ovaloids with four different width/height ratios, and for three different stress fields, are shown in Fig. 7.

The maximum boundary stresses around approximately ovaloidal openings do not occur on the axes of the openings, as in elliptical openings, but are shifted toward the junction of the semi-circular end and the straight sides.

STRESSES AROUND RECTANGULAR OPENINGS

Perfectly rectangular openings would produce infinitely large stress concentrations at the right-angle corners; therefore, rounded or filleted corners are assumed. (Mining conditions will normally result in a rounded corner in a rectangular opening.)

The tangential stress distribution around a rectangular opening with rounded corners has been studied by the photoelastic method.[4, 5]

Fig. 8 shows the boundary stresses around five rectangles with a constant ratio of fillet radius to short dimension for stress fields of three types. In all cases the stress concentration is maximum at the corners.

STRESSES AROUND INCLINED OPENINGS

Elliptical and rectangular openings having cross-sectional axes which are inclined at an angle to the horizontal and vertical are considered in this section.

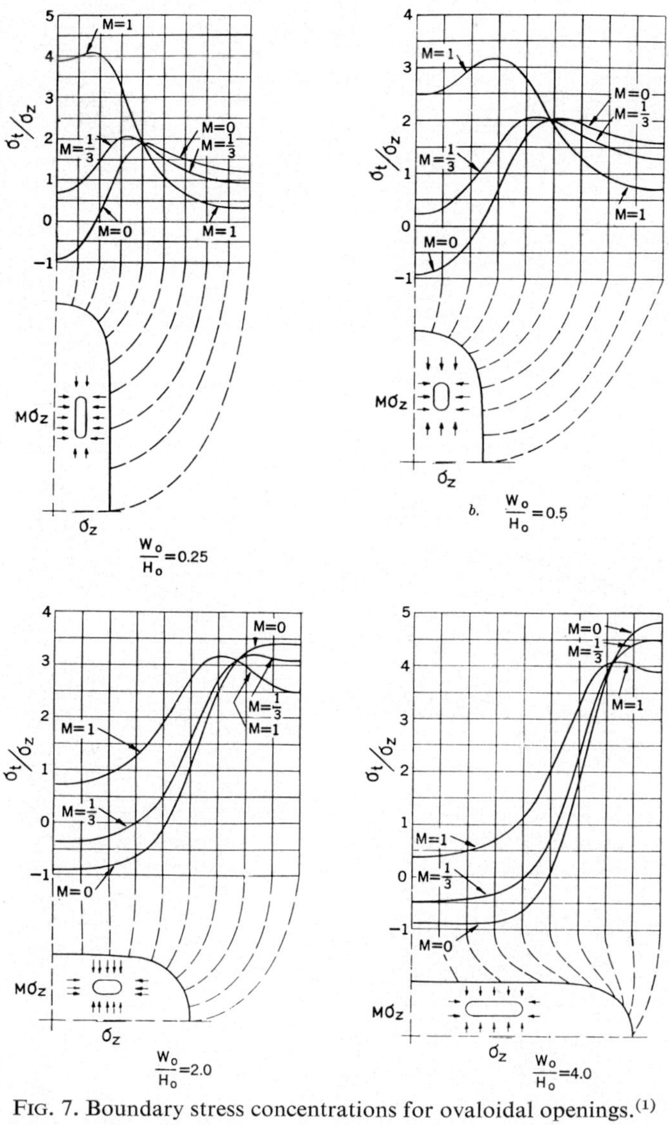

FIG. 7. Boundary stress concentrations for ovaloidal openings.[1]

Magnitudes of Boundary Stresses around Inclined Elliptical Openings

Panek[5] has computed critical boundary stress concentrations for elliptical openings for three width/height ratios at various angles of inclination and three types of stress fields. Fig. 9 defines the variables involved and Table 1 summarizes the results of the calculations.

The critical stress concentrations are at the ends of the major and minor axes for $\delta = 0$ or $90°$. For other values of δ the critical stress concentrations are near the ends of the major axis.

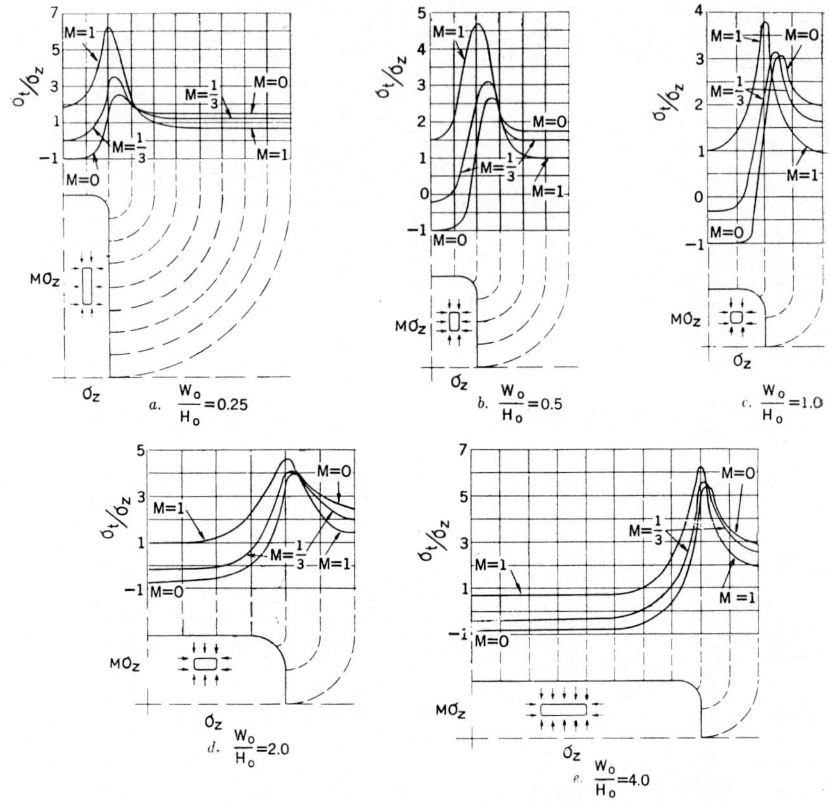

FIG. 8. Boundary stress concentrations of rectangular openings with rounded corners; ratio of fillet radius to short dimensions, 1 to 6.(1)

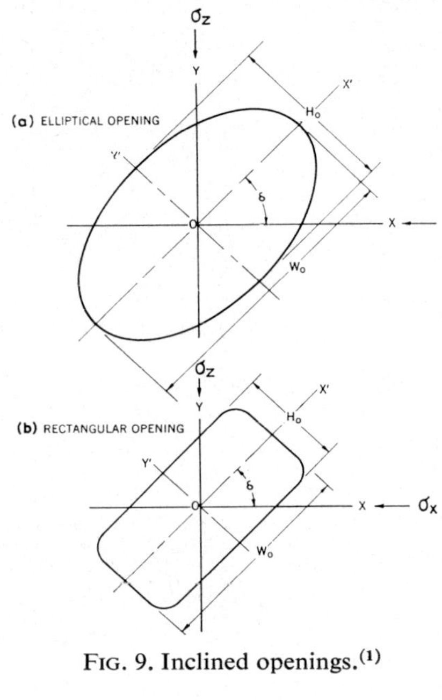

FIG. 9. Inclined openings.(1)

TABLE 1. CRITICAL STRESS CONCENTRATION FOR ELLIPTICAL OPENINGS AS FUNCTION OF ANGLE OF INCLINATION OF MAJOR AXIS TO HORIZONTAL[5]

W_o/H_o	δ, °	$M=0$		$M=1/3$		$M=1$
		Tension	Comp.*	Tension	Comp.*	Comp.*
1	0–90	−1.0	3.0	0	2.7	2.0
2	0	−1.0	5.0	−.33	4.7	4.0
	22.5	−1.0	4.6	−.29	4.3	4.0
	45	−1.1	3.7	−.13	3.5	4.0
	67.5	−1.1	2.5	†	2.4	4.0
	90	−1.0	2.0	†	1.7	4.0
3	0	−1.0	7.0	−.44	6.7	6.0
	22.5	−1.1	6.3	−.43	6.1	6.0
	45	−1.3	4.6	−.29	4.7	6.0
	67.5	−1.2	2.7	†	3.0	6.0
	90	−1.0	1.7	†	1.33	6.0
4	0	−1.0	9,0	−.50	8.7	8.0
	22.5	−1.2	8.0	−.52	8.0	8.0
	45	−1.5	5.8	−.42	6.1	8.0
	67.5	−1.4	3.0	†	3.7	8.0
	90	−1.0	1.5	†	2.0	8.0

* Compression.
† No critical tensile stress concentration.

Magnitudes of Boundary Stresses around Inclined Rectangular Openings

Panek[5] has also determined, by means of photoelastic studies, the critical stress concentrations around rectangular openings with rounded corners. Table 2 summarizes his results for various angles of inclination and width/height ratios. The ratio of the fillet radius to the height is $\frac{1}{6}$.

For $M = 0$ or $\frac{1}{3}$, the critical compressive stress concentrations are maximum for $\delta = 45°$. For $M = 1$ the critical compressive stress concentration is virtually independent of the angle of inclination. Critical tensile stress concentrations are large for $M = 0$ and for angles of inclination of 22.5–67.5°. For $M = \frac{1}{3}$ the critical tensile stress concentrations are small and for $M = 1$ they are zero.

STRESSES AROUND SPHERICAL AND SPHEROIDAL OPENINGS

The openings previously considered have been those whose length was large compared to cross-sectional dimensions.

If the length of the opening is not large compared with the cross-sectional dimensions the problem must be considered in three dimensions and becomes much

TABLE 2. CRITICAL STRESS CONCENTRATION FOR RECTANGULAR OPENINGS WITH ROUNDED CORNERS AS FUNCTION OF ANGLE OF INCLINATION OF MAJOR AXIS TO HORIZONTAL[5]

W_o/H_o	δ, °	$M=0$		$M=1/3$		$M=1$
		Tension	Comp.*	Tension	Comp.*	Comp.*
1	0	−1.0	3.1	−0.3	3.1	3.8
	22.5	−.5	3.9	†	3.8	3.7
	45	−1.1	4.7	†	4.3	3.6
	67.5	−1.3	4.0	−.3	3.9	3.7
	90	−1.0	3.1	−.3	3.1	3.8
2	0	−.8	4.0	−.1	4.1	4.7
	22.5	−.7	5.0	†	4.7	4.6
	45	−1.6	5.7	†	5.2	4.5
	67.5	−1.4	4.5	−.1	4.5	4.6
	90	−1.0	2.7	−.2	3.1	4.7
3	0	−.8	4.6	−.4	4.7	5.2
	22.5	−.7	5.9	−.1	5.5	5.2
	45	−1.8	6.5	−.4	6.0	5.3
	67.5	−1.6	5.0	†	5.0	5.2
	90	−1.0	2.6	−.1	3.3	5.2
4	0	−.9	5.4	−.4	5.6	6.2
	22.5	−1.0	6.5	−.1	6.0	5.9
	45	−2.0	7.1	−.5	6.6	5.9
	67.5	−1.9	5.5	†	5.5	5.9
	90	−1.0	2.5	†	3.5	6.2

* Compression.

† No critical tensile stress concentration.

more difficult to solve mathematically. Only two problems of this type have been solved, the sphere and the spheroid.[6]

Fig. 10 shows the circumferential principal stress at the center of the roof of a spheroidal opening as a function of its width/height ratio, and Fig. 11 shows the tangential principal stress at the mid-height of the wall surface as a function of the width/height ratio.

The critical stresses for tunnels with the same cross-sectional ratios are also plotted for comparison purposes.

The stress distribution around spherical and spheroidal openings depends upon the values of Poisson's ratio whereas the stress distribution around openings which are long are independent of the elastic constants.

Usually the maximum boundary stresses around spherical and spheroidal openings are 30–40 per cent lower than around long openings having the same cross-sectional shape.

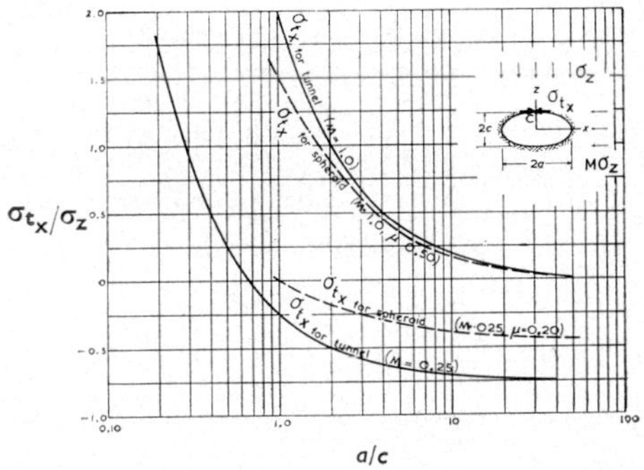

FIG. 10. Circumferential principal stresses, σ_{t_x}, at center of roof of cavity in homogeneous stress fields.[6]

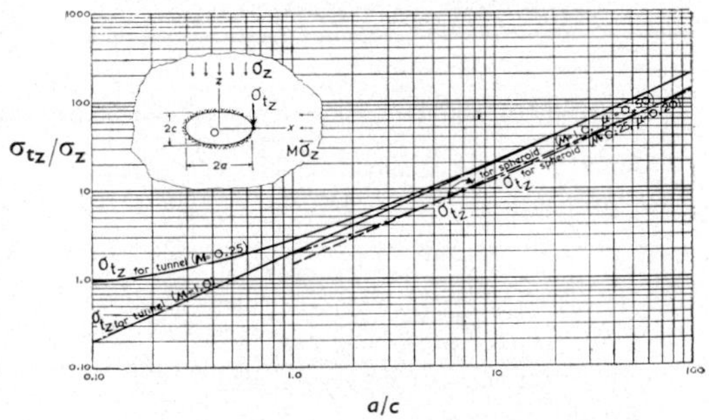

FIG. 11. Circumferential principal stress, σ_{t_x}, at mid-height of wall surface of cavity located in homogeneous stress fields[6].

STRESSES AROUND GALLERIES

Fig. 12[6] shows the stress distribution around an opening which has a cross-sectional shape typical of the galleries provided for inspection purposes in large concrete dams. Stresses were determined photoelastically and the stress distribution shown is for an assumed M value of 0.25. This cross-section is typical of that of many tunnels and of many mine openings, such as adits, drifts, and crosscuts.

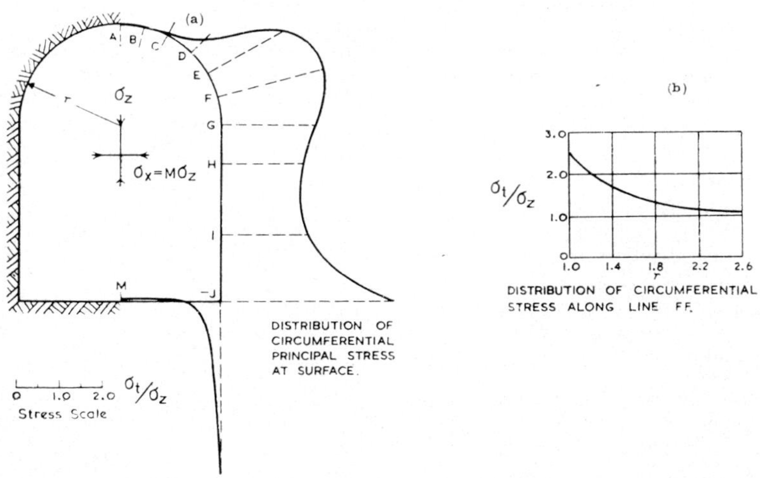

FIG. 12. Stress distribution about an inspection gallery in homogeneous stress fields ($M = 0.25$).[6,7]

It will be noted that maximum stress concentration occurs at the sharp corner at the junction of the wall and the floor. Fracturing of rock at this point would not seriously affect the stability of the ordinary tunnel or mine opening because the fractured rock would not be a hazard and would not require support as would rock which broke from the roof or walls.

SUMMARY – STRESS CONCENTRATIONS FOR VARIOUS SHAPES OF OPENING IN VARIOUS STRESS FIELDS

The results of the preceding studies are summarized in the set of critical stress curves shown in Figs. 13, 14, 15 and 16. These curves show the maximum stresses induced around openings of various cross-sections and indicate which cross-sections are the most likely to be self-supporting for any given width/height ratio in various stress fields.

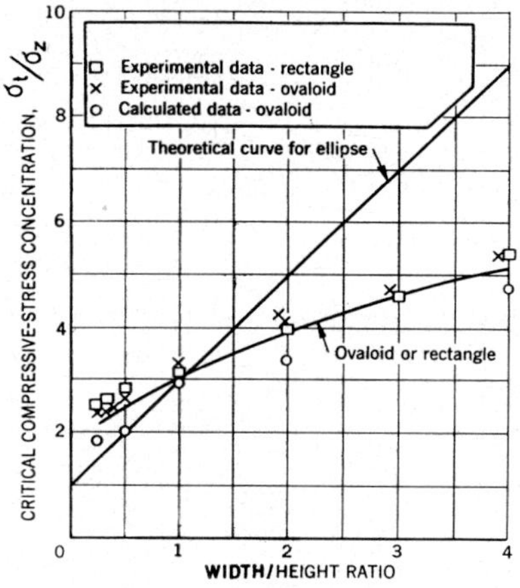

FIG. 13. Critical compressive stress concentration for openings of various cross-sections; unidirectional stress field, $M = 0$.(1)

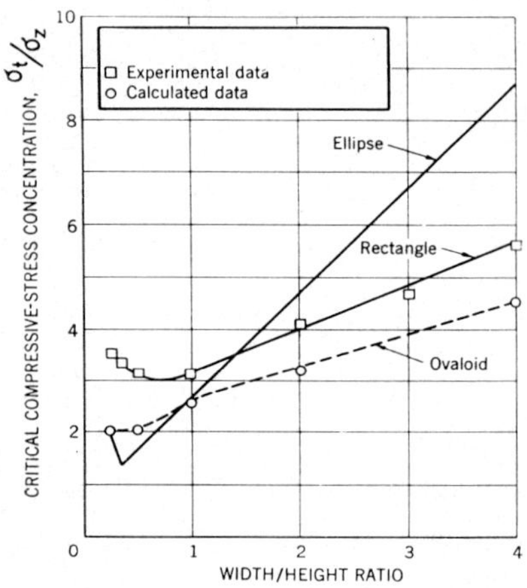

FIG. 14. Critical compressive stress concentrations for tunnels of various cross-sections; two-dimensional stress field, $M = \frac{1}{3}$.(1)

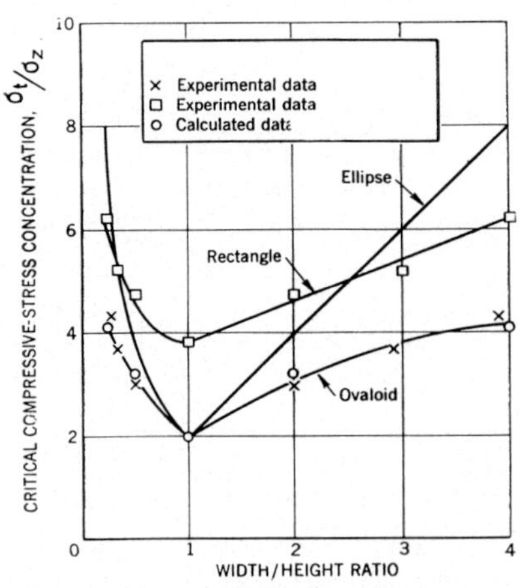

FIG. 15. Critical compressive stress concentration for tunnels of various cross-sections; hydrostatic stress field, $M = 1$.(1)

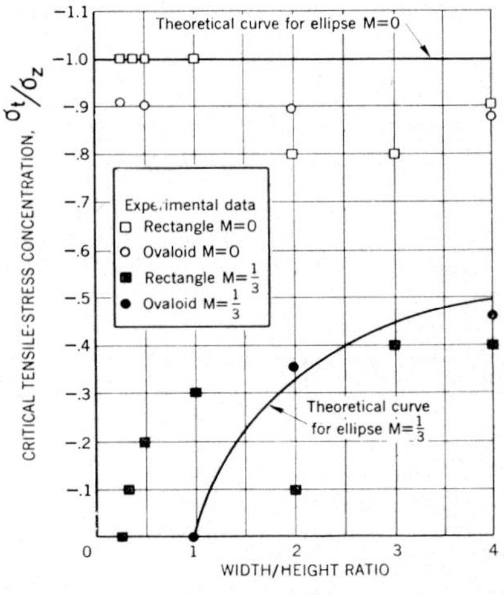

FIG. 16. Critical tensile stress concentration for tunnels of various cross-sections; two types of stress fields, $M = 0$ and $M = \frac{1}{3}$.(1)

SHEAR STRESSES

The foregoing sections have dealt with the stresses tangential to the boundaries of openings of various shapes. It has been noted that tensile stresses, when they occur, are considered critical because most rock is very weak in tension.

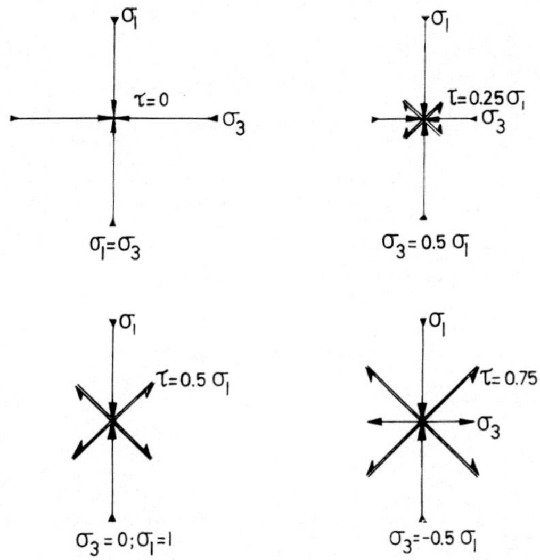

FIG. 17. Maximum shear stresses resulting from various combinations of the principal stresses. Computed from the formula

$$\tau = \frac{\sigma_1 - \sigma_3}{2}.$$

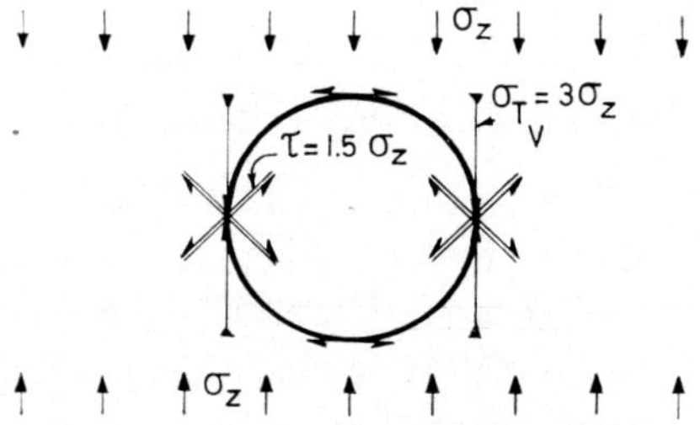

FIG. 18. Principal stresses and shear stresses at the boundary mid-height of a circular opening.

In general, it appears that tensile stresses will usually be produced in the roofs of narrow openings only when lateral constraint is lacking and the natural (primitive) stress is a unidirectional vertical stress. The author believes that such stress conditions (no lateral constraint) must occur very rarely in nature and that the condition of tangential tensile stress above narrow openings must also occur rarely (see Chapter 2, Part B).

It appears therefore that most rock failure around single openings occurs by fracture under excessive compressive stress. There are two types of fracture which occur when a brittle material is compressed. These are:

(1) Failure and spalling of thin slabs from the rock surface due to transverse tensile forces in the "skin" of the opening. The nature and origin of these tensile forces are described in more detail in Chapters 3 and 4 (Part B) and will not be elaborated on here.

(2) Failure of the rock in the highly compressed zones by shearing. In contrast to the tension slabbing which is a surface phenomenon, shear fracturing may extend for considerable distances into the walls, or roof and floor, of an opening. (The process is described in more detail in Chapters 3 and 4.)

Fig. 17 illustrates the shear stresses which are induced by various combinations of the principal stresses. The magnitude of maximum shear stress at any point may be computed from the equation

$$\tau = \frac{\sigma_1 - \sigma_3}{2} \tag{12}$$

Maximum Shear Stress

Maximum shear stress occurs on planes which make angles of 45° with the major principal stress. As shown in Fig. 17 and indicated by equation (12), the magnitudes of such shear stresses are functions of the difference between the principal stresses. Since one principal stress is zero at the boundary of the opening the shear stresses will be maximum at the opening boundary and will decrease away from the boundary. Maximum shear stress will occur at the opening boundary at that point where the compressive stress is maximum.

Orientations of Shear Fractures around Single Openings

It has been noted that the maximum shear stresses make angles of 45° with the principal stresses. However, fracturing does not take place along these planes but along the planes of maximum *effective shear* stress. These planes make angles of $45°-\Phi/2$ with the major principal stress (where Φ is the angle of internal friction).

Fig. 19–26 show graphically the magnitudes and alignments of the principal stresses and the effective shear stresses which are induced by the construction of circular, or nearly circular, openings in various types of stress fields. These effective shear stresses, τ_{es}, have been computed from equation (8), Chapter 4. The equation is:

$$\tau_{es} = \frac{\sigma_1 - \sigma_3}{2} \sin (90° - \Phi)$$

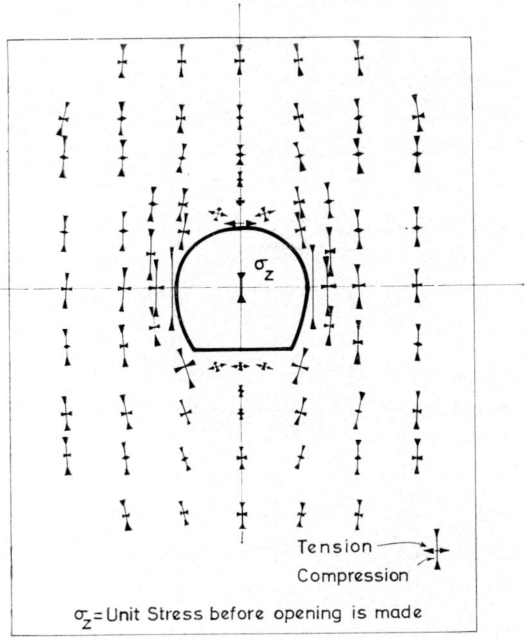

FIG. 19. Magnitudes and directions of the principal stresses around an opening located in an elastic material which is subjected to a uniform vertical compressive stress and which is unrestrained laterally.[8]

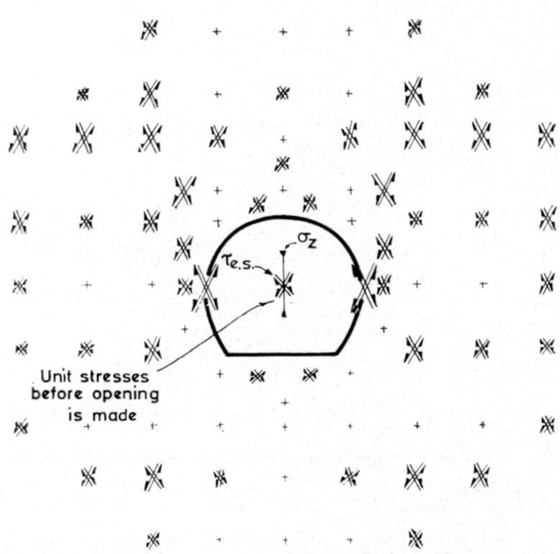

FIG. 20. Magnitudes and directions of the maximum effective shear stresses at the same points shown in Fig. 19 (Φ assumed as 40°).

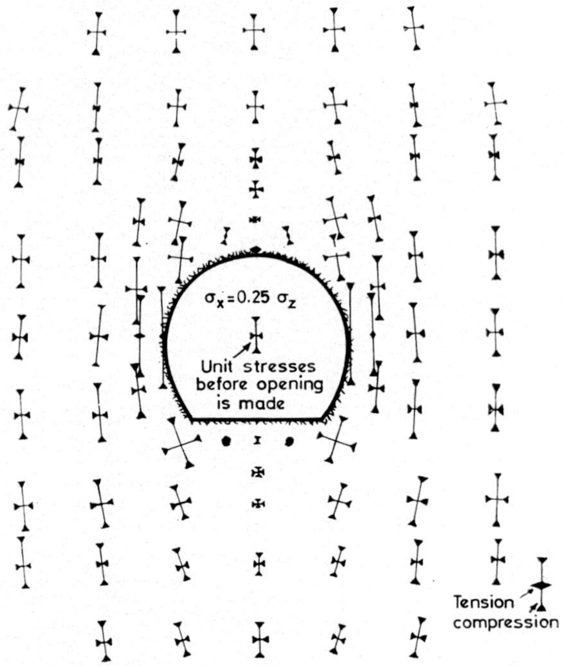

FIG. 21. Magnitudes and directions of the principal stresses around an opening located in an elastic material which is subjected to a uniform vertical stress and is restrained laterally so that $\sigma_x = 0.25\,\sigma_z$.[8]

FIG. 22. Magnitudes and directions of maximum effective shear stresses at the same points shown in Fig. 21 (Φ assumed as 40°).

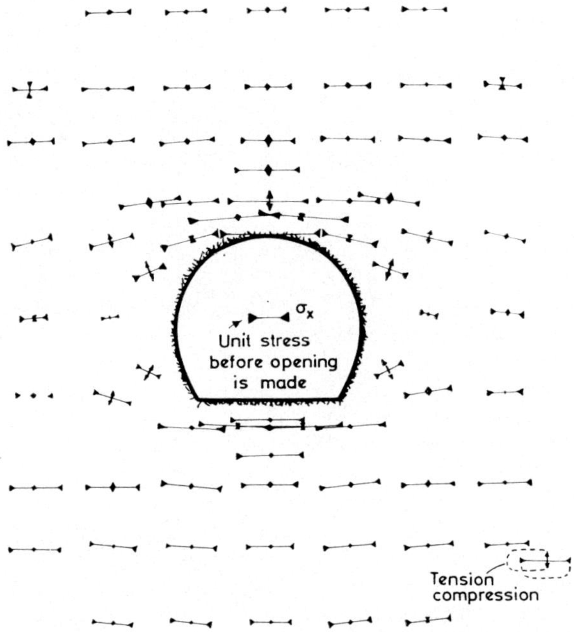

FIG. 23. Magnitudes and directions of the principal stresses around an opening located in an elastic material which is subjected to a uniform horizontal compressive stress and which is unrestrained vertically.[8]

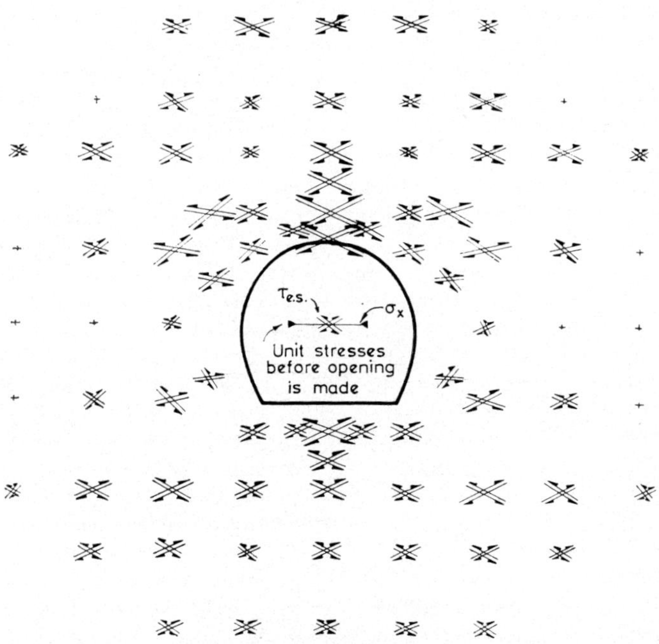

FIG. 24. Magnitudes and directions of maximum effective shear stresses at the same points shown in Fig. 23 (Φ assumed as 40°).

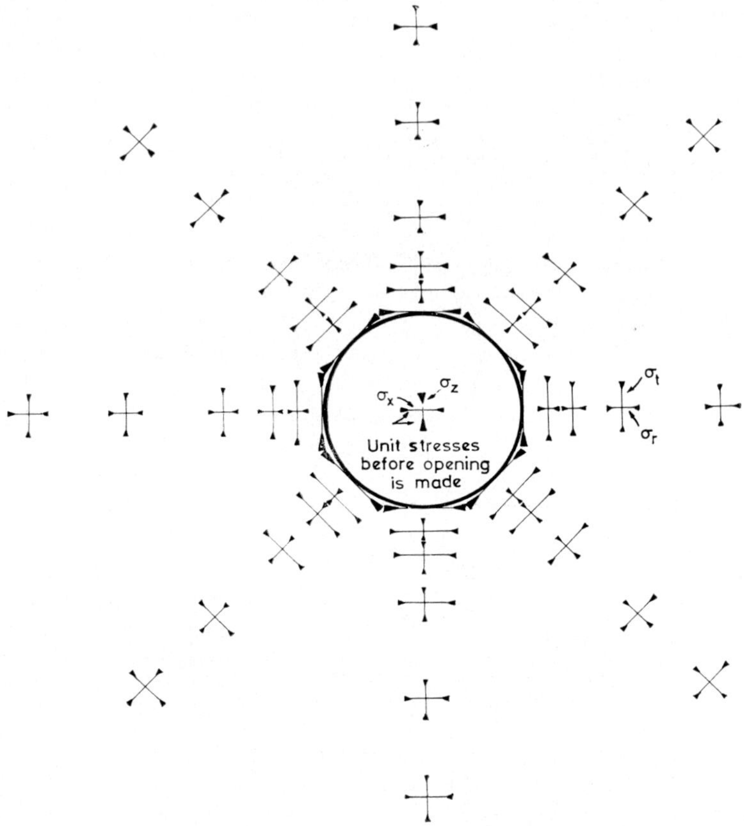

FIG. 25. Magnitudes and directions of the principal stresses around an opening located in an elastic material which is subjected to "isotropic" stress.

FIG. 26. Magnitudes and directions of maximum effective shear stresses at the same points as shown in Fig. 25 (Φ assumed as 40°)

Shear Stresses around Wide Openings — Photoelastic Stress Analysis

As used herein the term "wide opening" is used to designate an underground excavation whose width is greater than two or three times its height. Thus the term would include wide rooms and development headings as well as stopes and the mined-out areas behind longwall faces.

Fig. 27 (a) shows the typical pattern of alternating light and dark loops which is obtained when a transparent plastic model is subjected to loading and examined under polarized light. The principles of photoelastic stress analysis are discussed in Chapter 11 (Part B).

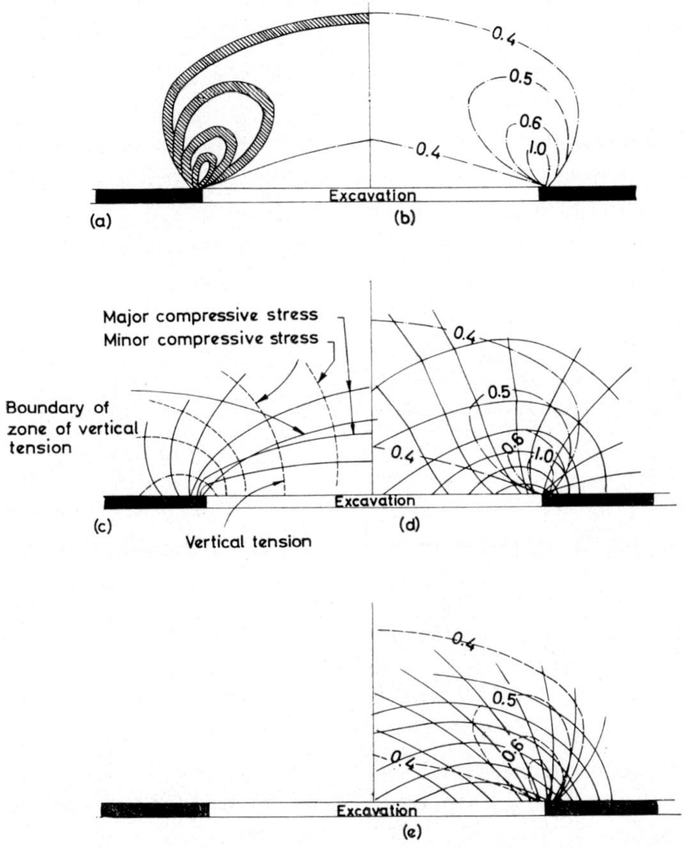

FIG. 27. (a) and (b) Contours connecting all points of equal magnitudes of shear stress, as shown by photoelastic experiments. (Isotropic stress field.) (c) Directions of the principal stresses. (Adapted from Hackett[9]. (d) Directions and relative magnitudes of the maximum shear stresses. (e) Directions of probable shear failure for rock with an angle of internal friction (Φ) of 40°.

The loops are isochromatic lines. At every point on such a line the shearing stresses have equal magnitudes; that is, the lines are "contours" of equal shear stresses. In Fig. 27(b) the inner loop represents all points for which the shearing stress has a relative magnitude of 1.0. The next loop outside represents all points for which the shearing stress has a relative magnitude of 0.6, etc.

Points where the loops converge are points of high stress concentration. Stress concentrations commonly occur at corners and acute angles. The concentration of stress as shown at the ends of the openings in Fig. 27 is typical.

This type of fringe pattern indicates that stress concentrates in the vicinity of the face, but it does not tell us the directions of the shear stresses or the directions of the resulting fractures.

The direction of an isochromatic line at any point has no definite relationship to the orientation of the shear stress at that point.

Stress Fringes around an Opening

Fig. 27(a) shows photoelastic stress fringes typical of those in a transparent model subjected to isotropic compression; that is, a model subjected to horizontal and vertical stresses of equal magnitude.

In Fig. 27(b) the dashed loops are contours connecting points of equal shear stress. The maximum shear stress occurs in the zone enclosed by the 1.0 curve; outside this loop the shear stress decreases to the relative magnitudes indicated by contours marked successively by curves of 0.6, 0.5 and 0.4.

The stress distribution shown in illustrations (a–d) was deduced analytically by Hackett [(9)]. Similar photoelastic stress patterns are used frequently in the literature to illustrate stress distribution around openings in elastic materials.

Directions of Maximum Effective Shear Stresses are as Important as Magnitudes

Fringe patterns such as that illustrated in Fig. 27 (a) and (b) are of limited value in predicting the orientation of fractures at the boundary of an opening in brittle material such as rock. Such stress contours show only the relative magnitudes of the shear stresses at various points and have no relationship to the directions of such stresses. Thus they fail to give any information as to the orientation of the maximum *effective* shear stress relative to the orientation of the exposed rock surface. This is one of the most important of the factors which determine the stability of an opening.

Principal Stress Patterns

In order to visualize the nature of fracture around an opening it is desirable to derive a stress net showing the directions of the principal stresses. Such a net is shown in Fig. 27(c). The solid lines show compressive stresses and the dotted lines show tensile stresses.

Derivations of the directions of the principal stresses may be carried out analytically, or by means of photoelastic processes. Either process requires a great deal of time and effort and, in spite of the value of such derivations in visualizing potential fracture patterns, they are seldom carried out.

Orientation of Shear Stresses and of Fracture Planes

Directions of the maximum shear stresses may be sketched in from the principal stress net. The maximum shear stresses make angles of 45° with the principal stresses. Directions of the maximum shear stresses are sketched in Fig. 27 (d).

In actuality the maximum shear stresses show only approximately the directions of potential fracture planes because it is the maximum *effective* shear stresses which control the directions of fracture rather than the maximum shear stresses.

In Fig. 27 (e) the directions of the maximum *effective* shear stresses are sketched. Such stresses make angles of $45° - \Phi/2$ (see Chapters 3 and 4) with the major principal stress. In this example the rock is assumed to possess an angle of internal friction, Φ, of 40°. Thus the maximum effective stress is oriented at 25° from the major principal stress and shear fracture occurs along the planes of maximum effective stress.

Referring to Fig. 27 (e) we see that shear fracture is most likely to occur in the vicinity of the face and that fracture planes will be inclined at some angle between 40 and 90° with the horizontal.

Effects of Shear Fracture (Induced Cleavage)

The orientation and the dip of the shear fractures (induced cleavage) determine how seriously they will affect the stability of the opening. Generally, when viewed in plan, they will be oriented parallel with the long dimension of the opening. When viewed in a section cut normal to the face the cleavage planes will be inclined to the horizontal. Those fractures which dip at angles steeper than about 60° may not seriously weaken the roof since the fractured elements may still support themselves in the manner of a linear arch. When fractures dip at angles of less than about 40°, however, they form a series of relatively thin, flat-lying plates which have no self-supporting ability whatever. In such cases much timber is required for support.

STRESSES AROUND MULTIPLE OPENINGS

A single opening in a uniform stress field disturbs the stress for a distance of approximately twice the diameter from the opening. Thus, when two openings are less than two diameters apart, the stress distribution around one opening affects the stress distribution around the other.[1]

Because of the complexity of the problem only a few examples involving relatively simple shapes have been treated analytically.

The stress distribution around multiple openings such as ovaloids and rectangles with rounded corners have been determined from photoelastic models.

Row of Circular Openings

The stress distribution around an infinite row of equal-sized circular holes equally spaced in an infinitely wide plate subjected to a uniform stress, either normal to, or parallel to, the line of holes has been studied theoretically and the results of the study are summarized in Figs. 28, 29, 30, and 31.

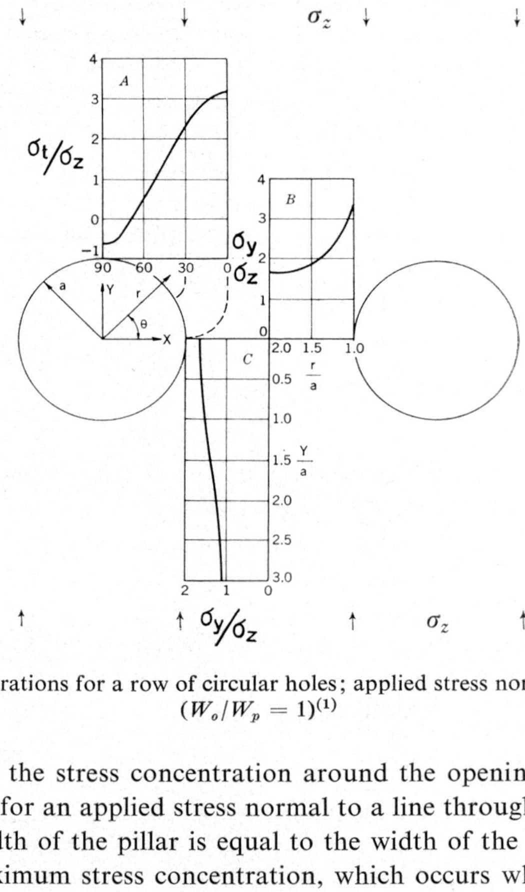

FIG. 28. Stress concentrations for a row of circular holes; applied stress normal to line of centers. $(W_o/W_p = 1)$[1]

Figure 28 shows the stress concentration around the opening and through the center of the pillar for an applied stress normal to a line through the centers of the holes when the width of the pillar is equal to the width of the opening. It will be noted that the maximum stress concentration, which occurs where the horizontal axis intersects the boundary of the hole, is 3.29, while the average stress concentration in the pillar along the horizontal axis of the holes is about 2. It can be inferred from this that the average stress in the pillar increases more rapidly than the maximum stress and that the average stress in the pillar will approach the maximum stress for large stress concentrations.

Figure 29 shows the stress concentrations around the opening and in the pillar when the applied stress is parallel to the line of centers and

Width Opening/Width Pillar = 1.

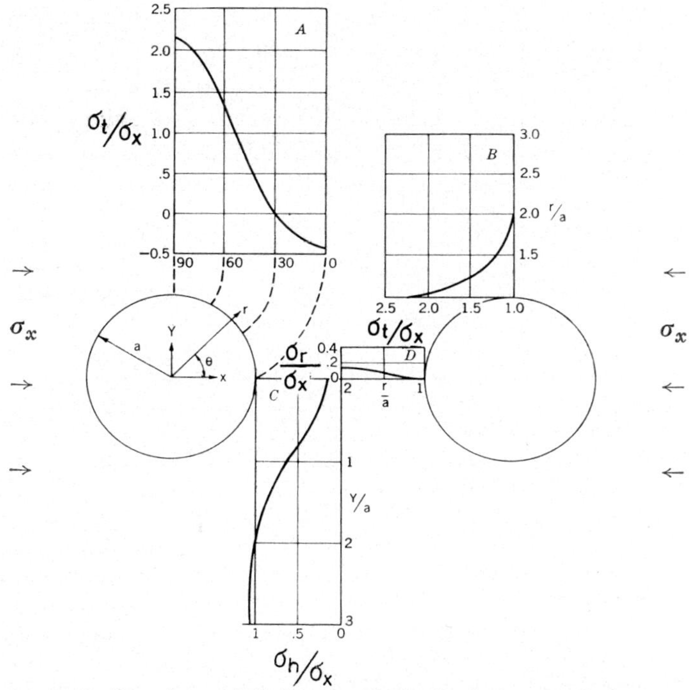

FIG. 29. Stress concentrations for an infinite row of circular holes; applied stress parallel to line of centers ($W_o/W_p = 1$).[1]

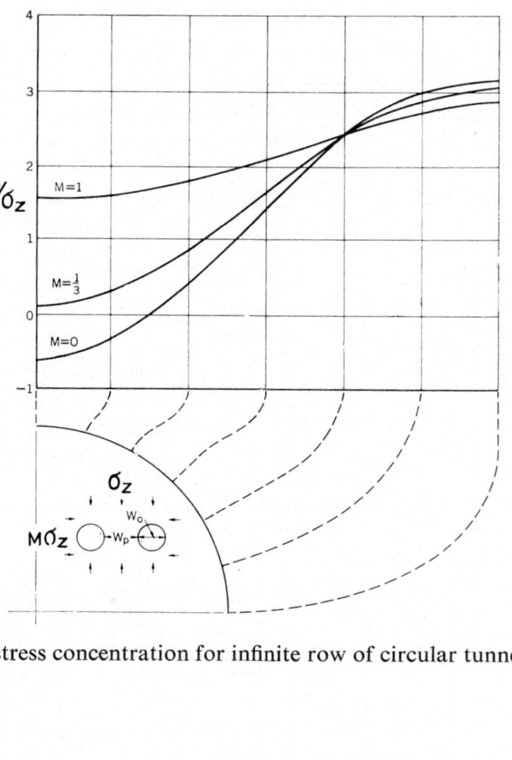

FIG. 30. Boundary stress concentration for infinite row of circular tunnels ($W_o/W_p = 1$).[1]

Figure 30 shows the boundary stress concentrations around one of a row of circular holes spaced one diameter apart in a two-directional stress field. The results are plotted for M values of 0, $\frac{1}{3}$, and 1.

If these stresses are compared with the stresses around a single circular opening as shown in Fig. 3 it will be noted that the maximum boundary stress concentrations on the sides of the multiple holes are larger, and stress concentrations at the top and bottom are smaller for a row of circular holes than for the single opening.

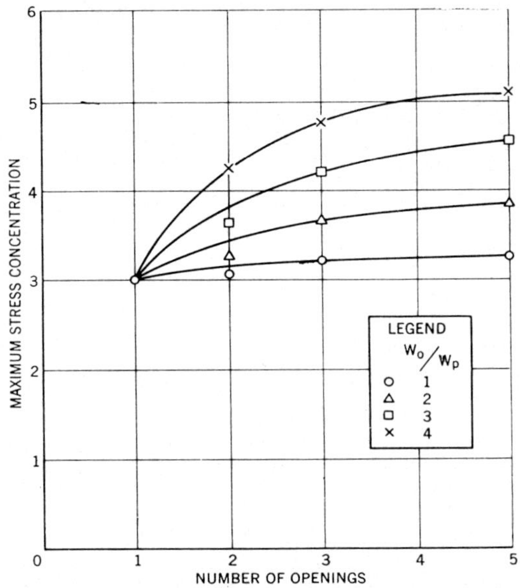

FIG. 31. Increase in maximum stress concentration with number of circular openings; applied stress normal to line of centers.[1]

Figure 31 shows the results of photoelastic stress determinations around a finite number of circular holes. The curves indicate that as the number of holes increases the maximum stress concentration approaches an upper limit, which corresponds to that for an infinite row of holes.

Row of Ovaloidal Openings

Stress distributions around a row of five ovaloidal holes in a plate subjected to stress normal to the line of centers of the holes has been studied by the photoelastic method.[10]

Figure 33 shows the stress concentrations at points where the horizontal axis intersects the boundaries of the holes, for various width opening/width pillar ratios. As might be anticipated, the maximum stress occurs at the boundary of the middle hole and stress concentrations increase as the width opening/width pillar ratio increases.

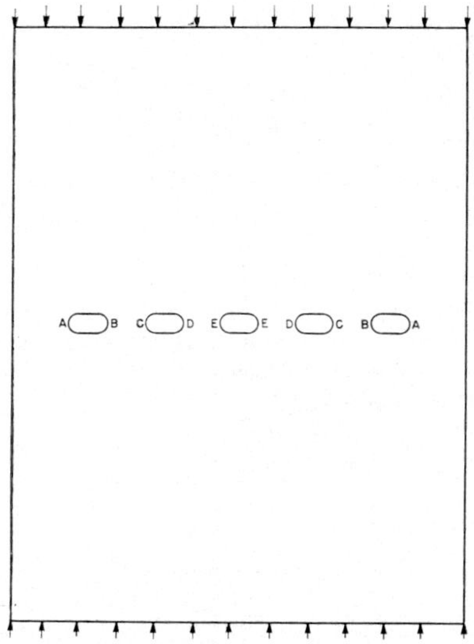

FIG. 32. Plate with five ovaloidal openings (height to width ratio of 0.5); applied stress field perpendicular to line of centers.[10]

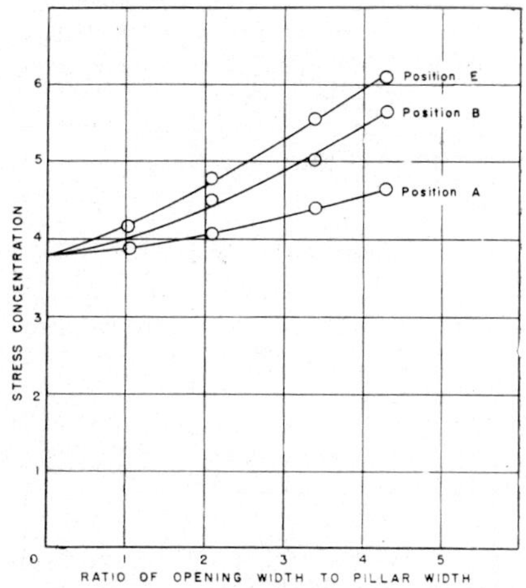

FIG. 33. Stress concentration as a function of opening width to pillar width ratio; plate with five ovaloidal openings (height to width ratio of 0.5); applied stress field perpendicular to line of centers.[10]

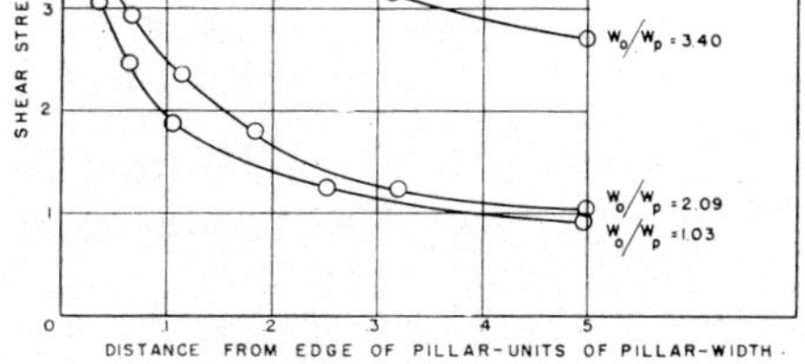

FIG. 34. Shear stress distribution in central pillars; plate containing five ovaloidal openings (height to width ratio of 0.5); applied stress field perpendicular to line of centers.[10]

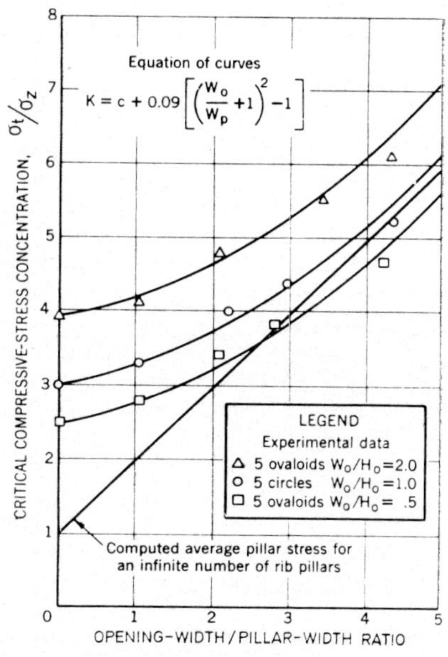

FIG. 35. Critical compressive stress concentration for multiple openings; unidirectional stress field, $M = 0$.[1]

Figure 34 indicates that for increasing values of W_o/W_p the average stress in the pillar increases more rapidly than does the stress at the opening boundary.

Figure 35 illustrates the increase in critical compressive which takes place with increased W_o/W_p ratios for circles and ovaloids of different W_o/H_o ratios.

Fig. 37 illustrates the rate of increase of pillar stress which takes place with increased percentages of recovery of the mineral being mined. Stresses increase very rapidly for recovery percentages in excess of 75 per cent.

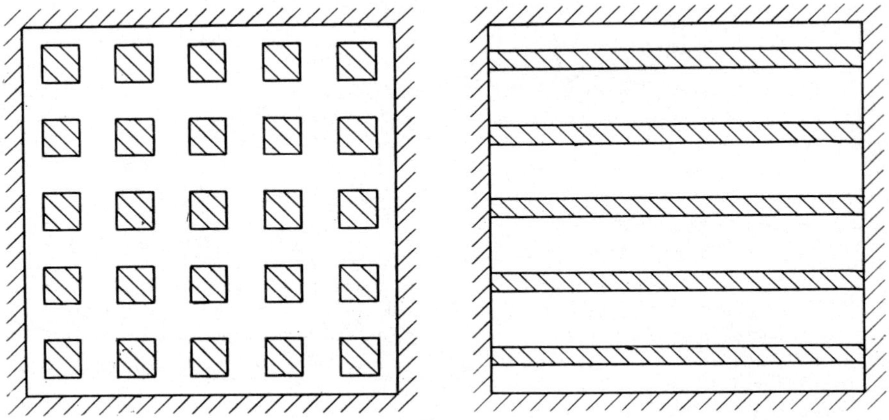

FIG. 36. Comparison of rib and square pillar support for 75 per cent recovery.[10]

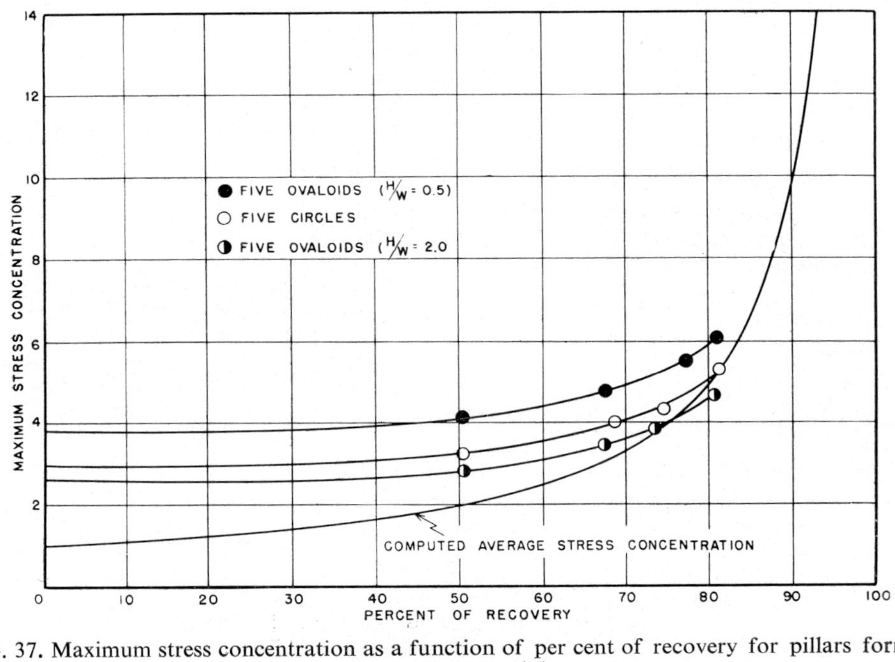

FIG. 37. Maximum stress concentration as a function of per cent of recovery for pillars formed by five openings.[10]

The following equation has been derived from experimental data:[11]

$$K = c+0.09\left[\left(\frac{W_o}{W_p}+1\right)^2-1\right] \tag{13}$$

where:

K = maximum stress concentration in pillars;
c = maximum stress concentration around a single opening for a unidirectional stress field;
W_o = width of the opening; and
W_p = width of the pillar.

Equation (13) can be used to compute the maximum stress concentration in rib pillars for a series of parallel openings. Values of c for different shaped openings can be obtained from the curves in Fig. 13. Equation (13) has been evaluated for circles and ovaloids having height/width ratios of 0.5 and 2.0, and the corresponding maximum stress concentration in the pillars as a function of opening to pillar-width ratio is shown in Fig. 35 together with experimental data. Figure 35 also shows the average pillar stress concentration as a function of W_o/W_p. The average pillar stress is obtained by assuming that any one rib pillar uniformly supports the weight of the rock overlying the pillar and one-half the opening on each side of the pillar. Thus,

$$\sigma_p = \left(1+\frac{W_o}{W_p}\right)\sigma_z \tag{14}$$

where:

σ_p = average pillar stress;
σ_z = applied vertical stress;
W_o = width of opening; and
W_p = width of pillar.

Pillar Contour

The theory of elasticity predicts that the stress concentration at a corner, such as that formed by the junction of the pillar and roof, is an inverse function of the radius of curvature of the corner. Examples of four assumed pillar contours are shown in Fig. 38 and the resulting boundary stress concentrations are shown in Fig. 39.[11] Examination of these figures indicates that as the radius of curvature increases the stress concentration decreases.

These diagrams indicate that in designing mine openings which will be subjected to severe stress conditions sharp corners should be avoided if possible. In the case of stoping excavations, drifts, etc., in massive rock the corners at junction of back and walls usually tend to assume a somewhat rounded configuration as a result of mining operations.

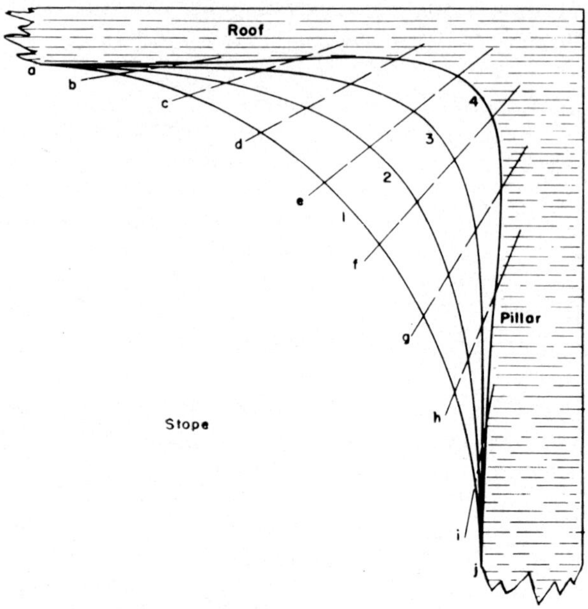

FIG. 38. Examples of four pillar contours[11]

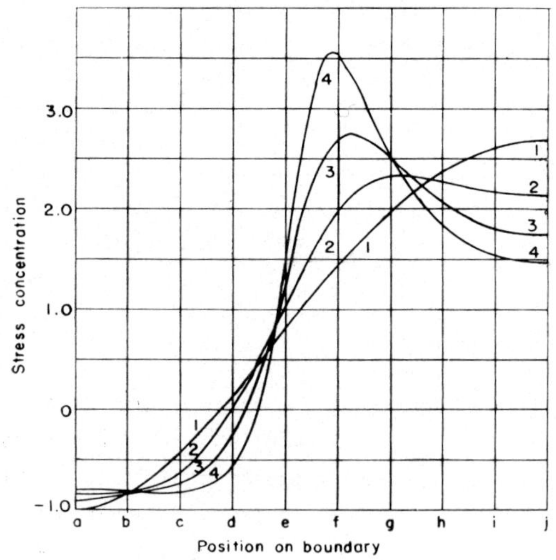

FIG. 39. Boundary stress concentrations for the contours shown in Fig. 38.[10]

In mining bedded deposits by conventional methods, however, it is not usually feasible to leave a rounded corner at junction of rib and roof because the corner rock tends to separate at a bedding plane and drop off, especially if the opening is made by blasting as in conventional mining.

Continuous mining machines of the boring type cut smoothly contoured openings without sharp corners, and such openings stand exceptionally well with a minimum amount of support.

Summary

The following generalizations can be made regarding the stress distribution around systems of equal-size openings equally spaced in a unidirectional compressive stress field:

(1) The maximum boundary compressive stresses occur on the sidewalls of the openings; that is, along the edge of the rib pillars between the openings.

(2) The boundary stress at the top and bottom of the openings is tension and is approximately equal in magnitude to the applied stress. However, this tensile stress concentration decreases rapidly with applied confining side pressures.

(3) The pillar stress concentration increases with increasing opening/pillar width ratios, but the stress distribution in the pillar becomes more uniform. Thus, the average stress in the pillar is nearly equal to the maximum stress for opening/pillar width ratios greater than 4.

(4) As the number of openings in a row increases the maximum stress concentrations in the central pillar approach an upper limit. This upper limit corresponds to the stress concentration for an infinite row of openings, and for practical purposes five openings or more are considered an infinite number.

SINGLE OPENINGS IN BEDDED ROCK

Roof Considered as a Gravity-Loaded Beam with Clamped Ends[11]

The stresses in a single gravity-loaded (loaded by its own weight only) roof slab overlying an underground opening may be approximated by the stresses developed in a uniformly loaded beam clamped at both ends, provided that the thickness is small compared with its lateral dimension, the length of the opening is more than twice its width, the roof slab extends over the entire opening and is composed of uniform homogeneous rock, and no bond exists between this slab and the rock above.

The standard equations for deflection, and for shear and tensile stresses in a clamped-end beam are:

$$D_{\max} = \frac{wL^4}{384EI}$$

$$\tau_{\max} = \frac{3wL}{4}$$

$$T_{\max} = \frac{Mc}{I}$$

$$M = \frac{wL^2}{12} \text{ (at the ends of the beam)}$$

$$M = \frac{wL^2}{24} \text{ (at the center of the beam)}$$

$$c = \frac{t}{2}$$

$$I = \frac{bt^3}{12}$$

where:

D_{max} = maximum deflection;
τ_{max} = maximum shear stress;
T_{max} = maximum tensile stress;
L = length of span (shortest lateral dimension of slab);
t = thickness of slab;
E = Young's modulus;
w = load per unit length of beam;
b = width of the beam;
γ = density (weight per unit volume);
A = cross-sectional area of beam;
γ_a = apparent density = w/A;
R = modulus of rupture (outer fiber tensile strength in bending);
F = factor of safety = ratio of critical stress to working stress.

Considering only a beam of unit width ($b = 1$) and setting $w = \gamma t$ the equations for deflection, and for shear and tensile stresses reduce to the following:[(11)]

$$D_{max} = \frac{\gamma L^4}{32Et^2} \quad (15)$$

$$\tau_{max} = \frac{3\gamma L}{4} \quad (16)$$

$$T_{max} = \frac{\gamma L^2}{2t} \quad (17)$$

The maximum deflection will occur at the center of the beam and the maximum shear and tensile stresses at the ends of the beam. At the center of the beam the shear is zero, and the tension is one-half the maximum value; thus, the point of initial failure would be expected to occur at the ends of the span, rather than at the center. Distribution of tension, compression and shear is shown in Fig. 40.

The magnitude of the maximum shear stress varies directly as the span length; the maximum tensile stress varies directly as the square of the length of the span and inversely as the thickness. The ratio of these stresses, obtained from equations

(16) and (17), is

$$\frac{\text{Maximum tensile stress}}{\text{Maximum shear stress}} = \frac{2L}{3t}$$

When beams are long compared to their thickness, stresses developed in tension are higher than in shear. For example, the ratio of tension to shear in a beam 8 ft thick and 60 ft long is 5:1. Thus, if the tensile to shear strength of the rock is also 5:1, both of the stresses must be considered in design calculations. However, for oil shale, as well as for most rock, the ratio of the tensile strength (modulus of rupture) to the shear strength is less than 5:1, and, therefore, the effect of shear stresses can be neglected, with little error.

Multiple-member Roof Beams

When the immediate roof consists of two or more members (see Fig. 41) and the lowest member is loaded by the upper members, the increase in load on the lowest member can be considered as an increase in density of the lowest member. This "apparent" density is given by:[13]

$$\gamma_a = \frac{E_1 t_1^2(\gamma_1 t_1 + \gamma_2 t_2 + \ldots + \gamma_n t_n)}{E_1 t_1^3 + E_2 t_2^3 + \ldots + E_n t_n^3} \tag{18}$$

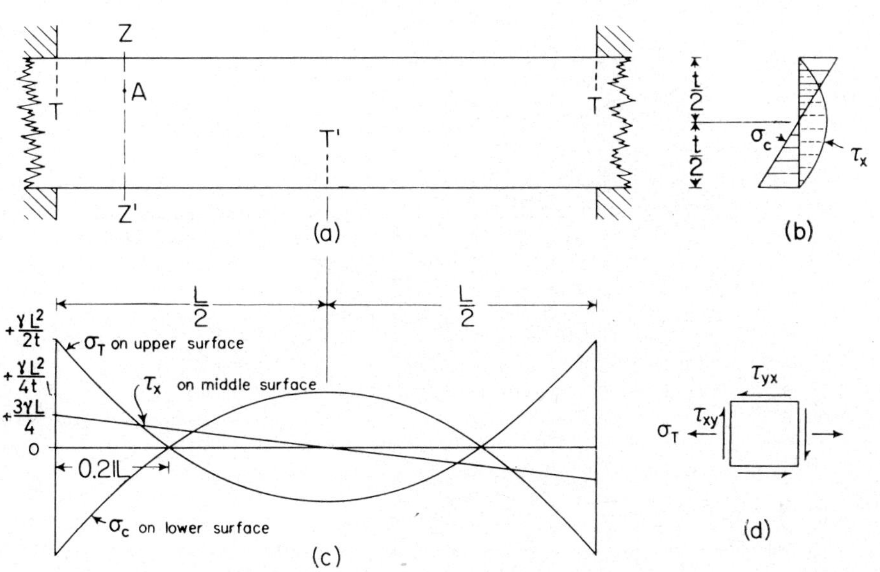

FIG. 40. (a) Gravity-loaded beam. (b) Bending and shearing stresses on section Z–Z'. (c) Distribution of outer-fiber bending stress and middle surface shearing stress across the span. (d) Stresses acting on an element at A. Failure in flexure by tension cracks, T, T'.[12]

where:

γ_a = apparent density of lowest member;
$\gamma_1, \gamma_2, \gamma_n$ = density of each member;
E_1, E_2, E_n = modulus of elasticity of each member;
t_1, t_2, t_n = thickness of each member.

If γ in equations (15) and (17) is replaced by γ_a, these equations can be used to determine the maximum tensile stress and maximum deflection of the lowest member of the multi-member immediate roof.

Most roof formations are not perfectly elastic and may contain defects within the layers, such as joints, fractures, faults, and fissures. Also, the roof may contain horizontal planes of weakness below which wholly detached roof layers will form.

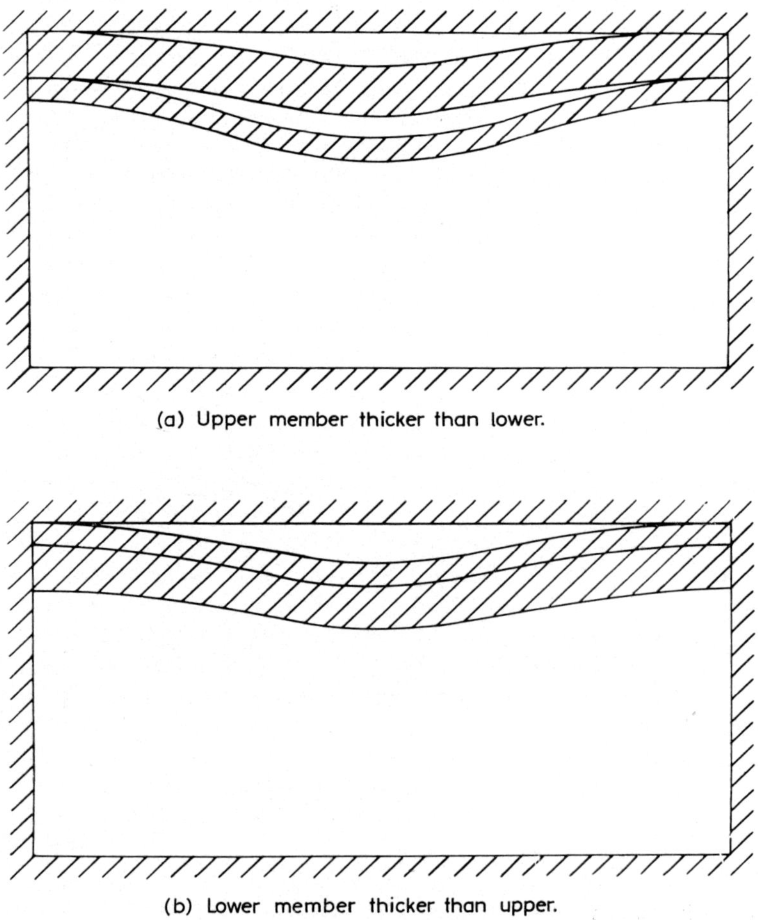

(a) Upper member thicker than lower.

(b) Lower member thicker than upper.

FIG. 41. Two-member beam: (a) Upper member thicker than lower. (b) Lower member thicker than upper.[11]

In some instances the rock exhibits a degree of weakening with time (called time deterioration), which may cause a change in the number of layers, the physical properties of the rock or any other factor that would affect the stability of the roof with time. Therefore, when equation (17) is used for design purposes, relatively high stress dividers are used to allow for the effect of these unknown conditions. Furthermore, the value of the maximum outer-fiber tensile strength of the roof is considered the same as the value of the modulus of rupture of samples from the roof measured in the laboratory. For design purposes, equation (17) can be re-written as

$$L = \sqrt{\left(\frac{2Rt}{\gamma_a F}\right)} \tag{19}$$

where:

R = modulus of rupture of the layer;
F = stress divider (safety factor) which is the ratio of the critical stress to the working stress.

Tests can be made in full-scale experimental stopes to evaluate the degree of bond across the vertical defects (joints, etc.), the degree of bond of lateral planes of weakness, the elastic behavior of the roof, the value of the physical properties of the roof rock in place and the time deterioration of the roof.

Application of Theory to Roof-span Calculations

For a room that is long compared to its width (length/width ratio of 2.0 or more), the safe roof span L_S can be determined from equation (19), provided that the value of the modulus of rupture of the roof rock can be substituted for the tensile stress T and that a safety factor F (ratio of the critical stress to the working stress) be introduced to provide allowance for minor geologic defects or any other unknown factors.

The use of equation (19) as a design equation is valid, provided that (1) the roof spans are long compared to the thickness, and (2) there are no major geologic defects, such as open faults or fractures that intercept the span.

Wright and Bucky prepared a chart [II] showing the relationship between the safe spans, tensile strength of the roof rock and roof thickness. In preparing this chart a specific gravity of 2.5 (density = 156 pcf) and a safety factor of 8 were used. Assuming the condition of a roof whose thickness is 8 ft and tensile strength is 3000 psi, the safe roof span is 72 ft. For a roof span of 50 or 60 ft under the same conditions the safety factor would be 18 and 12, respectively. (See Fig. 15, Chapter 6.)

If the roof is considered as a gravity-loaded plate (length/width ratio less than 2.0) the safe roof span may be calculated from the following equation

$$L_p = \left(\frac{Rt}{\sigma\gamma\, BF}\right)^{1/2} \tag{20}$$

where B is a constant whose value depends on the length/width ratio and also on Poisson's ratio. If we take Poisson's ratio to be 0.33 then B varies from 0.0513 to 0.0829 for length/width ratios from 1/1 to 2/1 respectively.

If we substitute identical values for R, t, γ and F in equations (19) and (20) the ratios of safe span lengths for plates, L_p, and beams, L, for values of plate length/width ratios of 1/1, 1.25/1, and 1.5/1 are found to be $L_p/L = 1.27$, 1.12, and 1.05 respectively.

Example Calculation[13]

A sample evaluation of equation (19) is presented below.

$$L = \sqrt{\left(\frac{2Rt}{\gamma_a F}\right)}$$

Let

$R = 2600$ psi (Modulus of rupture)

$$\gamma_a = \gamma = 2.69 \times \frac{62.4 \text{ pcf}}{1728 \text{ cicf}} = 0.097 \text{ pci.}$$

$F = 4$,

$t = 12$ in.

Then

$$L = \sqrt{\left(\frac{2(2600)(12)}{0.097\times 4}\right)} = 401 \text{ in.} = \frac{401}{12} = 33.4 \text{ ft.}$$

In the above calculation, the immediate roof is presumed to consist of a single layer; if the immediate roof consists of two or more layers, γ_a may not equal γ. As an example, consider that the immediate roof consists of two layers that have the following physical properties and thicknesses:

Layer 1 (lowest layer):

Modulus of elasticity $= 10\times 10^6$ psi
Specific gravity $= 2.50$.
Thickness $= 20$ in.

Layer 2 (second layer from roof horizon):

Modulus of elasticity $= 5\times 10^5$ psi.
Specific gravity $= 2.77$.
Thickness $= 10$ in.

Equation (18) now appears as

$$\gamma_a = \frac{E_1 t_1^2(\gamma_1 t_1 + \gamma_2 t_2)}{E_1 t_1^3 + E_2 t_2^3},$$

and

$$\gamma_1 = 2.50 \times 62.4/1728 = 0.090 \text{ pci.}$$

$$\gamma_2 = 2.77 \times 62.4/1728 = 0.100 \text{ pci.}$$

$$\gamma_a = \frac{(10\times 10^6)(20^2)(0.09\times 20 + 0.1\times 10)}{(10\times 10^6)(20^3) + (5\times 10^6)(10^3)},$$

$$= \frac{(10^7)(400)(1.8+1)}{(80\times 10^9) + (5\times 10^9)},$$

$$= \frac{(4\times 2.8)(10^9)}{(85)(10^9)},$$

$$= 0.132 \text{ pci.}$$

This value of γ_a must then be substituted in equation (19) to find L for the two-layer roof.

MULTIPLE OPENINGS IN BEDDED ROCK[1]

This section will consider only those openings situated in bedded rock in which the rock is divided into beds which are thin compared with the span of the openings and in which the bond between beds is weak.

In most mine openings in bedded rock the roof is formed at a bedding plane and is relatively smooth and flat. Because of the weak bond between beds, the roof rock will either immediately, or after a time, become detached from the overlying rock and form a layer (or layers) loaded only by gravity. The layer, or layers, is called the immediate roof and the overlying roof is called the main roof. The stresses in the main roof can be computed by procedures outlined previously under the section, Design of Single Openings in Massive Rock.

The stresses in the immediate roof can be computed from the theory of a clamped beam or plate provided that:

(1) Within each layer the thickness is uniform.

(2) The flexure of the beam or plate is caused by gravity; that is, that there are no end thrusts or other external forces.

(3) The rock in the layers is elastically perfect, isotropic and homogeneous.

(4) The ends of the layers are rigidly clamped by the overlying rock and the sidewalls.

(5) There are no fractures in the roof strata. (Vertical fractures parallel to the directions of the span of the assumed roof beams would not affect stability, however.)

For multiple openings whose lengths are great in comparison to their spans the roof design is about the same as that for single openings of the same type.

For multiple openings supported by square or rectangular pillars the stresses in the roofs can be computed from the theory of slabs clamped by supporting columns.

In the following discussion the cross-section of the opening is considered to be square or rectangular and the pillars are assumed to be square or rectangular in plan view. It is assumed that the horizontal planes of weakness in the pillars do not change the stress distribution due to vertical compression, that the differences in the strength of the rock from bed to bed do not change the overall pillar strength significantly, and that the average compressive strength can be used as a basis for design.

Single-layer Roof

For rows of parallel openings separated by rib pillars, the stress concentration in the main roof decreases as the distance between the openings decreases. Moreover, the stress concentration in the main roof is small compared with the stress in the immediate roof owing to flexure; therefore, equation (19) can be used for design.

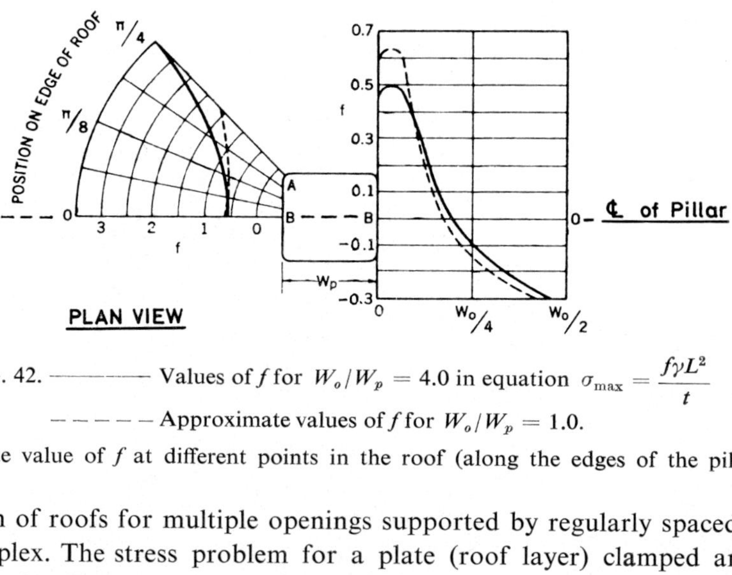

FIG. 42. ——— Values of f for $W_o/W_p = 4.0$ in equation $\sigma_{\max} = \dfrac{f\gamma L^2}{t}$

– – – – – Approximate values of f for $W_o/W_p = 1.0$.

Approximate value of f at different points in the roof (along the edges of the pillar).(1)

The design of roofs for multiple openings supported by regularly spaced pillars is more complex. The stress problem for a plate (roof layer) clamped and supported by randomly spaced columns (pillars) has no known theoretical solution. The stress problem for a plate clamped (and supported) by regularly spaced square columns with rounded corners has been solved. This problem can be considered representative of a roof layer supported on regularly spaced pillars. For the square pillar, the stresses in the layer are maximum at the corners (Fig. 42, point A). For example, for a room-and-pillar system in which $W_o/W_p=1$ or $W_o/W_p=4$, maximum compressive and tensile stresses are about two to six times greater, respectively, than the corresponding stresses computed for a clamped beam having the same span and supported by rib pillars. The maximum stress in the layer at the centerline of the pillars (Fig. 42, point *B*) is about the same as that for a clamped beam or plate.

Design of Pillars for Multiple Openings

Where relatively large areas are mined with a system of rooms and pillars, the average pillar stress is about equal to the maximum stress. The average pillar stress depends upon the stress before mining, σ_z, and the extraction, R_E; the average stress is related to the stress before mining by

$$\sigma_p = \frac{A_t}{A_p}\sigma_z \tag{21}$$

or

$$\sigma_p = \left(\frac{1}{1-R_E}\right)\sigma_z \tag{22}$$

where:

σ_p = average pillar stress;
σ_z = vertical applied stress;
A_t = total area;
A_p = pillar area; and
R_E = extraction ratio.

The extraction ratio is given by

$$R_E = \frac{A_t - A_p}{A_t} = \frac{A_m}{A_m + A_p} \tag{23}$$

where A_m = mined area.

If the average stress in the pillar, σ_p, in equations (21) and (22) is replaced by the compressive strength of the pillar, C_p, and divided by a safety factor, *F*, the following design equations are derived:

$$A_p = \frac{F\sigma_z}{C_p}A_t \tag{24}$$

and

$$R_E = 1 - \left(\frac{F\sigma_z}{C_p}\right) \tag{25}$$

where C_p = compressive strength of pillar.

For openings in bedded rocks, the roof span is computed from equations previously given. Therefore, the pillar width (or area) is determined for a given roof span (or mined area and pillar strength). For example, consider the following cases:

(1) Regularly spaced rib pillars, (2) regularly spaced rectangular pillars or square pillars, and (3) randomly spaced pillars with irregular shapes. Plans of these various systems of pillars are shown in Fig. 43 (a), (b), and (c).

For rib pillars, case (1), where the roof and pillars are the same length, the ratio A_m/A_m+A_p can be replaced by $L/(L+W_p)$, and the design equation is:

$$W_p = \frac{FL\sigma_z}{C_p - F\sigma_z} \tag{26}$$

For rectangular pillars, case (2), A_p in equation (24) is replaced by W_pL_p (where L_p is the length of the pillar) and

$$A_t = (L+W_p)(L+L_p),$$

where L = span between pillars in both directions.

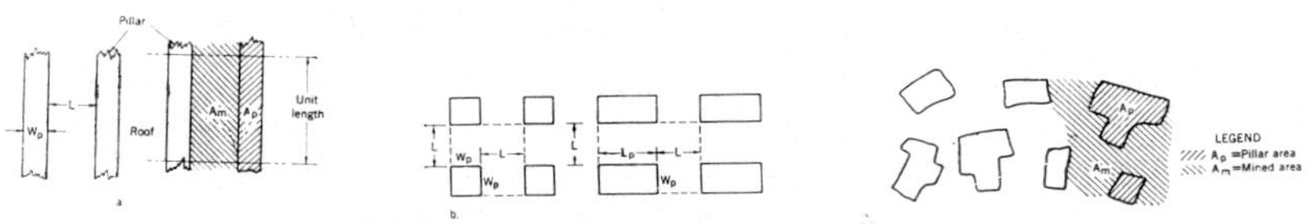

(a) (b) (c)

FIG. 43. Examples of different pillar patterns.[1]

(a) Rib pillars. (b) Square and rectangular pillars, equally spaced. (c) Irregular pillars, irregularly spaced.

Solving for W_p *(See* Fig. 43b),

$$W_p = \frac{L}{\sqrt{\left[\frac{kC_p}{F\sigma_z} + \left(\frac{k-1}{2}\right)^2 - \left(\frac{1+k}{2}\right)\right]}} \tag{27}$$

where $k = L_p/W_p$, the ratio of the pillar length to width. For square pillars, k in equation (27) becomes $K = L_p/W_p = 1$.

For randomly spaced, irregularly shaped pillars, case (3), the ratio of the pillar area to the mined area can be computed and the size and shape of the pillar adjusted accordingly; the ratio is

$$\frac{A_p}{A_m} = \frac{F\sigma_z}{C_p - F\sigma_z} \tag{28}$$

For first approximations from equations (26), (27), and (28), the compressive strength of the pillar, C_p, can be considered the same as the compressive strength of the rock, which had a height/diameter ratio of 1 as determined in the laboratory. If the height/width ratio of a pillar is less than 2 or greater than 0.5, the corrected compressive strength C_s is used, as given by the following equation:

$$C_s = C_1\left[0.778 + 0.222\left(\frac{d}{h}\right)\right] \tag{29}$$

where:

C_1 = compressive strength for specimen having $d/h = 1$;
C_s = compressive strength of specimen having $d/h = \pm 1$;
d = diameter of specimen;
h = height of specimen.

If the height/width ratio of a pillar is greater than 2:1, the compressive strength of the pillar should be determined from samples that have the same height/width ratio.

llustrative Problem

As an example of pillar design, consider the following problem:

Depth of opening (ft)	1000
Height of opening (ft)	30
Safe roof span (ft)	30
Average compressive strength of rock (psi)	11,500
Average weight-density of rock above opening (pcf)	144

From equation (14),

$$\sigma_z = \frac{1000 \times 144}{144} = 1000 \text{ psi}$$

$$\sigma_p = 11{,}500 \text{ psi}$$

and, for a safety factor of 4, the extraction ratio is

$$R_E = \frac{(11{,}500 - 4000)}{11{,}500} = 0.65$$

From equation (26) the width of rib pillars is

$$W_p = \frac{4 \times 30 \times 1000}{11{,}500 - (4 \times 1000)} = 16 \text{ ft}$$

For regularly spaced square pillars, the square pillar width from equation (27) is

$$W_p = \frac{30}{\sqrt{\left(\dfrac{11{,}500}{4 \times 1000} - 1\right)}} = 43 \text{ ft}$$

For the rib pillar the height/width ratio of the pillar is about 2:1 and for the square pillar, about 1:1.5. Because the height/width ratio is not greater than 2:1, or less than 1:2, the corrected strength can be computed from equation (29). The corrected strength, for $d/h = 1/1.5$, is $C_s = 10{,}000$ psi and, for $d/h=2$, $C_s = 14{,}000$ psi. Because the ratio of the corrected compressive strength, C_s, to the uncorrected strength, C_1, is small compared with the safety factor of 4, the uncorrected compressive strength can be used with little error.

BIBLIOGRAPHY

1. OBERT, L., DUVALL, W. I., and MERRILL, R. H., Design of underground openings in competent rock, *U.S. Bur. Mines Bull.* 587, 1960.
2. TIMOSHENKO, S., *Theory of Elasticity*, McGraw-Hill, New York, 1934, pp. 75–81.
3. GREENSPAN, M. Effect of a small hole on the stresses in a uniformly loaded plate, *Quart. Appl. Math.* Vol. 2, No. 40, 1945, pp. 60–71.
4. DUVALL, W. I., Stress analysis applied to underground mining problems, Part 1. Stress analysis applied to single openings, *U.S. Bur. Mines R.I.* 4192, 1948.
5. PANEK, L. A., *Stresses about Mine Openings in a Homogeneous Rock Body*, Edwards Bros., Ann Arbor, Michigan, 1951, pp. 1–50.
6. TERZAGHI, K., and RICHART, F. E., JR., Stresses in rock about cavities, *Geotechnique*, Vol. 3, No. 2, June, 1952, pp. 57–90.
7. PHILLIPS, H. B. and ZANGAR, C. N., *Stresses around a Gallery — Determined by the Photoelastic Interferometer*, U.S. Department of Interior — Bureau of Reclamation, Engineering monograph No. 12, April, 1952.
8. LIVINGSTON, C. W., Research at The Colorado School of Mines in subjects related to the mechanics of rock failure, *Trans Am. Geophys. Union* Vol. 32, No. 2, 1951, pp. 268–278.
9. HACKETT, P., An elastic analysis of rock movements caused by mining, *Trans. Inst. Mining Engrs., London*, 1958–1959, Vol. 118, Part 7 (Paper 3748).
10. DUVALL, W. I., Stress analysis applied to underground mining problems. Part II. Stress analysis applied to multiple openings and pillars, *U.S. Bur. Mines. R.I.* 4387, November, 1948.
11. MERRILL, R. H., Design of underground mine openings, Oil Shale Mine, Rifle, Colorado, *U.S. Bur. Mines R.I.* 5089, December, 1954.
12. PANEK, L. A., Principles of reinforcing bedded mine roof with bolts, *U.S. Bur. Mines R.I.* 5156, March, 1956.
13. MERRILL, R. H., Roof span studies in limestone, *U.S. Bur. Mines R.I.* 5348, July, 1957.
14. MERRILL, R. H. and MORGAN, T. A., Method of determining the strength of mine roof, *U.S. Bur. Mines R.I.* 5406, 1958.
15. MERRILL, R. H., MORGAN, T. A., and STEHLIK, C. J., Determining the in-place support of mine roof with rock bolts, White Pine Copper Mine, Michigan, *U.S. Bur. Mines R.I.* 5746, 1961.

CHAPTER 2

NATURAL (PRIMITIVE) STRESSES IN UNDISTURBED SOIL AND ROCK

VERTICAL (GRAVITY) STRESS OR "DEPTH PRESSURE"

IT IS generally assumed, for purposes of stress calculation, that vertical compressive stress (due to the weight of overlying rock and soil) increases approximately in proportion to the depth below the surface, and in proportion to the density (weight per unit volume) of the overlying soil and rock.

Some authorities question the strict application of this rule in steeply dipping strata since it is possible that the stiffer beds (with higher modulii of elasticity) may tend to carry more of the load than weaker beds. However, in order to simplify the presentation herein, vertical stresses computed on theoretical bases will be taken as being the product of depth and rock density.

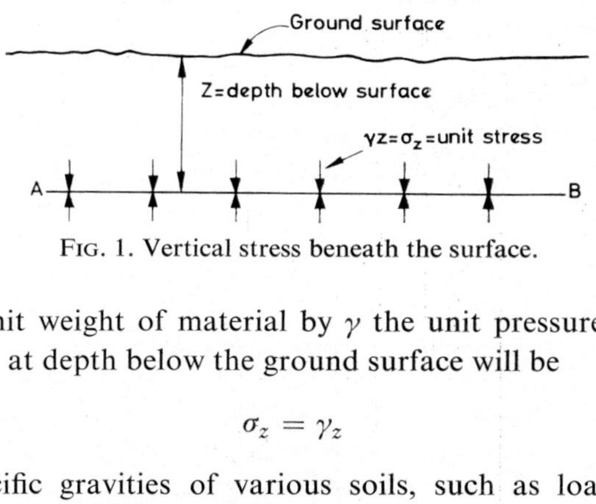

FIG. 1. Vertical stress beneath the surface.

Signifying the unit weight of material by γ the unit pressure on the horizontal plane A–B in Fig. 1 at depth below the ground surface will be

$$\sigma_z = \gamma_z \tag{1}$$

The apparent specific gravities of various soils, such as loam, weathered rock products, clay, silt, sand, gravel, boulders, etc., range from about 1.5 to about 2.0 (95–125 pcf). The presence of water in a soil increases the effective weight which then consists of weight of solid material plus the weight of water in the voids.

Sedimentary strata, such as shales and sandstones of the coal districts, are ordinarily considered to have a specific gravity of approximately 2.3 (144 pcf) which gives a vertical pressure of about 1 psi per ft of depth.

Igneous and metamorphic rocks, granites, gneisses, etc., are somewhat heavier than the sedimentary rocks. The most common of these rocks possess specific gravities in the neighborhood of 2.7 (170 pcf) or about 1.2 psi per ft of depth. Some of the basic igneous rocks, such as basalts, gabbros, peridotites, and their metamorphic equivalents possess specific gravities in the neighborhood of 3.0 or more (about 190 pcf) although these are not as common as the lighter rock varieties.

LATERAL STRESS

Elastic Material

Assuming that a natural uniform vertical compressive stress $\sigma_z = \gamma z$ exists on the plane *A–B*, Fig. 1, then the problem is to determine the lateral stresses σ_x and σ_y at any point in the plane. Orogenic (tectonic) stresses will be neglected for the moment.

In an elastic material the horizontal stresses σ_x and σ_y engendered by vertica stress in a confined elastic material are given by the equation

$$\sigma_x = \sigma_y = \sigma_z \left(\frac{1}{m-1}\right) \tag{2}$$

where m is Poisson's number. (Poisson's number is the reciprocal of Poisson's ratio which is the ratio of lateral deformation to longitudinal deformation under longitudinal stress.)

Poisson's number as determined by tests on laboratory specimens of various types of rocks varies from about 20 to about 2. For the development of theory and for computation purposes m is usually taken to have a value of 4 or 5 for most

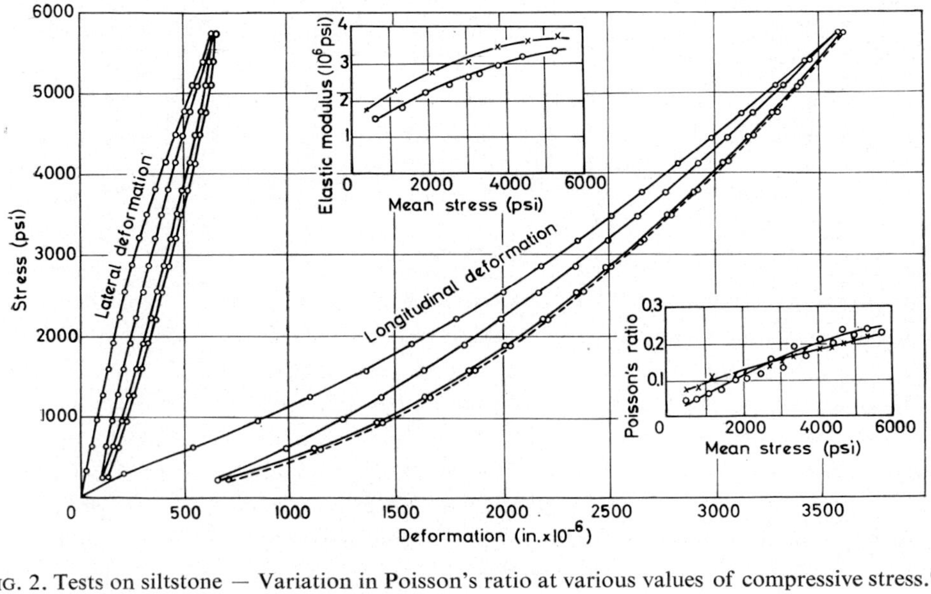

FIG. 2. Tests on siltstone — Variation in Poisson's ratio at various values of compressive stress.[1]

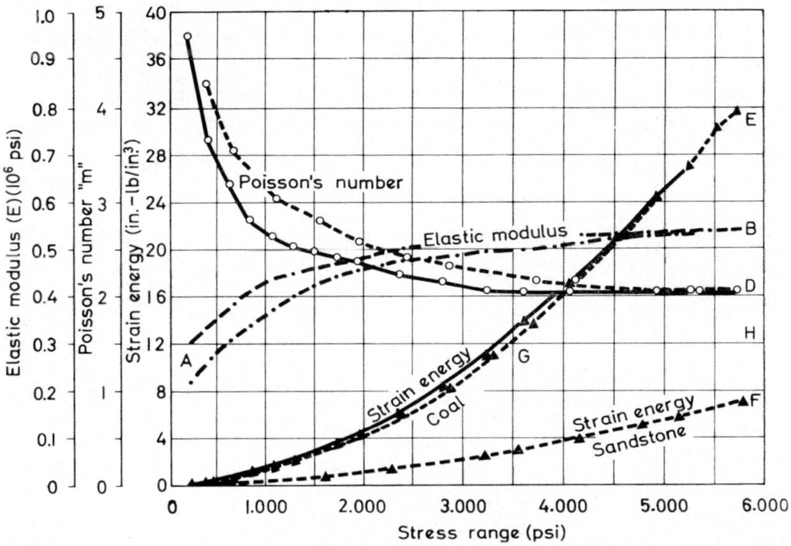

FIG. 3. Tests on coal – Variation of Poisson's number, elastic modulus, and stored strain energy with varying compressive stress.(1)

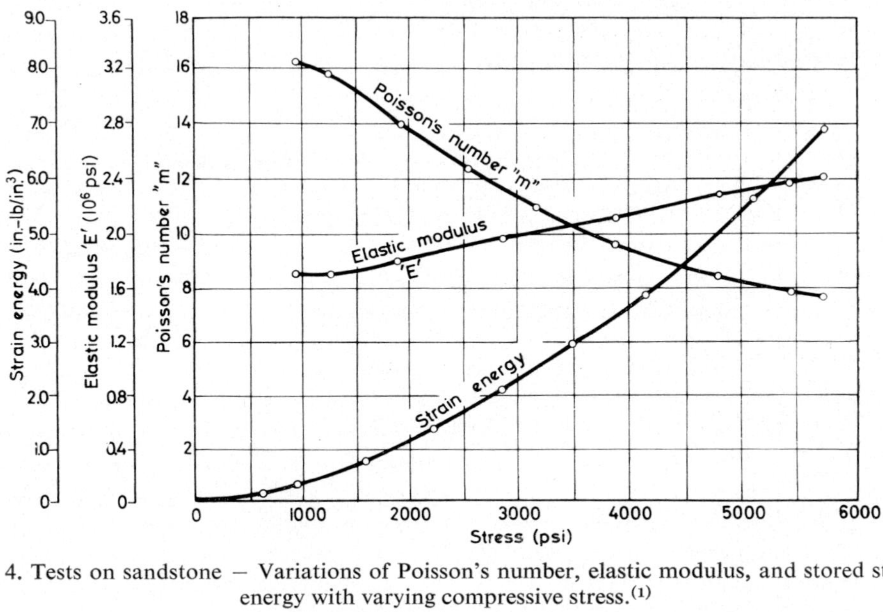

FIG. 4. Tests on sandstone – Variations of Poisson's number, elastic modulus, and stored strain energy with varying compressive stress.(1)

rocks. In comparison both rubber and steel have Poisson's numbers in the range from 3 to 4, while that for concrete is about 5–6.

Using a value of $m = 4$ in equation (2) we obtain

$$\sigma_x = \sigma_y = 0.33\, \sigma_z$$

A value of $m = 5$ gives

$$\sigma_x = \sigma_y = 0.25\,\sigma_z$$

Thus, neglecting tectonic stresses, the natural lateral stress in rock is usually taken, for computation purposes, as a third to one-quarter of the value of the vertical stress at any given point. Laboratory tests have demonstrated that the value of m for a given rock decreases with increasing values of confining compressive stresses. This is illustrated in Figs. 2, 3, and 4.[(1)] Thus, at high pressures at great depths the value of m will approach two. This implies a state of "hydrostatic stress" or

$$\sigma_x = \sigma_y = \sigma_z.$$

MATERIAL IN THE "CLASTIC" STATE OF STRESS

Cohesionless material, in which stress equals or exceeds the value of its shear resistance may be said to be in the "clastic state" of stress. Such material obtains its stability, that is, it maintains equilibrium between vertical and horizontal stresses, by reason of its internal friction. This consists of the friction between surfaces of the individual particles composing the mass, as well as an interlocking effect of the same particles. The coefficient of internal friction is designated as tan Φ, and Φ is called the "angle of internal friction". The relationship between σ_x, σ_y and σ_z in material which is in the "clastic state" is given by

$$\sigma_x = \sigma_y = i\sigma_z \tag{3}$$

where $i = \dfrac{1}{j} = \dfrac{1-\sin\Phi}{1+\sin\Phi}$

As an example of stress equilibrium in a material in the "clastic state" of stress take the case of a dry sand. Depending upon its porosity and upon grain size and shape, the angles of internal friction, Φ, of such sands normally ranges between 35 and 40°. The corresponding values of σ_x and σ_y are:

Φ	σ_x and σ_y
30°	$0.33\sigma_z$
35°	$0.27\sigma_z$
40°	$0.22\sigma_z$
45°	$0.17\sigma_z$

Fractured brittle materials, such as rock which has been fractured by excessive stress and does not possess any appreciable cohesion, may be in the clastic state of stress. Other pulverulent materials which may exist in the clastic state of stress are the clean sands, gravels and boulders.

MATERIAL IN THE "PLASTIC" STATE OF STRESS

Material which possesses cohesion after shearing may be said to be in the "plastic state" of stress when the applied stress is equal to the value of its shear resistance. The additional resistance furnished by cohesion reduces the magnitude of the lateral stresses, σ_x and σ_y induced by σ_z, thus

$$\sigma_x = \sigma_y = i\sigma_z - 2iC \frac{\cos \Phi}{1-\sin \Phi} \tag{4}$$

where C = cohesion.

Clays, and granular or fragmented materials containing appreciable amounts of clay, or fine silt, to act as binder, are typical of materials which may exist in the plastic state of stress.

NATURAL STRESSES OBSERVED IN SEDIMENTARY ROCKS

The theoretical natural stress conditions outlined in the preceding section may exist in those sedimentary rocks which have not been subjected to tectonic forces (lateral thrust) since they were laid down. Such rocks include a large proportion of coal-bearing strata as well as a large proportion of the sedimentary strata associated with the production of petroleum. These include shales, mudstones, siltstones, and sandstones in various degrees of consolidation.

Generally, these beds lie flat, or dip at relatively flat angles, and evidence of regional lateral comression such as thrust faulting, or tight folding, is lacking.

In some British coal mines the presence of moderate natural lateral stress, such as would be inferred from equation (2), is evidenced by horizontal shearing of the shale roofs of "bords" (roadways driven at right angles to the cleat).[2] Such fracturing occurs in the shale roofs of seams at depths greater than 700 ft. The fracturing may be induced in a desired location by advancing a narrow vertical slot or "nicking" ahead of the face. Shear fracturing then takes place at the location where the nicking exposes the roof.

Fracturing of this type in ventilation and haulageways can be reduced by first driving ahead a "pressure relief" bord midway between locations where main bord roadways will be driven. Roof supports are withdrawn and the roof of the pressure relief bord is allowed to collapse. The roofs of bords subsequently driven then remain sound and unbroken, testifying to the fact that lateral stress has been relieved by expansion of strata toward the pressure relief bord.

STRESSES DEDUCED FROM HYDRAULIC FRACTURING PROCESSES

The petroleum industry has devoted considerable study and experimentation to determining the actual magnitudes of horizontal stresses in sedimentary rocks. However, there is considerable disagreement among groups who have studied the subject. The results of a fairly recent and rather thorough study of the subject by Harrison, Kieschnick and McGuire are presented hereafter.[3]

The study is based on a process employed by the petroleum production industry which is known as "formation fracturing". This process has for its purpose the increase of productivity of oil wells by pumping into a well a fluid at pressure sufficiently high to cause fracturing of the walls of the hole and extension of the fractures into the rock.

The pressure required to fracture twenty-seven wells, located in the Permian basin of West Texas and New Mexico, with depths from 3500 ft to 12,000 ft, were found to lie between the values of $0.44\,Z$ and $0.74\,Z+500$ (where Z = vertical depth). The increase of vertical stress would be about 1 psi per ft of depth due to the weight of these rocks. Thus the depth pressure, σ_z, at any point (in psi) would be $1 \times Z$. At 3500 ft depth it would be $\sigma_z = 3500$ psi, approximately and at 12,000 ft it would be $\sigma_z = 12{,}000$ psi, approximately.

The authors of the above study point out that fracturing occurs at pressure less than depth pressure, σ_z, therefore such fracturing cannot take place by vertical separation along horizontal planes because pressures exceeding σ_z would be required to separate the rock along these planes, and raise it against the weight of overlying rock. Therefore, they say that fracturing occurs along vertical tension fractures radiating from the wall.

Poisson's number for these rocks lies between 3.7 and 5.5. Then from equation (2) the natural horizontal stresses induced by vertical stress should be

$$\sigma_x = \sigma_y = \sigma_z\left(\frac{1}{m-1}\right) \tag{2}$$

and horizontal stress would range from

$$\left(\frac{1}{5.5-1}\right)\sigma_z \quad \text{to} \quad \left(\frac{1}{3.7-1}\right)\sigma_z \quad \text{or} \quad 0.22\sigma_z \quad \text{to} \quad 0.37\sigma_z$$

By elastic theory the internal pressure required to rupture a thick cylinder is:

$$p_i = 2\sigma_r + S_t = 2\left(\frac{1}{m-1}\right)\sigma_z + S_t$$

where σ_r is radial compressive stress, in this case $\sigma_{x,\,y}$ and S_t is tensile strength of the rock. Thus the pressures in the well required to produce rupture of the walls would lie between:

Minimum; for $m = 5.5$ $\quad p_i = 2(0.22\sigma_z) = 0.44\sigma_z$
Maximum; for $m = 3.7$ $\quad p_i = 2(0.37\sigma_z)+500 = 0.74\sigma_z+500$

(Taking the minimum tensile strength (S_t) of rock as zero, and maximum as 500 psi.)

Thus the theoretical values based on the assumption that natural horizontal stress is related to the vertical stress by equation (2) seem to be in close agreement with values inferred from these measurements.

The same authors have charted the pressures which were required to fracture the formations around fourteen wells located in the lightly consolidated sands of Texas and Louisiana. The depths ranged from 5700 ft to 9000 ft. Values for internal pressure, p_i, required to produce rupture, were between 0.75 Z and 1.0 Z.

The authors explain the higher pressures required in these formations as being due to the poorly consolidated sands which, at depth, would be in the "clastic state" of stress. The authors infer that horizontal stress at depths where the "clastic state" of stress exists would be about 0.4 Z. This would be equivalent to an angle of internal friction, Φ, of about 25° in the formula for equilibrium of material in the clastic state.

$$\sigma_x = \sigma_y = i\sigma_z \tag{3}$$

The conclusions of the authors of the paper are:[(3)]

(1) The competent formations of the Earth's crust, even when arched, lack sufficient strength to alleviate the load from underlying formations, and the overburden load may be taken as approximately equal to 1 psi per foot of depth.

(2) Hard, well-consolidated rocks can be considered as existing in the elastic state of stress to the depths penetrated in present oil well operations. As a result, as indicated by theory and data from the field, such formations usually possess horizontal stresses in the neighborhood of 0.33 psi per foot of depth. (Occasional exceptions would be expected when horizontal stresses are significantly influenced by tectonic forces.)

(3) The soft shales and weakly cemented sands most frequently found in the Texas and Louisiana Gulf Coast area can be considered to exist in the plastic state of stress and to possess horizontal stresses in the neighborhood of 0.4 psi per foot of depth.

(4) Pressure induced wellbore fractures may be horizontal when the bottom hole fracturing pressure has exceeded the overburden load of about 1 psi per foot of depth. However, this seldom occurs below about 3000 ft, except possibly in the infrequent case of abnormally high horizontal stresses.

(5) Vertical or nearly vertical fractures are obtained at bottom hole pressures below the overburden load and frequently below 0.7 psi per foot of depth.

VARIATIONS IN POISSON'S RATIO

Figures 2 and 4 illustrate the tendency for Poisson's ratio to increase with increasing values of the applied compressive stress. This is particularly noticeable in the softer sedimentary rocks.

In the harder rocks the values of Poisson's ratio as determined by laboratory tests is usually less than the 0.20–0.25 used for calculation purposes in previous examples given.

Figure 4 shows the results of tests on sandstones samples in which Poisson's ratio increases from about 0.06 (Poisson's number = 16) at 1000 psi applied compressive stress to about 0.13 (Poisson's number = 7.7) at 5700 psi.

Figure 5 shows the values of Poisson's ratio determined for samples of strong sedimentary rocks from a Canadian coal mine. The values shown are the tangent values measured at a compressive stress of 6000 psi. Values of Poisson's ratio generally range from 0 up to about 0.25 but most specimens tested fall in the range from 0.05 to 0.15.

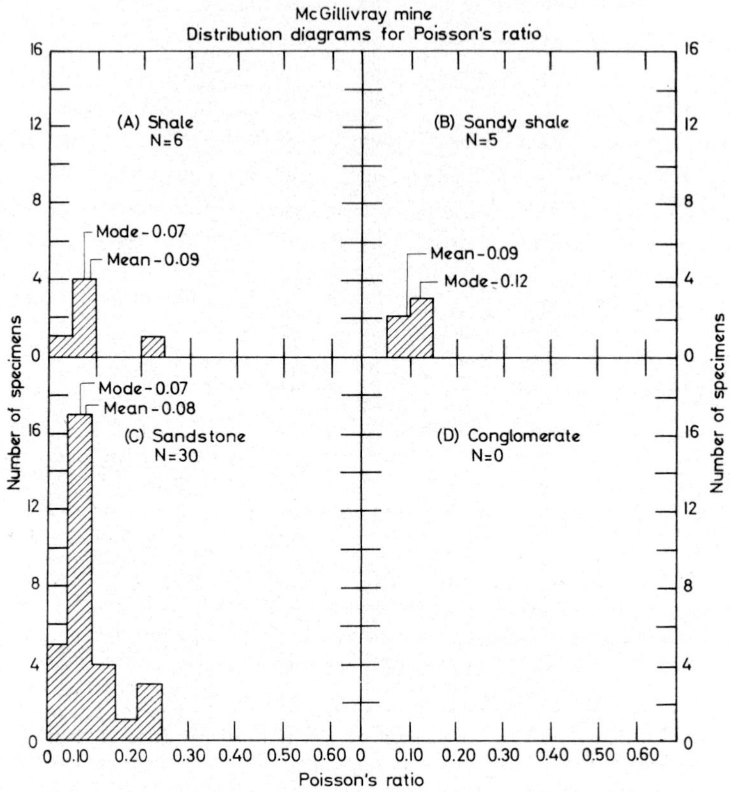

FIG. 5. Distribution diagram for Poisson's ratio — rocks from the McGillivray Mine, Canada.[20]

EFFECTS OF COMPACTION PROCESSES

Corlett and Emery[18] have pointed out that natural lateral strains in sedimentary rocks may not be related to the vertical strains by Poisson's ratio but more probably are the results of mechanical processes which operated during the deposition, compaction and cementation of the rock. They suggest that the magnitudes of lateral strains may be determined by the wedging action which occurs as individual grains of sediment are forced between other grains by compaction processes. The lateral strain would also be a function of the grain shape and roughness, and as the nature of the deposited sediment varied, so would the relative lateral strain vary from layer to layer.

Water contained in the pore spaces of the sediment would also exert pressure on the individual grains. Such hydrostatic pressure would increase by 0.43 psi for each foot of depth beneath the surface and would be instrumental in determining the state of stress in the rock.

There are examples on record which indicate that in some localities the stresses in apparently undisturbed sedimentary rocks greatly exceed the stresses which would be created by existing depths of cover.

One such instance occurred during the development of an open pit mine in Ontario. It was necessary to strip up to 160 ft of overlying rock before the ore-bearing formations were reached. During the removal of the Palaeozoic limestone the pit had been developed to the approximate dimensions of 1000 ft $\times$ 2000 ft $\times$ 50 ft deep when quite suddenly the bottom of the pit cracked and a portion heaved (buckled upward) to a maximum height of 8 ft within a few minutes. Calculations, based on certain simplifying assumptions, indicated that the lateral natural stress required to cause this buckling of the limestone strata might be on the order of 2000 psi.

It is probable that the lateral stresses in the limestone were residual from a time when the depth of cover was much greater.

In summarizing, it appears that equation (2) may be used as a convenience for estimating lateral stresses in undisturbed sedimentary rocks but that the values obtained by use of this equation will generally be low and that actual stresses may have any magnitude from about $\frac{1}{3}\sigma_z$ up to values several times that of σ_z.

(Results of stress measurements in some sedimentary rocks are given in the Appendix to this chapter.)

LATERAL STRESSES OF TECTONIC OR RESIDUAL ORIGIN

The previous section dealt with natural vertical stresses existing beneath the Earth's surface and, neglecting residual and tectonic stresses, outlined the theoretical lateral reactions which would be induced by these vertical stresses in various types of material.

Heim long ago (1878) proposed the theory that the state of stress within the Earth's crust, even in elastic materials, in hydrostatic; that is, that the natural lateral stresses are equal to the natural vertical stresses. He theorized that when subjected to vertical stress for long periods of time, a slow adjustment takes place in the rock which allows the lateral stresses to approach the values of the vertical stress.

This theory has received considerable acceptance in the civil engineering field, in connection with the design of tunnels, particularly pressure tunnels. However, until recently very few attempts had been made to measure the actual natural subsurface stresses. Recent measurements indicate that lateral stresses deviate considerably from the "hydrostatic" state proposed by Heim. Natural stresses in intact igneous and metamorphic rocks have been measured at a number of locations in various parts of the world. Minimum recorded lateral stress in such intact rocks has been about $0.6\,\sigma_z$[6] and maximum about 8 σ_z[17]. In many cases the lateral stress considerably exceeds the vertical, and such conditions indicate that forces other than the existing depth pressure, σ_z, have been active. Such natural lateral stress may result either from lateral compressive stresses such as produce overthrusts and mountain building (tectonic forces) or they may be residual from the time when the rock was located at a greater depth beneath the surface, before erosion removed a large portion of the overlying rock.

The term "hydrostatic stress" implies, to this author, a state of fluidity or semi-fluidity wherein the stresses at a point are equal because the material lacks shear strength. Yet it is entirely possible to have equal stresses applied at a point in an elastic material (that is, $\sigma_x = \sigma_y = \sigma_z$), while at the same time the material has a Poisson's number greater than 2. The author prefers to use the term "isotropic stress" for the condition of equal horizontal and vertical stresses in material which is in the "elastic state" of stress. The term "hydrostatic stress" will be reserved for the condition in which stresses are equal because a material lacks shear strength.

EVIDENCE OF EXISTENCE OF LATERAL RESIDUAL AND/OR TECTONIC STRESSES

Prior to recent quantitative measurements of lateral subsurface stresses many qualitative observations were on record of violent stress fracturing of rock around mine development openings and civil works tunnels and in mine workings located at relatively shallow depths beneath the surface. Such fracturing around single openings could only have been the result of high lateral subsurface stresses existing before the opening was made.

"Rock bursts" are a type of violent stress fracturing typified by sudden explosive fracturing of rock in roof, walls, or face of a tunnel, or other opening. The fracturing is accompanied by the projection of flakes or slabs from the rock surface. The action is usually accompanied by snapping, popping, or explosive sounds.

As is explained in Chapter 8, page 323, such fracturing results when shear stresses in the rock exceed the shear strength. The resulting fracturing allows the loosened fragments to expand suddenly, with the result that they project themselves from the rock surface into the opening.

High stresses are required to produce rock bursts around single openings, and at shallow depths where vertical stress is small, the presence of such stress fracturing indicates the existence of large lateral stresses. In some isolated instances bursts may be produced by localized stress along tight folds, or along faults, but where bursting persists over any appreciable distance it is an indication that large natural lateral stress exists over a considerable area, that is, stresses are residual, or tectonic in nature.

In the granite quarries of Vermont and Georgia one type of rock burst occurs in the rock at a few tens of feet beneath the surface.(4) When released on three sides the rock expands as if released from compression and corners spall off the square blocks. In addition, the hard, massive rock, when perforated with rows of drill holes, expands laterally with sufficient force to crush the rock between adjacent holes. Lateral stresses in the rock, although they have not been measured, appear to range up to several thousand pounds per square inch. Since vertical stresses are practically non-existent at this depth the lateral stress here greatly exceeds the vertical.

In some mine workings in hard brittle rocks, such as granite, diorite and quartzite, minor rock bursts such as spitting or popping are a frequent occurrence at depths of a few hundred feet. Here again vertical stress cannot be sufficient to fracture strong rocks, and the presence of large lateral stresses is indicated.

Rock bursts have caused some difficulty during the construction of underground hydroelectric plants in Sweden. They are excavated in hard granite within a few hundred feet of the surface; therefore, any stresses sufficient to cause rock bursts can only be lateral stresses and obviously great exceed the vertical stress.

During the construction of Duchesne Tunnel in Utah (U.S.B.R.) considerable trouble was experienced with rock bursts. This occurred in hard quartzite at a depth of only a few hundred feet so obviously vertical stresses were not responsible.

Bird[(5)] tells of rock bursts during the construction of the East Delaware Aqueduct Tunnel. When the heading was in massive granite he saw violent and continuous rock bursting where the cover was less than 500 ft.

Long portions of the tunnel were in sedimentary rocks which had been subjected to lateral compressive stresses resulting in thrust faulting and sharp folding with high dips. Sandstones in this part were subject to violent bursting. In general, Bird reported fairly violent popping at less than 500 ft of depth although it did not appear to become general until the depth exceeded 1000 ft, a depth which is not, however, sufficient to create vertical stress sufficient to cause fracturing of a strong rock.

The foregoing instances have been cited to illustrate the existence of large lateral compressive forces, even at very shallow depths in the earth. It is evident that they often exceed the vertical stresses.

MEASURED ROCK STRAINS AND COMPUTED STRESSES

Techniques for the determination of absolute (total) strain and stress within rock *in situ* usually involve one of the following methods:

Strain Relief Method

A smooth hole is bored in the rock and a gage capable of measuring diametral deformation is inserted. The hole is then overcored with a large coring bit so that the cylinder of rock containing the deformation measuring gage is free to expand. The change in the diameter of the hole when the rock cylinder is free to expand is a function of the original stress in the rock and its elastic modulus.

A series of recent stress determinations carried out by the U.S. Bureau of Mines,[16] using strain relief techniques, in a trona mine, a salt mine and at other locations in granite, tuff and limestone is described in the Appendix to this chapter.

Strain Restoration Method

The strain restoration method involves:

(a) Installation of strain gages on the rock surface.

(b) Cutting of a slot in the rock between the strain gages so that the surface rock is free to expand.

(c) Installation of a "flat jack" (hydraulic pressure cell) in the slot.

(d) Application of hydraulic pressure to the flat jack until the rock is restored to its original state of strain.

The original stress in the rock is presumed to be equal to the final pressure in the flat jack.

MEASURING CHANGES IN ROCK STRAIN

Techniques for determining only the *change* in strain in rock adjoining mine workings, as such workings advance, require only that suitable strain measuring devices be affixed to rock surfaces, or inserted in boreholes, and their readings noted at suitable intervals.

Results of strain and stress measurements on coal mine and in uranium mine roof rocks are described in Chapter 11, page 426. These measurements were made by means of a recently developed technique which involves the application of "patches" of a photoelastic plastic material directly to the rock surface. Stress measurement methods and apparatus are described in detail in Chapter 10 of this volume.

MEASURED ROCK STRAINS AND STRESSES

The earliest attempts to determine rock stress by the strain relief method were carried out by Lieurance for the U.S. Bureau of Reclamation at Hoover (Boulder) Dam in 1932.[6,7]

These measurements were carried out in a tunnel at a depth of 150 ft below the original elevation of the bottom of Boulder Canyon in a sedimentary rock locally known as Dam Breccia. The axis of the tunnel was approximately at right angles to the centerline of the canyon.

Results indicated that the horizontal stress existing in the rock beneath the canyon floor prior to excavation of the tunnel was equal to almost three times the vertical pressure. This could be explained by the fact that the rock slopes at both sides of the canyon floor rose steeply to a height of more than 1000 ft, and the horizontal stresses may have been residual from the original depth of cover as indicated by the present topography.

Gorge Powerhouse Penstock Extension Tunnel

Subsequently the technique for cutting a groove and measuring strain were improved and simplified and the method was applied by the U.S. Bureau of Reclamation in the Gorge Powerhouse penstock extension tunnel (power system of the City of Seattle, Washington). This tunnel is located a depth of about 200 ft beneath the crest of a gentle ridge, although the canyon wall nearby rises to much greater heights. The rock consists of a sound granitic gneiss with typical jointing into huge rectangular blocks.

Figure 6[(7)] shows a plan and profile of the test locations.

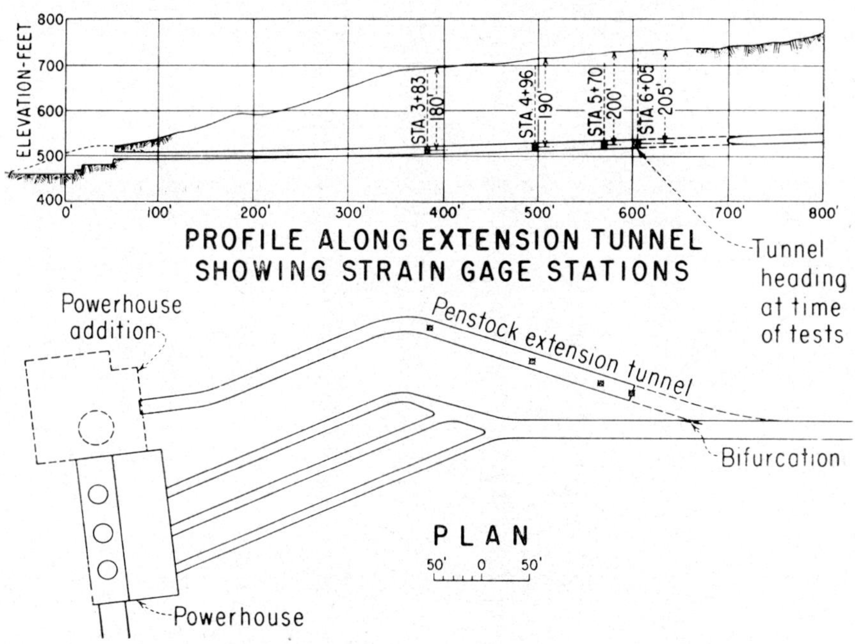

FIG. 6. Profile along the Gorge Powerhouse Extension Tunnel showing strain gage stations.[(7)]

Figure 7 shows the measured stresses represented graphically on the developed inside surface of the tunnel. The stress distribution varied rather erratically from profile to profile, and the stress along the direction of the tunnel axis varied from point to point in each profile.

In analyzing these measured values Terzaghi and Richart[(6)] note that the stress diagram which most nearly fits with theoretical considerations gives a minimum circumferential stress of about 1050 psi at the top of the tunnel and a maximum of about 3750 psi near the spring-line of tunnel. This corresponds to a natural stress ratio (σ_x/σ_z) or M value of about 0.6. That is, the natural lateral stress was about

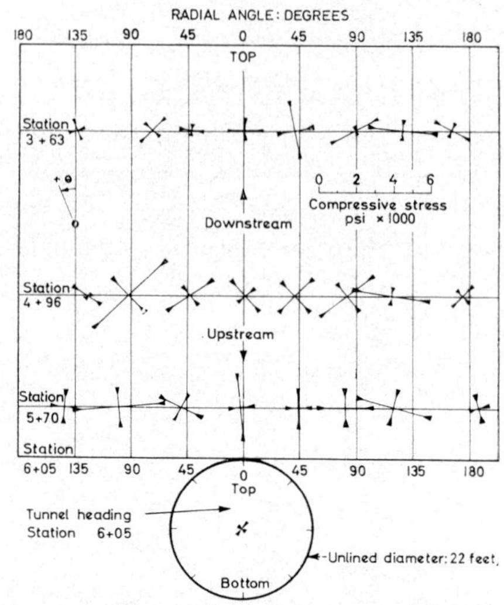

FIG. 7. Residual principal stresses plotted on the developed inside surface of the Gorge Powerhouse Extension Tunnel.(6)

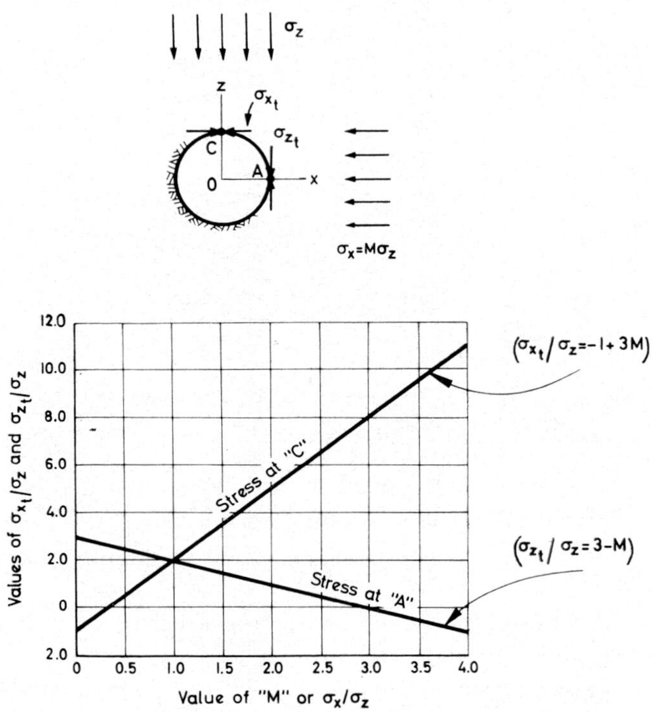

FIG. 8. Relationship between the tangential principal stresses at the boundary of a circular tunnel and the natural stresses in the rock mass. (Adapted from Terzaghi and Richart(6).)

0.6 the intensity of the natural stress in an approximately vertical direction. The maximum principal stress was inclined about 15° from the vertical.

From Figure 8 it can be seen that the natural vertical stress would be about 1400 psi, whereas the weight of the overburden can account for only about 230 psi.

An example of the use of the diagram of Fig. 8 follows:

Example: Assume that measurements have shown that rock stress at A is 2000 psi and at C rock stress is 1000 psi. Then $\sigma_{x_t} = \frac{1}{2}\sigma_{z_t}$. Enter the diagram of Fig. 8 and by inspection find the location where a vertical line intersects C with a value one-half of that of line A. This vertical line intersects the M axis at a value of 0.7. Then $\sigma_x = 0.7\sigma_z$.

To find numerical value of σ_z transpose the equation

$$\sigma_{z_t} = \sigma_z(3-M) \quad \text{to}$$

$$\sigma_z = \frac{\sigma_{z_t}}{(3-M)}$$

Then

$$\sigma_z = \frac{2000}{(3-0.7)} = \frac{2000}{2.3} = 880 \pm \text{psi}$$

Notch Stresses

A theory has recently been proposed which explains the existence of large lateral stresses at certain locations, as well as excessive vertical stresses, as due to the concentrating effects of "notches" in the topography. However, it does not appear adequate to completely explain the presence of a measured vertical stress with an apparent magnitude which is more than six times the theoretical vertical stress, σ_z, due to the weight of the overlying rock. It would seem that such excess stress, if existing over a considerable area, would find relief by simple upward expansion of the rock toward the surface.

Restrained Deformation Recovery

The presence of apparent vertical stresses in excess of those predicted by theory may possibly be explained by the phenomenon of "restrained deformation recovery" which was first discovered and recorded by Emery[(21)].

Emery noted that large strain energies were present on the inside of a drill core (quartzite) which had been cored 5 years previously. When the core was sliced longitudinally and strain relaxation was measured on the new surface, it was found that there was little further relaxation near the edges but that there was a high relaxation near the center. The outer $\frac{1}{4}$ in. of core volume appeared to have restrained the internal portion progressively inward. Emery has observed this same phenomenon in many other rocks.

Emery says: "Such relaxation is a function of the surface area of the specimen. Hence a $\frac{7}{8}$-in. diameter drill core could be expected to be in a different rheological condition from a 1–$\frac{1}{8}$-in. diameter core taken otherwise identically after any given time interval".

It is conceivable that this same type of restraint may operate on a much larger scale and that the rock in the interior of a mountain, for example, is restrained in a manner similar to that in the drill core so that the primitive strains at any given point will never be fully recovered until erosion has removed all but a few feet, or a few tens of feet, of the overlying rock. Thus we might find residual strains in the rock at depths of a few hundred feet which would correspond to vertical stresses imposed by a much greater depth of rock cover.

Effect of Anisotropy

It is probable that anisotropies in rock, such as bedding, gneissic banding, foliation, schistocity, and cleavage may cause the development of localized zones of high stresses which exceed those computed from the weight of overburden.

Such localized zones could exist also as residuals from previously greater depth of overburden and would be prevented from relieving themselves upward by the restraining effects of adjacent rock.

Modulus of Elasticity

Habib[8] has demonstrated that the modulus of elasticity of rock measured in place by means of pressure applied by a jack may be as little as about one-fifth of that obtained by measurement on laboratory specimens. Part of the discrepancy between measured vertical stresses and theoretical vertical stresses might be accounted for in this manner.

The important fact brought out by the stress measurements, and the analysis by Terzaghi and Richart, is that the ratio of the initial lateral stress to the initial vertical stress was $\sigma_x = 0.6\,\sigma_z$, approximately. Thus, the lateral stresses were larger than would be indicated by equation (2)

$$\left(\sigma_x = \sigma_z\left(\frac{1}{m-1}\right)\right)$$

and the presence of tectonic or residual lateral stresses in the granitic gneiss is indicated.

Prospect Mountain Tunnel

The strain-relief method was also employed by engineers of the U.S. Bureau of Reclamation for measuring rock stresses in the Prospect Mountain Tunnel to determine whether reinforcing steel was needed in the concrete lining. This was a pressure tunnel, 5650 ft long, with a maximum depth of overburden of 900 ft. The rock was, in general, a good quality massive grey granite with occasional pegmatite dikes.

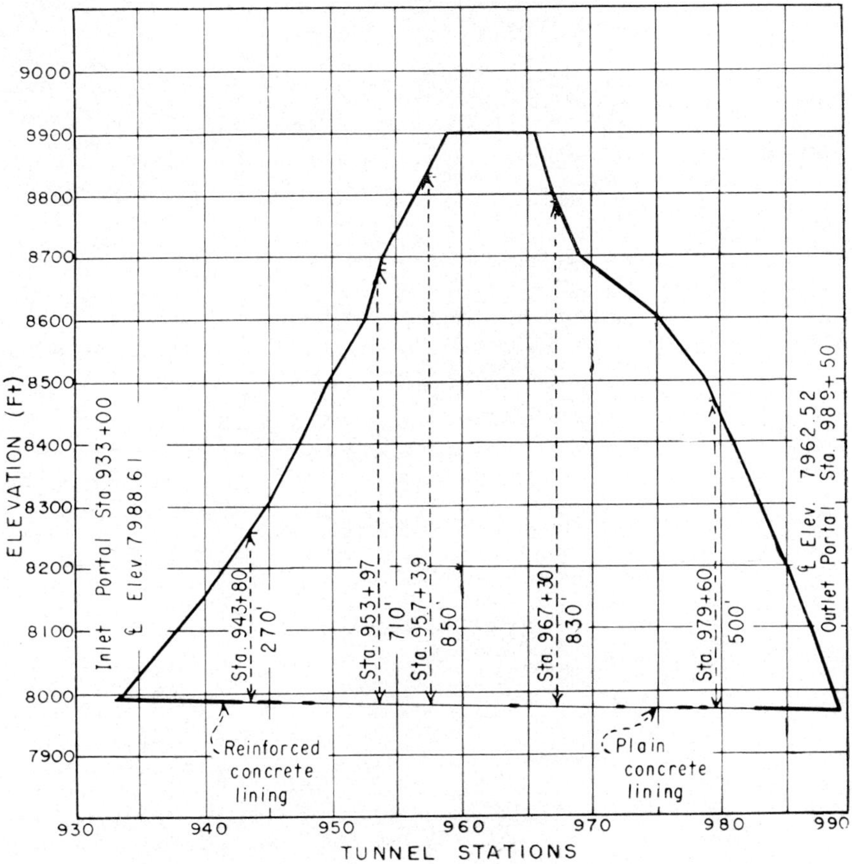

FIG. 9. Profile of Prospect Mountain Tunnel.(7)

Figure 9(7) shows a profile of Prospect Mountain Tunnel with locations of the five measuring stations, and Fig. 10(b)(7) shows, graphically, the magnitudes and directions of principal tangential rock stresses plotted on the developed inner surface of the tunnel. Figure 10(a) shows radial locations of measuring points.

Quoting Oleson:(7)

The maximum principal stress ranged from 1090 psi, to 11,840 psi, compression. The direction of the maximum principal stress occurs more generally transverse to the tunnel axis than parallel to it. A general correlation exists between the maximum stress and depth of cover although relationship indicates only a trend, and is not linear.

For a maximum depth of cover of 900 ft it is apparent that the magnitude of existing rock stress is much higher than can be accounted for by the weight of the overlying rock plus the

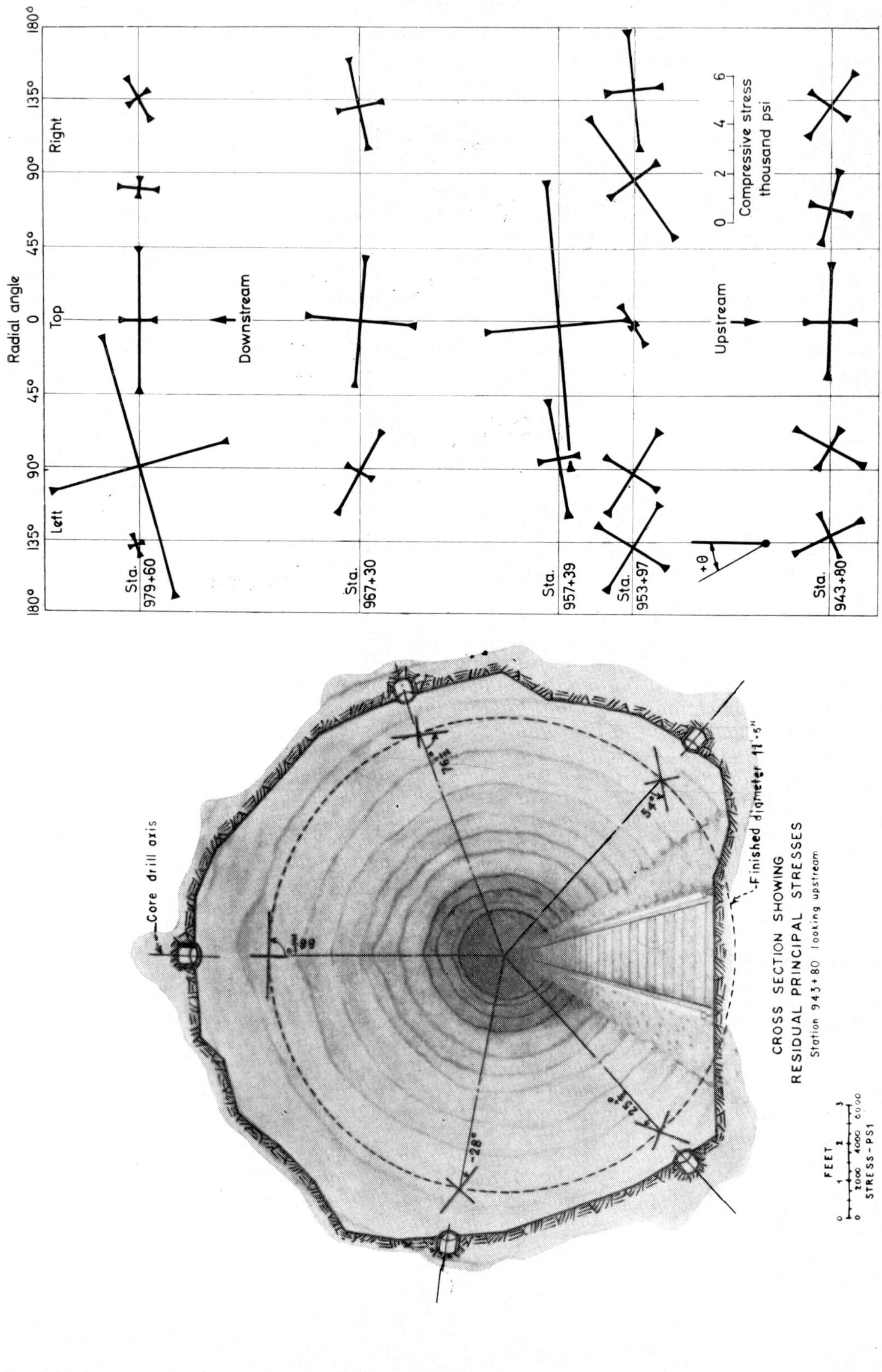

FIG. 10 (A) Tunnel cross-section showing the orientations and the magnitudes of the principal residual stresses.[7] (B) Computed magnitude and

stress concentration produced by drilling the tunnel. Even after eliminating these stress concentrations the vertical stresses were still 2 to 6 times greater than could be accounted for by the overburden.

Examination of Fig. 10(b) indicates that, as a rough approximation, and taking an overall average, the circumferential stresses at the crown were of the same order of magnitude as those at the sides, and were roughly twice the magnitudes of the longitudinal tangential stresses.

These are the same relative stress concentrations which are produced when a circular opening is excavated in an elastic material which is subjected to isotropic stress (see Chapter 1). Thus, it would seem that the original rock stress was essentially isotropic and in general ranged from about 2000 to 6000 psi.

Under 900 ft of rock cover the vertical stress due to weight of overlying rock should be about 1000–1100 psi.

The important relatioship indicated by the tests in this instance is that, in a sound igneous rock, the initial lateral stresses were approximately equal to the initial vertical stress, that is

$$\sigma_x = \sigma_y = \sigma_z$$

Oleson says[7] "As a result of these tests and subsequent analysis and inspection, the tunnel rock was judged to be of sufficient soundness that reinforcement could be omitted from 61 per cent of the total length of tunnel. The actual amount of reinforcing steel saved as a result of the investigation was 833,000 lb."

Tunnel and Power Stations (Australia)

The results of rock stress measurements made during the construction of the tunnels and underground power stations in the Snowy Mountains of Australia have been reported by Moye.[13]

Tangential compressive stress in the surface layer of rock around the excavation was measured in horizontal and vertical directions at several sites, using an adaptation of the "flat-jack" method, which is described in Chapter 10. The method consists essentially of cutting a slot into the rock surface, which results in an expansion of the rock into the slot. Pressure is then applied with a flat hydraulic jack to the walls of the slot to restore the rock to its initial position. The jack pressure, corrected for the shape of the slot, is assumed to equal the tangential pressure in the rock face before the slot was cut.

The next step consisted of making a two-dimensional photoelastic analysis on model openings of the same shape as the tunnels where the measurements were made. From the analyses stress concentration factors were derived which were applied to the measured stresses. These were assumed to give the magnitudes of the original stresses in the directions of measurements, although these would not necessarily be the principal stresses.

Figure 11 shows the results of stress measurements at three localities.

The natural stresses at both T.1 and T.2 show similar features. The compression in the vertical direction is greater than that due to the weight of the rock vertically above the sites. It is probable that these high values may be partially due to the V-notch effect of the adjacent valley, in causing realignment and concentration of stresses. These power stations were excavated in granite and granitic gneiss.

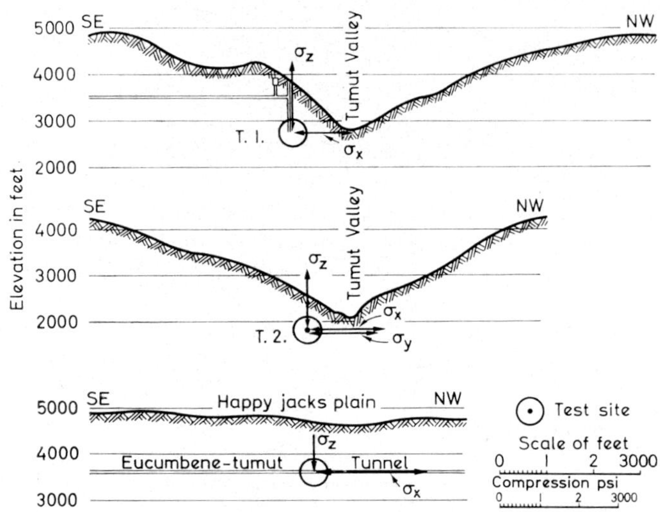

FIG. 11. Natural rock stresses at three localities in the Snowy Mountains, Australia.
σ_z vertical compression; σ_x horizontal compression; σ_y horizontal compression in a direction perpendicular to σ_z and σ_x.(13)

To determine the natural stresses at a location unaffected by topography a site was selected in Eucumbene–Tumut Tunnel, at a depth of 1100 ft below a relatively level plain. The tunnel was nominally circular, 24 ft in diameter and driven through broadly jointed granite which was generally very sound, although at intervals, long after completion of excavation, there is slight but widespread spalling of thin slabs from the surface of the tunnel along fresh fractures.

The average natural compressive stress measured in Eucumbine-Tumut Tunnel in the horizontal direction was about 2.6 times the stress in the vertical direction. The vertical stress was about 0.8 of the stress due to the weight of overlying rock.

It is interesting to note that the horizontal stress was measured parallel to the longitudinal axis of the tunnel only and that this longitudinal axis of the tunnel nearly parallels the long axis of the major geologic structures of the area.

In the case of natural stresses in mines of the Kolar Gold Fields the natural lateral stresses of greatest magnitude have a direction normal to the long axes of geologic structures of the area.

Powerhouses in Portugal

Measurements made by the "strain relief" method on a tunnel in granite at Canicada powerhouse (Portugal)[14] indicated that the stress distribution in the rock was essentially isotropic before the tunnel was excavated.

Measurements made in the Picote Powerhouse, also in granite,[14] gave essentially the same results on the upstream side of the powerhouse but on the downstream side stresses as high as 6000 psi were observed. The high localized stresses were attributed to the presence of a nearby vertical fault.

Kolar Gold Mines

Recent strain measurements in the mines of the Kolar Gold Field, Mysore State, India, have indicated the presence of lateral compressive stresses which greatly exceed vertical stresses, even at great depth.[9]

Magnitudes of pre-existing earth stresses computed from strain-relief measurements made on periphery of a drift on the fortieth level at a vertical depth of 3646 ft showed an initial vertical stress of 5820 psi and a lateral stress of 6700 psi. Taking the density of the rock as 190 pcf the vertical pressure should increase at the rate of about 1.32 psi per foot of depth. At a depth of 3646 ft, σ_z should be $3646 \times 1.32 = 4800$ psi.

Other measurements made include measurements of lateral strain in country rock 1000 ft from the "reef" (vein) and normal to the plane of the reef. Indicated stress at a depth of 4199 ft was 11,800 psi. Thus, the indicated lateral stress was more than twice the theoretical vertical stress for this depth.

Other strain measurements in the foot wall of a stope at a depth of 7929 ft, indicated a lateral stress of 13,600 psi in the immediate footwall which built up to 15,800 psi at a distance of some 120 ft into the country rock. Since these veins at depth dip nearly vertically the stresses indicated were essentially lateral stress. Again, the indicated lateral stress greatly exceeded the theoretical vertical stress, which at that depth would be $7929 \times 1.32 = 10{,}500$ approximately.

An east–west natural lateral stress of almost 8 tons per square inch was deduced from strain measurements at a depth of about 9000 ft where the theoretical vertical stress would be about 11,800 psi.[10]

The gold-bearing quartz veins of the Kolar field occur in a narrow strip of Pre-Cambrian hornblende schist which is surrounded by a complex of gneissic and granitic rocks.[11] The veins have a strike varying only a few degrees from north and south, and dip toward the west. Only two of the quartz veins, the Champion and the Oriental, are of economic value. The Champion Lode (the principal producer) dips westerly at an angle from 40 to 50° near the surface, gradually changing to vertical at depth. In places the quartz attains widths of 20 ft, but average width is between 3 and 4 ft.

The presence of large lateral stresses in the rocks of this field had long been suspected because the first rock burst in the field took place at a depth of only 500 ft. The direction of the maximum lateral stress is apparently normal to the strike of the vein. This is illustrated by deformation measurements in a concrete-lined circular shaft of nominal diameter 12.5 ft located at a depth just short of 10,000 ft. Periodic measurements of shaft diameters after completion of the concrete lining revealed that during an eight-week period the north–south diameter increased by 0.21 in. while the east–west diameter decreased by 0.45 in. while the north west–south east and the northeast–south west diameters remained relatively constant. The deformation was due to time–strain effects only, since no other workings were active in the same vicinity. Direction of deformation indicated that the east–west stress greatly exceeded the north–south stress; that is, the maximum lateral stress is normal to the strike of the veins, and therefore normal to the long axes of the workings.

Mines in Sweden

Notch Effects

In Sweden, Hast[12, 17] has, over a period of years, perfected a method for measuring rock stresses by means of magnetostrictive gages. Results from measurements in Swedish mines indicate the presence of large natural lateral stresses. These results are in line with results of measurements of lateral stresses which have been cited previously.

Figure 12 shows the results of one of a series of measurements in the iron mines at Grangesberg. The orientation of the principal stresses appears to conform to the trajectories of stresses around a notch in an elastic stressed material, and the magnitudes of lateral stresses, which here exceed the vertical stresses, may be partially explained as the natural realignment of stress underneath the excavation.

By applying the equation for stress concentration around a right-angle notch in elastic material

$$\sigma_x = \sigma_x\left(1+2\sqrt{\left(\frac{a}{\varrho}\right)}\right)$$

where σ_{x_t} = maximum compressive tangential stress at bottom of notch;
σ_x = natural rock stress (lateral);
a = vertical depth of notch;
ϱ = radius of the bottom of the notch.

Hast was able to secure a fair agreement with observed stresses on the assumption that the lateral stresses existing before the excavation was made had the magnitude $\sigma_x = \sigma_z$.

However, this relationship does not entirely explain the high lateral stresses at some other locations and, in general, it appeared that natural lateral stress exceeded the vertical stress at many points.

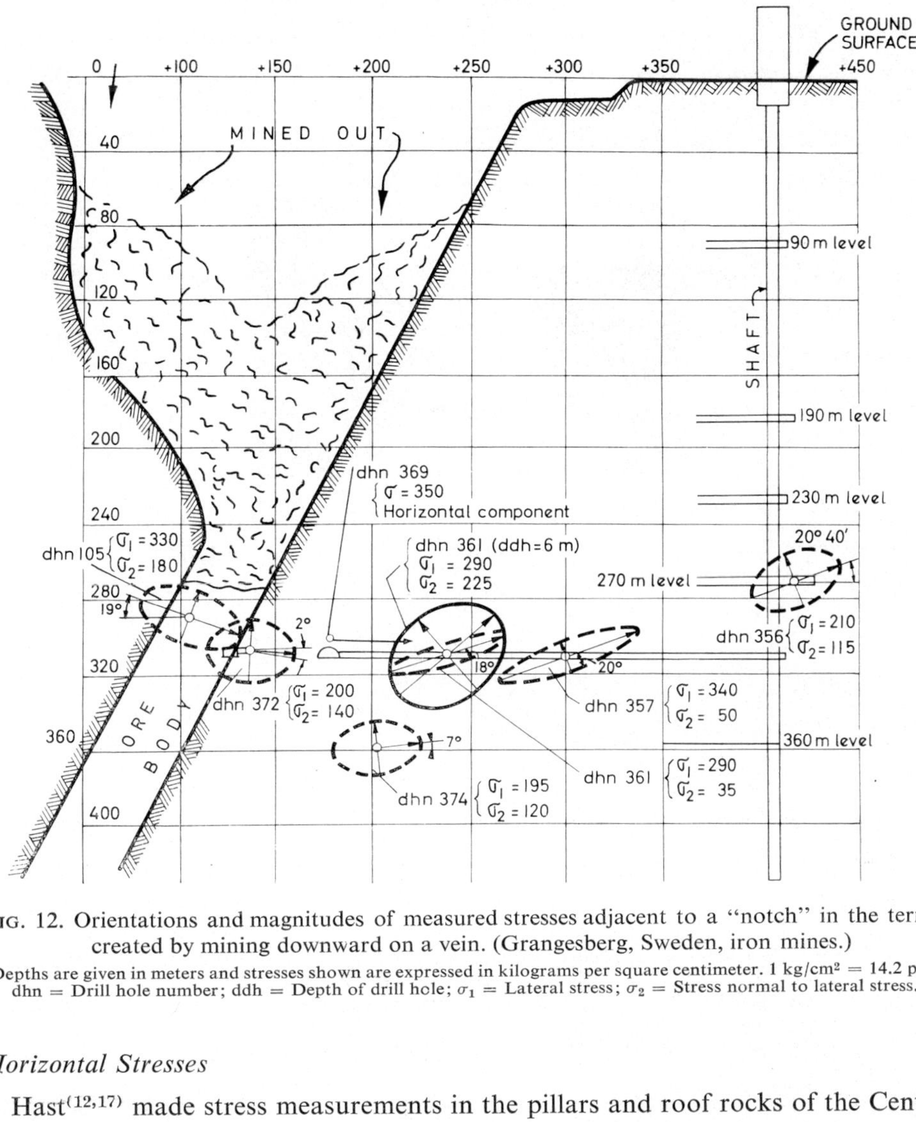

FIG. 12. Orientations and magnitudes of measured stresses adjacent to a "notch" in the terrain created by mining downward on a vein. (Grangesberg, Sweden, iron mines.)

(Depths are given in meters and stresses shown are expressed in kilograms per square centimeter. 1 kg/cm² = 14.2 psi[12] dhn = Drill hole number; ddh = Depth of drill hole; σ_1 = Lateral stress; σ_2 = Stress normal to lateral stress.)

Horizontal Stresses

Hast[12,17] made stress measurements in the pillars and roof rocks of the Central Mine in the Laisvall lead mines in Central Northern Sweden.

Geology

The mining operations are carried on at different places and in different horizons of a locally mineralized bed of sandstone. In this area, borings disclosed the following sequence of nearly horizontal strata:

Shallow top layer (residual soil or recent sediments); 34 m mica or quartz schists; 20 m various types of mylonite, including arkose and syenite mylonite; 6 m schists; 5 m schist conglomerate; 36 m sandstone; 3 m arkose; granite basement.

In the Central Mine the seat of the mining operations is located in the lowest portion of the sandstone stratum. The sandstone is described as "hard and strong", but its continuity is disrupted by numerous horizontal, more or less discontinuous, clay or shale partings. The vertical distance between these planes of weakness is

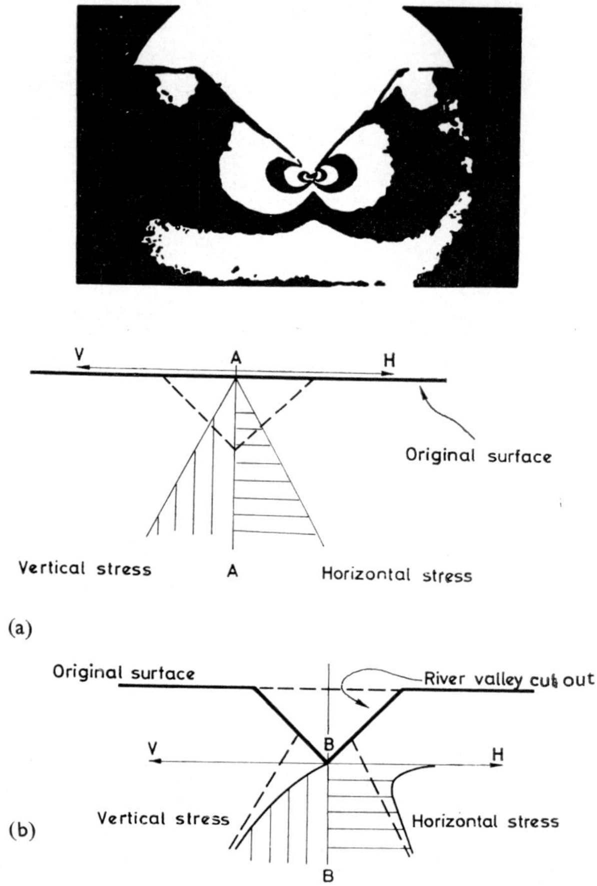

FIG. 13. (a) Stress distribution along *A–A* before cutting of valley. (b) Stress distribution along *B–B* after cutting valley.

Effect of a V-notch in a gravity stress field.(19)

commonly several metres, though in a few places it is less than half a meter, with detrimental effects on the stability of the roof. The sandstone is also weakened by two–sets of joints, at right angles to the bedding planes, striking approximately north–south and east–west. Most of the east–west joints are tight whereas many of the north–south joints are open. In some of these, the open spaces between the walls are occupied by gouge-like material, and the other north–south joints are commonly water-bearing.

The geological survey of the area has disclosed the following relationships between the strata encountered in the borings in the Central Mine. The sandstones overlying the granite basement as well as the schists above the mylonite are of Cambrian age. The mylonite originated during the Caledonian upheaval, when the schists were pushed over the sandstones in a south-easterly direction. The base of the schist represents the thrust plane. The mined-out spaces are located at a considerable depth below this plane, in apparently intact sandstone. The joints exposed on roof and floor intersect the direction of the overthrust at approximately 45°. Their age is not known.

Measured Stresses

Figure 14 shows the location of the roof points 1–5 and the floor points 26 and 27 where stresses have been measured. At all points the measurements showed that the maximum horizontal stress is several times greater than the overburden pressure which prevailed at the point before the mining operations were started, and the difference between the maximum and the minimum value of the horizontal stress at each point is rather important. The maximum values of the horizontal stresses, σ_{x_t}, encountered at elevations of about 2.0 m above the roof ranged be-

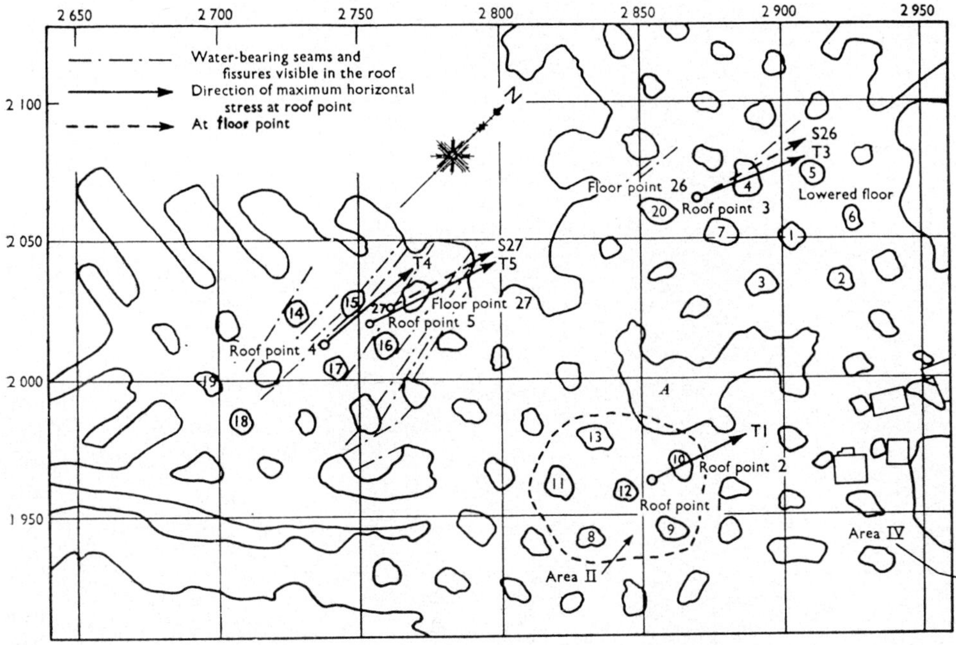

FIG. 14. Locations of pillars and observation points in the Central Mine, Laisvall, Sweden. Pillars in which stress measurements were made are numbered. The arrows labelled *T* and *S* indicate directions of the maximum horizontal stress.[12, 17]

tween 140 and 360 kg/cm² whereas the initial overburden pressure, σ_z, at the same elevation, prior to the beginning of the mining operations was about 25 kg/cm². Hence $\sigma_{x_t}/\sigma_z = 5.6\text{–}14.5$. The corresponding M-values range between about 3 and 8. These high M-values indicate that the initial maximum horizontal stresses were very much greater than the overburden pressure. The prevalent direction of the maximum stress was north–south, which is the direction of the open and water-bearing joints.

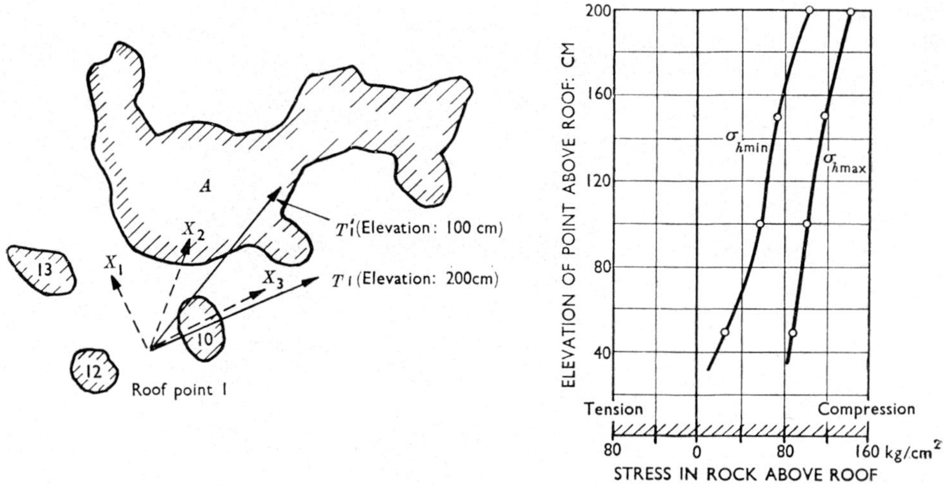

FIG. 15. (a) Location of roof point 1 and adjacent pillars. X_1 to X_3 indicate directions in which stresses were measured. $T1$ and $T1'$ show direction of maximum horizontal stress at two different elevations above the roof. (b) Intensity of maximum and minimum horizontal stress at different elevations above the roof.[12, 17]

Close to the rock surface exposed in the mine the orientation of the maximum horizontal stress appears to be influenced by the location of the boundaries of the mined-out spaces. This possibility is suggested by the results of the stress measurements above roof point 1, located half-way between pillars 10 and 12. Figure 15(a) shows the location of the point with respect to the pillars, and Fig. 15(b) the maximum $\sigma_{h,\,\max}$ and the minimum $\sigma_{h,\,\min}$ horizontal stress in the rock at different elevations above the roof. According to Fig. 15(b) the intensity of both stresses increases with increasing elevation above the roof. The results of the individual gage readings, in three different directions, left no doubt that the direction of the maximum stress $\sigma_{h,\,\max}$ also changed with increasing elevation above the roof. In Fig. 15(a) the line $T1'$ indicates the direction of the maximum horizontal stress at an elevation of 100 cm above the roof and the line $T1$ at an elevation of 200 cm. The line $T1'$ is directed towards the large body of rock A in Fig. 15(a) whereas $T1$ is parallel to the prevalent direction of the maximum horizontal stresses at all the other points at an elevation of 200 cm as shown in Fig. 14.

The measured values of the largest horizontal stress at a depth of 200 cm beneath the floor were 170 kg/cm^2 at point 26 and 140 kg/cm^2 at point 27 (*see* Fig. 14). The corresponding M-values are approximately 3.6 and 2.8. In Fig. 14 it can be seen that the direction of the largest horizontal stresses at floor points, indicated by the dash lines S 26 and 27, is roughly parallel to that of the largest horizontal stresses at an elevation of 200 cm above the roof ($T1 \times T3 \times T5$) and to the direction of the open and water-bearing joints. Both above and below the mined-out space, the minimum measured stress ranges between 0.4 and 0.6 times the maximum stress at the same point.

STRESS DETERMINATIONS AT THE SITE OF A PROPOSED UNDERGROUND POWERHOUSE (CALIFORNIA)

Recent stress determinations carried out by the State of California Division of Water Resources and the U.S. Bureau of Mines at the site of a proposed underground hydroelectric plant indicated that magnitudes of lateral stresses were about equal to vertical stresses at corresponding points in the fine-grained amphibolite.

The stress determinations were made at a depth of about 350 ft beneath the surface and about 900 ft from the adit portal.

Measurements were made at the walls of drifts about 6 ft wide and 7 ft high by means of borehole deformation meters and also by means of flat jacks, and the results obtained by the two different methods were in reasonable agreement.

Outside the zone of stress concentration the magnitude of the calculated major principal stress varied from about 300 to 900 psi. The principal stresses were generally inclined from the vertical and horizontal. (See Chapter 12 for details of these stress measurements.)

LOCALIZED STRESSES

In rocks which have been intensely folded and deformed, stresses may vary from point to point. A tight fold, for instance, may be under considerable stress as also may the rock in the vicinity of a fault. Rock bursts may occur in isolated locations due to such local stress concentrations, while openings in the remainder of the rock mass experience no difficulty.

The author recalls one such rock burst (at a mine in the Bingham Canyon District, Utah) which occurred when a mine cross-cut at a depth of about 2800 ft exposed a tightly folded quartzite bed. The bent quartzite strata, some hours after the face had exposed it, sprang apart, somewhat in the manner of a clock spring which breaks when wound too tightly. Force of the rock burst was sufficient to bend upward the mine rails for a distance of several feet back from the face. No other rock bursts were experienced in this vicinity.

Bursting conditions were encountered throughout a large part of the Delaware Aqueduct during construction.[5] However, Bird cites an example of an isolated stressed bed encountered during the sinking of a shaft 1550 ft deep which was lined with 2 ft of concrete. The lower several hundred feet of shaft was sunk through alternating beds of shale and sandstone with a dip of less than 10°. One sandstone bed, 8 ft thick, was located just above the tunnel. Within 2 weeks this bed had begun to expand and to crush the concrete shaft lining. It was the only bed encountered in the shaft which exhibited any pre-stressing and a similar sandstone bed just below it showed no strain where it was cut by the pump station 50 ft from the shaft.

Isolated localities of high stress appear to be rather rare, however, and most high lateral stresses seem to be of district-wide, or regional extent. For instance, the presence of high lateral stress was in evidence over a large part of a 33-mile length of the Delaware Aqueduct, as indicated by bursting and spalling of roof rock.

As another example, high lateral stresses are in evidence in all the mines of the Kolar Gold Field, in India. These mines extend for a strike length of approximately 5 miles.

STRESSES IN THE VICINITIES OF FAULTS

In many localities the existing rock stresses adjacent to a fault considerably exceed the stresses prevailing in the rock a short distance from the fault.

At the Sunnyside Coal Mines, in Utah,[15] "bumps" ("rock bursts" or "coal bursts"; that is, the violent failure of rock or coal under excessive stress) have occurred in connection with the driving of development headings (single openings) under as little as 500 ft of cover. Although the Sunnyside Mines had many bursts which can be ascribed to overstressed pillars, the bursts in development headings have frequently occurred far away from any locality of previous extraction operations and have been noted to occur in the vicinities of faults.

Two possible explanations present themselves:

(1) The possibility that the fault system is still active, and occasional slight fault movement strains the adjacent rock and causes violent shear failure in nearby workings.

(2) The rock adjacent to the fault is already in a state of strain and is therefore highly stressed, and fails violently when penetrated by an opening.

In some instances the first supposition seems to apply, as for example, on November 21, 1957 [15] indications of movement were discernible in No. 1 Mine rock tunnel where it intersected a block formed by a strike fault, and a dip fault. Along the same fault zone, but 13,000 ft distant in No. 2 Mine, disturbance was noted immediately adjacent to fault planes. In both localities cover was less than 500 ft. In another instance, on December 13, 1957, a bump manifested itself along a distance of 7000 ft and was simultaneously recorded by men working at four points along different fault planes.

On the other hand, instances can be cited which appear to support the existence of an increase in the magnitudes of natural rock stresses around development headings as they approached faults. In one case a tunnel face became very active and fragments of rock flew from the face as it approached within 30 ft of a fault. After the fault was penetrated, however, no further disturbance was noted. In another instance a slope encountered violently bursting ground after penetrating a fault with an 18-ft displacement. With difficulty, because of bursting conditions, the slope was driven 840 ft at which point it encountered another fault. After penetrating this latter fault no further disturbance was noted.

It is possible that the manifestations described above are the result of both an active fault system, and a pre-existing stressed zone adjacent to the fault. A burst at one location along a fault could upset equlibrium and trigger movement at distant locations along the fault.

Figure 16 shows the stress conditions along a fault.

"Rock bursts" and "bumps" are discussed more extensively in Chapter 8.

STRESSES ADJACENT TO FAULTS

The existence of a fault is evidence that, during its history, the rock at that location has been subjected to stresses which have exceeded its shear strength. When the rock "shears" to form a fault a certain amount of "elastic rebound" occurs and a portion of the original localized stress is relieved.

Along those faults which have long been inactive it is probable that most of the stress has been relieved by "time yield" of the rock. Along relatively recent faults, or those which are still active, evidence indicates that the rock retains considerable strain and stress.

Faults in hard brittle rocks such as granites, quartzites, and strong sandstones, etc., are more likely to retain strain and stress than those of the softer rocks such as shales or siltstones.

"Tight" faults are more likely to retain strain and stress in adjacent rock than those faults which contain wide zones of fault clay and breccia.

An irregular fault plane is more likely to retain strain and stress in the adjacent rocks than is a smooth, regular fault plane.

Figure 16 (a), (b), and (c) shows the manner in which a fault may form, and the distribution of stresses in the adjacent rocks. The square element shown is for the purpose of demonstrating the directions of deformations in the rock.

Maximum shortening occurs along axis *A–B* and the compressive stress in the rock will be maximum parallel to this axis. Tensile stress is maximum along axis *CD*. The effect of depth pressure is neglected in this illustration. It would have the effect of rotating the maximum compressive stress to a more nearly vertical position.

Figure 16 (d) shows the locations of stress concentrations around a tunnel which is driven parallel to the fault. This is the simplest case. Obviously, there are innumerable other situations possible, depending upon the direction of movement along the fault and the direction and shape of the opening approaching the fault.

With a knowledge of the dip and direction of movement along a fault the orientation of the maximum stresses may be deduced as in the foregoing example.

The same reasoning applies to folds, which are, in many cases, merely the early stages in the formation of faults.

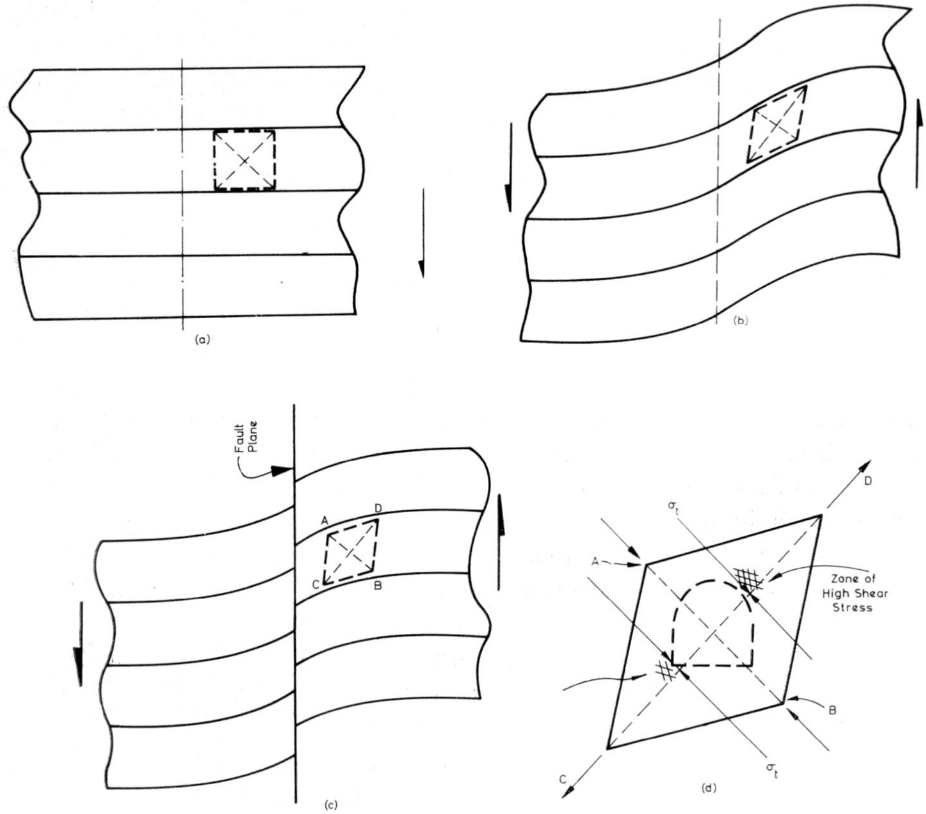

FIG. 16. (a) Unstressed rock. (b) Deformation under shear stress. (c) Fracture and recovery of deformation by elastic rebound. (d) Stresses around a tunnel driven parallel to the fault.

LATERAL STRESS VARIATION WITH DEPTH

The measured stress values cited previously show a general tendency to vary with the depth of overlying rock. It was noted that the vertical stress varied roughly with depth, although generally the measured values of vertical stress exceeded the values to be expected from weight of overburden, σ_z. It was noted also that the measured lateral stress values varied, in a rough way, with depth, although no definite linear relationship could be noted.

In many cases the magnitudes of measured lateral stresses exceeded the vertical stress at the same point. It is evident that such stresses are tectonic or residual, and are not induced by the existing depth pressure.

Since they frequently exceed but at the same time tend to vary with depth, it appears that the tectonic or residual lateral stresses may be the remnants of stresses which were originally much greater, and which found some means of partially relieving themselves as erosion removed the overlying rock. Possibly the rock is working toward the conditions of "hydrostatic" stress postulated by Heim, and if given enough time the excessive lateral stress would be reduced to the same magnitude as vertical stress. Thus, the present stresses may represent a stage in transition to "isotropic" stress conditions.

ORIGINS OF TECTONIC AND RESIDUAL STRESSES

Residual lateral earth stresses may result from several causes:

(1) Lateral compressive stresses such as result in regional deformation and produce folding and thrust faulting in all types of rocks.

(2) Intrusive igneous rocks solidify at depths of thousands of feet below the surface. The hydrostatic stresses in the molten mass will be partially preserved in the solidified mass and, as erosion removes the overlying rock vertical stresses are relieved. However, lateral stresses can be only partially relieved by the vertical expansion.

The same conditions would apply to a metamorphic rock which may be plastic or semi-liquid at some period during its formation.

(3) Some authorities[(5)] ascribe some lateral stresses to re-crystallization in metamorphic rocks or original crystallization in igneous rocks (with an accompanying volume increase).

Similarly, the volume increase attending the formation of serpentine is thought to be the cause of large lateral stresses in serpentinized rocks.

NATURAL STRESS CONDITIONS TO BE ANTICIPATED

The magnitudes of lateral stresses which may be encountered in various types of rocks may be summarized as follows.

In soils the state of stress equilibrium is usually determined by the angle of internal friction. Generally, no tectonic stresses will exist in such material.

In sedimentary rocks which are undisturbed by horizontal stresses equation (2) may be expected to give the magnitude of the minimum lateral stress which may be encountered.

In folded and overthrust sedimentary rocks the lateral stress may have a magnitude greater than that indicated by equation (2) and, in fact, may have any value compatible with the strength of the rocks.

There seems little doubt that "hydrostatic" stresses existed in metamorphic and igneous rocks during the stages of their formation when they were in plastic state. If such a rock is still intact and relatively unfractured and unweathered then it is likely that residual lateral stresses will still exist within it. If the rocks have been subjected to subsequent faulting then lateral stresses may have been partially relieved. If the rock shows open jointing or other tension fractures then it is improbable that any lateral restraint remains.

In the extrusive igneous rocks it is improbable that lateral stresses will be encountered as these rocks are not subject to vertical compressive stresses during solidification and, upon cooling, they ordinarily form numerous tension fractures.

APPENDIX

THE following paper is an account of methods used and results obtained in a recent series of measurements made at underground installations to determine the stresses existing in the rocks at those locations.

The paper, by Leonard Obert, appeared in *Mining Engineering*, for August 1962, and is reprinted here in its entirety by permission of the American Institute of Mining, Metallurgical and Petroleum Engineers.

IN SITU DETERMINATION OF STRESS IN ROCK

Underground research by the Bureau of Mines has led to development of a method for determining the absolute magnitude and direction of stress in rock. This work has also shown that by making measurements at a distance from the surface of underground openings, the magnitude and direction of the stress field (i.e. the state of stress in the rock before mining) can be ascertained.

The structural stability of any mine or underground opening in rock is dependent on the stress field, that is, the state of stress in rock before mining, the stress distribution in the rock created by the opening, and the *in situ* strength of the rock. In the past mine design problems have been considered by assuming that the stress field is due only to the weight of the overlying rock, the stress around the opening can be calculated from elastic theory using laboratory-determined elastic constants, and the *in situ* strength of the rock can be determined from laboratory tests on small specimens. Solutions to hypothetical problems using this procedure are given in a report by Obert, Duvall, and Merrill.[22] Because of the uncertainties of these assumptions, this method can give only a rough approximation to the solution of actual structural problems. Obviously, this approach to determining the stability of underground openings would be greatly improved if the rock stress could be measured *in situ*, and it is undoubtedly for this reason that in recent years many investigators have spent considerable effort on the development of procedures for measuring rock stress.

In this report a method is described for determining *in situ* the absolute stress in rock. The method requires the measurement of the deformation of a borehole before and after the rock is stress-relieved. The magnitude and direction of the applied stresses are calculated from the borehole deformation and modulus of elasticity of the rock. Also, the results of absolute stress determination made in the walls and pillars of underground openings in five rock types are presented. From these results a number of conclusions are drawn regarding the accuracy and limitations of this method and the problem of stress measurements in general.

In this investigation rock stress was determined by measuring the deformation (change in diameter) of a borehole before and after the hole was stress-relieved. It has been shown that the borehole deformation in a biaxial stress field is related to the magnitude and direction of the applied stresses in the plane perpendicular to the axis of the hole (Fig. 17) by the equations*. (See also Chapter 10, page 390.)

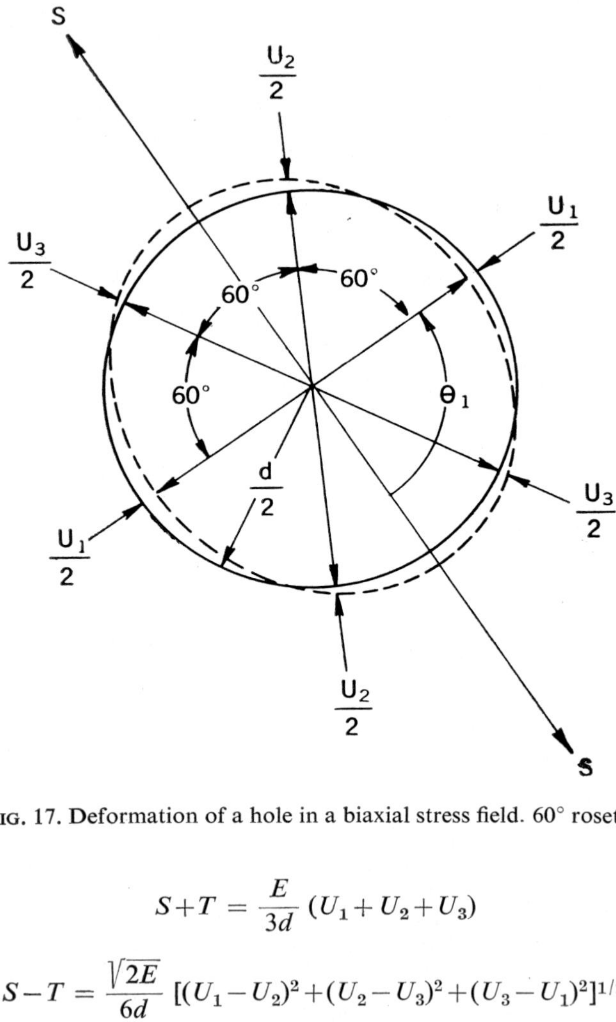

FIG. 17. Deformation of a hole in a biaxial stress field. 60° rosette.

$$S+T = \frac{E}{3d}(U_1+U_2+U_3) \tag{5}$$

$$S-T = \frac{\sqrt{2}E}{6d}[(U_1-U_2)^2+(U_2-U_3)^2+(U_3-U_1)^2]^{1/2} \tag{6}$$

$$\tan 2\theta_1 = \frac{\sqrt{3}(U_3-U_2)}{2U_1-U_2-U_3}, \tag{7}$$

* In the derivation of these equations it is assumed that the medium is perfectly elastic, isotropic, and in plane stress. For, a review of the theory and empirical verification, see Ref. (23).

where

S, T = maximum and minimum stress in the plane perpendicular to the axis of the borehole, psi

E = modulus of elasticity of the rock, psi

d = diameter of a borehole, inches

U_1, U_2, U_3 = borehole deformation at a 60° separation (60° deformation rosette), inches. U is positive for increasing diameter

θ_1 = angle from S to U_1, measured in the counterclockwise direction, degrees

$$\text{if } U_2 > U_3,\ \theta_1 \text{ is between 90 and 180°;}$$

$$\text{if } U_2 < U_3,\ \theta_1 \text{ is between 0 and 90°;}$$

$$\text{if } U_2 = U_3, \text{ and if (a) } U_1 > U_2,\ \theta_1 = 0°;$$

$$\text{(b) } U_1 < U_2,\ \theta_1 = 90°.$$

All borehole deformation measurements were made with a USBM borehole deformation gage[(24)]. This gage has a sensitivity of about 20 μ-in. per inch, corresponding to a stress sensitivity of 13.3 psi for a rock in which $E = 3.0 \times 10^6$ psi. The gage is designed to fit in a $1\frac{1}{2}$-in. diameter hole.

Two methods of stress-relief were employed in these investigations. In the first method stress-relief in soft rock was effected by cutting to a depth of 10 ft around, or at least on three sides of the measurement area with a universal cutting machine. A stress-relief block is shown in Fig. 18.

In the second method the stress-relief was made by overcoring the $1\frac{1}{2}$-in. diameter gage hole with a larger bit, normally a 6-in. outside diameter bit that cuts a $5\frac{5}{8}$-in. diam core. A photograph of the overcoring drill and a section of stress-relief core is shown in Fig. 19.

Because only a limited number of gages were available for placement in a cutter relief block, the gages were oriented to measure the vertical deformation. The gages were usually placed at a depth of 0.5, 1, 2, 3, 5 and 8 ft from the collar of the hole. With a cutter depth of 10 ft it was found that the maximum depth at which a gage could be fully relieved was 8 ft. The stress was calculated, assuming a vertical uniaxial stress field, by the equation,

$$S = \frac{UE}{3d}, \tag{8}$$

where U is the vertical borehole deformation produced by the stress-relief.†

The modulus of elasticity, E, was determined in from either stress–strain or borehole deformation‡ measurements on samples of rock.

† If the actual stress-field is biaxial, the stress value calculated on uniaxial basis will be low by about 11 to 16%.

‡ See footnote on page 118.

FIG. 18. Stress-relief block made with a universal cutting machine.

FIG. 19. View of overcoring drill and section of stress-relief core.

The overcoring stress-relief procedure was as follows:

(1) A $1\frac{1}{2}$-in. diameter EX gage hole (pilot hole) was diamond drilled to the desired depth (usually between 10 and 20 ft). The orientation of the hole was determined with respect to the spacial attitude of the underground opening.

(2) The gage was placed in the hole at a point 5 in. from the collar of the hole and oriented to measure in a specified direction normal to the axis of the hole.

(3) An initial deformation reading was taken.

(4) The borehole gage was overdrilled to such a depth that further drilling produced no further change in the deformation. The depth was usually about 1 in. past the point of measurement.

(5) A final deformation reading was taken. The borehole deformation is the difference between the initial and the final readings.

(6) Next the gage was moved into the hole 5 in. past the previous overcoring depth, oriented at 60° clockwise with respect to the initial direction of measurement, and the overcoring procedure repeated.

(7) The gage was again placed 5 in. past the previous overcoring depth, oriented at 60° counterclockwise from the initial direction of measurement, and overcored.

(8) This cycle was continued until a total depth of 24 in. had been drilled (which was the length of the overcoring core barrel). The core was then pulled, the gage reinserted in its next cyclic position, and procedure continued.

(9) To obtain the medium stress the rosette readings were continued until such a gage hole depth was reached that no further change in the borehole deformation was produced.

(10) Stress–strain or borehole deformation curves obtained in the laboratory from sections of the stress-relief core or sections of solid core taken from the same area were used to determine the modulus of elasticity of the rock.

(11) The magnitude and direction of the maximum and minimum stress in the plane normal to the axis of the borehole were calculated from the rosette data and the modulus of elasticity from equations (5), (6), and (7).

In this method it is not imperative that the gage hole and the $5\frac{5}{8}$-in. diameter core be kept concentric. However, to keep the $5\frac{5}{8}$-in. diameter coring bit on a reasonable course, the driller would guide the drill with respect to the direction of the gage hole—a procedure that is purely art. Using this method, stress-relief measurements have been made at depths up to 20 ft. The overcoring procedure has superseded the cutter relief method because the latter was limited to soft rock and the cutting equipment was exceedingly cumbersome. This overcoring procedure is similar to that employed by Hast.[(25)]

Stress Determinations in Trona

Absolute stress determinations were made in the walls and pillars of the trona mine operated by the Intermountain Chemical Co., Westvaco, Wyo. In this mine the trona bed lies about 1500 ft from the surface, varies in thickness from 7 to 11 ft and is nearly horizontal. Except for an occasional parting, the mineralized bed shows only a slight bedding plane weakness and virtually no fracturing. Both the strength and elastic properties vary slightly in different directions. This mineral also had a comparatively linear stress–strain (or stress-deformation) curve, and the

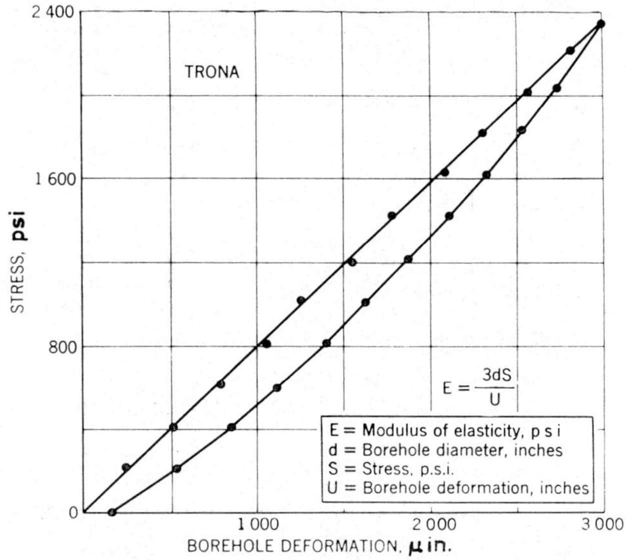

FIG. 20. Trona: representative stress-deformation curve.

area under the increasing and decreasing stress curve (hysteresis loop) was small. The average modulus of elasticity, E, was $3.30 \pm 0.3 \times 10^6$ psi. A representative stress-deformation curve is shown in Fig. 20.

This deposit is mined by both the conventional drilling and blasting and by continuous mining machines. The stoping system is a semi-retreating room and pillar method in which the pillars are split and recovered on the retreat, permitting the roof to cave. Although roof bolts and bolted lagging are used to support the roof, no other structural support is used except occasional light stulls.

The first stress determinations were made using the cutter relief procedure. Six or seven holes varying from 6 in. to 8 ft in depth were drilled in a 5×5 ft wall or pillar surface. A borehole deformation gage was inserted in each hole at a specified depth, oriented to measure vertical deformation, and an initial deformation reading taken. The block was then undercut, overcut, and either cut in one or both of the vertical sides. Usually deformation readings were taken throughout the cutting process. However, the borehole deformation used for calculating the stress was determined only from the initial and final readings. Using this procedure, measurements were made in four machine-mined and five conventionally mined sites.

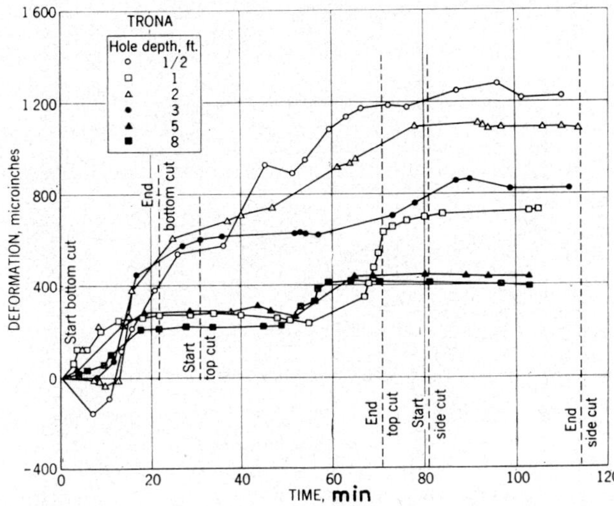

FIG. 21. Trona: representative borehole deformation data, cutter-relief.

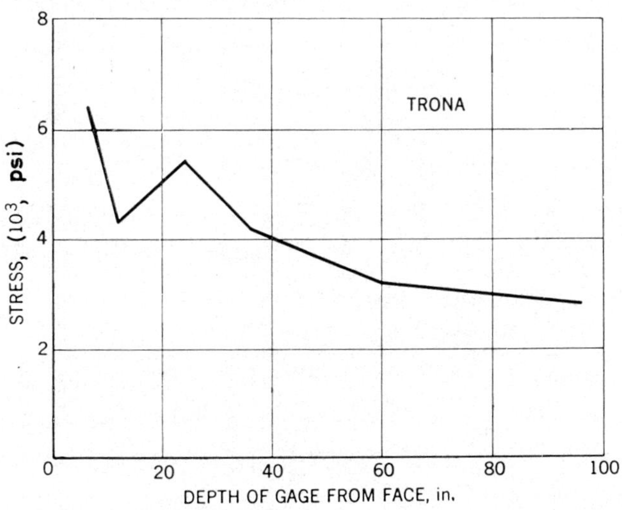

FIG. 22. Trona: stress vs. distance from face, average four machine-mined sites.

Representative borehole deformation data and the average stress results for machine-mined areas are given in Figs. 21 and 22.

Overcoring relief measurements were made at five machine-mined and two conventionally mined sites. As trona is soluble, the bit was cooled and the cutting cleared with air. Representative borehole deformation data from a machine-mined area are given in Fig. 23, and the corresponding stress curve in Fig. 24. The direction of the maximum and minimum stresses, which were nearly vertical and horizontal, are also indicated in the upper part of Fig. 24.

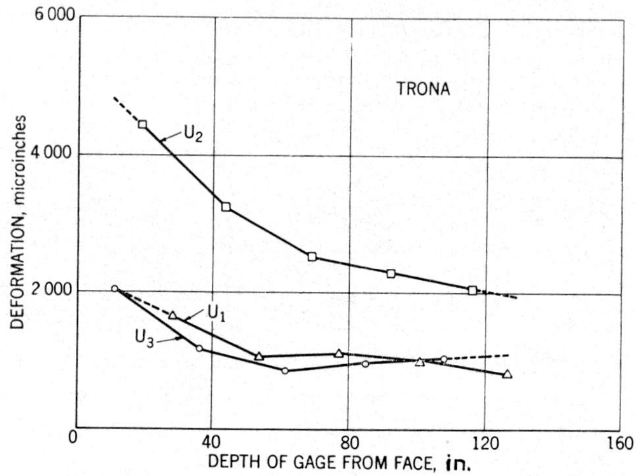

FIG. 23. Trona: representative borehole deformation data, borehole relief.

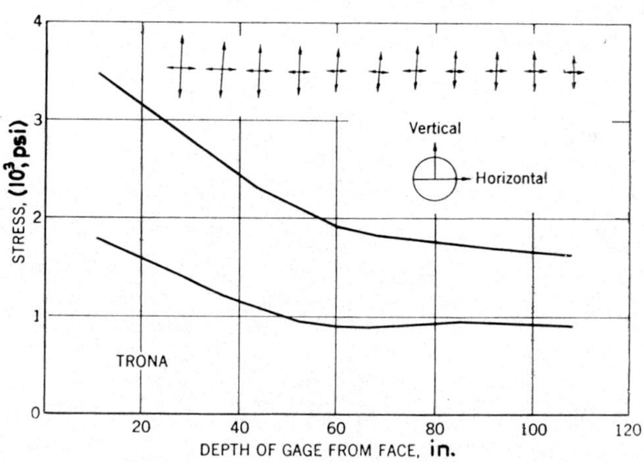

FIG. 24. Trona: stress vs. distance from a machine-mined face, borehole relief.

The results from the 16 tests listed are given in Tables 1, 2, 3, and 4. In Tables 3 and 4 the direction of the maximum and minimum stresses, S and T, were found to be almost vertical and horizontal in each instance.

Summarizing these results:

(1) The vertical stress at (or near) the wall or pillar surfaces ranged from 2620 to 8200 psi. At a depth of 8–10 ft from the wall or pillar surface the vertical stress

TABLE 1. VERTICAL STRESS VS. DEPTH FROM A CONVENTIONALLY MINED FACE-UNDERCUTTER RELIEF

Hole depth, in.	Stress, psi				
	Site 3	Site 4	Site 5	Site 6	Site 7
6	5950	5150	4250	5250	3000
12	9700	4800	3300	5050	2800
24	5600	4700	2650	4700	2600
36	4600	2400	4550	3700	2750
48	3000	2800	2850	3250	2400
72	1800	1500	1600	2460	2500
96	2200	1600	1850	2250	1850

TABLE 2. VERTICAL STRESS VS. DEPTH FROM MACHINE-MINED FACE-UNDERCUTTER RELIEF

Hole depth, in.	Stress, psi			
	Site 9	Site 10	Site 11	Site 12
6	8200	5200	5900	8200
12	4200	4550	4000	4400
24	5650	—	5300	5200
36	4950	—	4200	3500
60	5000	3000	2400	2450
96	4650	—	2250	1400

TABLE 3. MAXIMUM AND MINIMUM STRESS VS. DEPTH FROM CONVENTIONALLY MINED FACE-OVERCORING RELIEF

Hole depth, in.	Stress, psi			
	Site 1		Site 2	
	S	*T*	*S*	*T*
12	—	—	2600	950
24	—	—	2400	830
48	2600	1750	2000	630
72	2500	1700	1900	570
96	2150	1700	1850	550
120	2300	1650	1700	330

TABLE 4. MAXIMUM AND MINIMUM STRESS VS. DEPTH FROM MACHINE-MINED FACE-OVERCORING RELIEF

Hole depth, in.	Stress, psi									
	Site 2		Site 3		Site 4		Site 5		Site 7A	
	S	*T*	*S*	*T*	*S*	*T*	*S*	*T*	*S*	*T*
12	2750	1400	3500	1800	3600	2000	3700	2200	2900	1750
24	2500	1100	3050	1500	2700	1600	3000	1900	2550	1650
48	2200	850	2200	1000	1800	1050	2300	1350	2050	1450
72	2050	610	1800	900	1550	860	2100	1100	1900	1350
96	2050	560	1700	920	1500	890	1950	970	1750	1150
120	2050	780	1500	820	1350	980	1950	930	1600	1200

ranged from 1415 to 4640 psi. This large site-to-site difference was due to the fac that some sites were close to the cave line and others distant from the cave line

(2) In all instances the smallest vertical stress was greater than the calculated weight of the overlying cover, which is about 1450 psi for a depth of 1500 ft.

(3) In all overcoring relief tests the horizontal stress was compressional and ranged from about $\frac{1}{3}$ to $\frac{2}{3}$ of the vertical stress.

(4) The stress concentration, that is, the ratio of the stress at any point in the borehole to the average stress near the end of the hole, ranged from about 1.5 to 5.0. Theoretically, the stress concentration depends primarily on the geometry of the opening (s), and for this stoping system it is estimated that the stress concentration should be 4 or greater. Thus, there appeared to be a stress relaxation at the wall and pillar surfaces in some instances.

(5) In general, the stress determinations were less erratic in the machine-mined than in the conventionally mined sites, a factor that was probably caused by the presence of blasting fractures near the surface of conventionally mined sites.

(6) In general, the borehole relief method was easier to effect and gave better results (more uniform stress curves) than those obtained using the cutter relief method. In the cutter relief method there was a tendency for the block to fracture during relief. There was also a tendency for the borehole relief cores to fracture at high stress levels. This fracture problem is discussed in more detail in a previous publication.[(26)]

Stress Determinations in Salt

Stress determinations in salt were made by the overcoring method in the Carey Salt Co. mine, Winnfield, La., as a complement to the Atomic Energy Commission tests made at this site. This mine is situated in a salt dome at a depth of about 800 ft from the surface. This salt is comparatively pure and relatively free from joints or fractures. The salt crystals range in size from $\frac{1}{4}$ to $\frac{3}{4}$ in., and there is some evidence that interstitial gas is entrapped within the salt.

The tests were made in the end of a stope 50 ft wide and 20 ft high, which was situated on the boundary of mining. Near the test site an area had been mined by the room and pillar method with an extraction of about 50 per cent.

The modulus of elasticity of the salt was determined from samples taken in the test area. A representative stress–strain curve is given in Fig. 25. In all stress–strain

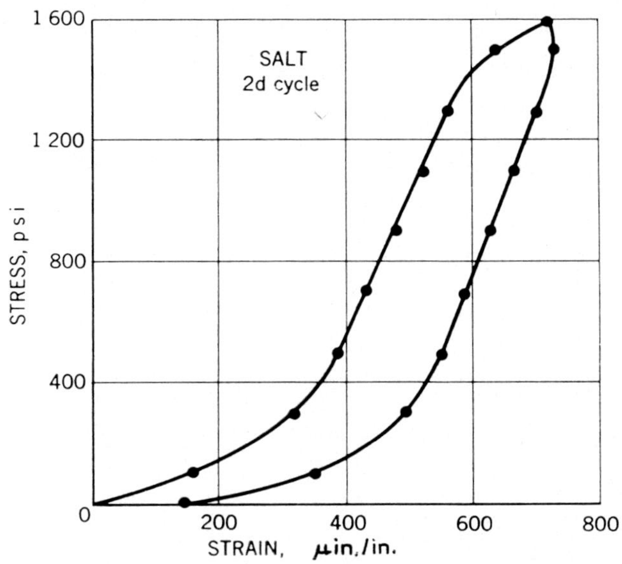

FIG. 25. Dome salt: representative stress–strain curve.

measurements in this salt the virgin increasing stress curve was not similar to the corresponding decreasing stress curve. The second, third and fourth cycle curves showed a general agreement, but in each the slope of the increasing and decreasing curves changes rapidly at low stress levels. This deviation from linearity may have been caused by the presence of microscopic cracks that developed during the coring of the salt or possibly to microscopic voids between crystal planes.†

Representative borehole deformation data from one of the horizontal holes drilled in this test area are given in Fig. 26, and the corresponding magnitude and direction of the maximum and minimum stresses normal to the axis of the hole are given in Fig. 27. Note that the value of the maximum and minimum stresses differ by less than 15 per cent over the length of the hole from 60 to 200 in., indicating that in this salt the stress is nearly hydrostatic. Although the direction of the

† Both the deviations from linearity and the area between the increasing and decreasing curves for this salt is greater than that for a dome salt from Hockley, Tex., or for bed salt from Kansas and Michigan.

stresses appears to follow some definite pattern, this result cannot be considered too significant because, as the maximum and minimum stresses become equal, their directions become indeterminate.

The numerical value of the stresses given in Fig. 27 was calculated using a modulus of elasticity of 3.5×10^6 psi, which was the average of the secant modulus between 0 and 1600 psi for four samples taken at this site. However, there is considerable uncertainty regarding how to select a modulus of elasticity for use in these stress

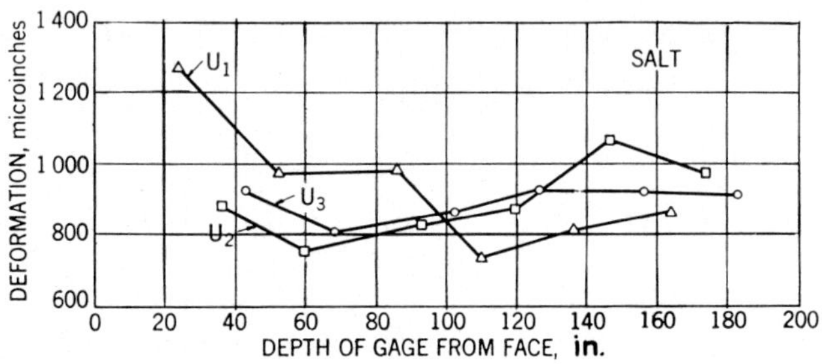

FIG. 26. Salt: representative borehole deformation data.

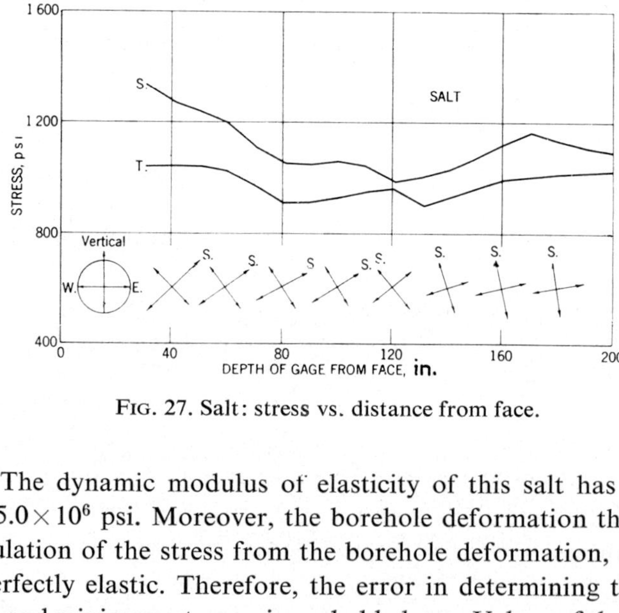

FIG. 27. Salt: stress vs. distance from face.

calculations. The dynamic modulus of elasticity of this salt has been accurately measured at 5.0×10^6 psi. Moreover, the borehole deformation theory, which permits the calculation of the stress from the borehole deformation, assumes that the medium is perfectly elastic. Therefore, the error in determining the magnitude of the maximum and minimum stresses is probably large. Values of the maximum stresses measured at this site are in reasonable agreement with the average stress calculated on the basis of a vertical gravity load, which was equal to 880 psi for a depth of 800 ft.

Two conclusions of importance were obtained from these tests. First, the magnitude of the maximum and minimum stresses are very nearly equal, and hence, the stress in the salt is nearly hydrostatic; second, there was virtually no stress concentration no the surface of the underground opening. Both of these conclusions are in agreement with results obtained in other salt mines.

Stress Determinations in Granite

Tests in granite were made in an underground excavation at the AEC test site in Nevada. This granite was comparatively strong but jointed, the joint spacing ranging from about 4 in. to over 2 ft. The laboratory-determined modulus of elas-

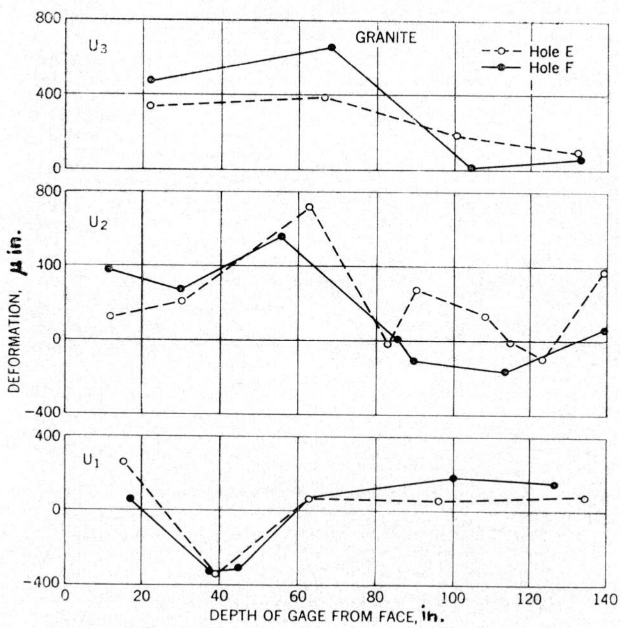

FIG. 28. Granite: representative borehole deformation.

ticity was 10.0×10^6 psi, the stress–strain curve was almost a straight line, and the area between loading and unloading curves was virtually zero.

The test area was in the end of a drift 8×8 ft in cross-section. This drift was 820 ft from the surface, and there were no excavations of sufficient size in the area to affect the direction and magnitude of the stress in this site. Stress-relief measurements were made in two horizontal holes 24 in. apart and 140 in. long drilled in both the north–south and east–west directions and in two vertical holes of approximately the same spacing and length.

Representative borehole deformation results from the two holes (holes *E* and *F*) drilled in the east–west direction are given in Fig. 28. The calculated direction and magnitude of the corresponding maximum and minimum stresses from hole *E* are given in Fig. 29.

The results from hole *E*, as well as from other holes at the site, were similar in that, with increasing depth from the drift surface, both the magnitude and direction of the maximum and minimum stresses varied erratically and between wide limits. However, the correspondence between the deformation results from holes *E* and *F* (Fig. 28) indicates a good reproducibility, especially where the joint spacing was large, as it was in the first 80 in. of the hole (estimated at 2–3 ft). Where the joint spacing was small, as it was from 80 to 140 in. (estimated at 6–12 in), the deformation and the corresponding stress measurements were more erratic.

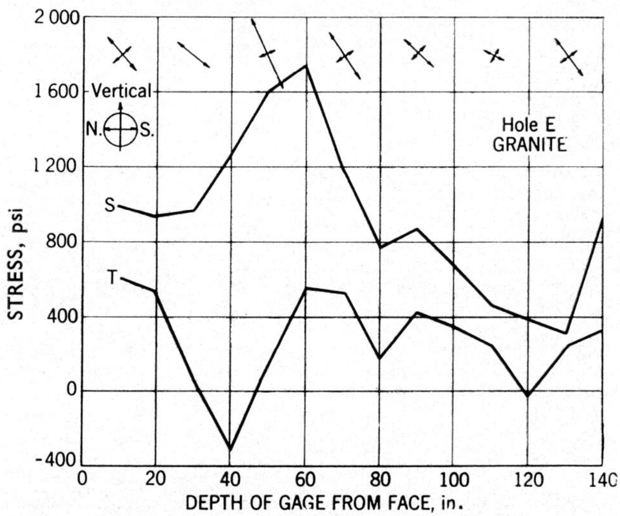

FIG. 29. Granite: stress vs. distance from face, hole E.

The stress from point to point in the hole varied over such wide limits that no conclusion can be made regarding the stress concentration. The average of the maximum stress, *S*, over the 140 in. of measurement was 935 psi. As the depth of the drift was 820 ft from surface, the gravity stress field would be about 900 psi; hence, there is no indication of any large tectonic stress. However, the average direction of the stress was not vertical, but at an angle of 23° from the vertical.

From these measurements in fractured granite it is evident that a single stress measurement made either on the surface or at any point near the surface would be meaningless.

Stress Determinations in Tuff

The stress determinations in tuff were made in an underground excavation at the AEC test site in Nevada. This tuff is a relatively soft, weak rock containing nodules of chert and some soft components, possibly bentonitic. The rock contained occasional recemented or partially recemented joints.

The test site was in a drift 8×10 ft in cross-section and was approximately 1200 ft below surface. The properties of the tuff at this site were affected by moisture; hence, all pilot and strain–relief holes were drilled with air. Within the stress limits of this test, the modulus of elasticity of the tuff was 0.5×10^6 psi. The stress–strain

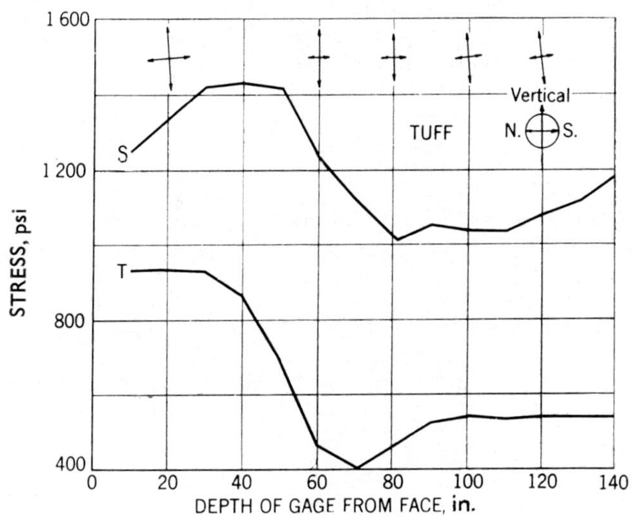

FIG. 30. Tuff: stress vs. distance from face.

curve was virtually a straight line, and the area between the loading and unloading curves small.

Borehole deformation measurements were made in a horizontal hole drilled normal to the axis of the drift to a depth of 140 in. The magnitude and direction of the maximum and minimum stresses normal to the axis of the hole are given in Fig. 30. These data indicate that:

(1) The average magnitude of the maximum stress, S, from a hole depth of 80–140 in. was 1075 psi, a value comparable with the calculated vertical gravity stress at this depth, which was 1050 psi.

(2) The minimum stress, T, was approximately one half of S, which implies a Poisson's ratio of approximately 0.33.

(3) Within the first 60 in. of the test hole there was some indication of a surface stress concentration. For the maximum stress, S, the maximum concentration was about 1.3 times the average stress in the back of the hole. The stress concentration

dropped off rapidly near the surface of the opening, possibly because this drift was mined by drilling and blastings.

(4) The direction of the stress field was comparatively uniform from point to point in the hole, and the maximum angular deviation from the vertical was 17°. Because the maximum stress, S, was approximately vertical and because the average maximum stress was about equal to the gravity load, it can be presumed that tectonic stresses were absent.

The results indicate that this method is satisfactory for determining the stress in tuff and probably in other soft rocks that are reasonably elastic in the stress range encountered *in situ*.

Stress Determinations in Limestone

The Limestone mine, Columbia Southern Chemical Co., Barberton, Ohio, is approximately 2300 ft from the surface. This limestone deposit occurs in a succession of almost horizontal beds that range from 2 to 18 ft in thickness. The beds are formed by well defined and persistent partings, which contain some clay and organic material. Within the beds the limestone is comparatively free from joints and

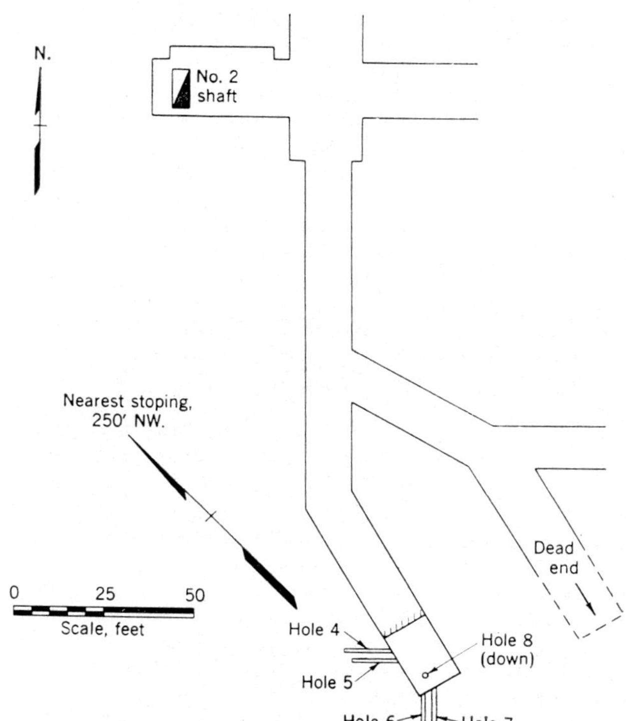

FIG. 31. Test area of limestone mine, Barberton, Ohio.

fractures, but there are occasional well-bouded stylolitic partings and bands of chert nodules. The average crushing strength of the specimens from the test areas was 2100 psi, and the modulus of elasticity, determined from static stress–strain curves, ranged from 8.0×10^6 psi to 9.7×10^6 psi. The stress–strain curves were linear, and the area within the hysteresis loop was small.

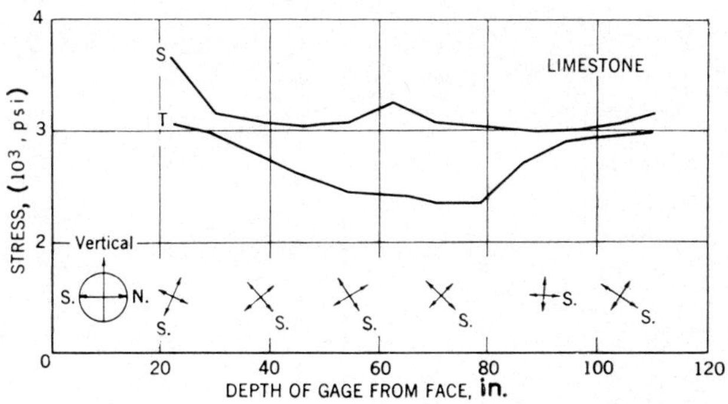

FIG. 32. Limestone: stress vs. distance from face, average from holes 4 and 5.

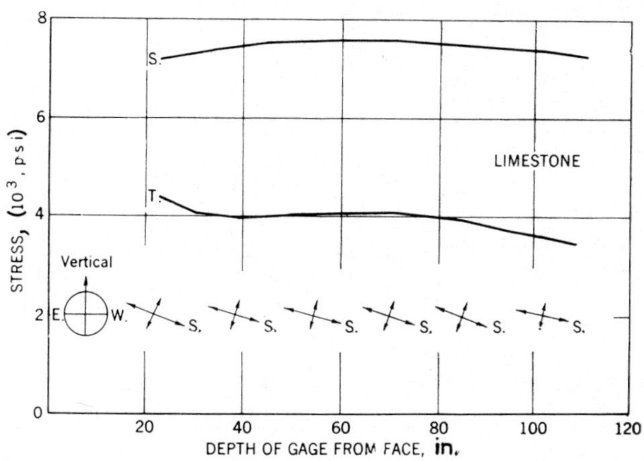

FIG. 33. Limestone: stress vs. distance from face, average from holes 6 and 7.

The stress determinations were made in a deadend drift of about 250 ft from the nearest stoping area (Fig. 31). At this site two parallel horizontal stress-relief holes 24 in. apart and 120 in. long were drilled in both the east–west and north–south directions, and one hole 120 in. long was drilled in the vertical direction. Generally, no borehole deformation data could be obtained in the first 24 in. from the face because of blasting fractures. In the remainder of each hole no difficulty was encountered in the drilling, and the core lengths ranged from 16 to 24 in.

The stress results from holes 6 and 7 and holes 4 and 5 were averaged, and the results are given in Figs. 32 and 33. The results from hole 8 are given in Fig. 34.

The stress data from holes 4 and 5 show some indication of a stress concentration greater than 1 between 24 and 48 in. from the face. For 48–112 in. the maximum stress was almost constant and equal to approximately 3000 psi. This is comparable with the calculated vertical gravity stress of 2700 psi. From 90 to 112 in. maximum and minimum stresses were almost equal. The direction of the stresses varied considerably, but this determination is subject to a large error when S and T stresses are so nearly equal.

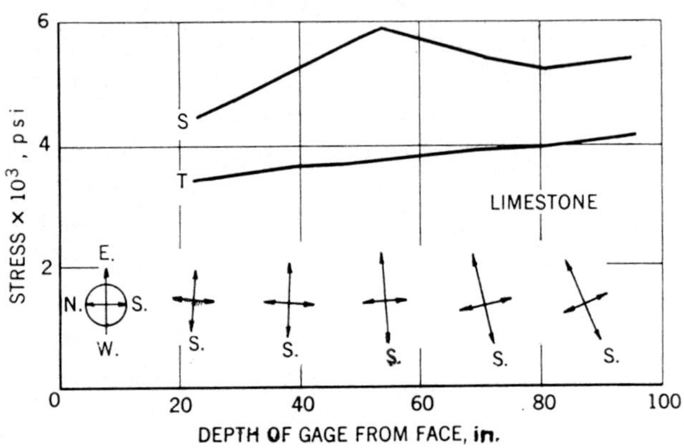

FIG. 34. Limestone: stress vs. distance from face, hole 8.

In holes 6 and 7 the maximum and minimum stresses were relatively constant from 24 to 112 in. The maximum stress was about 7500 psi, almost twice the minimum stress and in approximately the east–west horizontal direction. This stress is presumably of tectonic origin and was not anticipated in this flat-bedded limestone. The minimum stress, T, was about 4000 psi and in the near-vertical direction. This stress is 1300 psi larger than the calculated vertical gravity stress.

The results from hole 8 show that the maximum stress from 60 to 100 in. was about 5500 psi and in the east–west horizontal direction. The minimum stress was about 4000 psi and in the north–south horizontal direction.

There is general agreement between both the magnitude and direction of the maximum and minimum stresses measured in the three orthogonal planes, which is illustrated in the idealized stress ellipsoid (Fig. 35), in which the direction and magnitude of the stresses from the three sets of holes have been averaged.

Stress determinations were also made in a mine pillar in the center of a large stoped area. Representative stress results are given in Fig. 36. The data (average from holes 11 and 12), show that the maximum stress was 8600 psi and vertical and that there was a strong strain-relief (decrease in the stress) near the surface. This pillar stress is over three times the calculated vertical stress before mining. The minimum stress was compressive and horizontal. The horizontal stress in the center of a tall pillar (height several times width) should be nearly zero. However, in this short pillar (height to width ratio $\frac{1}{3}:\frac{1}{2}$), the compressive stress was probably due to two factors. First, a compressional component due to a vertical load on a short pillar laterally constrained at top and bottom; and second, a compressional component due to the horizontal stress in the floor.

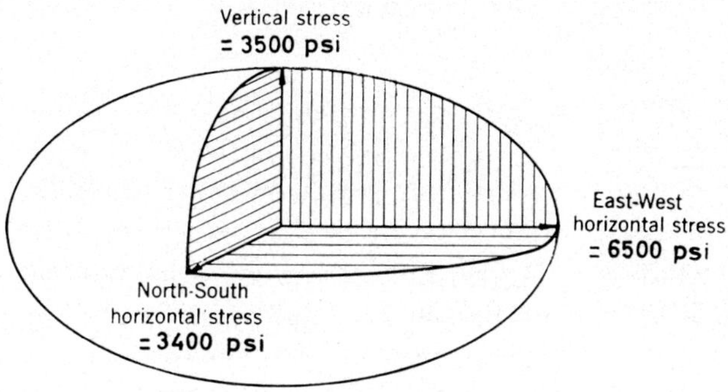

FIG. 35. Limestone: stress ellipsoid.

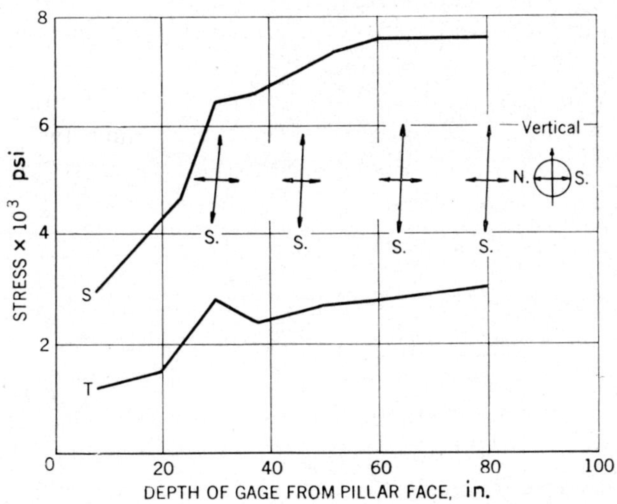

FIG. 36. Limestone: stress vs. distance from pillar face.

From the foregoing tests at this mine, it can be concluded that the borehole deformation method is satisfactory for determining the stress in this almost perfectly elastic limestone at least at stress levels as high as 9000 psi.

Conclusions

The test results described in the foregoing indicate that:

(1) The borehole deformation method of determining the absolute in situ stress in rock is satisfactory for use in rock that is relatively isotropic and for which

stress-strain curves are linear (within the limits of stress being measured) and the hysteresis loop is closed.

(2) The sensitivity of the gage is such that accurate measurements can be obtained in rock having a modulus of elasticity as high as 10×10^6 psi or greater.

(3) For rock that is relatively anisotropic or that has a nonlinear stress–strain curve, or a large area under the hysteresis loop, or a loop that does not close at zero stress, or that creeps at high stress or after removal of the applied load, the accuracy of absolute stress determinations is decreased in proportion to the extent of these deviations. Thus, in the tests described in the foregoing, the absolute stress determinations in granite, trona, limestone and tuff are considered sufficiently accurate for the intended engineering purposes, whereas the absolute stress determinations in salt are not. However, even for this dome salt (which exhibited extremely unfavorable stress–strain characteristics) information was obtained regarding the relative magnitude of the stresses and the stress concentrations.

(4) In all tests the stress concentrations near the surface of the openings were either absent or not as large as that calculated from stress theory.

(5) In general, maximum and minimum stresses in the walls of the test areas were compressional and of such relative magnitude that a complete constraint and a Poisson's ratio of 0.33 or greater were indicated.

(6) In rock containing natural fracture or jointing, or near surface fracture from blasting, the stress from point to point varied erratically and between wide extremes. Hence, in fractured or jointed rock, information regarding the magnitude or direction of the stress field can only be obtained by making a large number of observations.

BIBLIOGRAPHY

1. Phillips, D. W., Tectonics of mining, *Colliery Eng.*, Vol. 25, June, August, September, October, November, 1948.
2. Alder, H., Potts, E. L. J., and Walker, A., Research on strata control in the northern coalfields of Great Britain, *International conference about Rock Pressure and Support in the Workings, Liege*, 1951, pp. 106–120, Paper A8.
3. Harrison, E., Kieschnick, W. F., Jr., and McGuire, W. J., The mechanics of fracture induction and extension, *Trans. Am. Inst. Mining Mat. Petrol. Engrs.* Vol. 201, 1954, pp. 252–263, T. P. 3916.
4. White, W. S., Rock bursts in the granite quarries at Barre, Vermont, *U.S. Geol. Survey*, Circular 13, 1946.
5. *Rock Bursts — A Symposium*, American Institute of Mining and Metallurgical Engineers, Tech. Paper No. 1468, May, 1942, *Mining Technol.*
6. Terzaghi, K. and Richart, F. E., Jr., Stresses in rock about cavities, *Geotechnique*, Vol. 3, June, 1952, pp. 57–90. (Also Harvard Soil Mechanics Series, No. 41.)
7. Oleson, O. J., Measurement of residual stress by the strain-relief method, *Quart. Color. School Mines*, Vol. 52, No. 3, July, 1957, pp. 183–204.
8. Habib, M. P., Determination du modul d'élasticité des roches en place (The measurement of modulus of elasticity of *in-situ* rocks), *Ann. Inst. Tech. Batiment et Trav. Publ.* September 1950, Sols et Fondations No. 3, pp. 27–35.
9. Isaacson, E. De St. Q., Research into the rock burst problem on The Kolar gold field, *Mine & Quarry Eng.* December, 1957, pp. 520–526.

10. Cowlin, W. R., and Isaacson, E. De St. Q., Planning and research necessitated by rock bursts in an underground shaft, *Mine & Quarry Eng.*, February, 1959, pp. 75–82.
11. Kaw, J. M., The Kolar gold field, *Mine & Quarry Eng.*, July, 1956, pp. 258–268; August, 1956, pp. 306–316.
12. Hast, N., Bergtrycksmatningar i gruvor (The measurement of rock pressure), *Jerkontorets Ann.*, Vol. 141, No. 10, 1957, pp. 633–744.
13. Moye, D. G., Rock mechanics in the investigation and construction of T. 1 underground power station, Snowy Mountains, Australia. Paper presented at the Annual Meeting of the Geological Society of America at Symposium sponsored by the ASCE-GSA Joint Committee on Engineering Geology, St. Louis, Missouri, Nov. 6 to 8, 1958.
14. Serafin, J. L., Discussion — earth pressures on structures and tunnels, *Proceedings of the Fourth International Conference on Soil Mechanics and Foundation Engineering*, London, 1957, Vol. III, pp. 227-229.
15. Peperakis, J., Mountain bumps at the Sunnyside Mines, *Mining Eng.*, September, 1958, pp. 982–986.
16. Obert, L., *In situ* determination of stress in rock, *Mining Eng.*, August, 1962, pp. 51–58.
17. Terzaghi, Karl, Measurement of stresses in rock, *Geotechnique*, Vol. XII, No. 2, June 1962, pp. 105–124.
18. Emery, C. L., and Corlett, A. V., Prestress and stress redistribution in rocks around a mine opening, *Trans. Can. Inst. Mining Met.* Vol. 62, 1959, pp. 186–198.
19. Lang, T. A., Theory and practice of rock bolting, *Trans. Am. Inst. Mining, Met. & Petrol. Engrs.* Vol. 220, 1961, pp. 333–348.
20. Hardy, H. R., *Physical Properties of Mine Rock Under Short Period Uniaxial Compression*, Part 1, *Results of McGillivray Mine, Coleman, Alberta,* Report No. F.R.L. 243 of the Fuels Division, Canada Department of Mines and Technical Surveys, Mines Branch, Ottawa, June, 1957.
21. Emery, C. L., Testing rock in compression, *Mine & Quarry Eng.* May, 1960, pp. 196–202.
22. Obert, L., Duvall, W. I., and Merrill, R. H., Design of underground openings in competent rock, *U.S. Bur. Mines Bull.* 587, 1960.
23. Merrill, R. H. and Peterson, J. R., The deformation of a borehole in rock, *U.S. Bur. Mines R.I.*, 5881, 1961.
24. Obert, L., Merrill, R. H. and Morgan, J. A., Borehole deformation gage for determining the stress in mine rock, *U.S. Bur. Mines R.I.* 5978, 1962.
25. Hast, N., *The Me surement of Rock Pressure in Mines*, 52, No. 3, Series C, P. A. Norstedt & Sons, Stockholm, 1958.
26. Obert, L., Effect of stress relief and oth er chnges in stress on the physical properties of rock, *U.S. Bur. Mines. R.I.* 6053, 1962.

CHAPTER 3

TYPES OF ROCK FRACTURE

GENERAL CONSIDERATIONS

There is no general theory which satisfactorily explains the manner of fracture of rocks under all stress conditions. Mohr's theory of failure has long been in use in the field of soil mechanics and has proven satisfactory as a means of predicting the behavior of soils subjected to compressive loading.

The use of Mohr's failure theory has been extended into the field of rock mechanics and it has been used as an approximate method for predicting the structural behavior of rocks which fail under excessive stress.

Some materials, in some stress ranges, deviate from the requirements of Mohr's failure theory, however, and investigators continue to search for a general theory which will cover all phases of failure of brittle materials.

ROCK FRACTURE UNDER TENSILE STRESS

Because most rocks possess little tensile strength, an appreciable proportion of rock failure occurs under tensile stress. Roof failure under tensile stress occurs most commonly in those underground openings which are very wide and have long, flat roof spans. The roof fractures in the same manner as does an overloaded beam. The lower portion is in tension and the rock simply pulls apart.

The fracture of a beam of rock occurs when the flexural strength (modulus of rupture) of the rock is exceeded. The "modulus of rupture" is the strength of the material when considered as the outer fiber of a beam, and it is usually about three to six times the simple tensile strength.

FRACTURE INITIATION

When a rock specimen is subjected to compressive stress which is raised slowly to the breaking strength of the rock it may be deduced that failure of the specimen will begin at the points of weakness. In the absence of pre-existing fractures the weakest links in a granular material are the contact planes along the individual grain boundaries. Thus we may expect that, as stress increases, the individual grains composing the rock will begin to become detached, one from the other, along the directions of the most highly stressed planes.

The separation cracks at the grain boundaries begin to extend themselves for short distances into the rock, and as stress increases certain of these short cracks join to form through-going cracks which mark the final failure of the specimen.

ROCK FRACTURE UNDER COMPRESSIVE STRESS

There are three theories which are of interest in explaining the mode of fracture of a rock subjected to excessive compressive stress. No one of these theories satisfactorily explains the behavior of a rock over all stress ranges. It appears that each theory may apply within a certain range of magnitudes of the major principal stresses, and that the range of validity of any one of the theories may depend also upon the nature of the pre-existing flaws in the particular rock under consideration.

Brandtzaeg's Theory

Fracturing at right angles to the direction of the major principal compressive stress is postulated, because of lateral tensile stress induced by the compressive stress.

Griffith's Theory

It is predicted that an overstressed brittle material will fracture because of tensile stresses set up at the boundaries of flaws in the material. Flaws having certain preferred orientations with respect to the principal stresses are more highly stressed than others.

Mohr's Theory

This predicts that an overstressed brittle material will fracture by shearing along planes which make acute angles with the direction of the major principal stress.

Brandtzaeg's Theory

Experimenters[1, 2] have observed that a specimen of brittle material which is compressed between pressure plates with lubricated surfaces splits along planes parallel to the direction of the applied stress. This type of fracture is illustrated in Fig. 1 (a). Brandtzaeg's theory explains this type of fracture[3] as being due to internal tensile stress normal to the principal applied compressive stress. The forces inducing such transverse tensile stress are illustrated in Fig. 1 (b). The laws of statics require that the shear stresses on opposite sides of any element such as *A* be equal and opposite. These shear stresses, created by compression, tend to stretch the element laterally and create transverse tensile stress.

Brandtzaeg's theory does not give quantitative results which can be used in mathematical analyses, but we may investigate further such stress conditions by assuming that the specimen is homogeneous and contains numerous randomly oriented elliptically shaped flaws. The magnitude of the stress at the end of an elliptical crack which is oriented parallel with a unidirectional applied stress is given by (see Fig. 2 (a)).

$$\sigma_t = -\sigma_1$$

where σ_t is the tangential stress at the end of the crack and σ_1 is the applied stress. Thus the stress at the end of the opening is equal in magnitude and opposite in sign to the applied stress. Since, in this case, the applied stress is compressive, the induced stress is tensile or negative.

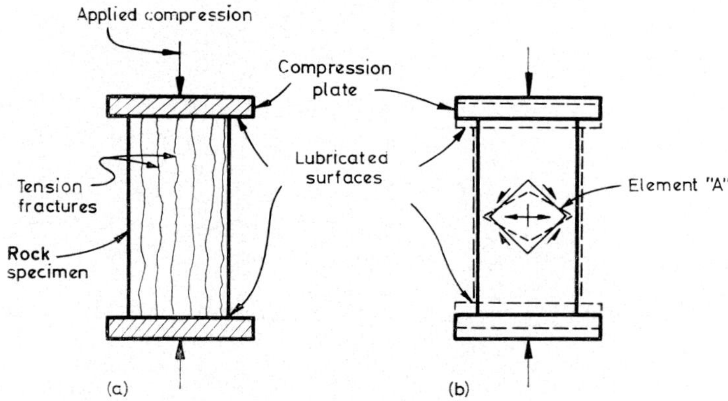

FIG. 1. Transverse tensile stress created by applied compressive stress.

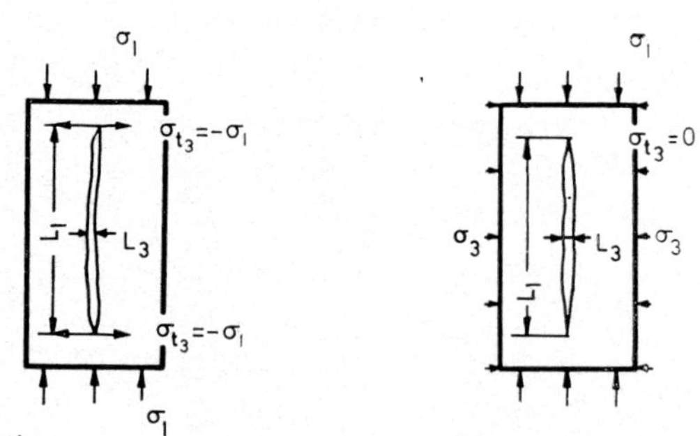

FIG. 2. Transverse tensile stress neutralized by very small lateral confining pressure.

If the specimen is unconfined laterally and the tensile stress exceeds the tensile strength of the material, then the crack will be extended indefinitely. However, as will be shown, only a very small uniform lateral stress need be applied to the material to completely neutralize the tensile stress at the ends of the crack.

The magnitude of the stress σ_t at the ends of the long axis of an elliptical opening in a material subjected to stress, σ_1, parallel to its long axis and stress, σ_3, normal to its long axis (see Fig. 2 (b)) is:[(9)]

$$\sigma_t = \sigma_3\left(2\frac{L_1}{L_3}+1\right)-\sigma_1 \tag{1}$$

where σ_t is the stress induced at the ends of the long axis, L_1, and L_3 is the length of the short axis of the ellipse.

In the case of a newly formed crack the ratio of its length to its width will be very great, that is, L_1/L_3 will be on the order of several hundred, or several thousand. Assuming a crack with a length of 1000 times its width, then:

$$\sigma_t = \sigma_3\left(2\frac{1000}{1}+1\right)-\sigma_1$$

If the lateral confining pressure is zero, then

$$\sigma_t = -\sigma_1$$

that is, the tensile stress at the ends of the crack is equal to the applied unidirectiona compressive stress. If, however, a lateral confining stress exists, then in this example:

$$\sigma_t = 2001\,\sigma_3-\sigma_1$$

and when all tensile stress is neutralized

$$\sigma_t = 0$$

and

$$\sigma_1 = 2001\,\sigma_3$$

and

$$\sigma_3 = \frac{\sigma_1}{2001}.$$

Thus, for the crack in this example a very small uniform stress, σ_3, applied normal to the long axis of the crack will neutralize tensile stress at the ends of the crack and stop its extension. If the compressive stress applied to the ends of the specimen is 2000 psi, extension of the crack would be prevented by a uniform confining pressure, σ_3, of only 1 psi. In an actual underground opening such confining pressure might be applied to the rock surface by support. Even in the absence of support, however, a certain amount of lateral restraint is present because of friction at the upper and lower compression surfaces. Such lateral restraint is greatest near the top and the bottom of the free rock face and may prevent tension cracks from extending all the way from roof to floor.

This type of tension crack may be observed at the rock walls of tunnels, drifts and cross-cuts, and in coal ribs and coal faces. In many cases the orientation of existing natural fractures or of incipient fractures created by blasting is such that they are easily extended by these tensile stresses which are induced by compression.

Such fractures naturally form near a free surface. In the rock, away from the free surface, lateral confining stress tends to build up and to prevent this type of tension fracture. Other types of fracture than these may form instead. These may be explained by Griffith's theory or by Mohr's theory.

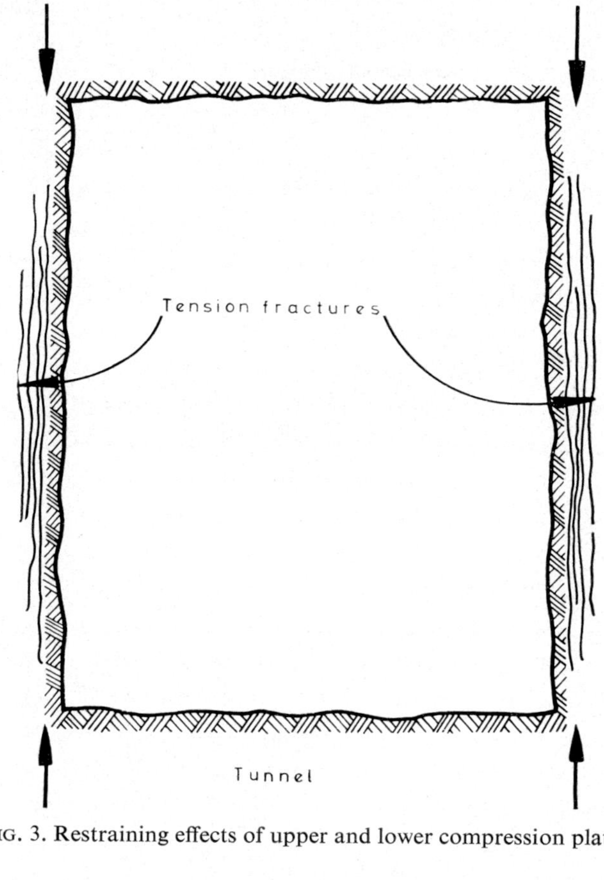

FIG. 3. Restraining effects of upper and lower compression plates.

Griffith's Theory

Griffith[(4-7)] derived expression for stresses around an elliptically shaped hole or crack in a biaxial stress field. He showed that even when the major and minor principal stresses, σ_1 and σ_3, were both compressive the stress along the boundary of the crack was tensile at some points, so that tensile fracture could take place. The maximum tensile stress occurs at, or very close to, the tip of the crack. Griffith's theory predicts the orientation of the most severely stressed crack in material with a variety of cracks. In a body containing a large number of cracks oriented at random, but not so closely spaced that the stress distribution around an individual crack is seriously disturbed by the presence of the others, it is the crack inclined at the angle θ which would initiate fracture. Griffith's theory does not, however, predict how the crack will propagate, and if the crack grows it may not follow the plane of the initial flaw.

Orientation of Cracks

Griffith derived equations for the principal stresses required to cause fracture, and for the tensile strength of the material and the orientation of the most severely stressed cracks. The equation for orientation of the most severely stressed crack is given by:

$$\cos 2\theta = \frac{\sigma_1 - \sigma_3}{2(\sigma_1 + \sigma_3)} \tag{2}$$

where compressive stress is taken as positive (+) and where θ is the angle between crack and the major principal stress.

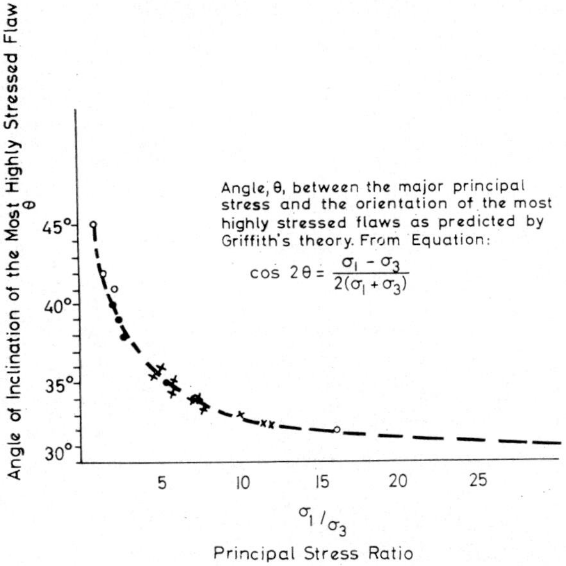

FIG. 4. Angle, θ, between the major principal stress and the inclination of the most highly stressed flaw as predicted by Griffith's theory. From equation

$$\cos 2\theta = \frac{\sigma_1 - \sigma_3}{2(\sigma_1 + \sigma_3)}$$

Figure 4 shows the theoretical relationship between the angle of inclination of the most severely stressed flaw and the principal stress ratio σ_1/σ_3. If σ_1/σ_3 = infinity; that is, if $\sigma_3 = 0$, then the predicted angle of inclination would be 30°. At the other extreme, if $\sigma_1/\sigma_3 = 1$, then the predicted angle of inclination would be 45°.

For the conditions in which we are primarily interested; that is, rock subjected to two principal stresses neither of which is tensile, the angle of inclination of the most severely stressed flaw should vary between limits of 30 and 45°, depending upon the principal stress ratio. If we could assume that the angle of fracture coincided with this angle of inclination, the flattening of Mohr's rupture envelope at high confining pressures (corresponding to an increase of the fracture angle) would be predicted by Griffith's theory, and one is tempted to consider this proof of the validity of Griffith's theory. However, the angle of propagation of a macro-fracture may not correspond to the inclination of the critical Griffith crack as given by the equation because a crack in a field of compression does not propagate in its own plane. Rather, a macroscopic crack or shear fracture grows from a series of microscopic *en echelon* cracks.

Griffith's theory also predicts that at the moment of fracture of a brittle material the principal stresses will have the following relationships when values of σ_3 are small: [7]

$$\sigma_1 = Q+3\sigma_3$$

where Q is the unconfined compressive strength. That is, for every 1 psi of lateral confining pressure applied, the compressive strength of the sample will be increased by 3 psi. The theory also predicts that at very high values of confining stress the coefficient 3 in the preceding equation decreases to 1.

Tensile Strength

According to Griffith's theory the relationship between tensile strength, T, and principal stresses is given by

$$T = \frac{(\sigma_1-\sigma_3)^2}{8(\sigma_1+\sigma_3)} \tag{3}$$

when $\sigma_3 = 0$, then

$$\sigma_1 = 8T$$

Thus the theory implies that the compressive strength of an unconfined test specimen is equal to eight times its tensile strength. Test data available to date on tensile strengths of brittle materials are not adequate either to prove or to disprove the validity of this prediction.

Modified Griffith Theory

The McClintock–Walsh modification of the Griffith theory [8] is based on the premise that Griffith cracks will partially close under the influence of compressive stresses and that frictional forces will be developed across the crack surfaces. The modified Griffith theory predicts that fracture will take place under the folloning conditions:

$$f(\sigma_3+\sigma_1-2\sigma_c)+(\sigma_1-\sigma_3)\,(1+f^2)^{1/2} = 4\,T_o(1-\sigma_c/T_o)^{1/2}$$

where σ_1 and σ_3 are the extreme principal stresses, T_0 is the tensile strength and f is the coefficient of friction at the crack surfaces, and σ_c is the stress normal to the crack surface which is required to close it.

It is difficult to test this equation experimentally because of uncertainty as to the exact values to be used for the coefficient of friction, and values to be used for σ_c.

Griffith Cracks

Although they are not obvious to the naked eye the most likely origins for Griffith cracks in brittle materials are the flaws along grain boundaries. Griffith predicted(5) that the tensile strength of a material should be inversely proportional to the square root of the length of the pre-existing flaws.

By assuming that flaws occur along grain boundaries and that the lengths of such flaws are directly proportional to grain diameters a rough correlation between rock tensile strength and average grain size has been found to exist.

Mohr's Theory

Mohr's failure theory postulates that of all the planes having the same magnitude of normal stress the one along which failure is most likely to occur is that which has the maximum shearing stress.

It has been previously noted that the maximum shearing stresses occur on planes which make angles of 45° with the directions of the maximum principal stresses. In material which is under compressive stress, however, there are stress components acting normal to all potential shear surfaces. These normal stress components induce frictional forces on potential shear surfaces (in accordance with Coulombs law, $s = \sigma_n . f$) so that actual shearing does not take place along the 45° planes of maximum shear stress, but occurs along planes which are inclined at an angle of $45 - \Phi/2$ to the direction of the major principal stress, (Φ is the angle of internal friction of the material).

Mohr devised a graphical method for determining the magnitudes and directions of the stresses at a point in a stressed material. This device, known as Mohr's stress circle, is a convenient method for determining the magnitudes of the shearing stresses on any plane. Mohr's stress circle and Mohr's theory of failure are discussed in detail in Chapter 4.

PROBABLE MECHANISM OF BRITTLE FRACTURE

It appears that the mode of fracture of a rock subjected to compressive stress depends upon (a) the number and nature of pre-existing flaws in the rock and (b) the conditions of confinement; that is, the ratio between the major and minor principal stresses.

The author suggests that rock fractures originate and propagate as follows.

(1) Rock at a free surface may fracture by the spalling off of thin slabs if there are pre-existing cracks, joints, foliation, cleavage, cleats, or other flaws parallel to the free surface. Such fracture and spalling would take place in accordance with Brantzaeg's theory which postulates that transverse tensile stresses are created at the ends of any openings (cracks) in a brittle material which is subjected to compression. These tensile stresses act at right angles to the direction of the applied compression and tend to extend any pre-existing cracks in a direction parallel to the applied compression.

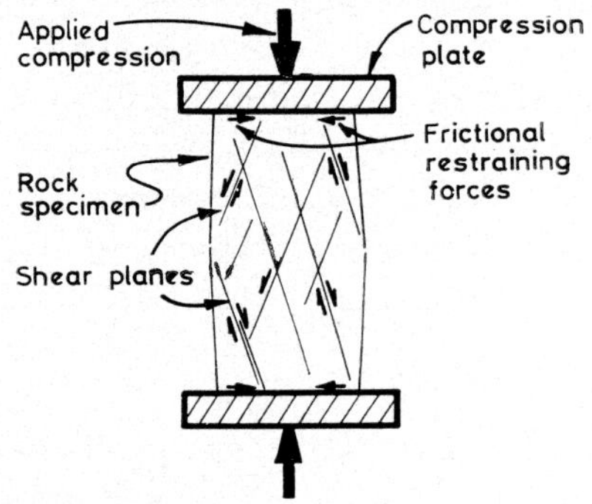

FIG. 5. Rock fracture produced by shear stress induced by compression.

However, these tensile stresses may be nullified by a very small uniform pressure applied laterally to the rock, and the restraint exerted by the rock in the roof and the floor of a mine opening may be sufficient to confine this type of fracture to the mid-height of the rib and to prevent such spalling from extending very far into the rock from the original free surface.

(2) Griffith's theory may account for the initiation of fracturing in rock subjected to confining compressive stresses. Griffith cracks originate from those flaws whose orientations are determined by the ratios of the principal stresses. These cracks propagate along directions which are a compromise between the orientation of the most highly stressed flaw and the direction of the major principal stress.

A macroscopic fracture then forms by coalescence of a number of *en echelon* Griffith cracks.

It is possible that, as the Griffith cracks coalesce to form a macroscopic fracture, the material begins to behave as predicted by Mohr's theory and the fracture becomes a shear fracture which is oriented in accordance with the requirements of Mohr's theory, and that resistance to movement along this shear plane is in accordance with Coulomb's law; that is, $s = \sigma_n f$. Movement would be along the plane of maximum effective shear stress and this plane would be inclined at an angle of $45° - \Phi/2$ from the direction of the major principal compressive stress.

Effect of Anisotropy

In the foregoing discussion of Griffith's and Mohr's theories of fracture rocks have generally been considered to be homogeneous and isotropic, and it has been assumed that pre-existing planes of weakness are randomly oriented. However, most rocks contain planes of weakness, such as bedding planes, joints, schistocity, or cleavage. Rocks without such obvious planes of weakness may show some orientation of mineral constituents which make them structurally anisotropic and influence their strength and directions of fracture.

Tests on slate specimens have shown that the compressive strengths of specimens oriented with cleavage at angles of 30° with the major principal stress have only about 10 per cent of the compressive strength of specimens oriented with cleavage normal to the major principal stress.[10]

Such results are to be expected when the planes of weakness in a rock are oriented so that they nearly coincide with the planes of maximum effective shearing stress as predicted by Mohr's theory, and this serves to emphasize the fact that these planes of weakness must be taken into account in evaluating a rock for its structural ability to support itself over a mine opening. Adler has proposed a method for three-dimensional analysis by means of Mohr's diagram which takes into account any pre-existing planes of weakness in the rock.[11] [12]

SUMMARY

In summarizing, it appears from observations in underground excavations that fracturing from a free rock surface which is subjected to excessive tangential compression may occur in accordance with Brandtzaeg's theory. Transverse tensile stress induced in the rock extends pre-existing flaws, cracks, or weaknesses in the rock with the result that flat slabs spall from the surface. Near the top and the bottom of the free rock surface the lateral restraint of the roof and the floor is effective and prevents extension of fractures for the complete height of the exposed surface.

When confining, or restraining, lateral stresses are exerted on the rock surface, then the rock may fail in tension and shear in accordance with Griffith's theory and with Mohr's theory.

The principal recommendation for Griffith's theory is that it appears to offer a satisfactory mathematical explanation for the increase in the angle of fracture, θ, at higher confining pressures, which is observed in the triaxial testing of brittle materials. There is considerable doubt, however, that the directions of the macroscopic shear planes which are formed during fracture of rock under compression are controlled by the directions of the microscopic Griffith cracks which propagate from flaws in the material, and it appears logical to assume that the orientations of the shear planes themselves are influenced more by the requirements of Mohr's failure criterion than by the directions of the Griffith cracks, which may initiate the failure.

STRUCTURAL FAILURE OF ROCK AROUND MINE OPENINGS

In dealing with actual excavations in rock we are interested mainly in the lower pressure ranges, that is, in the pressures encountered within a short distance inside a free rock surface, and with the relatively small lateral stresses required to prevent fracturing from proceeding inward from the surface. In this range the results obtained by using Mohr's failure criterion are consistent and reproducible.

In the chapters which follow, Mohr's theory of failure will be used in analyzing the structural behavior or rocks around underground openings. This course of action is encouraged by the fact that once a rock has fractured and lost its cohesion it may be expected to behave in conformity with Mohr's theory. Under this scheme the internal movements will be shearing movements, and internal resistance will be frictional resistance. By employing Mohr's theory we may make the transition from cohesive material to non-cohesive material without changing the mathematical equations used, since the same equations serve for treating the two different states.

Chapter 4 deals with Mohr's theory as applied to the structural behavior of soils and rocks.

BIBLIOGRAPHY

1. TSCHEBOTARIOFF, G. P., *Soil Mechanics, Foundations and Earth Structures*, McGraw-Hill, New York, 1951.
2. LABASSE, H., Le pressions de terrains dans les mines de Houilles, *Rev. Universelle Mines*, January, 1949.
3. RICHART, F. E., BRANDTZAEG, A., and BROWN, R. L., A study of the failure of concrete under combined compressive stresses, *Univ. Illinois Bull. Eng. Expt. Stat*. Ser. 185, 1928.
4. GRIFFITH, A. A., *Phil. Trans. Roy. Soc. London*, A., Vol. 221, p. 163, 1920.
5. GRIFFITH, A. A., *Proc. First International Congress of Applied Mechanics*, *Delft*, *Holland*, 1924.
6. MURRELL, S. A. F., The strength of coal under triaxial compression, *Proc. of Conference on Non-metallic Brittle Materials*, London, April 1958.
7. CLAUSING, D. P., Comparison of Griffith's theory with Mohr's failure criteria, *Third Symposium on Rock Mechanics; Quarterly Color. School of Mines*, Vol. 54, No. 3, July 1959, pp. 285–296.
8. BRACE, W. F., An extension of the Griffith theory of fracture of rocks. *J. Geophys. Research*, Vol. 65, October 1960, pp. 3477–3480.
9. ISAACSON, E. DE ST. Q., *Rock Pressure in Mines*, Mining Publications Ltd., London, 1958, 212 pp., p. 31.
10. DONATH, FRED A., Experimental study of shear failure in anisotropic rocks, *Bull. Geol. Soc. Am.* Vol. 72, June, 1961, pp. 885–990.
11. ADLER, L., Failure in geologic material containing planes of weakness, *Trans. A.I.M.E.*, Vol. 226, 1963, pp. 88–99.
12. ADLER, L., Clarification of three-dimensional plane of weakness concepts, *Trans. A.I.M.E.*, Vol. 226, 1963, pp. 340–342.

CHAPTER 4

STRUCTURAL CHARACTERISTICS OF SOILS AND ROCKS

TYPES OF MATERIALS COMPOSING THE EARTH'S CRUST

Soils

The Earth's crust consists of the overlying soil mantle and the underlying rock. The depth of the soil mantle varies at different locations, and may range from a few inches up to several hundred feet, and occasionally it may exceed a thousand feet in thickness. The soil mantle is made up of various types of plastic, cohesive, and/or granular materials including organic material, clay, silt, sand, gravel, boulders, rock fragments, and mixtures of various proportions of these materials. In consistency they vary from soft plastic clays, possessing cohesion but little internal friction, to the clean sands and gravels which possess no cohesion but have a relatively high value of internal friction. These latter *cohesionless* granular or fragmental materials will be referred to hereafter as "pulverulent" or "clastic" materials.

Most of these soils occur as mixtures which contain enough clay or fine silt to act as a binding material and to give the material appreciable cohesion. Thus most naturally occurring soils derive their strength from two sources, cohesion and internal friction.

Rocks

Sedimentary rocks are simply lithified soils which possess greater cohesion, and usually a greater magnitude of internal friction, than their unconsolidated equivalents.

Igneous and metamorphic rocks consist of masses of mineral grains or crystals which are in intimate contact, and in addition are usually interlocked. They derive their strength from cohesion between grains, and from internal friction, and in general their strengths exceed those of the sedimentary rocks.

Thus, in common with all materials, the resistance of soil, or rock, to applied stresses may be separated into two phases: (1) cohesion between constituent grains; (2) friction between the grains.

We may therefore analyze the behavior of an element of soil, which is subjected to stress in the same manner as we analyze the behavior of an element of rock since each derives its strength from the same basic properties. The chief difference between them is in the magnitude of resistance of each to fracturing; for instance, a specimen of weak clay may have a compressive strength of only 20 psi while a strong granite will have a compressive strength of 20,000 psi or more.

TYPES OF SUB-SURFACE STRESS

Compressive stresses exist at every point under the Earth's surface prior to creation of an opening. Vertical compressive stresses are created by the weight of overlying materials, while horizontal compressive stresses may be induced by the weight of overlying materials, or may be residual from past geologic processes. In addition, residual shear stresses may be encountered in the vicinities of faults or tight folds.

Creation of an opening induces realignment of stresses around the opening and causes concentration of stresses at certain critical locations on the opening boundary. Magnitudes of compressive stresses at these locations are greatly increased, while, under certain conditions, tensile stresses may be created at other locations around the opening.

Most soils and rocks have little tensile strength. Such strength may range from zero for a fractured material, to about 10 per cent of its compressive strength for a strong intact rock. Thus, most underground openings derive part, or all, of their stability from the self-supporting nature of natural structures in which soil, or rock, is subjected to compressive stress. This self-supporting tendency is sometimes referred to as "ring action" around single openings, or as "arching action" around wide openings. It has been demonstrated, that the roofs or "backs" of many wide flat openings derive their stability from "arching action" rather than from "beam action".

The unconfined compressive strength of a non-cohesive granular (pulverulent) soil is very low, hence such a soil will not stand unsupported. When support is installed to apply moderate pressure to the inner surface of an excavation such soil derives additional compressive strength because of its internal friction; the increase in compressive strength of sand with an angle of internal friction of 37°, for example, will be almost four times the value of the stabilizing pressure exerted by the support. The most efficient support is that designed to apply just sufficient pressure to surface of soil around an excavation to increase the compressive strength of the material until it forms a self-supporting "ring" or "arch".

Materials possessing cohesion between grains vary in strength from clays, sands, gravels, and fragmented rock possessing slight cohesion, (because of the presence of moisture, clay, or other binding material) up through the sedimentary, to the igneous and metamorphic rocks which possess molecular cohesion between the constituent mineral crystals. The strengths of common mine rocks vary from the soft shales with compressive strengths of 3000–5000 psi, or so, up to granites and quartzites with compressive strengths of 20,000–25,000 psi, and amphibolite, with compressive strengths of 40,000 psi or more.

Granular or fragmented cohesive materials may stand unsupported at shallow depths, while in deep mines even the hardest rock fractures under the stresses concentrated around openings.

In all cases the material tends to be self-supporting, and support which is installed serves the purpose of inhibiting shear stresses which would otherwise cause further fracturing and deterioration of supported material.

PRINCIPAL STRESSES AND SHEAR STRESSES

The stresses on an element of material situated beneath the surface may be resolved into three principal stresses. These are at right angles to each other so that each of the principal stresses may be visualized as bearing on two opposite sides of a cube. When the three principal stresses are not equal then shear stresses are induced. Maximum shear stress is a function of the maximum difference between two principal stresses.

The relationship of two principal stresses at a point to the shear stress is shown graphically in Fig. 3.

In this illustration the magnitude of the largest principal stress, σ_1, is represented as unity and, σ_3, the smallest principal stress, is represented as some proportion of σ_1 by a line of appropriate length.

Since soil, and/or rock failure around many single openings occurs ultimately as a result of excessive compressive stresses it is of interest to study the behavior of an element of material subjected to compressive stress sufficient to cause failure.

STABILITY OF AN ELEMENT

An element of material, such as that shown in Fig. 2, which is subjected to excessive compressive stress fails by shearing along planes inclined to the direction of the maximum principal stress. According to Mohr's failure theory the angle of inclination of the shear planes is determined by the angle of internal friction for the material.

TRIAXIAL TESTS IN THE LABORATORY

A unit of soil, or rock, located beneath the Earth's surface is subjected to vertical pressure of the overlying material and, in addition to lateral pressures which are induced by the vertical pressures, or which are residual from past geologic processes.

The pressures on an element may be resolved into the three "principal stresses", each at right angles to the other two.

As a basis for predicting the behavior of such materials when structures are erected upon them (or within them) it is common practice to select samples which are

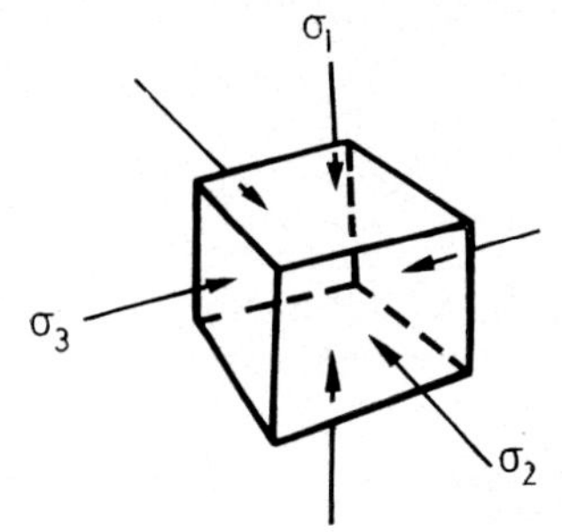

FIG. 1. Principal stresses on an elementary volume.

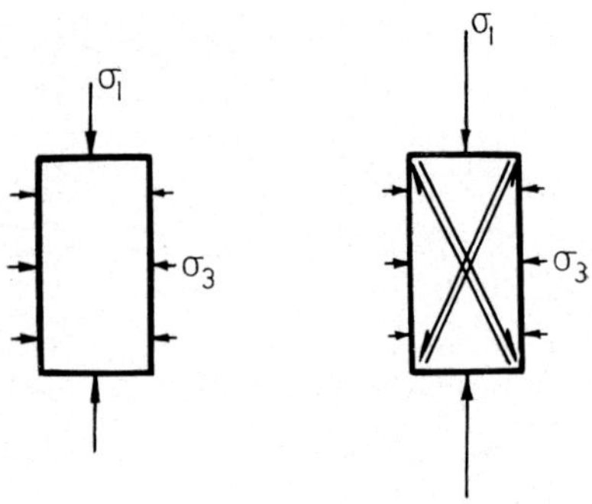

FIG. 2. Shear failure of an element when the maximum principal stress becomes excessive.

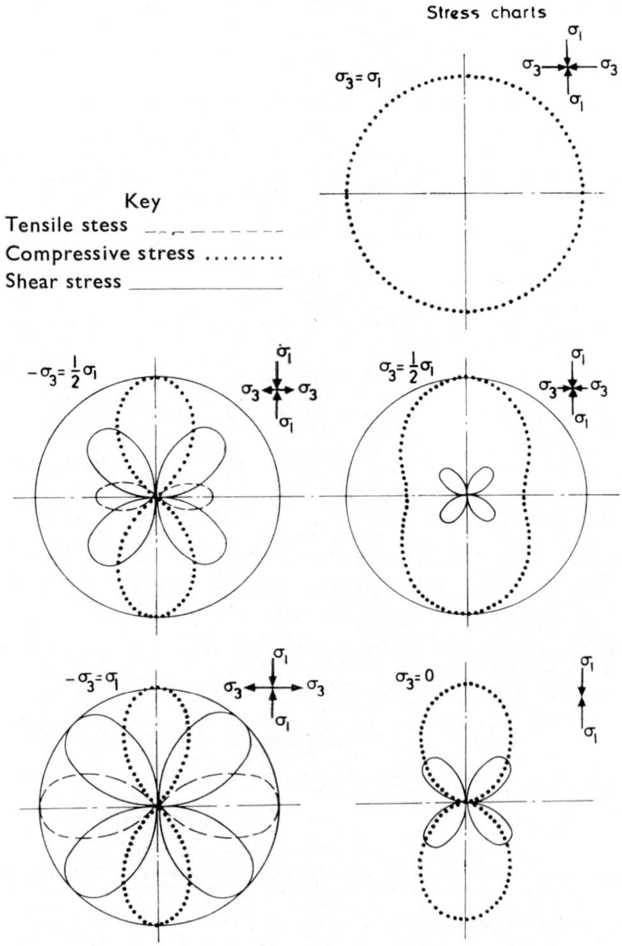

FIGS. 3. and 4. Stress charts showing the magnitudes of the shear stresses induced by various combinations of the principal stresses.[24]

then subjected to "triaxial-compression tests". A soil sample for such a test is enclosed in a rubber sleeve. The enclosed sample is then inserted into a fluid-filled cylinder and held between two pistons, which exert pressure parallel to the axis of the cylindrical sample. Controlled pressure is applied to the pistons and the sample is compressed until it shears or crumbles. Another sample of the same material is then tested in the same manner except that horizontal confining pressure is applied to the outside of the sample soil cylinder by means of a fluid surrounding it. (The sample is enclosed in a membrane to prevent the surrounding fluid from entering the pores.) The pressure required to produce failure of the sample will be higher

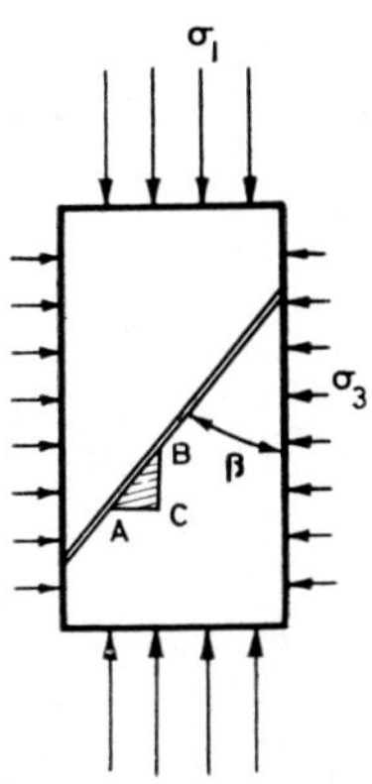

FIG. 5. Element subjected to major and minor principal stresses.

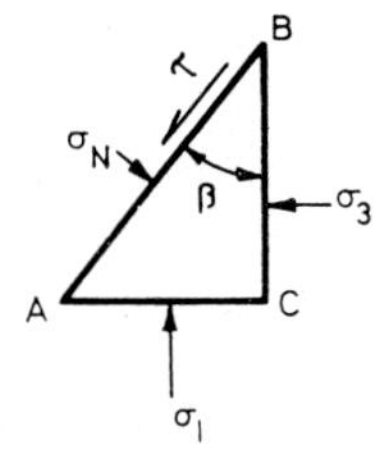

FIG. 6. Forces acting on an element.

than in the first test because the lateral confining pressure raises the compressive strength of the soil. A series of tests is made, each at a higher value of lateral restraining stress, and from the resulting information the magnitude of the components from which the soil, or rock, derives its strength, may be computed.

The basic elements of strength of a material such as soil or rock which fails in shear when subjected to excessive compressive stress are:

(1) Cohesion (c), or the resistance to shearing stress when no normal stress exists on the shear plane. Cohesion is considered to be derived from the same source as tensile strength, that is, from attraction between molecules of constituent minerals.

(2) Internal friction, or the resistance due to friction of grain on grain, plus the resistance due to interlocking of grains. Internal friction is designated as $\tan \Phi$ and Φ is called the angle of internal friction.

This information, properly projected, can be used to predict the behavior of soil, or rock, upon which structures are to be built, or within which excavations are to be made.

Triaxial tests on soil samples are routine. For rocks, however, such testing apparatus is expensive because the compressive strengths of rocks such as shales and sandstones may run from 5000 to 15,000 psi, while the strengths of stronger granites and quartzites may run 20,000–25,000 psi.

Referring to Fig. 6, which represents an element, *ABC*, along any plane making an angle β with the direction of the maximum principal stress, σ_1 (Fig. 5) the major and minor principal stresses can be resolved into a normal stress σ_N acting at right angles to plane *AB*, and a shearing stress τ, acting parallel to plane *AB*.

Taking the area of plane *AB* as unity and

$$F_1 = \text{total force acting on plane } AC$$
$$F_3 = \text{total force acting on plane } CB$$
$$F_N = \text{total force acting normal to } AB$$

then the normal forces acting on the element *ABC* are

$$F_1 = \sigma_1 \sin \beta$$
$$F_3 = \sigma_3 \cos \beta$$

and

$$F_N = F_1 \sin \beta + F_3 \cos \beta$$

Substitution for F_1 and F_3 gives:

$$\begin{aligned} \sigma_N &= \sigma_1 \sin^2 \beta + \sigma_3 \cos^2 \beta \\ &= \cos^2 \beta(\sigma_3 - \sigma_1) + \sigma_1 \\ &= \frac{\cos 2\beta + 1}{2}(\sigma_3 - \sigma_1) + \sigma_1 \\ &= \frac{\sigma_1 + \sigma_3}{2} - \frac{\sigma_1 - \sigma_3}{2} \cos 2\beta \end{aligned} \tag{1}$$

Likewise the tangential force (F_t) can be resolved into two forces:

$$F_t = F_1 \cos \beta - F_3 \sin \beta$$

(The vectors are subtractive rather than additive because they are acting in opposite directions.)

Since area of plane *AB* is unity:

$$\begin{aligned} \tau &= \sigma_1 \cos \beta \sin \beta - \sigma_3 \cos \beta \sin \beta \\ &= \frac{\sigma_1 - \sigma_3}{2} \sin 2\beta = \frac{\sigma_1 - \sigma_2}{2} \sin (90° - \Phi) \end{aligned} \tag{2}$$

The unit shearing stress and the unit normal stress are now given in terms of the known stresses σ_1 and σ_3 and trigonometric functions of the angle 2β.

Shear stress is maximum when $\sin 2\beta = 1$ or when $\beta = 45°$. Thus planes inclined at 45° to the direction of the maximum principal stress, σ_1, sustain maximum shear stress.

However, shear failure takes place along planes of maximum *effective shear stress* rather than along planes of maximum stress. In other words, according to Mohr, of all planes having the same normal component of stress, failure will take place along the plane having the greatest shear stress.

On any plane on which there is a shear stress, τ, there is also a normal stress acting at right angles to the plane. This normal stress, σ_N, on potential shear planes induces a resisting force proportional to $\sigma_N \tan \Phi$, where $\tan \Phi$ = the coefficient of internal friction. Thus the inclination of the plane along which shear failure occurs is determined by the coefficient of internal friction for the material.

Before proceeding further with the investigation of stress relations at failure we will present a graphical method for determining stresses at a point.

MOHR'S DIAGRAM OR MOHR'S STRESS CIRCLE

Values of σ_N and τ may be computed from equations (1) and (2). These equations may also be represented graphically by "Mohr's Diagram" or "Mohr's Stress Circle" (Mohr, 1871), which is shown in Fig. 7.

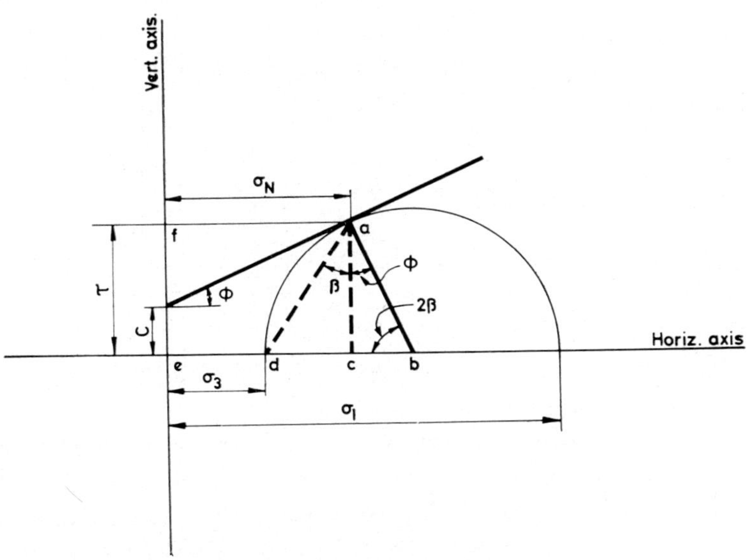

FIG. 7. Mohr's Stress Circle.

The maximum (σ_1) and minimum (σ_3) principal stresses are plotted on the horizontal axis of the diagram and a circle erected whose diameter is equal to the difference in principal stresses ($\sigma_1 - \sigma_3$). Then

$$\text{distance } db = \frac{\sigma_1 - \sigma_3}{2} = ab = \text{radius of circle},$$

and

$$\text{distance } ac = ab \sin 2\beta = \frac{\sigma_1 - \sigma_3}{2} \sin 2\beta$$

$$= \frac{\sigma_1 - \sigma_3}{2} \cos \Phi$$

This is the same as equation (2) for shear stress; so distance ac is equivalent to τ.

$$\text{Distance} \quad eb = \frac{\sigma_1 + \sigma_3}{2}$$

$$\text{and distance} \quad cb = \frac{\sigma_1 - \sigma_3}{2} \cos 2\beta$$

$$= \frac{\sigma_1 - \sigma_3}{2} \sin \Phi$$

and

$$\sigma_N = eb - cb = \frac{\sigma_1 + \sigma_3}{2} - \frac{\sigma_1 - \sigma_3}{2} \sin \Phi \tag{1a}$$

which is the same as equation (1); so distance fa represents normal stress (σ_N) on the shear plane.

The foregoing formulas and diagrams give stress magnitudes on any plane AB making an angle β with the major principal stress, σ_1.

STRESS CONDITIONS AT FRACTURE

It is generally assumed that the magnitude of resistance to failure, s, on failure plane is given by Coulomb's friction law:

$$s = c + \sigma_N \tan \Phi \tag{3}$$

where s is total unit shearing resistance, c is cohesion per unit area, and $\sigma_N \tan \Phi$ is the frictional resistance to shearing.

To relate the four quantities, σ_1, σ_3, τ, and $\tan \Phi$ we combine equations (1), (2), and (3), to obtain the following, which gives stress conditions at failure:

$$\frac{\sigma_1 - \sigma_3}{2} \sin 2\beta = c + \tan \Phi \left[\frac{\sigma_1 + \sigma_3}{2} - \frac{\sigma - \sigma_3}{2} \cos 2\beta \right] \tag{4}$$

Referring to Mohr's circle, at failure $2\beta = 90° - \Phi$ and equation (4) becomes

$$\frac{\sigma_1 - \sigma_3}{2} \cos \Phi = c + \tan \Phi \left[\frac{\sigma_1 + \sigma_3}{2} - \frac{\sigma_1 - \sigma_3}{2} \sin \Phi \right]$$

Then

$$\sigma_1 = 2c \frac{\cos \Phi}{1 - \sin \Phi} + \sigma_3 \frac{1 + \sin \Phi}{1 - \sin \Phi} \tag{5}$$

when $\sigma_3 = 0$, σ_1 = unconfined compressive strength and representing the unconfined compressive strength by Q

$$Q = 2c\,\frac{\cos\Phi}{1-\sin\Phi} \tag{6}$$

and

$$\sigma_1 = Q + \frac{1+\sin\Phi}{1-\sin\Phi}\,\sigma_3 = Q + j\sigma_3 \tag{7}$$

TRIAXIAL STRESS RELATIONS

It is desirable to relate the major and minor principal stresses, σ_1 and σ_3 to the shear stress, τ, and to the internal friction, tan Φ.

The foregoing equations relate the principal stresses σ_1 and σ_3 to the "cohesion" or "shear strength", c and the coefficient of internal friction, tan Φ, of the material.

It can be seen from examination of Mohr's diagram that the value of tan Φ determines the angle which shear failure planes make with the direction of maximum principal stress, and that this angle, β, is equal to $45° - \Phi/2$. The value of tan Φ also determines the rate at which the compressive strength of a soil or rock increases when lateral restraining pressure is applied.

Figure 10 shows, among other functions, the curves representing functions $\cos\Phi/(1-\sin\Phi)$ and $(1+\sin\Phi)/(1-\sin\Phi)$, $= j$ for various values of Φ. When the unconfined compressive strength and angle of internal friction of a sample are known, these equations give the stress relationships at the moment when unit stress equals strength of the sample and shear failure occurs. The relationships are expressed analytically and it will be convenient to use them later in determining whether shear failure will occur around an opening, and in computing theoretical lateral support pressures required to stabilize the material around an opening.

The same values obtained by use of the equations may be obtained graphically by constructing a series of Mohr's circles representing stress conditions for each test.

THE RUPTURE ENVELOPE

The "rupture envelope" is a line drawn tangent to a series of Mohr's circles. In actuality such a line for a uniform homogeneous material is a curve, concave downward. Figure 8 shows results of a series of tests on concrete. The tangent to the stress circle becomes flatter with increasing confining stresses; that is, the value of the internal friction becomes less as the confining pressure increases, and it is probable that at sufficiently high confining pressures all materials behave as liquids; that is, the internal friction approaches zero.

For homogeneous materials, such as the concrete tested by the U.S. Bureau of Reclamation, the rupture envelope may be sufficiently regular in outline to be expressed by a mathematical formula. However, soil and rock when tested under triaxial loads do not usually give a curve which is sufficiently regular in outline that the rupture envelope can be expressed by a simple equation, and in any case deviations between samples may be so large that such refinements are not warranted.

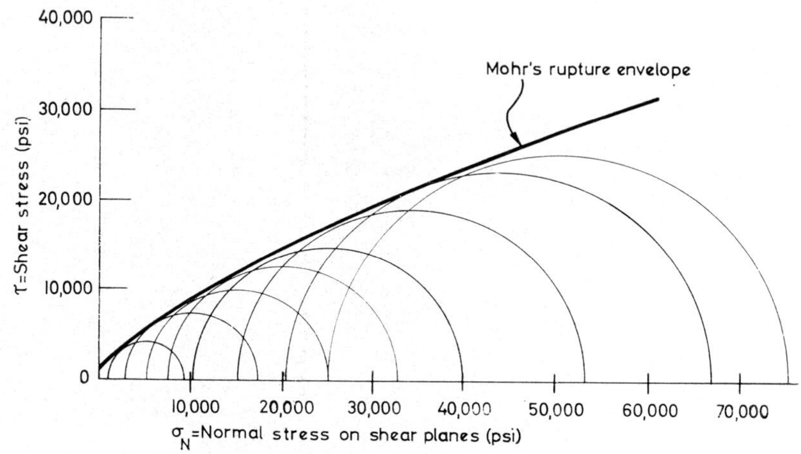

FIG. 8. Results of triaxial tests on concrete cylinders.(1)

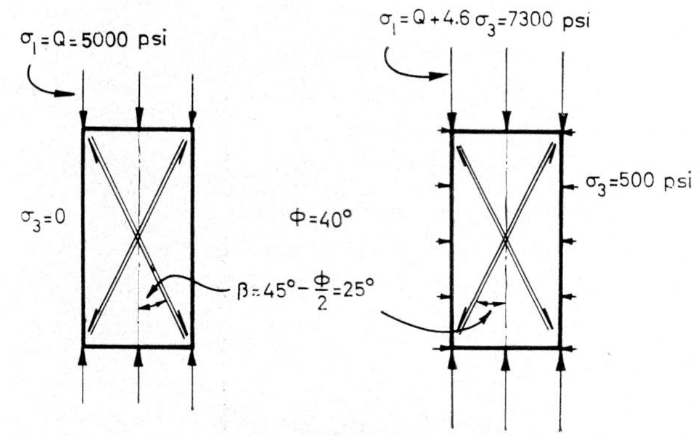

FIG. 9. Effect of lateral restraining stress on compressive strength of rock specimen; taking Φ as 40°.

Therefore, in further discussions and computations herein the rupture envelope will be taken as a straight line. (That is, σ_1 varies linearly with σ_3.) This assumes that Φ is constant over a wide stress range, and while not strictly true this assumption gives values which are close enough for practical purposes, particularly if computations apply over a limited stress range, such as is usually the case for a particular problem.

STRENGTH CALCULATIONS

Equation (7) can be used to compute confined compressive strength, Q, of soil or rock providing that values for Q (unconfined) and Φ are known. Taking for example a sandy shale with a compressive strength of 5000 psi and $\Phi = 40°$, then from Fig. 10, $j = 4.6$ and from equation (7) $\sigma_1 = Q + 4.6\,\sigma_3$ at fracture and

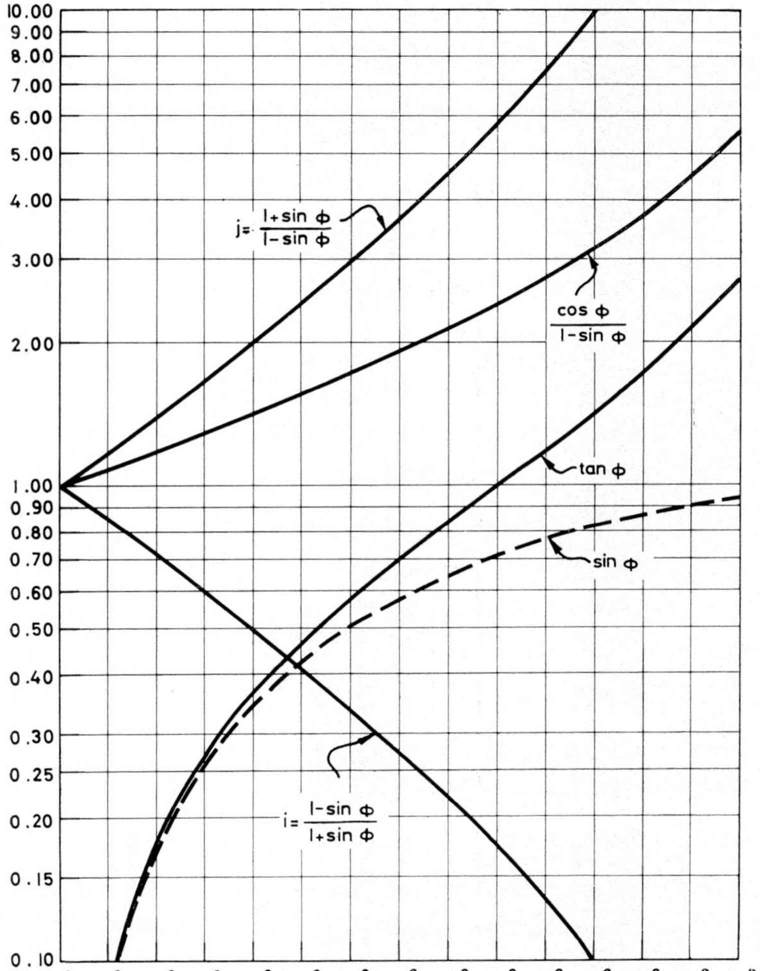

FIG. 10. Graphs of trigonometric functions commonly used in soil and rock stress computations.

FORMULAS FOR STRESS RELATIONSHIPS IN TRIAXIAL LOADING

σ_1 = major principal stress;

σ_3 = minor principal stress;

τ = shear stress;

σ_N = normal stress on the shear plane;
c = cohesion or shear strength;
Q = unconfined compressive strength;
$\tan \Phi$ = coefficient of internal friction.

$$\sigma_N = \frac{\sigma_1+\sigma_3}{2} - \frac{\sigma_1-\sigma_3}{2} \sin \Phi \tag{1a}$$

$$\tau = c+\sigma_N \tan \Phi = \text{shear stress at failure} \tag{3}$$

$$Q = 2c\times \frac{\cos \Phi}{1-\sin \Phi} = 2\,c\times \tan^2\left(45^\circ + \frac{\Phi}{2}\right) \tag{6}$$

$$\sigma_1 = Q+j\sigma_3 = \text{stress relations at failure} \tag{7}$$

$$\tau_{\text{effective}} = \frac{\sigma_1-\sigma_3}{2} \sin (90^\circ - \Phi) \tag{8}$$

$$T = \frac{c}{\tan \Phi} = \text{tensile strength} \tag{9}$$

STRESS RELATIONSHIPS AND TRIGONOMETRIC EQUIVALENTS FOR STRESS COMPUTATIONS

i = reciprocal of the "flow value"

$$i = \frac{\sigma_3}{\sigma_1} = \frac{\sigma_x}{\sigma_z} = \frac{\sigma_y}{\sigma_z} = \frac{1-\sin \Phi}{1+\sin \Phi} = \tan^2\left(45^\circ - \frac{\Phi}{2}\right) = \frac{1}{j}$$

j = "flow value" for materials in the clastic or plastic state of equilibrium

$$j = \frac{1}{i} = \frac{\sigma_1}{\sigma_3} = \frac{\sigma_z}{\sigma_x} = \frac{\sigma_z}{\sigma_y} = \frac{1+\sin \Phi}{1-\sin \Phi} = \tan^2\left(45^\circ + \frac{\Phi}{2}\right) = \left(\frac{\cos \Phi}{1-\sin \Phi}\right)^2$$

Powers for computing radial and tangential stresses:

$$j-1 = \frac{2 \sin \Phi}{1-\sin \Phi} = \frac{2}{1-\sin \Phi} - 2 = \frac{1+\sin \Phi}{1-\sin \Phi} - 1$$

$$j-2 = \frac{3 \sin \Phi - 1}{1-\sin \Phi}$$

Identities for use in equations for relating compressive strength and shear strength:

$$\frac{\cos \Phi}{1-\sin \Phi} = \tan\left(45^\circ + \frac{\Phi}{2}\right) = \tan \Phi + \sqrt{(1+\tan^2\Phi)}$$

if lateral restraining stress $\sigma_3 = 500$ psi is applied to the rock then $\sigma_1 = 5000 + (4.6)(500) = 7300$ psi at fracture.

In other words, if an element of the material fractures under a compressive force equal to 5000 psi when unrestrained laterally, then the same element is subjected to lateral restraining pressure equal to 500 psi it will sustain a compressive stress equal to 7300 psi before shearing.

This simple example illustrates the effect of underground support which is properly applied. The most efficient support for single openings is that designed and installed to exert a uniform pressure on the soil or rock surface in the zones where compressive stresses are maximum in order to inhibit shear fracturing induced by excessive compressive stress.

The same principle applies to materials such as sand, which have no cohesion. In this case pressure applied to the sand by structural support installed in a tunnel, for example, increases the resistance to movement along diagonal shear surfaces and tends to create in the sand a self-supporting arch, or ring.

If the internal cohesion, c, of a material, and its angle of internal friction, Φ, are known, then its unconfined compressive strength may be computed from formula (6)

$$Q = (2c)\frac{\cos\Phi}{1-\sin\Phi}$$

Values of $\cos\Phi/1-\sin\Phi$ may be conveniently obtained from Fig. 10 or Table 1. For a rock with $c = 1000$ psi and $\Phi = 45°$ then $Q = (2)(1000)(2.41) = 4820$ psi.

TABLE 1

TRIGONOMETRIC VALUES FOR USE IN SOIL MECHANICS COMPUTATIONS

Φ	tan Φ	sin Φ	$\frac{1-\sin\Phi}{1+\sin\Phi}$ i	$\frac{1+\sin\Phi}{1-\sin\Phi}$ j	$\frac{2\sin\Phi}{1-\sin\Phi}$ $j-1$	$\frac{\cos\Phi}{1-\sin\Phi}$	$\frac{1-\sin\Phi}{3\sin\Phi-1}$ $\frac{1}{j-2}$
0°	0.00	0.00	1.00	1.00	0.00	1.00	0.00
5°	0.09	0.09	0.84	1.19	0.19	1.09	−1.23
10°	0.18	0.17	0.70	1.42	0.42	1.19	−1.73
15°	0.27	0.26	0.59	1.70	0.70	1.30	−3.33
20°	0.36	0.34	0.49	2.04	1.04	1.43	25.30
25°	0.47	0.42	0.41	2.46	1.46	1.57	2.18
30°	0.58	0.50	0.33	3.00	2.00	1.73	1.00
35°	0.70	0.57	0.27	3.70	2.70	1.92	0.59
40°	0.84	0.64	0.22	4.60	3.60	2.14	0.38
45°	1.00	0.71	0.17	5.83	4.83	2.41	0.26
50°	1.19	0.77	0.13	7.55	6.55	2.75	0.18
55°	1.38	0.82	0.10	10.05	9.05	3.17	0.13
60°	1.73	0.87	0.07	13.93	12.93	3.73	0.08
65°	2.14	0.91	0.05	20.27	19.27	4.51	0.05
70°	2.75	0.94	0.03	32.34	31.34	5.67	0.03
75°	3.73	0.97	0.02	57.82	56.82	7.59	0.02
80°	5.67	0.98	0.01			11.43	
85°	11.43	0.99	0.00			22.90	
90°	∞	1.00	0.00				

SUMMARY: STRESS RELATIONSHIPS – SUB-SURFACE STRESSES

From equations (1a) and (2) in the foregoing discussion stress magnitudes may be computed. Equation (3) gives magnitude of resistance to failure. By combining the three equations we obtain equation (7) which gives stress relations at the in-

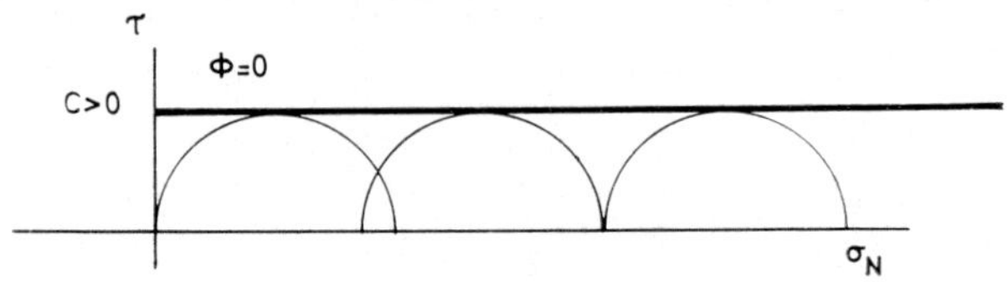

FIG. 11. Typical Mohr's diagram for a material without internal friction. This is typical of some soft saturated clays which possess cohesion but no internal friction.

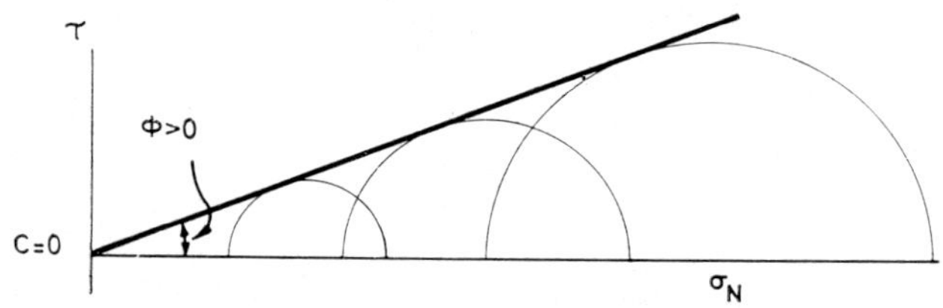

FIG. 12. Typical Mohr's diagram for a pulverulent material. Typical of granular or fractured materials without cohesion such as clean sands or gravels.

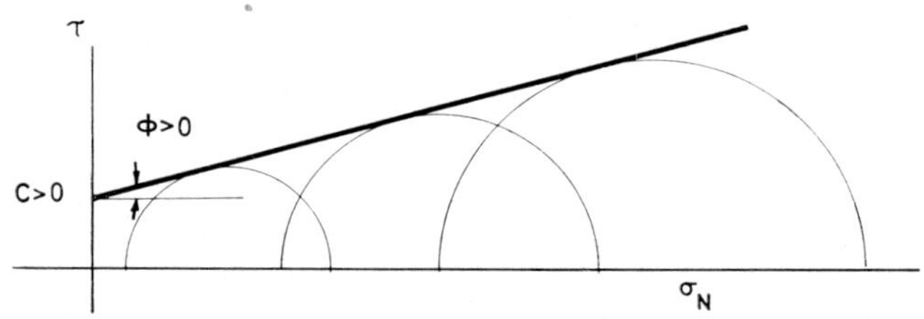

FIG. 13. Typical Mohr's diagram for a material possessing both cohesion and internal friction This is typical of most naturally occurring materials.

stant that shear failure occurs. These equations may be represented graphically by Mohr's stress diagram. Mohr's envelope represents a line tangent to circles for conditions of rupture, and any circle which touches the envelope represents stress conditions which will produce shear failure.

THE "ELASTIC STATE"

When an element is subjected to stresses such that Mohr's circle does not touch the rupture envelope then rupture does not take place and it may be presumed, for practical purposes, that the element is in the "elastic state". This implies that strain is proportional to applied stress, and when stress is released the element will return to its original dimensions. Examination of Mohr's diagrams for soils, as well as rocks, shows that all materials, except liquids and unconfined cohesionless materials, behave elastically to some extent. In the case of weak soil, a pressure of a few pounds per square foot may be sufficient to produce rupture, but up to the point of rupture the material behaves elastically. As long as the material surrounding an underground opening is not stressed beyond its elastic limit, it will require no support. Therefore, we determine the strength of a material, compare its strength with the stress applied to it, and if stress exceeds strength we know that rupture takes place and the broken material passes into a state in which it does not recover from the applied stress. Such a state may be either the "clastic state" (cohesionless) or the "plastic state" (cohesive), depending upon the composition of the material.

THE "CLASTIC STATE" AND THE "PLASTIC STATE"

It has been shown that shear failure occurs when the principal stresses on an element of soil, or rock, satisfy the equation:

$$\sigma_1 = (2c)\frac{\cos\Phi}{1-\sin\Phi}+j\sigma_3$$

The fractured material then obeys the laws for material in the "clastic state" or in the "plastic state". If $c = 0$, such as for a dry clean sand, then the equation becomes:

$$\sigma_1 = j\sigma_3$$

Most soils, and all rocks, in their undisturbed state have a value of cohesion, c which is greater than 0. Once such a material is fractured however, its c value drops and for most brittle substances, becomes zero. Many soils and rocks which contain appreciable quantities of clay may retain some cohesion after fracturing, but even so the major part of the resisting strength of granular or broken materials under compressive stress is derived from internal friction.

GENERAL RANGE OF SOIL AND ROCK STRENGTHS

Materials in the Earth's crust vary from the soft muds and clays with a Φ value, and c value both nearly zero, through the firmer clays with Φ from 10 to 20°, and c = 5–10 psi; clean sands and gravels with c values of zero and Φ values from 35 to 45°; soft shales with Φ values of 25–35° and Q = 2000–5000 psi, to granites and quartzites, with Φ from 45 to 55° and Q from 20,000 to 25,000 psi.

Triaxial Tests on Typical Soils

Clays and Silts

Table 2 and Fig. 14 and 15 show results of some of the tests on soft soils, tabulated by Golder and Skempton.[(2)] The soils all possessed varying degrees of cohesion but some possessed little or no internal friction.

TABLE 2. CLAY SHALES AND SILTSTONES

Description	Saturation (%)	Liquid limit (%)	Plastic limit (%)	Moisture content (%)	$^*\sigma_3$ (psi)	$^*\sigma_1-\sigma_3$ (psi)	c (psf)	$\Phi°$
[10] (a) Portishead. Stiff red clay shale Kemper Marl.	81	41	27	20	0	30	1440	23
					30	71		
					60	110		
[10] (b) Portishead. Stiff red and green clay shale, Kemper Marl.	100	29	23	19	0	38	1600	27
					20	69		
					40	109		
					60	133		
[10] (c) Newport, Mon. Hard fissured red siltstone. Devonian.	91	30	17	9	0	23.0	900	28
					20	49.7		
					40	90.0		
					60	125.0		
[10] (d) Newport, Mon. Firm red weathered siltstone. Devonian.	100	32	16	10	0	24.5	1440	16
					20	47.2		
					40	55.0		
					60	57.5		
[10] (e) Severn Estuary. Firm red clay shale. Kemper Marl.	100	30	17	15	0	41.5	2300	16
					20	62.0		
					40	78.0		
					60	85.5		
[10] (f) Waun-y-Gilfach. Laminated grey clay shale. Coal measures.	86	Not known		9	0	17.6	1600	25
					20	63.1		
					40	90.7		
					60	121.5		
[10] (g) Dormanstown. Hard grey shale	90	36	22	18	0	13.5	700	22
					40	60.5		
					60	86.5		

* Note: Values of σ_1 and σ_3 in Table 2 are given in psi while values in Figs. 14 and 15 are given in psf.

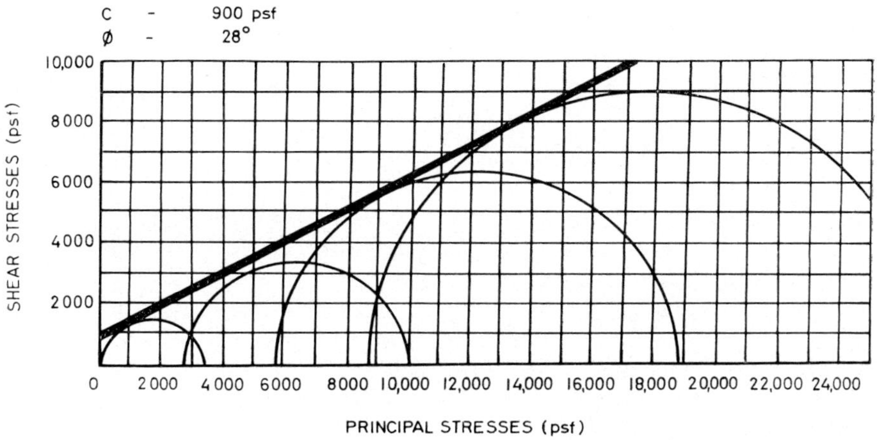

FIG. 14. Mohr's diagram for sample *c* in Table 4.2.[2]

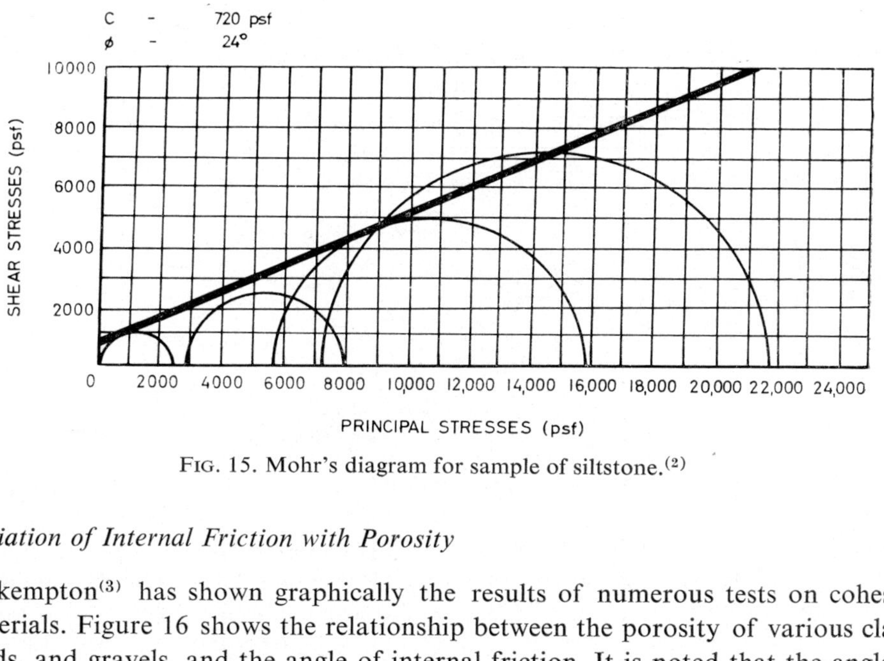

FIG. 15. Mohr's diagram for sample of siltstone.[2]

Variation of Internal Friction with Porosity

Skempton[3] has shown graphically the results of numerous tests on cohesive materials. Figure 16 shows the relationship between the porosity of various clays, sands, and gravels, and the angle of internal friction. It is noted that the angle of internal friction characteristically increases with decrease in porosity.

Quick Triaxial Tests

When the less pervious materials, such as clays, fine silts, and some soft shales, which contain appreciable amounts of water, are subjected to triaxial tests the values obtained for C vary with the speed of application of the load. As load is applied to such a specimen, part of the stress across potential shear planes is borne by contacts between the individual grains composing the soil and part is borne by the water trapped in the pore spaces. The more water contained in a given volume the more compressive stress is carried by the water and the less will be the cohesive strength (and the compressive strength) of the sample.

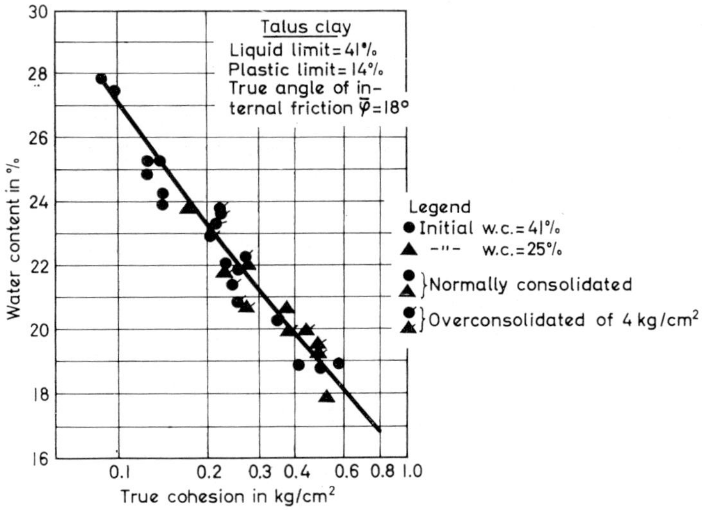

FIG. 17. Relationship between water content and true cohesion in a clay.[4]

In granular materials such as sand and gravel, or brecciated rock, the pore water escapes almost immediately and all stress is borne by contacts between individual grains or fragments. In clays and fine silts considerable time is required for pore water to work its way to the surface and escape. Therefore, if pressure is applied slowly, over a period of several hours, the contained water will have an opportunity to escape and the sample will show greater strength, that is, greater values for cohesion than if pressure is applied within a period of a few minutes. This latter procedure, known as the "quick triaxial test", is the one most commonly employed. It shows soil strength conditions which would be initially encountered when a tunnel, or other underground opening, is excavated in wet clays, or silts, before they have had a chance to drain, and to gain strength through consolidation.

Fig. 17 shows variation of true cohesion of clay with variation in water content.[4]

After an underground excavation has stood for some time unlined, any saturated clays surrounding it will have had a chance to drain and as a result the strengths of such clays will increase providing that no additional water is supplied to the clay.

Clay Shales

Fig. 19 shows results of direct shear tests on clay shales with the shear strength plotted against the normal pressure on the shear plane. The average slope of the lines shown is about $\frac{1}{2}$; that is, $\tan \Phi = 0.5$ and average Φ would be about 26°.

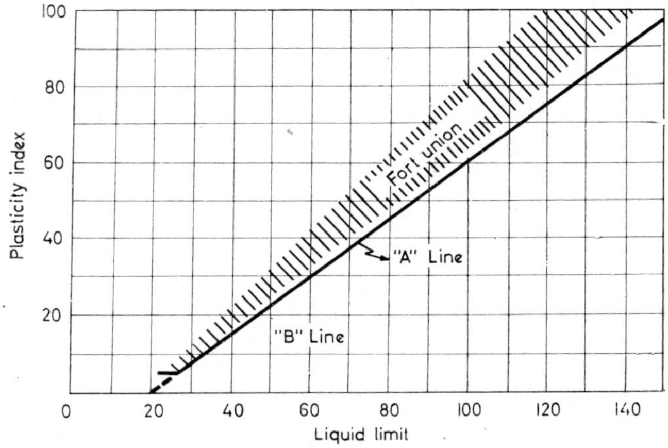

FIG. 18. Plasticity chart for Fort Union clay.(5)

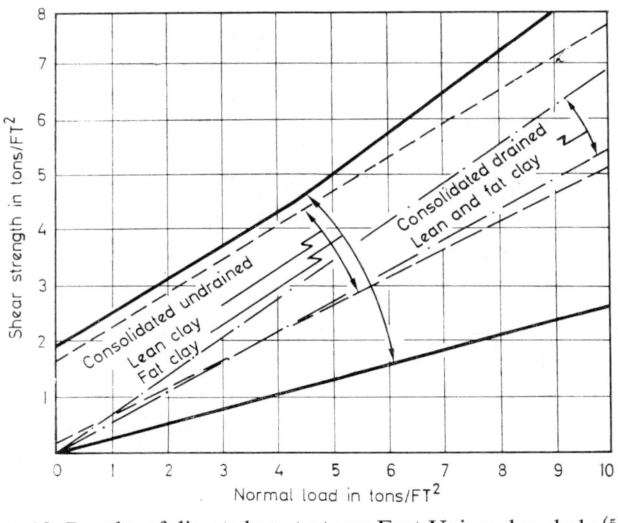

FIG. 19. Results of direct shear tests on Fort Union clay shale.(5)

The cohesion ranged in value from 0 psf up to about 4000 psf (28 psi); thus an average value would be about 2000 psf (14 psi).(5)

These rocks barely qualify as shales, but would seem to fit the description of tightly consolidated clays. The average compressive strength as computed from the data in Fig. 19 would be about 6400 psf or 44 psi. This compares with compressive strengths on the order of several hundred pounds per square inch and more often of several thousand pounds per square inch, for the weakest shales which are ordinarily encountered in mining operations.

Geological evidence as well as compression tests, indicated that these clays had been subjected to a preconsolidation load on the order of 80–100 tons psf (1100–1400 psi) which would correspond to a depth of overburden of about 1500 ft.

Sand

In Fig. 20[6] are shown the results of triaxial tests on samples of sand which were compacted to give different porosities. The coefficient of internal friction increases from about tan 33 to 41° as the porosity decreases from about 45 to about 39 per cent.

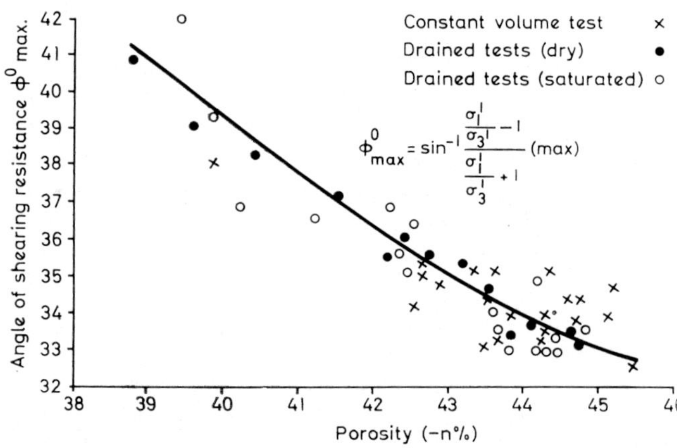

FIG. 20. Change of Φ with porosity, triaxial tests.[6]

Typical Values from Strength Tests on Rocks

Typical results for triaxial compression tests on a sandstone are shown in Figs. 21 and 22. Fig. 21 is a plot of Mohr's diagram for a saturated sandstone. It will be noted that the unconfined compressive strength of this sandstone is about 5300 psi and that the slope of the line which approximates a tangent to Mohr's circles indicates an angle of internal friction of about 47°.

Fig. 22 is a plot of the principal stress relations at failure for a saturated sandstone.

In Table 3, which was compiled by Wuerker[7], are summarized the results of tests on some typical rocks. Among other things, this table gives an equation for Mohr's envelope which is valid within the limits indicated in column 11 of that table.

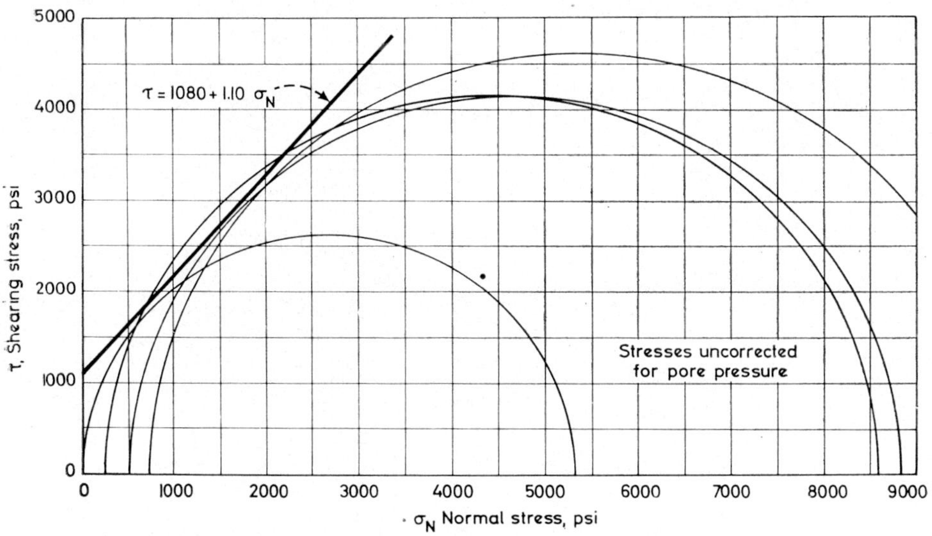

FIG. 21. Mohr's diagram for a saturated sandstone.

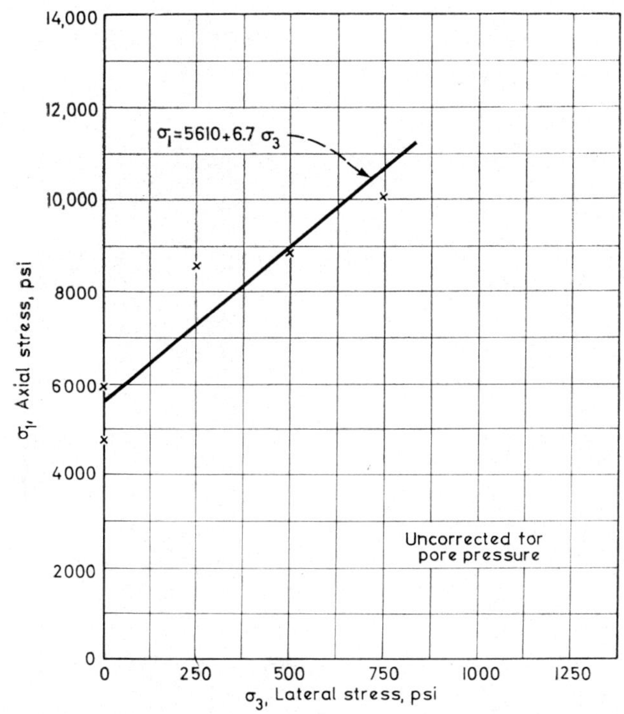

FIG. 22. Principal stress relations at failure for a saturated sandstone.

TABLE 3. STRENGTH DATA OF SOME ROCKS(7)

1	2	3	4	5	6	7	8	9	10	11	
										Validity up to X, psi	
Reference*	Type of rock	Compressive strength 10^3, psi	Modulus of elasticity (in compression) 10^6, psi	Tensile strength, psi	Modulus of elasticity (in tension) 10^6, psi	Shear strength, psi	Equation of Mohr's envelope	Ratio of compressive to tensile strength	Ratio of moduli of elasticity (compressive to tensile)	Point of tangency	Safe extrapolation
61.1 a	Chert (dolomitic), Tenn.	32.3	12.63	990	12.08	2900	$Y=2900+2.8X$	32.6	1:1	1000	2000
61.1 b	Chert (See also jaspilite)	29.3	8.15	820	—	2550	$Y=2550+3.0X$	34.7	—	750	1500
	Coal (Upper Silesia)	2.8	—	—	—	750	$Y=750+0.8X$	—	—	2800	3800
	Coal (Cwmtillery)	1.6	—	300	—	200	$Y=200+1.2X$	5.4	—	1900	2300
	Coal (Barnsley, hard)	6.2	—	—	—	1600	$Y=1600+0.8X$	—	—	1000	2000
42.2	Epidosite (mottled), Va.	49.9	13.7	660	—	3600	$Y=3600+4.5X$	75.6	—	500	1500
1.3	Granite, Vt.	33.2	4.41(d)	1030	—	2900	$Y=2900+2.8X$	32.2	—	900	1800
47.1	Granite (gneisses), Ga.	28.0	3.17	410	1.49	1720	$Y-1720+4.2X$	68.3	2.13:1	400	1000
47.2	Granite (gneisses), Ga.	30.9	3.08	560	—	2100	$Y=2100+3.6X$	55.2	—	500	1200
55.1	Granite, Colo.	25.5	3.94	590	—	2000	$Y=2000+3.4X$	43.2	—	600	1300
55.2 a	Granite, Colo.	23.0	5.54	520	—	1850	$Y=1850+3.6X$	44.2	—	500	1000
55.2 b	Granite, Colo. (parallel to bedding)	23.4	6.13	810	—	2200	$Y=2200+2.7X$	28.9	—	800	1600
55.3	Granite, Colo.	25.3	3.94	500	—	1850	$Y=1850+3.7X$	50.6	—	500	1000
51.2	Greenstone (amygdaloidal), Pa.	29.1	8.82	380	—	1700	$Y=1700+4.5X$	76.6	—	350	800
P-5620-A	Greywacke (Sub), Calif.	7.9	1.8	700	—	1200	$Y=1200+1.7X$	11.3	—	700	1400
8.4	Jaspilite, Soudan mine, Minn. (See also chert)	99.0	—	1100	7.28	5650	$Y=5650+5.1X$	90.0	—	900	2000
1.2	Limestone (fossiliferous), Ind.	10.9	6.41(d)	522	—	1250	$Y=1250+2.4X$	20.9	—	700	1400
59.1 a	Limestone (fossiliferous), Ohio	21.3	9.5	350	—	1320	$Y=1320+3.9X$	60.8	—	300	600
59.1 b	Limestone (parallel to bedding)	26.1	8.75	1380	—	3050	$Y=3050+2.2X$	19.0	—	1100	1800
1.1	Marble, Md.	30.8	7.15(d)	863	—	2650	$Y=2650+2.9X$	35.7	—	800	1600
41.4	Marlstone, Colo.	25.0	4.17	480	4.2	1700	$Y=1700+3.5X$	52.1	1:1	400	1000
	Salt, Potash-, N. M.	4.2	—	402	—	650	$Y=650+1.6X$	10.5	—	350	800
	Salt, Rock-, Retsof, N. Y.	2.2	1.35	85	0.468	210	$Y=210+3.2X$	25.8	2.9:1	400	800
1.4	Sandstone, Ohio	10.4	0.87(d)	120	—	560	$Y=560+4.7X$	86.7	—	100	250
38.4	Sandstone, Pa.	14.8	2.0	230	0.29	900	$Y=900+4.0X$	64.4	69:1	200	500
66.2	Sandstone, Pa.	12.6	1.2	130	—	680	$Y=680+5.2X$	97.0	—	200	500
	Sandstone (Pennsylvanian), Ill.	2.3	0.72	216	0.197	380	$Y=380+1.7X$	10.6	3.7:1	200	500
61.2 c	Shale (carbonaceous), Tenn.	16.3	2.01	110	—	850	$Y=850+7.7X$	148.0	—	200	500
	Shale (carbonaceous), Ill.	5.2	1.09	1538	4.98	1420	$Y=1420+0.6X$	3.4	1:2.2	1200	2000
59.3 b	Siltstone (wet, parallel-bedding), Ohio	5.0	12.6	440	—	750	$Y=750+1.7X$	11.4	—	500	1000
	Concrete	3.0	—	350	—	550	$Y=550+1.6X$	8.5	—	300	600

* Unprefixed numerals refer to group numbers used by USBM for labeling samples. Numbers with prefix P indicate samples tested by U. S. Bureau of Reclamation, Denver.

EFFECT OF MINERALOGICAL COMPOSITION ON THE STRENGTHS OF SOME SEDIMENTARY ROCKS

The Nature of Bonding Materials

Regarding a typical sandstone, Emery[14] says:

> The sand initially consisted of primary minerals derived by weathering from parent rocks. Original sands have porosities ranging up to 35 per cent but this drops to about 15 per cent upon compaction and cementing depending on packing geometry and bond volume. Most bond minerals are secondary and are deposited after the sand sorting movements have been virtually eliminated by compaction.
>
> Sands may become quartzitic aggregates if bonded with silica. Such bonding is derived usually from the decomposition of unstable silicates such as felspars or from corrosion of other quartz grains, and is deposited probably in colloidal form on the quartz grains. Quartz crystal growth occurs and in thin sections original grains may be seen outlined by inclusions of clay minerals, limonite, bubbles, etc., within the new quartz layer. Bonding with silica can occur by the solution of quartz grains at their points of contact due to high local pressure.
>
> Calcareous and dolomitic cements are also derived from felspar decomposition as well as from limestone sources. Sandstones cemented by limonite are brown and by hematite are red.
>
> Argillaceous bonds occur by the deposition of clays on the grains. The kaolinite group, montmorillonite group, and secondary micas are all secondary minerals and tend to deposit in thin platy layers on the quartz grains. Strong ionic bonds are formed between the plates and the quartz.
>
> All cements have characteristics akin to the cement in concrete. They tend to be amorphous or cryptocrystalline to some extent and exhibit some elasticity over short time intervals, but during finite times flow like viscous liquids. This is characteristic of Maxwell liquids.

Influence of Quartz Content on Strength

Price[13] studied the relationship of the compressive strengths of certain coal measure rocks to their mineralogical compositions by dividing them into two categories on the basis of the cement or matrix material.

The two categories were: (1) rocks with calcareous cement, and (2) rocks with clay mineral matrix. Each of these catagories includes the whole range of mudstones, siltstones, and sandstones. The cement and matrices in each category appeared to be fairly consistent from rock type to rock type, and it appeared reasonable, therefore, to relate the strengths of the coal measure rocks to their quartz contents.

Figure 23 shows the relationship between crushing strength and quartz content for those rocks with a calcareous cement. There is a marked relationship between the crushing strength and the quartz content of the rocks.

Figure 24 shows the relationship between strength and the quartz content of rocks with a matrix of clay minerals and with porosities of less than 3.5 per cent. In general the rocks with the higher quartz contents were stronger, but the relationship is far less well defined than that exhibited by the calcareous series. Price points out that the scatter can, in part, be attributed to the small variations in porosity of the series but that they were affected by another important variable factor, namely, compaction.

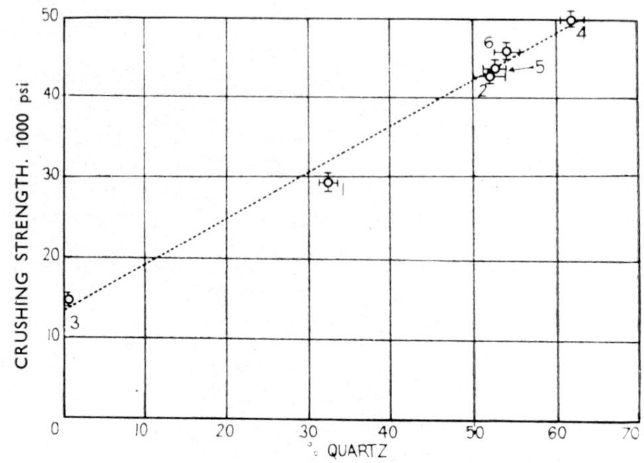

FIG. 23. Relationship between uniaxial strength and quartz content for rocks with calcareous cement.[13]

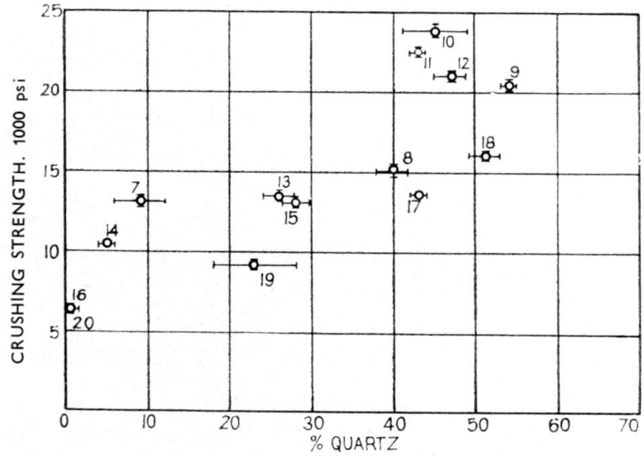

FIG. 24. Relationship between uniaxial strength and quartz content for rocks with clay-mineral matrices.[13]

Influence of Compaction on Strength

Compaction of sediments after deposition results in the fusion of the individual grains at their points of contact, and the greater the compaction the greater is the fusion and therefore the cohesion of the rock. In the South Wales coalfield there is an inverse linear relationship between volatile content of the coal seams and the depth of cover between the limits of 12 and 40 per cent volatiles, and the volatile content of a coal seam can therefore be used as a measure of the degree of compaction of its associated rocks.

Figure 25 shows the relationship between the quartz content of the various rocks and their "equivalent strength" for a standard degree of compaction.

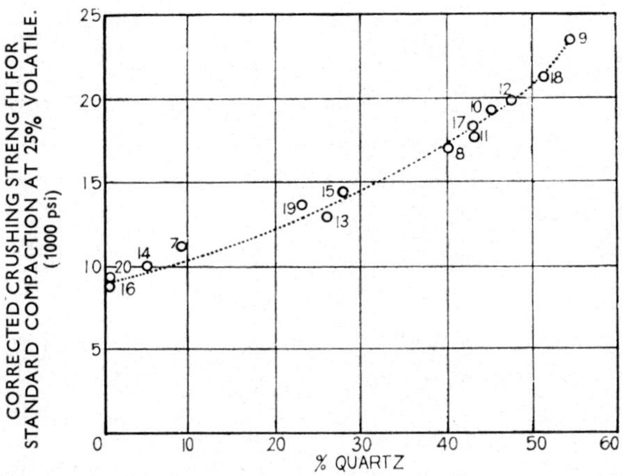

FIG. 25. Relationship between corrected strength and quartz content for rocks with clay-minera matrices.(13)

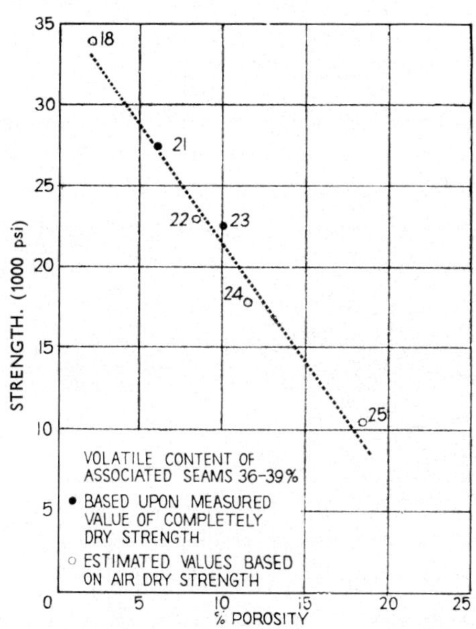

FIG. 26. Relationship between porosity and strength of completely dry rock. (Strength of rock corrected to 55 per cent quartz content.)(13)

Influence of Porosity on Strength

Price also studied the effect of porosity in rock strengths. Figure 26 shows the relationship of the estimated water-free compressive strength to porosity after corrections had been made for the effect of pore water content and quartz content.

The porosity of these rocks was attributed in part to their relatively large quartz content and to their large grain size.

Fracture Angles

In tests on some Canadian mine rocks, Hardy[8] found that mean values for rock specimens loaded to failure in unconfined compression tests were as follows:

		Standard deviation (%)
Shale	25°	20
Conglomerate	30°	7
Sandstone	33°	15

The angle of fracture is the angle between the shear planes and the direction of the applied compressive stress.

Specimens tested had approximately a 1:1 diameter/height ratio. It is possible that this ratio may have had some influence in determining angles of fracture.

Effects of Anisotropy on Compressive Strength

Most rocks contain planes of weakness such as the bedding planes in sedimentary rocks, the schistocity or gneissic banding in metamorphic rocks, and joints or cleavage planes which are common to all types of rocks.

The strength of such rocks varies as the orientation of the applied stress is varied with respect to the direction of the planes of weakness. The range of variation in compressive strength varies with the spacing of the planes of weakness, the strength of the boundaries across such planes, and the size of the rock specimen tested.

Slate specimens subjected to triaxial compression tests with the cleavage planes inclined at 30° to the direction of the major principal stress have been shown to have only one-tenth of the compressive strength of the same rock when the cleavage planes are oriented at right angles to the major principal stress.

Such rocks show maximum compressive strength when the planes of weakness are oriented either normal to, or parallel to, the direction of the major principal stress, σ_1. As the planes of weakness are rotated so that they approach the direction of the planes of maximum effective shear stress, the resistance to shear failure decreases. Any increase in the confining pressure σ_3 increases the rock resistance to shear failure proportionally.

Angle of Shearing

In a homogeneous, isotropic rock faults normally develop at approximately 20–30° to the direction of maximum compression. The angle of faulting in strongly anisotropic rocks varies considerably; it is dependent on the inclination of the anisotropy with respect to the principal stress directions. In slate deformed at low confining pressures, shear fractures developed parallel to cleavage for compression at inclinations up to 45° to the cleavage; the angle of shear fracture is strongly affected in the 60 and 75° orientations as well. Orientation of shear fractures is unaffected by cleavage only when maximum compression is perpendicular to the cleavage.

Summary

These test results point up the importance of natural planes of weakness in determining the ability of a rock to resist compressive and shear stresses. The effects of such weaknesses must be taken into account in assessing the ability of an anisotropic rock to support itself around a mine opening.

ROCK CREEP (TIME-DEPENDENT DEFORMATION OR STRAIN)

The foregoing discussion of the strength characteristics of soils and rocks considered these materials as being elastic solids up to the point of fracture. Thus the figures given for compressive strength in Table 3 were based on short-time compression tests. Under long-time tests the rocks would have progressed to final fracture by a process of slow deformation under stresses less than their short-time compressive strengths. Under low stresses the "creep" to final fracture would require infinite time. Under high stresses a rock may creep to final failure within a few minutes.

Elastic and Viscoelastic Deformation

Rock "creep" or "time strain" refers to the continuing slow deformation of a rock under a constant stress which is less than the yield stress. That phase of rock creep which is of immediate concern to the mining engineer is one phase of a much broader field of deformation studies which is known as rheology.

Rheology is the study of deformation and flow in materials, and the term includes all material at all temperature ranges.

In the earth sciences the geologist and the geophysicist are concerned with rock deformation at pressures up to tens of thousands of atmospheres and at temperatures up to the melting points of rocks. The mining engineer is primarily concerned with the more limited rock deformations covered by *elastic* and *viscoelastic* behavior which take place in the rocks surrounding mine openings. In such environments the temperatures are usually less than 100°F, (rarely over 140°F), and are not high enough to exert any practical effect on the deformation characteristics of mine rocks.

The stresses encountered in the rocks surrounding a mine opening are limited by the compressive strengths of the rocks. The compressive strengths of even the strongest rocks rarely exceed 40–50 thousand psi, and 20–30 thousand psi is a more normal range of strengths for the very strong rocks.

The range of pressures to which a mine rock may be subjected is also limited by the depths of the workings and by the geometry of the workings. The deepest mine workings, those in the Witwatersrand, South Africa, extend to depths of about 11,000 ft and the vertical overburden pressure at this depth may be taken as about 13,000 psi. It is unlikely that the peak stress around these deep openings exceeds the depth pressure by more than a factor of 3 or 4; that is, peak compressive stress in the rock around the deepest mine openings is probably not more than 40 or 50 thousand psi.

Thus, in contrast to the geologist and the geophysicist, the mining engineer is concerned with deformation of rocks at normal temperatures and at maximum stresses ranging from a few thousand pounds per square inch in the case of the weaker rocks up to possibly 40 or 50 thousand psi in the case of the strong rocks.

Rheology is a general term which covers large and small strains as well as the effects of heat and hydrostatic pressure as a function of time. The mining engineer is interested principally in the elastic and viscoelastic phase of rock deformation, that is, in the small strains at normal temperatures which can be described by equations containing elasticity and viscosity coefficients.

The Typical Creep Law

The form of the creep–time curve is basically the same for all materials and is illustrated in Fig. 27.[(15)] There are three sections to the curve. In the first (*BC*) the creep strain increases rapidly, but the strain rate decreases. Eventually the strain rate becomes constant over a section (*CD*). Finally the strain rate begins to increase again at *D* and the strain rapidly increases until fracture occurs at *E*. The gradient of *CD* depends on the stress and on the temperature. At low temperatures and stresses the gradient is small, and may approach zero, in which case the section *DE* may never be observed.

When the load is removed part of the creep strain recovers with time, the rate of strain recovery falling off in the same way that the rate fell off along section *BC*. This is also known as the elastic after-effect, or after-working.

Equations for Creep

Figure 28 illustrates the creep of a limestone which was subjected to uniaxial compressive stress of 1400 kg/cm^2 (about 19,800 psi) for over 500 days. At the end of this time the specimen had shortened by approximately 0.2 per cent. The short-

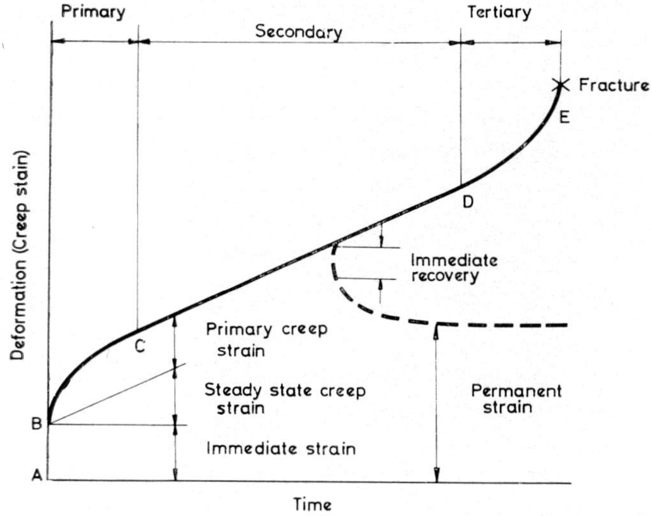

FIG. 27. Generalized creep and recovery curve.(15)

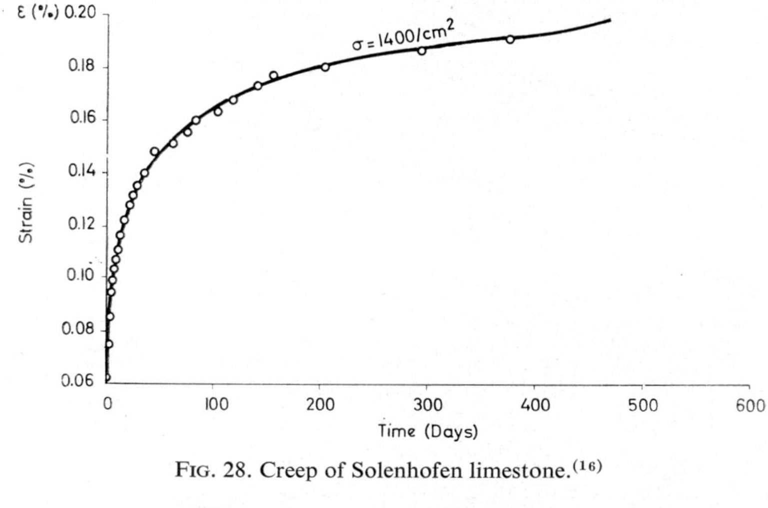

FIG. 28. Creep of Solenhofen limestone.(16)

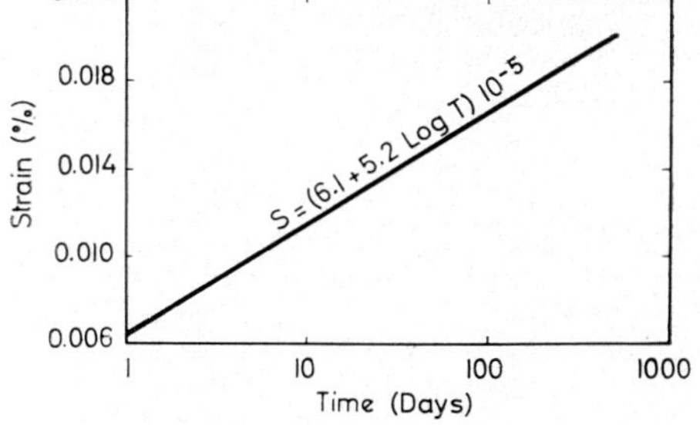

FIG. 29. Time strain curve for Solenhofen limestone under uniaxial compressive stress of 1400 kg/cm^2.(16)

time compressive strength of this limestone was about 2500 kg/cm^2 (about 35,000 psi) so the applied stress was about 56 per cent of the ultimate short-time compressive strength.

Figure 29 is a plot of the deformation of the limestone specimen against the logarithm of time. The equation has the form:

$$\varepsilon = \varepsilon_i + B \log t \tag{10}$$

where ε = total deformation, ε_i = the initial deformation, and B is a constant depending upon the stress, and the characteristics of the rock tested.

Equation (10) approximately fits the behavior of this rock under long term stress but should be modified when investigating rock deformation during the first few minutes, or hours, of stress application. For the study of short time deformation the term $B \ln t$ substituted for $B \log t$ in equation (10) defines a curve which more closely fits to the primary or transient creep curve.

A close approximation to the standard creep curve from A to D (Fig. 27) is given by the following equation:

$$\varepsilon = \varepsilon_e + B \ln t + At \tag{11}$$

where ε = total deformation, ε_e = elastic deformation, $B \ln t$ defines the primary (transient or short-time) creep and At defines secondary (steady state or long-time) creep.

Creep Variation with Stress

The rate of creep varies with the applied stress and the relationship is given approximately by the equation:

$$\dot{\varepsilon} = c\sigma^n \tag{12}$$

where $\dot{\varepsilon}$ is rate of creep, c is a constant, and n is an exponent whose value depends on the rock type and the stress range under consideration.

For low stress values n is about 1 but for higher stress values it may range up to 5 or more.

Creep Recovery

Upon release of stress a portion of the strain is recovered instantaneously, after which the rate of recovery decreases rapidly and eventually approaches zero. The total strain imposed on the rock is never fully recovered and a permanent "set" remains.

Typical creep and recovery curves are shown in Fig. 30. It will be noted that the shapes of the recovery curves are similar to the corresponding creep curves.

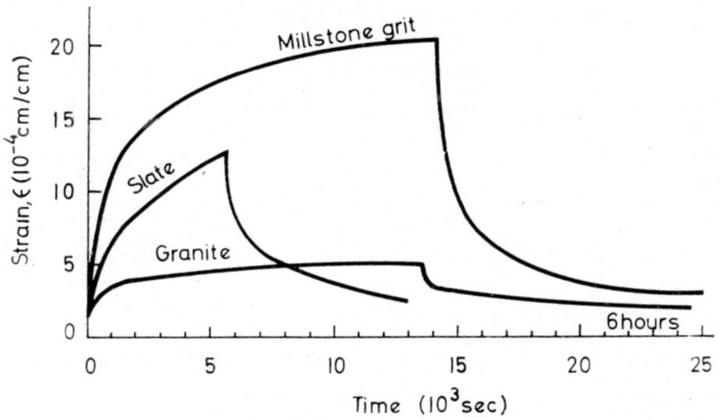

FIG. 30. Creep under constant stress and creep recovery curves. The Millstone grit and the slate were subjected to a stress of 1200 psi; the granite was subjected to a stress of 2340 psi.[27]

Deformation	Name	Mechanical Model
elastic	Hooke	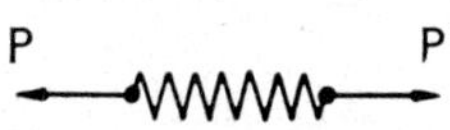
inelastic	Kelvin	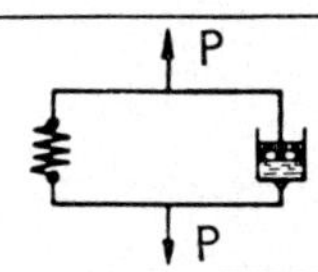
Kelvin	St. Venant	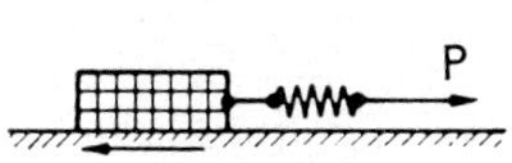 Friction
visco-plastic	Bingham	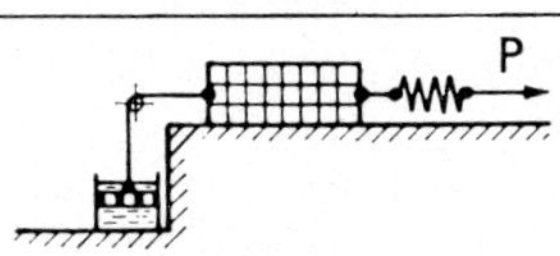
relaxing	Maxwell	
viscous	Newton	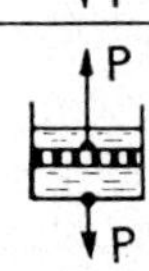

FIG. 31. Mechanical models to represent various modes of rock deformation.[18]

Rock Creep Considered as a Viscoelastic Phenomenon

Secondary or steady-state creep resembles viscous flow. Total deformation is the sum of the elastic deformation, the primary or transient creep, and the secondary or steady-state creep.

Mechanical Models of Creep Phenomena

Deformation phenomena may be more readily analyzed if they are represented by mechanical models. Figure 31 shows various mechanical models which have been proposed as representative of typical deformation processes.

The Maxwell model represents steady-state creep but not transient creep. The equation for deformation with respect to time is:[15]

$$\varepsilon = \frac{\sigma}{E}\left[1+\frac{E_1}{\lambda_1}\,t\right] \tag{13}$$

where σ = normal stress, ε = strain, E_1 = Young's modulus of the "spring", and λ_1 is a coefficient which depends on the viscosity of the fluid in the "dashpot".

The Kelvin–Voigt model represents transient (primary) creep and can also account for the delayed partial recovery of strain (elastic after effect) when stress is removed.

The equation for deformation of the Kelvin–Voigt model is:[15]

$$\varepsilon = \frac{\sigma}{E_2}\left[1-\exp\left(-\frac{E_2}{\lambda_2}\cdot t\right)\right] \tag{14}$$

The Burgers model is a Maxwell model linked to a Kelvin–Voigt model and represents the entire creep curve (Fig. 27) except the Tertiary creep. The equation is:

$$\varepsilon = \frac{\sigma}{E_1}\left[1+\frac{E_1}{\lambda_1}\cdot t\right]+\frac{\sigma}{E_2}\left[1-\exp\left(-\frac{E_2}{\lambda_2}\cdot t\right)\right] \tag{15}$$

The coefficients λ_1 and λ_2 in the equations are related to viscosity and it has been shown that such coefficients are related to the usual coefficient of viscosity, η, by the equation:

$$\lambda = 3\eta$$

It has been noted, however,[15] that the viscosity of a crystalline solid, instead of being a constant, depends upon the stress, so that in practice the coefficient of viscosity would be different for each value of the applied stress. This fact limits the usefulness of the foregoing equations.

It has also been noted[19] that the simple Burgers model with its single retardation time gives only an approximate description of the mechanical behavior of coal and that this material exhibits a spectrum of retardation times. This requires a series of *n* Kelvin units rather than the single one in the Burgers model, and makes the observation of a complete time–strain curve difficult.

Effect of Rate and Duration of Loading on Ultimate Strength

Griggs' experiments on Solenhofen limestone[17] indicate that there is a relationship between the ultimate compressive strength and duration of loading. The general shape of the ultimate compressive strength vs. time curve indicates that it becomes asymptotic to a certain value of ultimate strength at an infinite time. Therefore at applied stresses below this value the rock will not fail during finite time.[19]

Mechanism of Creep and Strain Recovery

The delayed deformation or "creep" of a rock under constant stress, and the delayed recovery of strain may be accounted for by the fact that rocks are composed of grains of different minerals with different deformation characteristics.

The sedimentary rocks (with the exception of limestone) consist primarily of grains of quartz which are glued together by calcareous cement or by argillaceous cements. The igneous and metamorphic rocks, on the other hand, consist of various polycrystalline granules held together by mechanical bonds.

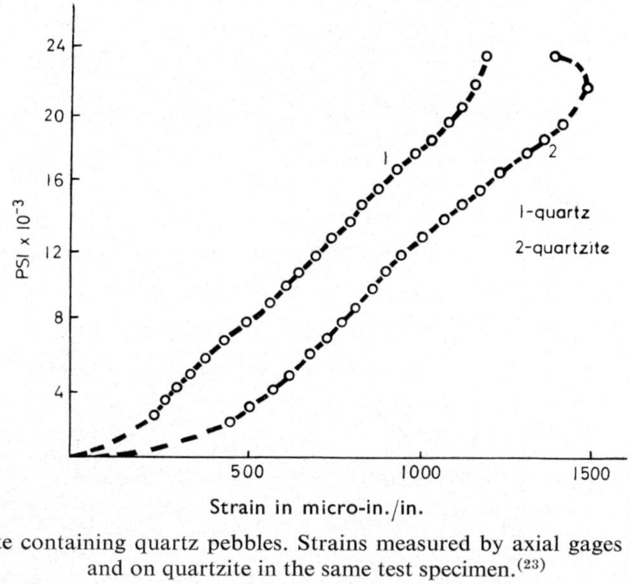

FIG. 32. Quartzite containing quartz pebbles. Strains measured by axial gages on quartz pebble and on quartzite in the same test specimen.[23]

When such an assemblage of mineral grains is subjected to stress, a mineral grain which is nearly perfectly elastic deforms immediately in proportion to the amount of stress transmitted to it. A neighboring mineral which has viscous characteristics yields continuously under stress until a sufficient proportion of the total load is transmitted to the more resistant grain and stress equilibrium is reached. If the load is then removed, the minerals which are more elastic attempt to expand immediately but are prevented from doing so by the restraining effects of the viscous minerals, or, in the case of sedimentary rocks, the "glue". This the strain recovery of the rock after the load is removed follows the same delayed strain pattern as when the load was first applied.

Hardy suggests[19] that under low values of applied stress (strain not in excess of 0.1 per cent) the Burgers viscoelastic model holds to a first degree of approximation and that at higher values of applied stress viscoelastic behavior is retained to some extent but micro-fractures begin to develop and to propagate.

If we take as an example a sandstone consisting of 50 per cent quartz grains and 50 per cent an argillaceous cement which acts as a "glue" and subject it to a stress of, say, 5000 psi, the rock will show an immediate deformation which will be inversely proportional to its modulus of elasticity under short-time loading. As the argillaceous glue yields, the quartz grains readjust themselves until they begin to make grain–grain contacts at certain points. As more and more of these contacts are made the resistance of the rock to deformation increases and the rate of creep continuously decreases.

If the applied load is now removed the quartz grains, being tightly compressed, expand immediately as much as the surrounding "glue" will allow. The quartz grains will still be stressed, however, and will continue to strain against the glue. As the glue gradually yields, the stress in the quartz grains decreases and rate of strain recovery of the rock decreases and approaches a zero value before the original strain is completely recovered.

Restrained Deformation Recovery

The fact that diamond drill cores and similar rock samples still retain internal strain energy for many years after being cored from the *in situ* rock was first noted and recorded by Emery, who describes the nature and causes of this phenomenon as follows:[20]

> Upon the removal of the piece from its natural restraint any completely elastic strain in the specimen as a whole will be instantaneously recovered. Such strain will be that due to the slight volumetric compression of the Maxwellian components and the similar and related portion of the volumetric compression of the elastic particles. The internal quartz-cement strain relationship will not immediately relax but will commence to do so on the surface. Exposed grains will begin to move, rotate, and expand at a rate controlled by the flow of the viscous cement. As the outer layer relaxes, the next layer has less restraint and can also relax but not as much as the surface layer unless the surface particles become unbonded and separate. Thus a strain gradient is built up through the specimen (analogous in some respects to a temperature gradient), and at some distance in, depending on the viscosity of the bond the original strains will remain substantially intact.
>
> Because the sandstone was strained largely before and during cementation the amount of instantaneous elastic strain relief is likely to be small when compared to the time-dependent strain relief. (In metamorphosed rocks, strained after cementing, one would expect higher volume strains and greater initial strain relief.)
>
> The author (Emery) first noted the large strain energies restrained in the inside of a drill core in 1958 in a specimen of quartzite which was then five years old. Upon slicing the core

longitudinally and measuring the strain relaxation on the new surface, it was found that there was little further relaxation near the edges but very high relaxation in the centre. The outer $\frac{1}{4}$ in. of core volume appeared to have restrained the internal portion progressively inward. The same phenomenon has since been observed in many rocks. An early prediction of it was made in a thesis at Queen's University, Kingston, Ontario.

Another example of such relaxation occurred in sandstone on a new surface from the interior of a 6 in. diamond drill core taken from the sub-Permian coal measure strata in Lincolnshire over one year previous to relaxation measurements.

Such relaxation is a function of the surface area of the specimen. Hence a $\frac{7}{8}$ in. diameter drill core could be expected to be in a different rheological condition from a $1\frac{1}{8}$ in. diameter core taken otherwise identically after any given time interval. One would therefore expect differing results in further tests on the two cores. Similarly one would expect differing results from tests on two cores identical except with respect to age, unless they were both very old.

On the other hand core drilled in the laboratory from a large rock sample, if used in tests within a few hours, would not be very different from the original fresh specimen.

Effect of Stress on Total Deformation

If the applied stress is relatively small, then the rate of rock deformation decreases asymptotically to a constant rate which may approach zero. At higher stresses the initial strain increases and the rate of creep increases gradually until at some stress value the secondary creep reaches ths tertiary phase and proceeds to fracture.

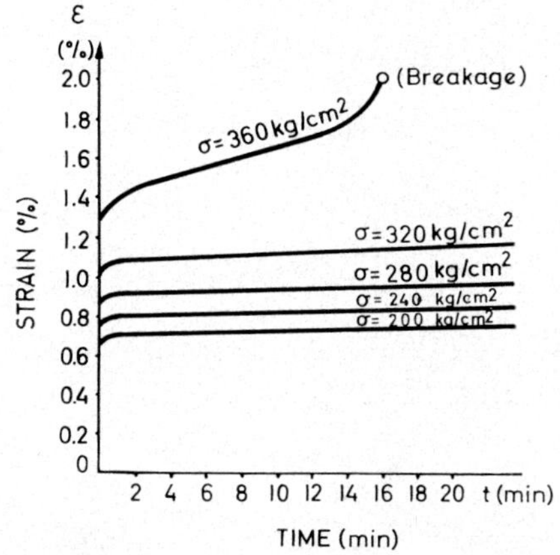

FIG. 33. Creep in rock salt under various stresses.(21)

Little information is available on the relative stress values which will cause a rock to progress through the various phases of creep to failure in a given finite time. Generally it appears that the stronger mine rocks approach "significant" creep rates when they are stressed to more than 60–75 per cent of their ultimate crushing strengths. ("Significant" creep rates would be those which might affect the stability of mine pillars or mine roofs over periods of 5–10 years.)

Figure 33(21) illustrates the creep rates of rock salt with each sample subjected to a constant load of the magnitude indicated for 20 min. When the stress was increased to 360 kg/cm^2 the salt crept to fracture in about 16 min. At lower stresses no fracture occurred during the period of the test, and there was a trend toward a reduced rate of creep for each increment of stress reduction.

EFFECTS OF ROCK CREEP ON STABILITY OF MINE WORKINGS

Stress Transfer

When a rock creeps under stress it relieves itself of a portion of the load which created the stress. This portion of the load is then transferred to other locations, and the stress peaks which may be created at the boundary of an opening when it is first excavated tend to decrease with time as the stress is spread over a broader base.

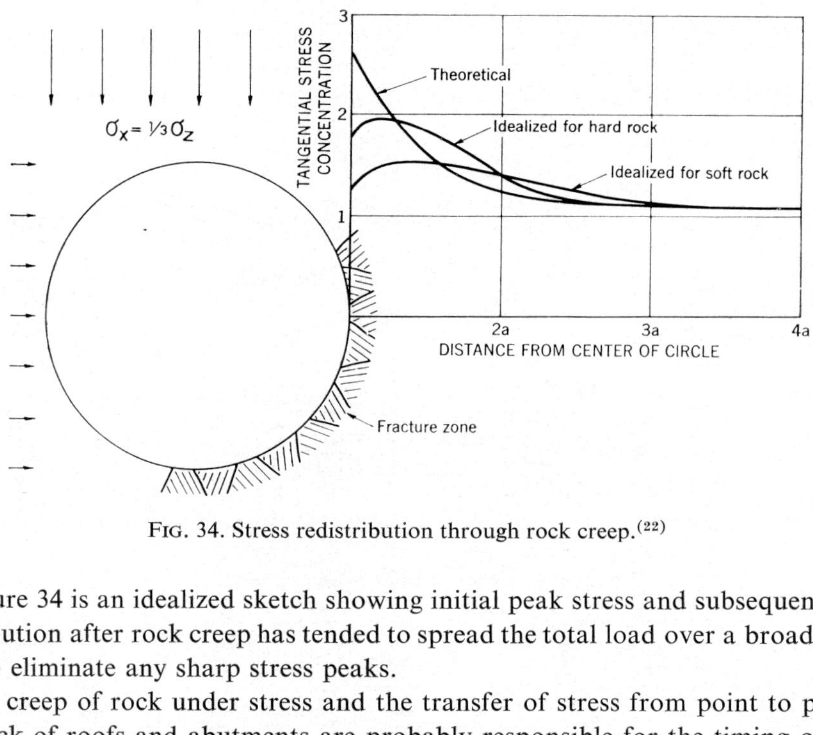

FIG. 34. Stress redistribution through rock creep.[22]

Figure 34 is an idealized sketch showing initial peak stress and subsequent stress distribution after rock creep has tended to spread the total load over a broader base and to eliminate any sharp stress peaks.

The creep of rock under stress and the transfer of stress from point to point in the rock of roofs and abutments are probably responsible for the timing of those rock bursts which are not actually triggered by blasting operations.

Practical Effects of Rock Creep on Mining Operations

The mining engineer is interested in predicting how long a roof span will stand without collapsing or how long a pillar will stand before it yields so that it no longer furnishes effective roof support.

In room-and-pillar mining systems in coal mines the pillars and the roof spans need only to stand for a few weeks or months before pillars are mined and the roof allowed to cave. Main entry pillars, on the other hand, may be in use for a period of several years and should be sized so that they do not yield excessively in that time.

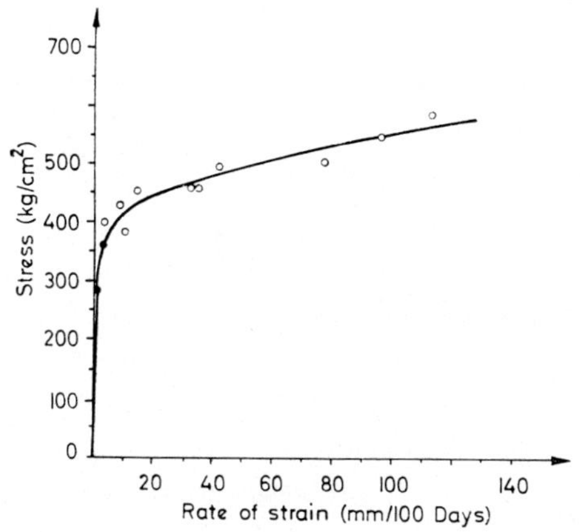

FIG. 35. Rate of lateral creep of pillars in potash mine as a function of the load on the pillar.[21]

Figure 35 shows the rate of transverse expansion of a pillar in a potash mine due to creep.

In metal mining operations the rates of progress in extraction operations are slower and a single large stope may require a year or more to complete.

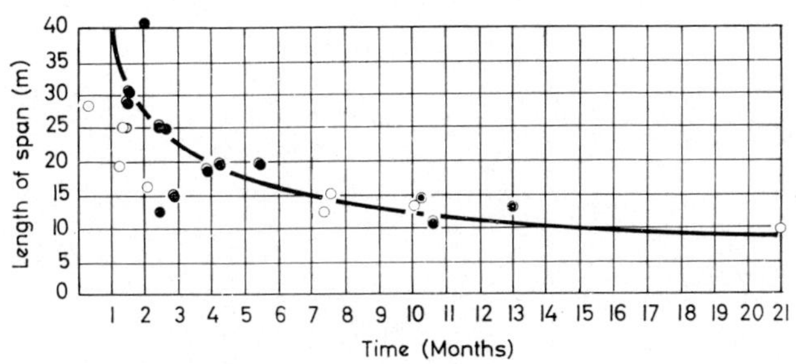

FIG. 36. "Standing time" of stope backs (roofs) as a function of stope spans in Russian iron mines.[21]

Figure 36 illustrates the effect of stope spans (in a Russian iron mine) on the stability of their backs (roofs). For the particular mine rock under consideration a stope roof with a span of about 30 m can be expected to stand for about $1\frac{1}{2}$ months, while a stope roof with a span of about 10 m can be expected to stand for about 15 months.

Figure 37 (a) shows the results of measurements of convergence of the sides of a 51-ft diameter tunnel excavated through horizontally bedded dolomite, sandstone, and shale in connection with the construction of the Sir Adam Back Niagara Generating Station No. 2, by Ontario Hydro. Tunnel excavation was done in two stages, and final deformation was restrained until after removal of the bench.

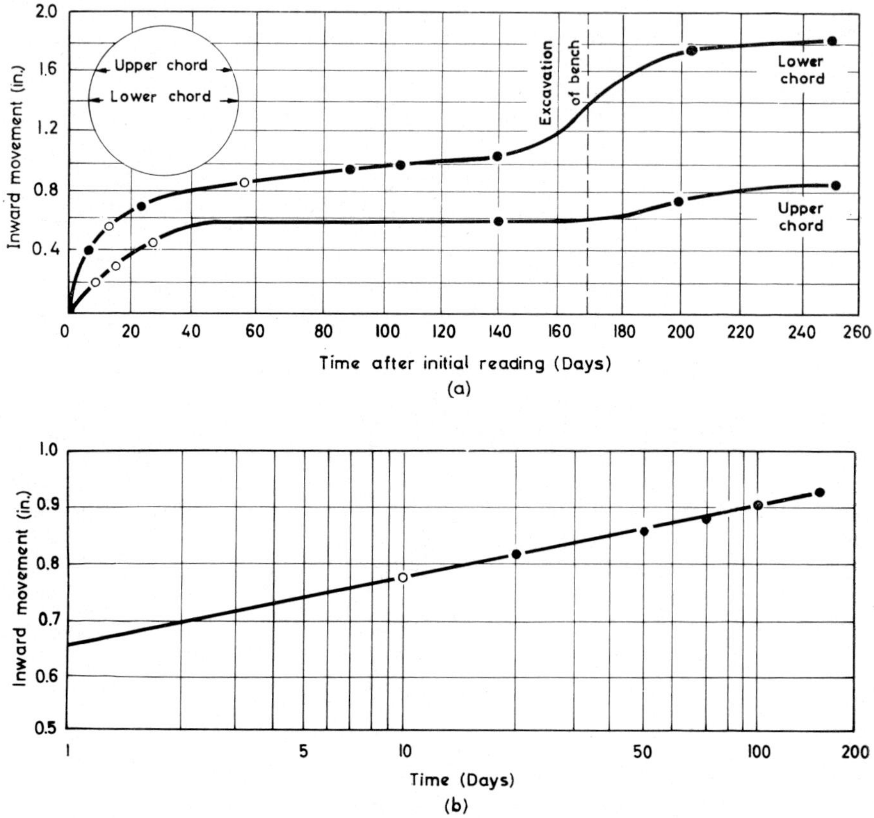

FIG. 37.

(a) Rock movement during tunnel excavation. (b) Inward movement of the rock walls at the spring line after bench removal.(25)

Figure 37 (b) indicates that rock creep was a logarithmic function of time. From the graph it was estimated that additional creep of about $\frac{1}{4}$ in. would occur in about 30 years.

By making certain simplifying assumptions, and assuming that the modulus of elasticity of the sedimentary rocks is about 5 million psi it has been calculated that the natural horizontal stress required to produce these deformations may be as much as 4800 psi. It is probable, however, that the modulus of elasticity of the rock *in situ* is much less than the assumed 5 million psi (see Chapter 10, in which are cited instances where the *in situ* modulus is only about one-fifth of the modulus as determined for laboratory specimens) and that the natural lateral stress is several orders of magnitude lower than that computed. (See Chapter 11, pages 451–461 for an account of measurements of *in situ* rock strain as a function of time.)

Predicting Rock Failure

A number of equations for deformation vs. time have been given previously herein. In addition to these, there are many more which have been proposed by various investigators for use in predicting time deformation of metals as well as of rocks and other brittle materials.

A review of these equations indicates that all are essentially empirical and that creep phenomena are very complex and that no one equation can exactly describe the behavior of any one rock throughout all stress ranges.

Within the range of interest of the mining engineer it is probable that one of the simpler logarithmic relationships, such as equation (10) or equation (11), is adequate for predicting the deformation which may be expected to take place in a given rock under a given stress over a period of years.

It is improbable that accurate predictions of rock creep for years in the future can be made on the basis of laboratory tests which last only a few minutes, hours, or days. Prediction of long-term time strain should be based on creep tests extending over periods of several months at the very least.

There appear to have been many short time laboratory tests made and recorded in the literature but there is a lack of long time tests from which predictions of long term rock deformation can be made.

Microseismic Activity as an Indication of Rock Creep

Figure 38 is a record of the rate of occurrence of microseismic noises during a 4-hr period after blasting in a mine. This record is typical of such activity and it

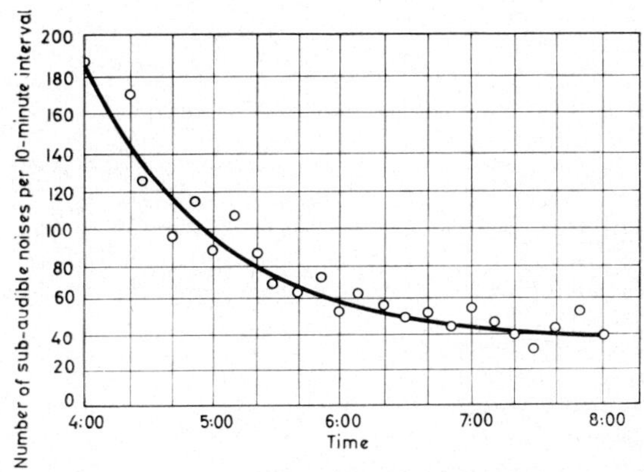

FIG. 38. Representative curve showing the rate of occurrence of subaudible noises during a 4-hr period after blasting.[26]

seems likely that there would be a correlation between microseismic activity and the rate of rock creep.

The writer suggests that it might be possible to correlate the rate of rock creep with the rate of occurrence of subaudible noises on laboratory specimens and to use this correlation to monitor the rate of rock creep in mine workings.

BIBLIOGRAPHY

1. Shearing strength of concrete under high triaxial stress — computation of Mohr's envelope as a curve, *U.S. Bureau of Reclamation Structural Research Laboratory Report* No. SP–23, Oct. 28, 1949.
2. Golder, H. Q. and Skempton, A. W., The angle of shearing resistance in cohesive soils for tests at constant water content, *Proceedings of Second International Conference on Soil Mechanics and Foundation Engineering, Rotterdam,* June 1948, Vol. I, pp. 185–192.
3. Skempton, A. W., A study of the immediate triaxial test on cohesive soils, *Proceedings of the Second International Conference on Soil Mechanics and Foundation Engineering, Rotterdam,* June 1948, Vol. I, pp. 192–196.
4. Bjerrum, L., Discussion, — (Shear strength of clays), *Proceedings of Third International Conference on Soil Mechanics and Foundation Engineering, Switzerland,* 1953, Vol. III, pp. 112–113.
5. Smith, C. K. and Redlinger, J. F., Soil properties of Fort Union clay shale, *Proceedings of the Third International Conference on Soil Mechanics and Foundation Engineering, Switzerland,* 1953, Vol. I, pp. 62–66.
6. Nash, K. L., The shearing resistance of fine closely graded sand, *Proceedings of the Third International Conference of Soil Mechanics and Foundation Engineering, Switzerland,* 1953, Vol. I, pp. 160–164.
7. Wuerker, R. G., The shear strength of rocks, *Mining Eng.* October, 1959, pp. 1022–1026.
8. Hardy, H. R., Jr., Physical properties of mine rock under short period uniaxial compression. Part 1. Results of McGillivray Mine, Coleman, Alberta, Report No. F.R.L. 243, Fuels Division, Canada Dept. of Mines and Technical Surveys, Mines Branch.
9. Panek, L. A., *Stresses about Mine Openings in a Homogeneous Rock Body,* Edwards Bros., Ann Arbor, Michigan, 1951.
10. York, B. and Reed, J. J., An analysis of mine opening failure by means of models, *Mining Eng.* July, 1953, pp. 705–710.
11. Livingston, C. W., Research at the Colorado School of Mines in subjects related to the mechanics of rock failure, *Trans. Am. Geophys. Union* Vol. 32, No. 2, pp. 268–278, 1951.
12. Mohr, F., Observations in shafts of ground movements due to mining, *International Strata Control Congress, Leipzig,* October, 1958.
13. Price, N. J., The compressive strength of coal measure rocks, *Colliery Eng.,* July 1960, pp. 283–292.
14. Donath, F. A., Experimental study of shear failure in anisotropic rocks, *Bull. Geol. Soc. Am.* Vol. 72, No. 6, June, 1961, pp. 985–990.
15. Murrell, S. A. F. and Misra, A. K., Time dependent strain or creep in rocks and similar non-metallic materials, *Trans. Am. Inst. Mining Met. Engrs.* Vol. 71, 1961–1962, Part 7, pp. 353–356. (Bulletin No. 665, April 1962.)
16. Griggs, D. T., Creep of rocks, *J. Geol.* Vol. 47, No. 3, 1939, pp. 225–251.
17. Griggs, D. T., Deformation of rocks under high confining pressures, *J. Geol.* Vol. 44, No. 5, 1936, pp. 541–577.
18. Salustowicz, A., New conceptions as to phenomena of stress and strain in rocks around mining excavations, *International Strata Control Congress, Leipzig,* October, 1958, pp. 1–13 (sponsored by Deutsche Akademie Der Wissenschaften Zu Berlin, Sektion Für Bergbau), Internationales Büro für Gebirgsmechanik.
19. Hardy, H. R., Jr., Time dependent deformation and failure of geologic materials, *Quart. Colo. School Mines,* Third Symposium on Rock Mechanics, Vol. 54, No. 3, July 1959, pp. 137–175.

20. Emery, C. L., Testing rock in compression, *Mine & Quarry Eng.* May 1960, pp. 196–202.
21. Hofer, K. H., The principles of creep in rock salts and their general significance to mining engineering, *International Strata Control Congress, Leipzig*, October, 1958, pp. 19–63, (sponsored by the Deutsche Akademie Der Wissenschaften Zu Berlin, Sektion Für Bergbau), Internationales Büro für Gebirgsmechanik.
22. Obert, L., Duvall, W. I., and Merrill, R. H., Design of underground openings in competent rock, *U.S. Bur. Mines Bull.* 587, 1960.
23. Corlett, A. V. and Emery, C. L., Prestress and stress redistribution in rocks around a mine opening, *Trans. Can. Inst. Mining Met.*, Vol. 62, 1959, pp. 186–198; and *Can. Mining Met. Bull.* June, 1959.
24. Kinsman, F. W., Charts for stress visualization, *Machine Design*, February, 1951, pp. 192–193.
25. Gibson, W. L., Some experiences in investigating stresses and forces in massive structures, *Trans. Can. Inst. Mining. Met.*, Vol. 59, April, 1956, pp. 141–153.
26. Obert, L. and Duvall, W., Use of subaudible noises for the prediction of rock bursts, Part II, *U.S. Bur. Mines R.I.* 3654, July, 1942.
27. Evans, R. H. and Wood, R. H., Transverse elasticity of natural stones, *Proc. Leeds Phil. & Lit. Soc.*, Vol. 3, Pt. 5, 1937, pp. 340–352.

CHAPTER 5

GROUND FAILURE AND PRESSURE ON SUPPORTS — SINGLE OPENINGS

THE BASIS OF EFFICIENT GROUND SUPPORT

Figure 1 illustrates a sub-surface opening which is excavated in ground subjected to vertical pressure, σ_z, due to overburden weight. Taking an element, A, located at the perimeter of the opening, the stress on the element will be something in excess of the uniform vertical stress, σ_z, because of stress concentration at the perimeter of the opening. Designating principal stresses on the element by σ_1 and σ_3 we see that at the edge of the opening

$$\sigma_3 = 0$$

Assuming that the material is a shale with an angle of internal friction of 37° then

$$\sigma_1 = (2c)\frac{\cos\Phi}{1-\sin\Phi}+j\sigma_3 \qquad \text{(Eq. 5, Ch. 4)}$$

where

$$j = \frac{1+\sin\Phi}{1-\sin\Phi}$$

taking values from Fig. 10, Chapter 4,

$$\sigma_1 = (2c)\,(2)+4\sigma_3$$
$$\sigma_1 = 4c+4\sigma_3$$

At the edge of the opening element A is completely unrestrained laterally, that is, $\sigma_3 = 0$, and assuming that $\sigma_1 > 4c$ (that is, the vertical compressive stress exceeds the effective resistance of the rock) then shearing proceeds into the solid rock until sufficient lateral stress develops to inhibit further progress of shearing. This occurs when shearing has progressed to element B, where σ_3 attains sufficient magnitude so that

$$\sigma_1 \leqq 4c+4\sigma_3$$

It is evident that if sufficient lateral pressure were exerted on the rock surface at the instant that the excavation was created, shear fracturing of the rock at the sides would be prevented entirely. This example illustrates the effect of support in limiting structural deterioration of soil, or rock, around an excavation.

In the case of a fractured rock which supports itself by "arching action" in the manner shown in Fig. 2 the effect of the stress in the arch is to produce failure by shear, as in the element *A*. The installation of support which exerts pressure against

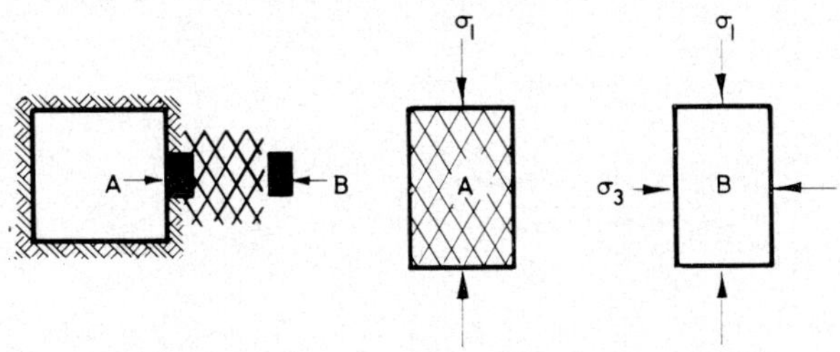

FIG. 1. Typical shear fracturing at the rib of a mine opening.

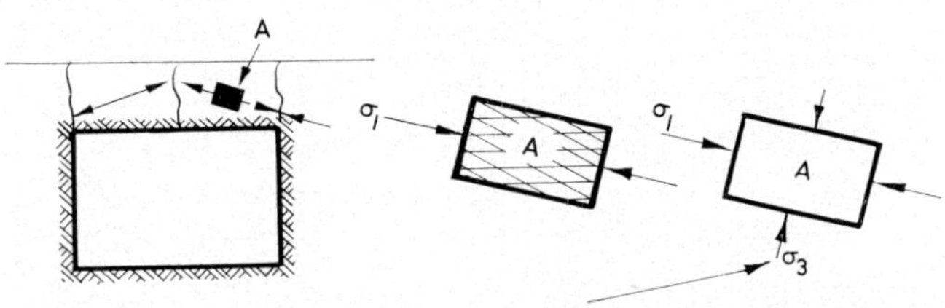

FIG. 2. Effect of support in stabilizing the "Linear arch".

the rock, normal to the major principal stress, increases the value of the quantity σ_3 and by inhibiting shear failure under compressive stress helps the material to support itself.

FRACTURES PRODUCED BY TRANSVERSE TENSION

In addition to the shear fractures which extend out to some distance from the opening boundary, and which are produced by tangential compressive stress, another type of fracture frequently occurs parallel to, and adjacent to, the free rock surface. This latter is the result of compressive stress acting parallel to planes of weakness in the rock. Such planes include bedding planes, shear planes, joints, and cracks created by blasting. The mechanics of fracture due to transverse tension are described in Chapter 3.

Weaknesses in the rock which form the basis for such tension fractures may include pre-existing natural flaws in the rock such as bedding planes, shear planes, or joints; or they may be induced flaws such as "induced cleavage" or fractures created by blasting. Thus, rock at a free surface may spall because of transverse tensile stresses at the ends of pre-existing rock flaws.

Exposed flat, or convex, surfaces are the most susceptible to this type of fracturing, although the orientation of pre-existing weaknesses is also of prime importance in determining which surfaces will spall.

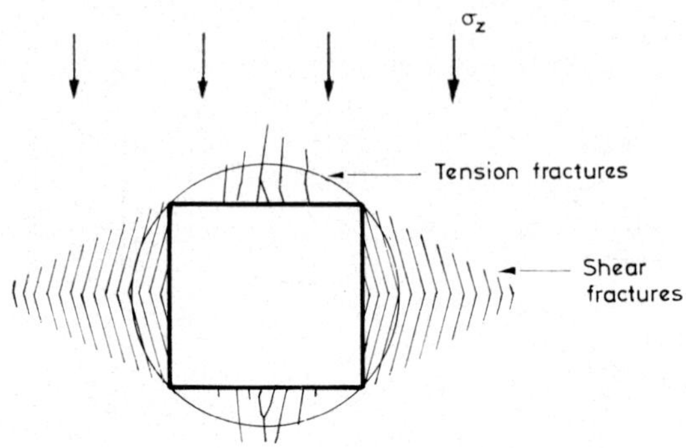

FIG. 3. Fracturing around an opening subjected to excessive vertical stress.

As compared with shear fracturing, such tension fracturing is of minor importance in affecting the stability of an opening. Shear fracturing extends much further out from the boundary of the opening and destroys the cohesion of large volumes of soil, or rock.

ROCK FRACTURING AROUND OVERSTRESSED SINGLE OPENINGS

Many tunnels are excavated to a modified square cross-section. When such an opening is located in overstressed soil or rock, failure tends to take place as shown in Fig. 3.

Unidirectional Stress

In the case of a square opening subjected to excessive stress in the vertical direction only, shearing of the rock takes place and extends into the rock until lateral stress is sufficient to stop further shearing. The shearing of the rock at the ribs allows a downward yield of the roof and has the effect of lengthening the "unsupported" roof span. Thus, the tensile stresses in the roof will be increased in magnitude as well as being extended across a greater width of the roof, and further vertically into the roof.

Similar fracturing takes place at the ribs and the back of an arched, or a circular, opening under similar vertical stress conditions. With the circular, or arched, opening, however, the configuration of the roof, or back, conforms more closely with the shape of the natural self-supporting arch (compression arch) and fractured material may partially support itself, while with the square opening the fractured material in the immediate roof has no "arch abutments" against which to support itself, and therefore tends to fall into the opening.

Isotropic Stress

If circular and square openings are excavated in material which is subjected to isotropic stress, the fracture patterns will be as shown in Fig. 4 (a) and (b). Since no tensile stresses exist around the peripheries under such conditions the only destructive stresses will be shear stresses. Shear fracturing extends out from the periphery along the spiral curves as shown; the direction of such fracture will at every point theoretically make an angle of $45° + \Phi/2$ with a radial line at that point.

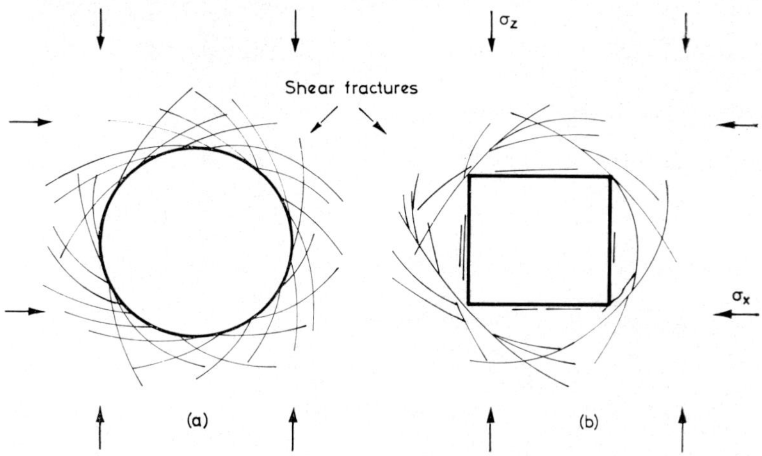

FIG. 4. Fracturing around openings subjected to excessive "isotropic" stress.

If the material is friable, then broken material will fall into the opening and the surrounding material will "squeeze" in all around the periphery, until shearing has extended far enough into the rock to achieve stability.

A square opening fractures first at the corners, where stresses concentrate, and from the corners shear planes extend into the surrounding material. If a circle is circumscribed through the corners of a square opening, as shown in Fig. 4 (b), it is observed that the shear planes are similar to those radiating from the periphery of the equivalent circular opening. The effect is as if the square opening first attempted to become a circular opening, and then the fracturing proceeded from the periphery of that circular opening.

The foregoing patterns of fracturing under isotropic stress are apparent in results of tests by York and Reed, who subjected plaster models to hydrostatic loading.[24]

Natural Arching of Tunnel Roofs

Openings with a high fracture arch may be produced under the following conditions of natural earth stresses:

(1) Excessive vertical stress with little lateral restraint. This is illustrated in Fig. 6. Under these conditions tension cracks extend vertically into the roof and rock may

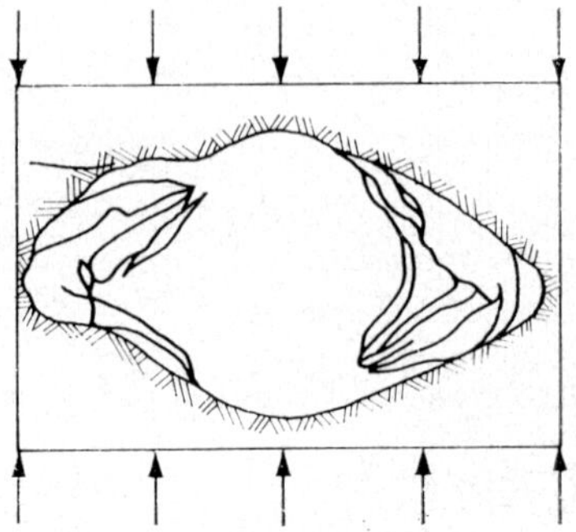

FIG. 5. Shear zones at the sides of a horizontal borehole subjected to vertical compressive stress. (After Mohr.[23])

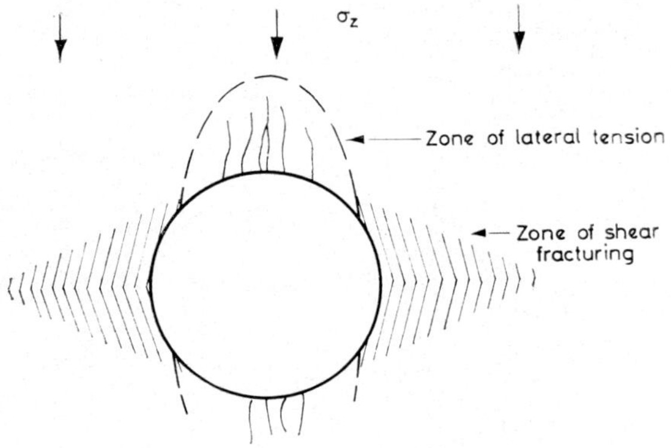

FIG. 6. Formation of an arched "back" due to lateral tensile stresses induced by unidirectional vertical stress.

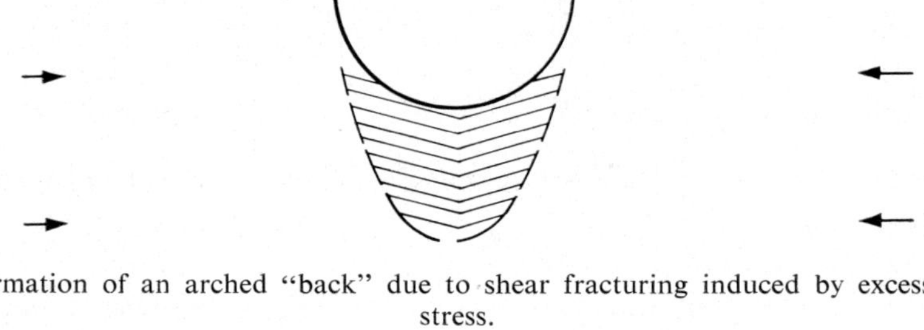

FIG. 7. Formation of an arched "back" due to shear fracturing induced by excessive lateral stress.

fall out, leaving a high arched opening. Eventually, if any natural lateral stress exists in the rock, stability is achieved when the vertical dimension of the opening is sufficiently large that the initial lateral compressive stress concentrates at the top of the arch to nullify tensile stress. The dimensions of the opening for such conditions may be calculated from equations (1) and (2), which are presented later in this chapter. If no lateral stress is present then the opening may fill completely with broken rock before stability is achieved.

(2) Excessive lateral stress with little vertical pressure. This produces shear fracturing which extends into the roof as shown in Fig. 7. Eventually, when fracturing extends far enough into the rock, if interlocking slabs are produced, they may create enough restraint to inhibit further shearing. If not, then the opening will fill with broken rock before stability is achieved.

Ground Failure around Openings in Interbedded Hard and Soft Strata

When an opening is excavated in bedded rock consisting of alternating hard and very soft strata there is a tendency for the soft strata in the high pressure zone to be squeezed laterally toward the opening (that is, to be extruded from between the harder strata) and to "flow" plastically into the opening. In order to find relief, the extruded strata may buckle and punch down through the top of the roadway supports.

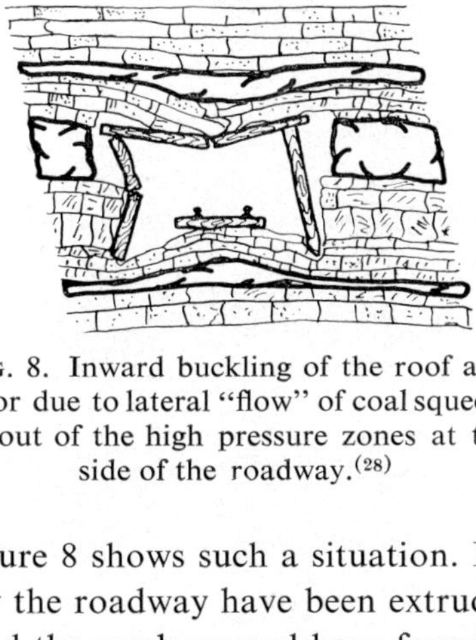

FIG. 8. Inward buckling of the roof and floor due to lateral "flow" of coal squeezed out of the high pressure zones at the side of the roadway.[28]

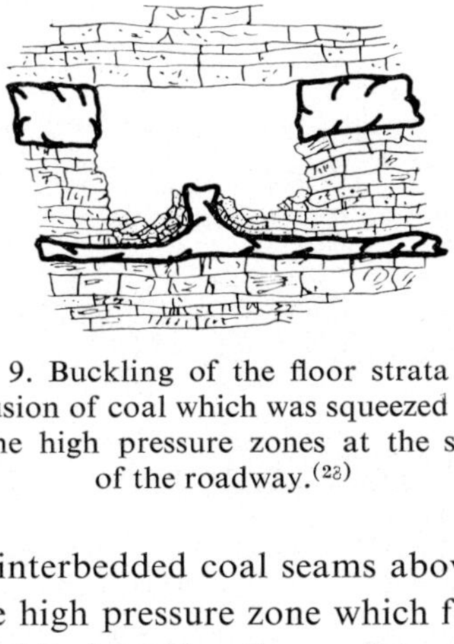

FIG. 9. Buckling of the floor strata and intrusion of coal which was squeezed out of the high pressure zones at the sides of the roadway.[28]

Figure 8 shows such a situation. In this case interbedded coal seams above and below the roadway have been extruded from the high pressure zone which formed around the roadway and have found partial relief by buckling the roof downward and the floor upward.

Figure 9 shows a roadway driven immediately above a soft coal seam. The coal "flowed" plastically and was extruded upward through the floor strata.

Figure 10 shows a roadway in the Campine coalfield which was completely closed when soft strata in the floor were extruded into, and completely filled, the roadway. In this case the thin-bedded schist and the soft coal seam were subjected to abutment

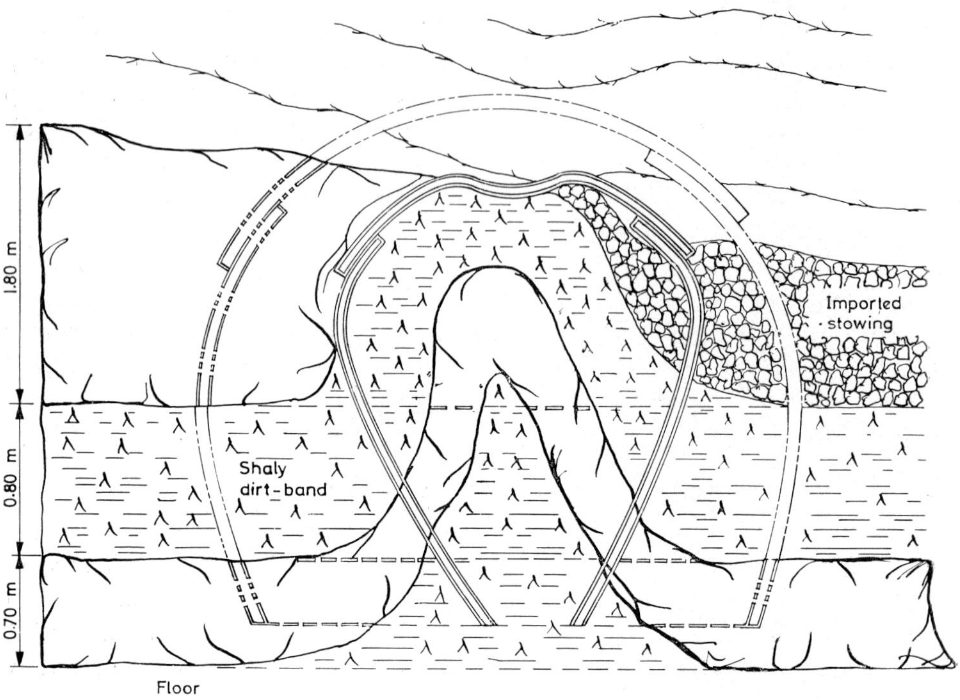

FIG. 10. Roadway completely closed by intrusion of coal and soft shale squeezed out of the high pressure zones at sides of roadway.[26]

pressure ahead of a longwall face and the ground stresses were further multiplied in the pressure zone around the opening.

This type of rock failure occurs frequently in openings driven through interbedded shales and sandstones. When very soft shale beds occur close to the roof or floor, the shale tends to "flow" laterally and to cause the floor to heave or the roof strata to buckle downward.

PRESSURES ON SUPPORTS FOR SINGLE OPENINGS

Classification of Stress States of Soil and Rock

A material composing the Earth's crust may be thought of as existing in one of the three following states of stress equilibrium:

(1) The elastic state.

(2) The clastic state.

(3) The plastic state.

(1) Materials in the elastic state are those which possess sufficient strength, or are so confined, that they yield elastically when stress is applied to them and recover

such yield when the stress is removed. In addition, the definition of elasticity requires that the yield or "strain" be proportional to the stress.

Materials in the elastic state are typified by those which possess some compressive strength when unconfined, such as common building materials, rocks, and cohesive soils. Each of these is in the "elastic state" so long as it is not stressed beyond its elastic limit.

A granular or fragmented (pulverulent) material may be considered to be in the elastic state provided that it is confined on all sides, and the applied stresses are so related that they do not produce shearing movements in the material. When this condition exists the material behaves elastically under stress; that is, each of the individual elastic fragments composing the material yields elastically as stress is transmitted from fragment to fragment but no movement occurs between the surfaces of adjacent grains or fragments.

(2) Materials in the "clastic state" include those which possess no significant cohesion. These "pulverulent" materials, when unconfined, yield under stress and do not recover any strain upon removal of the stress. Such materials are typified by cohesionless granular or fragmented materials such as brecciated zones in rock, clean sands, gravels, and boulders, and similar granular materials.

(3) Materials in the plastic state include those which yield under stress, do not recover the yield, but retain their cohesion during yield. The most common such materials are the moist clays, and materials containing enough clay to act as a binder, such as clayey sands, clayey gravels, zones of intermixed fault clay and breccia in rock, etc.

The "Clastic State"

The term "clastic state" is used herein to describe an equilibrium condition of "pulverulent" material that is material which has no cohesion. It has been used in this sense by Dr. Jacobi[(1)].

It has been customary in soil mechanics literature to designate only two conditions – the "elastic state" and the "plastic state". While these meanings may be clear to one well versed in soil mechanics, the author feels that the term "plastic" conveys to the average reader a picture of a soft yielding material such as moist clay. Therefore, the conditions covered by the term "plastic state" as ordinarily used in soil mechanics will be divided herein into the two conditions, the "clastic state" and the "plastic state." The author feels that the term "clastic state" is particularly needed in the field of rock mechanics where we are dealing, a great deal of the time, with the cohesionless fragmented (pulverulent) material which is produced when a brittle rock fractures under excessive compressive stress.

It should be noted that the term as used herein has no connection with the geological term "clastic" which implies a hard rock composed of fragments cemented together.

Letters and Symbols

In the analyses presented hereafter the Greek letters will be used to designate stresses and strains in soil and rock while the common alphabet will be used to designate stresses, strains, dimensions, and directions in structures and supports.

σ	designates unit stress in soil or rock;
p	designates unit stress in or on supports;
x	designates horizontal direction normal to the opening axis;
y	designates horizontal direction parallel to the opening axis;
z	designates vertical direction;
t	designates direction tangential to the boundary of the opening;
r	designates direction radial (normal to the opening boundary);
a, b, L, r	designate direction dimensions;
Φ	designates the angle of internal friction.

Thus:

σ_z	designates vertical stress at a point;
σ_x	designates horizontal stress at a point (normal to opening axis);
σ_y	designates horizontal stress at a point (parallel to opening axis).

The Elastic Boundary

Material around the boundary of an underground opening which requires no support is evidently in the elastic state. No pressure need be exerted against the boundary to prevent the material from fracturing and falling into the opening. This is true of most rock tunnels.

It is also true, although less evident, that there exists around every underground opening which has been stabilized by support an elastic boundary. This may be located at some distance from the periphery, its actual location being determined by the distance into the soil, or rock, which shearing must extend to establish equilibrium.

The material lying outside this elastic boundary is considered to be in the elastic state, while material inside the boundary is considered to be in the clastic state or in the plastic state.

The location of the elastic boundary is determined by the ratio between the radial stresses and the tangential stresses, and these in turn depend upon the shape of the opening, the magnitude of natural earth stresses, and the cohesion and internal friction of the soil or rock.

This elastic boundary may exist in material of any nature, that is, in fragmented or granular cohesionless material as well as in cohesive material. The only requirement for the establishment of the elastic state is that the minor principal stresses be large enough to prevent shearing due to the major principal stress.

Before proceeding to further consideration of equilibrium around underground openings we shall review the stress relationships existing around elliptical and around circular elastic boundaries.

Stresses at Boundaries of Circular and Elliptical Openings

The magnitudes of the stresses around openings in elastic materials are discussed in Chapter 1. Boundary stresses will be reviewed here briefly.

At the boundary of a hole in elastic material the radial stress, commonly designated by σ_r, is zero. Therefore, the only stress acting is the stress tangential to the boundary of the hole. Progressing away from the boundary the radial stress increases and the tangential stress decreases. Tangential stress is designated herein as σ_t.

Equations for Boundary Stresses

The general equations for stresses in elastic material around circular and elliptical openings were presented in Chapter 1. For our present purposes it is sufficient to determine the magnitude of stresses at the points where they are maximum and minimum. In the following discussions these points will be assumed to occur at the ends of the horizontal and the vertical axes. (See Fig. 11.) The following equations apply:[27]

$$\sigma_{t_x} = \sigma_x\left(2\,\frac{L_z}{L_x}+1\right)-\sigma_z \tag{1}$$

$$\sigma_{t_z} = \sigma_z\left(2\,\frac{L_x}{L_z}+1\right)-\sigma_x \tag{2}$$

When $L_z = L_x$ the ellipse becomes a circle and

$$\sigma_{t_x} = 3\sigma_x-\sigma_z \tag{3}$$

$$\sigma_{t_z} = 3\sigma_z-\sigma_x \tag{4}$$

(See Fig. 11.)

When σ_x and σ_z are of different magnitudes an elliptical opening may be so dimensioned that the resultant tangential stresses are equal; that is,

$$\sigma_{t_x} = \sigma_{t_z}$$

and the resultant tangential stress at the boundary is everywhere compressive and equal. The axis ratio required for such a condition is:[2, 27]

$$\frac{L_z}{L_x} = \frac{\sigma_z}{\sigma_x} \tag{5}$$

An elliptical opening may be so dimensioned (its z-axis lengthened) that tensile stress at the ends of the z-axis becomes zero. Then $\sigma_{t_x} = 0$ and

$$\frac{L_z}{L_x} = \frac{1}{2}\left(\frac{\sigma_z}{\sigma_x} - 1\right) \tag{6}$$

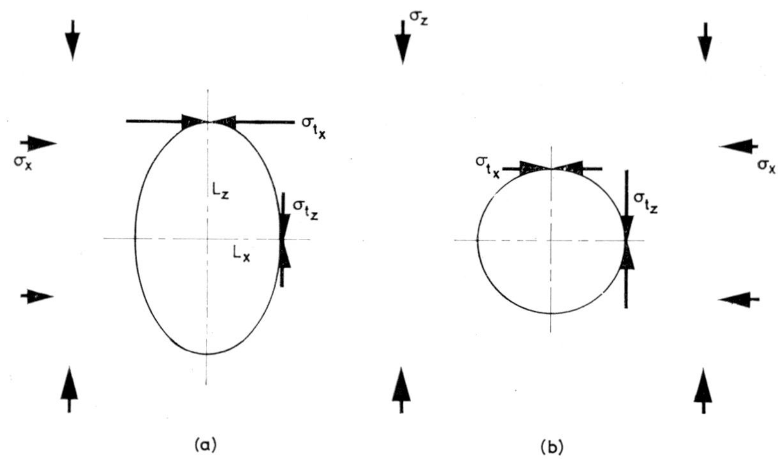

FIG. 11. Tangential stresses at elastic boundaries of elliptical and circular openings. Maximum and minimum stresses occur at the ends of the axis.

Square Openings

In the case of square openings stresses concentrate at the corners and the magnitude of such stress concentration depends upon the "sharpness" of the corner of the opening. The more rounded the corner, the less the magnitude of stress concentration.

The fracturing at corners is such that extension of corner fractures tends to create a roughly elliptical or circular zone (depending upon the ratio σ_x/σ_z). Stress conditions at the boundary of this zone then conform approximately to those previously described for elliptical and circular openings.

Stress Equilibrium around an Opening Bounded by Material in the Clastic State or in the Plastic State

With the aid of the equations for the magnitudes of stresses at boundaries of circular and elliptical openings we can compute the theoretical pressure on lining when the opening is stabilized; or, what amounts to the same thing, we can compute the pressure which the lining must exert on the periphery of the opening in order to prevent movement of the surrounding ground toward the opening.

Material in the clastic state, since it possesses no cohesion, will crumble and fall into an opening and obviously will not stand without support. Material in the plastic state will not stand without support to prevent its "squeezing" into the opening. If a material stands without support then it is in the elastic state. Most materials which possess cohesion after shearing are fairly soft, such as wet clays, and can remain in the elastic state only so long as the applied stresses are small, as in the case with openings at shallow depths.

If the tangential stress at the periphery of an opening does not exceed the compressive strength of the material then such material will be in the elastic state. If the compressive strength of the material is exceeded, then shearing progresses away from the boundary until sufficient radial stress is built up to inhibit further shearing.

Part of the radial stress is furnished by support but most of such radial stress is derived from the natural strength of the material itself which develops "arching" action, or "ring" action, due to compressive strength derived by virtue of its internal friction (see page 527, Chapter 12).

On the following pages will be outlined the mathematics of the basic principles involved in computing theoretical pressure on lining, or supports. The case which offers the simplest stress conditions is that of a horizontal section cut through a vertical circular shaft sunk in a granular cohesionless material; the material at the periphery of the shaft is therefore in the clastic state.

Notation:

p_i = pressure exerted by support on interior of opening;
a = final radius of the opening;
R = radial distance to a point in the clastic zone;
b = radial distance to boundary of clastic zone;
σ_r = radial stress (parallel to radii);
σ_t = tangential stress (parallel to tangents to opening);
σ_x = one principal stress in the horizontal plane;
σ_y = other principal stress in the horizontal plane.

The general equation for stresses in a horizontal plane around a circular opening in granular cohesionless material is:[(3–5)]

$$\sigma_{r_R} = p_i \left(\frac{R}{a}\right)^{j-1} \quad \text{or} \quad p_i = \sigma_{r_R}\left(\frac{a}{R}\right)^{j-1} \tag{7}$$

and

$$\sigma_t = j\sigma_r$$

and

$$j-1 = \frac{1+\sin\Phi}{1-\sin\Phi} - 1 \tag{8}$$

Values for this function may be read from Fig. 10, Chapter 4.

In Fig. 12 (b) the element 1, at the boundary of the opening, is subjected to tangential stress σ_t, which is the major principal stress and which tends to produce movement along the shear planes as shown. If the minor principal stress, p_i, is not large enough to inhibit movement along shear planes then shearing progresses outward from the periphery of the opening until stresses in the sheared zone are in equilibrium with those in the surrounding ground.

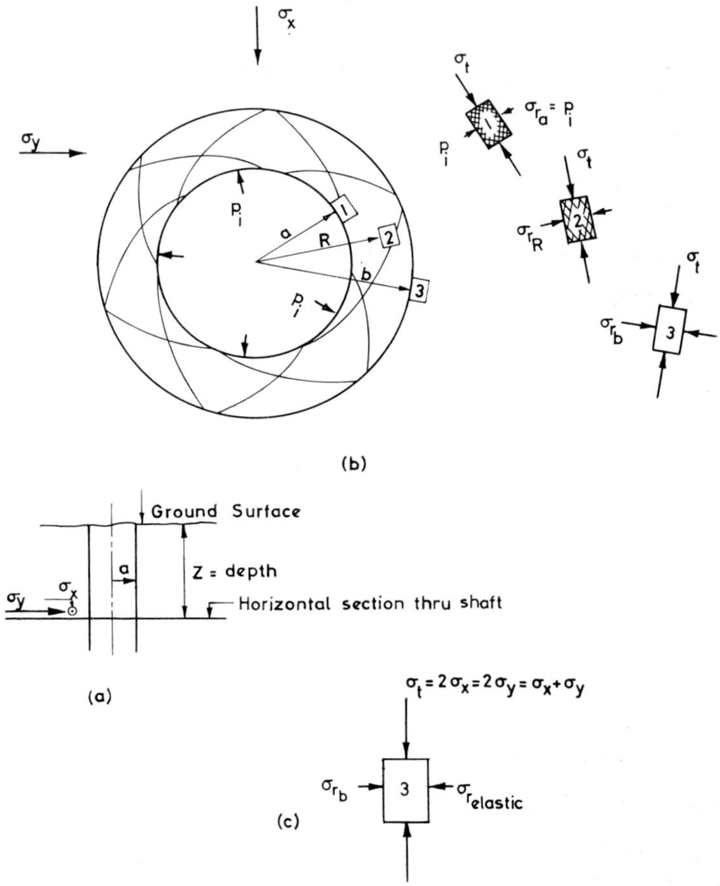

FIG. 12. Stress equilibrium in a horizontal plane around a vertical circular shaft.
(a) Vertical section through the shaft showing the stress relationships. (b) Equilibrium conditions in a horizontal plane around the shaft. (c) Stresses acting on element 3 at the boundary between the clastic and elastic zones

At the boundary of the sheared or clastic zone the element 3 is subjected to stresses as shown in Fig. 12 (c).

Just outside the boundary of the clastic zone the material is in the elastic state and the tangential stress will be (assuming isotropic stress):

$$\sigma_t = 2\sigma_x = 2\sigma_y = \sigma_x + \sigma_y$$

At the boundary of the clastic zone σ_{r_b} must have a magnitude sufficient to prevent further shearing.

The element 3 corresponds to a laboratory specimen subjected to triaxial compression, with σ_t the major principal stress and σ_r the minor principal stress. From Chapter 4 the stress relations at failure of such an element are:

$$\sigma_1 = Q + j\sigma_3 \quad \text{or} \quad \sigma_3 = (\sigma_1 - Q)i$$

where Q is the unconfined compressive strength of material and $i = {}^1/_j$. But element 3 is located at the circular interior boundary of the elastic zone and the major principal stress ($\sigma_t = \sigma_1$) is reduced when the minor principal stress ($\sigma_r = \sigma_3$) is applied to this circular elastic boundary. Designating the reduced tangential stress by σ_t', then

$$\sigma_t' = \sigma_t - \sigma_{r_b}$$

and at equilibrium

$$\sigma_{r_b} = (\sigma_t' - Q)i$$

thus:

$$\sigma_{r_b} = [(\sigma_t - \sigma_{r_b}) - Q]i$$

and

$$\sigma_{r_b}(1+i) = (\sigma_t - Q)i$$

or

$$\sigma_{r_b} = (\sigma_t - Q)\frac{1}{2}(1 - \sin\Phi) \tag{9}$$

When the material is cohesionless the unconfined compressive strength is zero, $Q = 0$, and

$$\sigma_{r_b} = \sigma_t\left(\frac{1}{2}\right)(1 - \sin\Phi) \tag{10}$$

To compute pressure on supports, substitute values of σ_{r_b} from the above equations in

$$p_i = \sigma_{r_b}\left(\frac{a}{b}\right)^{j-1} \tag{11}$$

or substituting for σ_{r_b}

$$p_i = \sigma_t\left(\frac{1}{2}\right)(1 - \sin\Phi)\left(\frac{a}{b}\right)^{j-1}$$

and for a circular shaft in a uniform horizontal stress field:

$$\sigma_t = 2\sigma_x = 2\sigma_y$$

Then:

$$p_i = 2\sigma_x\left(\frac{1}{2}\right)(1 - \sin\Phi)\left(\frac{a}{b}\right)^{j-1}$$

$$p_i = \sigma_x(1 - \sin\Phi)\left(\frac{a}{b}\right)^{j-1} \tag{12}$$

and

$$\sigma_x = \frac{p_i}{1 - \sin \Phi}\left(\frac{b}{a}\right)^{j-1} \tag{13}$$

STRESS EQUILIBRIUM IN A HORIZONTAL PLANE AROUND A VERTICAL CIRCULAR SHAFT

Pressure on Shaft Lining — Example Calculation (Considering Equilibrium Conditions in a Horizontal Plane)

For a numerical example take the simplest case, that of a circular shaft sunk in cohesionless sand above the water table. Equilibrium conditions in the horizontal plane will be computed. The following conditions are assumed:

z = depth = 100 ft;
γ = density = 100 pcf;
Φ = angle of internal friction = 37°;
a = radius of the opening = 5 ft;
b = radius of the clastic zone;
σ_z = vertical pressure on the plane = 100 × 100 = 10,000 psf.

Stress is assumed to be uniform in all directions on the horizontal plane: that is,

$$\sigma_x = \sigma_y$$

For 37° (from Fig. 10)

$$i = {}^1/_4 \qquad j = 4 \qquad j-1 = 3$$

For a cohesionless sand the maximum initial lateral stress existing in the sand prior to excavation of the shaft may be taken as

$$\sigma_x = \sigma_y = i\sigma_z = {}^1/_4\sigma_z$$

then

$$\sigma_x = \sigma_y = {}^1/_4\,(10{,}000) = 2500 \text{ psf}$$

and tangential stress due to external stress concentration at the boundary of the opening is:

$$\sigma_t = 2\sigma_x = 2\sigma_y = 2(2500) = 5000 \text{ psf}$$

and for equilibrium in cohesionless sand (equation 10)

$$\sigma_{r_b} = \sigma_t\left(\frac{1}{2}\right)(1 - \sin \Phi) = (5000)\,(0.2) = 1000 \text{ psf}$$

Assuming that the support yields sufficiently to allow shearing to progress to a distance of $7\frac{1}{2}$ ft from the center of the shaft, then

$$p_i = 1000\left(\frac{5}{7.5}\right)^3 = (1000)\,(0.30) = 300 \text{ psf}$$

If sufficient space is left, or if support yields enough to allow shearing to progress to 10 ft from the center of the shaft, then

$$p_i = 1000\left(\frac{5}{10}\right)^3 = 125 \text{ psf}$$

It should be noted that a slight movement between sand grains is required in order for the maximum shear resistance to be mobilized. This movement is accompanied by a volume increase. Thus, in order to reduce pressure on support, it is necessary to allow the sand to squeeze in uniformly around the periphery of the opening, within limits the more squeeze allowed the less the pressure on support, since shear resistance is mobilized over a larger volume of sand.

Obviously, this requires more excavation initially, and the most economical method is determined by balancing the cost of excavation against the cost of support.

The example given above has investigated stress conditions only in a horizontal plane. Around a vertical shaft in sand the stresses in vertical planes are often the most critical. Such cases are discussed later in this chapter.

Pressure on Supports — Openings in Cohesive Materials

Taking again the simplest case, that of stress equilibrium in a horizontal plane around a vertical circular shaft, the tangential stress at the boundary of a circular opening in material in the elastic state is:

$$\sigma_{t_a} = 2\sigma_x = 2\sigma_y = \sigma_x + \sigma_y$$

Assuming, for purposes of illustration, that σ_{t_a} exceeds the compressive strength of the material then shearing takes place and extends into the material until σ_r is sufficiently large to inhibit further shearing.

For a numerical example assume that the sand surrounding the shaft contains a minor amount of clay which binds the sand grains together and gives the sand an initial cohesive strength of about 500 psf. (Sand in the clastic zone is presumed to have lost all cohesion.) This corresponds to an unconfined compressive strength of about 2000 psf (when $\Phi = 37°$).

Assuming that other conditions remain the same as in the previous example, including the initial lateral compressive stress, we have (from equation (9))

$$\sigma_t = 5000 \text{ psf}$$

$$\sigma_{r_b} = (\sigma_t - Q)\frac{1}{2}(1-\sin\Phi) = (5000-2000)\frac{1}{5} = 600 \text{ psf}$$

If shearing progresses to $7\frac{1}{2}$ ft from the center of the shaft,

$$p_i = 600\left(\frac{5}{7.5}\right)^3 = 180 \text{ psf}$$

In order for the material to be self-supporting ($p_i = 0$) it would be necessary for it to possess an unconfined compressive strength, Q, of

$$Q = \sigma_t = 2\sigma_x$$

or

$$Q = 5000 \text{ psf}$$

This corresponds to (for $\Phi = 37°$) a cohesion of (from equation 6, Chapter 4)

$$c = \frac{Q}{2}\left(\frac{1-\sin\Phi}{\cos\Phi}\right) = \left(\frac{5000}{2}\right)\frac{1}{2} = 1250 \text{ psf} = 8.7 \text{ psi}$$

Such a soil would be classed as a strong soil.

EQUILIBRIUM CONDITIONS AT ZONE BOUNDARIES

Clastic-Elastic Equilibrium

Before proceeding further it is of interest to further examine graphically the stress conditions existing around an opening where clastic–elastic equilibrium has been established.

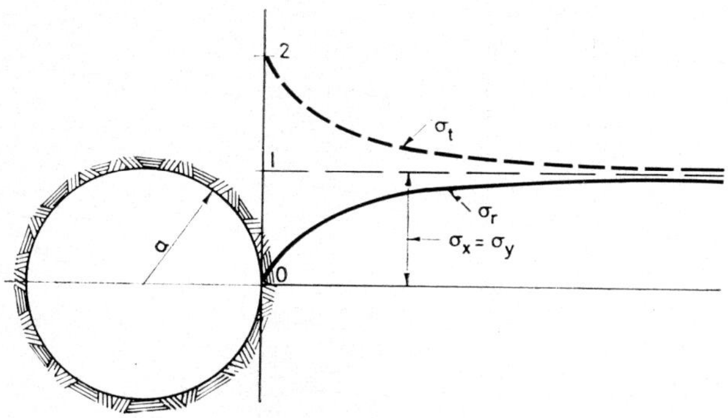

FIG. 13. Elastic equilibrium around a circular vertical shaft.

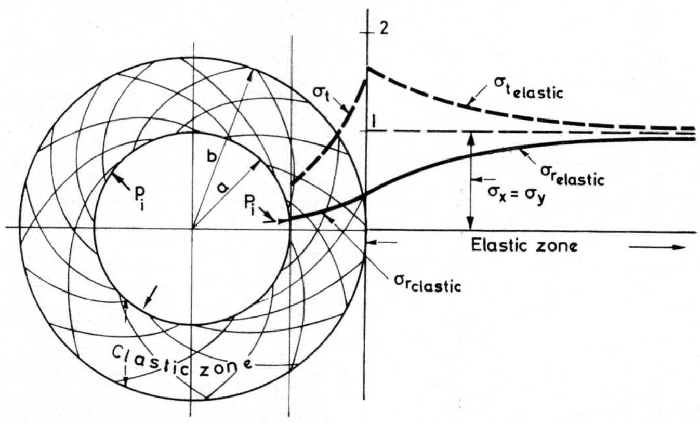

FIG. 14. Clastic–elastic equilibrium around a circular vertical shaft.

Figures 13 and 14 show graphically the magnitudes of the stresses around the same vertical shaft discussed previously. The stress conditions shown in Fig. 13 assume that the ground is strong enough to resist the tangential compressive stresses and is therefore in the elastic state at the boundary of the opening. Under these conditions the radial stress, σ_r, at every point on the boundary is zero, and the tangential stress at every point on the boundary is $\sigma_t = 2\sigma_x = 2\sigma_y = \sigma_x+\sigma_y$. Also at every point in a horizontal plane in the material surrounding the shaft the following stress relationship exists:

$$\sigma_t+\sigma_r = 2\sigma_x = 2\sigma_y = \sigma_x+\sigma_y$$

Figure 14 illustrates stress equilibrium conditions if the ground is weak and shearing extends outward from the perimeter until radial stress, $\sigma_{r,\text{ clastic}}$, is sufficiently large to establish equilibrium. At such points $\sigma_{r,\text{ clastic}}$ acting outward at the boundary of the clastic zone must equal $\sigma_{r,\text{ elastic}}$ (exterior) which acts inward at the boundary of the elastic zone. The radial stress applied to the elastic boundary reduces the magnitude of $\sigma_{t,\text{ clastic}}$ because of the requirement that the total stress on the boundary at equilibrium must be constant at $\sigma_x+\sigma_y$.

Although $\sigma_{t,\text{ clastic}}$ is shown in this instance as equal to $\sigma_{t,\text{ elastic}}$ at the boundary it can be seen that such conditions are not necessary for equilibrium, and, in fact, would not exist at the angle of internal friction, Φ_{clastic}, for material in the clastic zone differed from Φ_{elastic} for the material in the elastic zone.

The only requirement for equilibrium is that $\sigma_{r,\text{ clastic}}$ be sufficiently large with respect to σ_t to prevent further shearing of the material. It is then required that $\sigma_{r,\text{ clastic}}$ be equal to $\sigma_{r,\text{ elastic}}$ at the boundary where the two media are in contact

Plastic-Elastic Equilibrium

The foregoing analysis has considered materials within the sheared zone as being entirely without cohesion. It is apparent that a brittle material, after fracture, will consist of a number of "plates" or of fragments with little or no cohesion. On the other hand, a material containing some clay may retain some cohesion even after extensive shearing.

Referring to Mohr's stress diagram we recall that cohesion c is the shear strength of a material when the normal stress, σ_N, on the shear plane is equal to zero.

In order to represent cohesion by a value which may be conveniently introduced into the equations, c may be considered as being the result of an applied normal stress, σ_c, whose value is (from Mohr's diagram) [9]

$$\sigma_c = \frac{c}{\tan \Phi}$$

Then the following relationship pertains (from Mohr's diagram)

$$\frac{\sigma_1+\sigma_c}{\sigma_3+\sigma_c} = \frac{1+\sin \Phi}{1-\sin \Phi} = j$$

and

$$p_i = (\sigma_{r_b} + \sigma_c)\left(\frac{a}{b}\right)^{j-1} - \sigma_c \tag{14}$$

Taking for an example the conditions used previously, including a sand-containing clay which gives it an initial cohesion of 500 psf, we recall that σ_{r_b} required for equilibrium was 600 psf.

If we assume, for an example, that the sand loses most of its cohesion after shearing and in the plastic zone retains a cohesion of only 100 psf we can compute the resultant pressure on supports as follows:

$$\sigma_c = \frac{c}{\tan \Phi} = \frac{100}{\tan 37^\circ} = \frac{100}{0.75} = 133 \text{ psf}$$

and from equation (14)

$$p_i = (600 + 133)\left(\frac{5}{b}\right)^3 - 133$$

If shearing is allowed to progress to a distance of 7.5 ft from the center of the shaft

$$p_i = (733)\left(\frac{5}{7.5}\right)^3 - 133 = 87 \text{ psf}$$

This compares with $p_i = 180$ psf when the sand was presumed to possess no cohesion.

Vertical Shaft in Cohesionless Material — Stress Conditions on Vertical Cylindrical Surfaces

The preceding section has dealt with stresses in horizontal planes. For a vertical shaft which is sunk in a pulverulent material, such as a clean sand, above the water table, the stress conditions on vertical cylindrical surfaces concentric with the shaft must also be investigated.

An annular ring of sand outside the shaft perimeter tends to subside under its own weight, W shearing along the cylindrical surface which is represented by lines AB and CD, in Fig. 15. Such subsidence is resisted by the force Q acting upward and by shearing resistance acting along the cylindrical surface. The difference between forces W and Q must be carried by this shearing resistance.

Terzaghi[7] has derived the equations for stability under these conditions.

Figure 16 (from Terzaghi[7]) gives values of m_σ from which the theoretical pressure on the lining of a circular shaft may be computed. (The following notation differs from that originally used by Terzaghi because it has been changed to conform with notation used in the remainder of this book.)

z = depth of the point where pressure on lining is to be computed;
a = radius of shaft;
p_i = pressure on shaft lining (that is, radial pressure exerted by lining against the sand);
γ = unit weight (density) of sand;
m_σ = value from the chart;

$$n_1 = \frac{r}{a}.$$

The curve of Fig. 16 shows that for values of z/a more than about 16 (that is, 8 diameters) the pressure approaches a limiting value which is $p_i = (0.398)\gamma a$ at infinity.

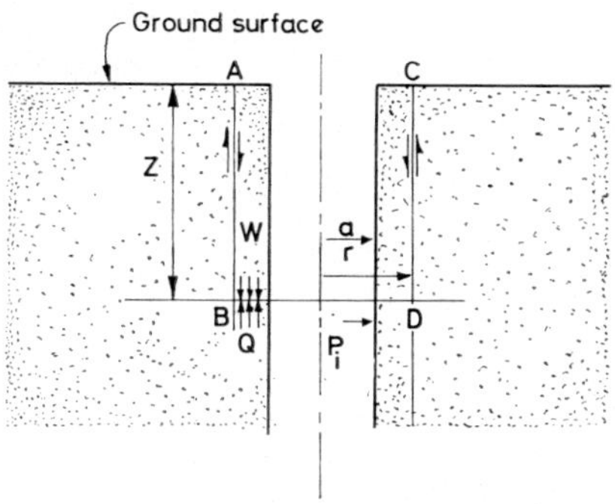

FIG. 15. Section through a vertical shaft in cohesionless sand — showing stresses acting in vertical planes.(7)

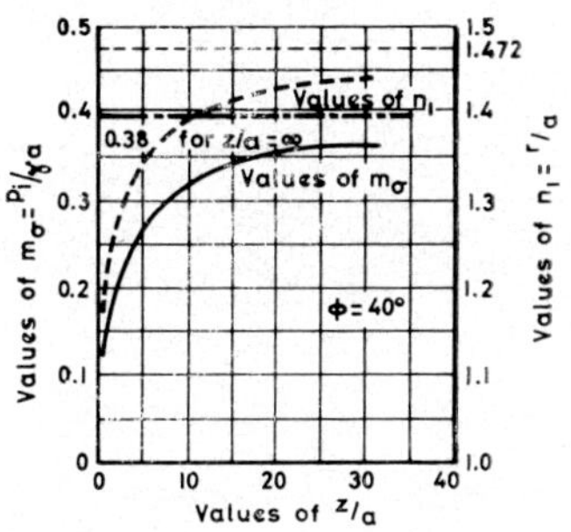

FIG. 16. Values for computing pressure on lining of a vertical shaft in cohesionless sand above the water table. (After Terzaghi.(7))

For a shaft 10 ft in diameter ($2a$ = 10 ft) the pressure on lining at a depth of 100 ft would be (assuming it to be above the water table, and taking Φ for the sand as 40°):

$$p_i = m_\sigma \gamma a = (0.36)\,(100)\,(5) = 180 \text{ psf}$$

The value of n_1 approaches a limiting value with value of $r/a = 1.47$; that is, the ratio between shearing stresses and normal stresses will be greatest at a distance of 1.47 a from the shaft centerline.

For a shaft located partially below the water table the pressure on the shaft lining will be the sum of the water pressure and the pressure due to the submerged weight of the sand.

Measured Pressure on Lining of Vertical Shaft in Plastic Clay

Peck and Berman[20] have reported on pressures measured on the lining of a vertical circular shaft which was sunk, in clay, to a depth of 72 ft below the ground surface (Chicago).

The unit pressures on lining were determined by means of a hydraulic jack which was used to measure the pressure required to force apart two semicircular segments of the steel lining.

The shaft was sunk from the basement of an old building and the shaft collar was about 13 ft below street level. For a depth of 30 ft below the basement floor the shaft diameter was 15.5 ft. For the remainder of the depth the diameter was 12 ft.

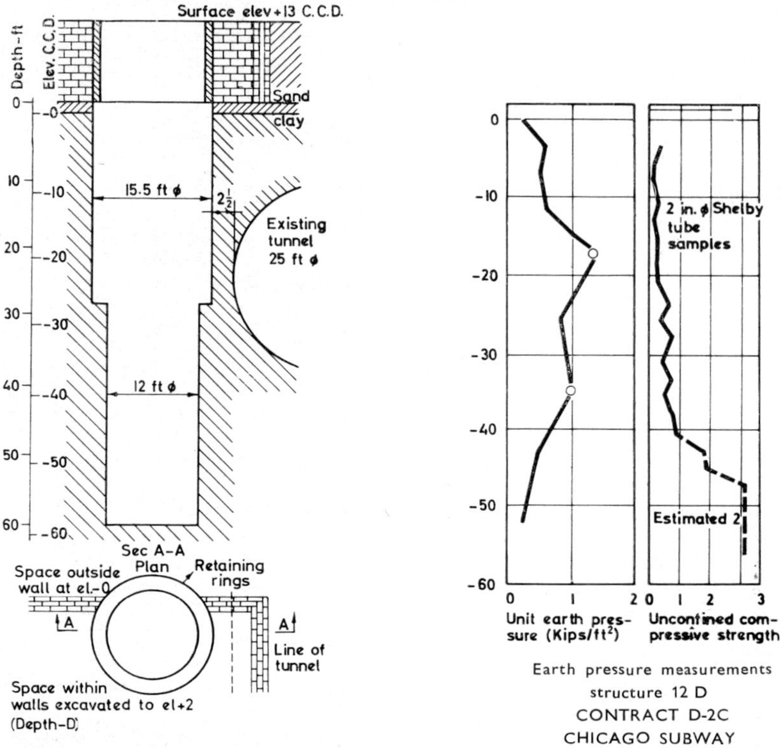

FIG. 17. Measured pressures on the lining of a vertical shaft in plastic clay.[20]

The shaft was lined with rings of 9-in. steel channels placed 2.75 ft center to center which were lagged across with steel plates. Any voids outside the lining were filled by driving wood blocks between the lining and the clay. Excavation was done carefully and voids more than 1 in. deep seldom existed, even before blocking.

Squeeze-in of the clay was probably reduced to a minimum, as evidenced by the fact that the maximum settlement of an adjacent building during excavation was only $\frac{5}{8}$ in.

Figure 17 shows the unit pressures, as measured, versus the depth, as well as the compressive strength of the clay at various depths.

The pressure increased with depth to a maximum of about 1400 psf at a depth of about 32 ft below the street surface. The depth pressure at this depth would be about $(32)\times(115) = 3700$ psf, whereas the pressure on the shaft lining was about 38 per cent of this value.

Below a depth of 32 ft the pressure on the lining decreased with depth, reaching a minimum value near the shaft bottom.

This phenomenon may be due to the transfer of lateral stress to the clay beneath the shaft bottom by shearing forces. Such behavior has been predicted by Terzaghi[7] on theoretical grounds.

EQUILIBRIUM AROUND SINGLE HORIZONTAL OPENINGS

The previous section has examined the simplest case of stress equilibrium that of stresses in a plane section at right angles to the axis of a vertical circular opening, with stresses in the plane section being of the same magnitude in any direction in a horizontal plane.

The case in which we are most interested, however, is that of a horizontal opening corresponding to a tunnel, drift, crosscut, or gallery. We shall examine the theoretical pressure on support required for equilibrium conditions around a horizontal circular opening. For such an opening two factors additional to those previously considered must be taken into account:

(1) The natural horizontal and the natural vertical stresses in the soil or rock around the opening may be of different magnitudes.

(2) In addition to the radial pressure, p_i, which must be exerted by the support to maintain equilibrium, supplementary pressure must be exerted by the support in the vertical direction to stabilize the material loosened by shearing action in the clastic (or in the plastic) zone.

A large proportion of this weight is carried by the natural "arching action" of the sheared material. The effectiveness of this arching action in material above the opening depends on the angle of internal friction of the fractured material. The greater the internal friction the more nearly does a fragmented material tend to support its own weight by "arching action". (See page 527, Chapter 12.)

Equation for Equilibrium around a Horizontal Opening

Labasse[6] has derived a general equation for equilibrium around a circular horizontal opening in a homogeneous material. The derivation follows:

Consider an infinitely small element in the ground around an opening. This element has the polar coordinates ω, r (Fig.18). The forces acting upon the faces of the element are shown. In order for the element to be in equilibrium it is necessary that the projections of these forces on the radius passing through the center of gravity, and the projections on the perpendicular to this radius, add up to zero.

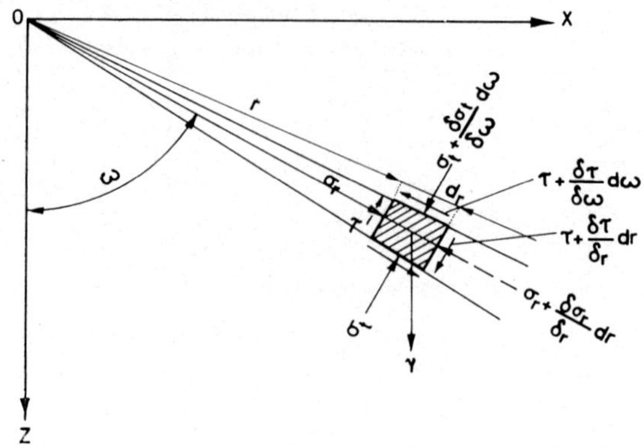

FIG. 18. Stress conditions for the equilibrium of an element in the ground around a circular horizontal opening.[6]

If we represent the specific weight of the rock by γ we may obtain the following relationships:

$$\sigma_r r\, d\omega - \left(\sigma_r + \frac{\delta\sigma_r}{\delta r}\, dr\right)(r+dr)\, d\omega + \sigma_t\, dr \sin\frac{d\omega}{2} +$$

$$\left(\sigma_t + \frac{\delta\sigma_i}{\delta\omega}\, d\omega\right) dr \sin\frac{d\omega}{2} + \tau\, dr \cos\frac{d\omega}{2} -$$

$$\left(\tau + \frac{\delta\tau}{\delta\omega}\, d\omega\right) dr \cos\frac{d\omega}{2} + \gamma \cdot r\, d\omega\, dr \cos\omega = 0 \qquad \text{(a)}$$

$$\sigma_t\, dr \cos\frac{d\omega}{2} - \left(\sigma_t + \frac{\delta\sigma_t}{\delta\omega}\, d\omega\right) dr \cos\frac{d\omega}{2} + \tau r\, d\omega -$$

$$\left(\tau + \frac{\delta\tau}{\delta r}\, dr\right)(r+dr)\, d\omega - \tau\, dr \sin\frac{d\omega}{2} -$$

$$\left(\tau + \frac{\delta\tau}{\delta\omega}\, d\omega\right) dr \sin\frac{d\omega}{2} - r\, d\omega\, dr\, \gamma \sin\omega = 0 \qquad \text{(b)}$$

Neglecting the infinitely small quantities beyond the second order, and taking

$$\cos\frac{d\omega}{2} = 1, \qquad \sin\frac{d\omega}{2} = \frac{d\omega}{2}$$

and dividing the two members of the equation by $dr d\omega$ the relationships become:

$$-\sigma_r - r\frac{\delta\sigma_r}{\delta r} + \sigma_t - \frac{\delta\tau}{\delta\omega} + r\gamma\cos\omega = 0 \qquad \text{(a')}$$

$$-\frac{\delta\sigma_t}{\delta\omega} - 2\tau - \frac{\delta\tau}{\delta r} \cdot r - r\gamma\sin\omega = 0 \qquad \text{(b')}$$

and in deriving (b′) with respect to ω we obtain:

$$\frac{\delta^2\sigma_t}{\delta\omega^2}+2\frac{\delta\tau}{\delta\omega}+\frac{\delta^2\tau}{\delta r\delta\omega}r+r\gamma\cos\omega=0$$

If we limit the problem to an investigation of stresses along the axes the first term may be neglected because it is nil along the axes and the third term may also be neglected because within the limits of variations of r one can neglect the variation of $\delta\tau/\delta\omega$ with respect to r. We can then write:

$$2\frac{\delta\tau}{\delta\omega}+r\gamma\cos\omega=0$$

and substituting this in the equation (a′) we obtain:

$$\sigma_r-\sigma_t+r\frac{\delta\sigma r}{\delta r}+\frac{3}{2}r\gamma\cos\omega=0$$

Along the axes the stresses σ_t and σ_r are the principal stresses. In a pulverulent material they are related by the expression:

$$\frac{\sigma_t}{\sigma_r}=\frac{1+\sin\Phi}{1-\sin\Phi}$$

whence the differential equation:

$$-\frac{2\sin\Phi}{1-\sin\Phi}\sigma_r+r\frac{\delta\sigma_r}{\delta_r}+\frac{3}{2}r\gamma\cos\omega=0 \qquad \text{(c)}$$

If we designate the pressure which the ground exerts on the support at equilibrium by p_i then the integral of equation (c) between the limits (a, r) and (σ_r, p_i) is

$$\sigma_r=p_i\left(\frac{r}{a}\right)^{\frac{2\sin\Phi}{1-\sin\Phi}}+\frac{3}{2}\gamma a\frac{1-\sin\Phi}{3\sin\Phi-1}\left[\left(\frac{r}{a}\right)^{\frac{2\sin\Phi}{1-\sin\Phi}}-\frac{r}{a}\right]\cos\omega \qquad (15)$$

This relationship is valid only for

$$\omega=0,\ 90,\ 180\quad\text{or}\quad 270°.$$

This general equation for equlibrium around a horizontal circular opening may be expressed in the form:

$$p_i=\sigma_{r_b}\left(\frac{a}{b}\right)^{j-1}-\frac{3}{2}\frac{\gamma a}{j-2}\left[1-\left(\frac{a}{b}\right)^{j-2}\right]\cos\omega \qquad (16)$$

$$\text{where } j-1=\frac{2\sin\Phi}{1-\sin\Phi}\quad\text{and}\quad j-2=\frac{3\sin\Phi-1}{1-\sin\Phi}$$

Values for these functions may be read directly from Fig. 10, Chapter 4.

Equation (16) is somewhat cumbersome; however, when each of its two terms is considered separately it is more easily understood.

$$p_{i_A} = \sigma_{r_b}\left(\frac{a}{b}\right)^{j-1} \tag{16a}$$

$$p_{i_B} = -\frac{3}{2}\frac{\gamma a}{(j-2)}\left[1-\left(\frac{a}{b}\right)^{j-2}\right]\cos\omega \tag{16b}$$

and at the elastic boundary:

$$\sigma_{r_b} = \sigma_t\left(\frac{1}{2}\right)(1-\sin\Phi) \tag{10}$$

Fig. 19 shows equilibrium conditions around a horizontal circular opening in ground where the natural (primitive) stresses are isotropic.

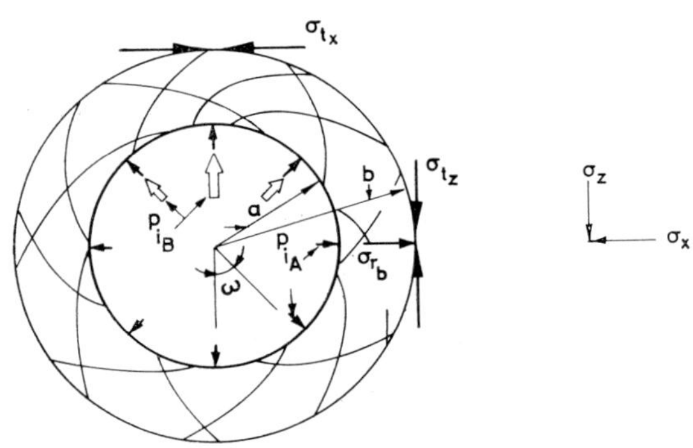

FIG. 19. Support pressures and equilibrium conditions around a horizontal opening.

Equation (16a) is the same equation (11) used previously in this chapter to determine stress conditions in horizontal planes around vertical circular shafts. This equation gives the radial support pressure which is required to neutralize the natural (primitive) ground stresses and to establish equilibrium at the boundary of the clastic zone. This pressure is represented in Fig. 19 by p_{i_A}.

Equation (16b) gives the magnitudes of the additional pressure which must be exerted by the support to stabilize the material in the clastic (or plastic) zone above the opening. This pressure is represented in Fig. 19 by the double arrows marked p_{i_B}. It is maximum at the crown of the opening and decreases to zero at the sides. This is in contrast to p_{i_A} which is uniform around the perimeter of the opening. The total pressure on support at any point is the algebraic sum of p_{i_A} and p_{i_B}.

Effect of Variation of ω

The pressure varies with the central angle, ω, and becomes maximum when $\omega = 180°$. The central angle, ω, is measured upward from the vertical and the value of $\cos\omega$ varies from $(+1)$ when $\omega = 0°$ to (-1) when $\omega = 180°$. The value $\cos\omega$

is zero when $\omega = 90°$ or $270°$. Therefore equation (16b) is negative and is subtracted from equation (16a) in the two lower quadrants, and is positive and added to equation (16a) in the upper quadrants. At the ribs, where $\omega = 90$ or $270°$ equation (16b) is zero.

Maximum negative value of equation (16b) occurs when $\omega = 0°$ and the maximum positive value when $\omega = 180°$; that is vertically overhead, as would be expected.

From the foregoing general equation the value of the theoretical radial pressure required for equilibrium may be computed at any point on the circumference by using the appropriate value of the central angle ω. However, we actually need to know only the maximum pressure, which occurs vertically overhead, and the pressures at the ribs (ends of the horizontal axis).

Examination of equation (16b) shows that when $b = a$, that is, when no shearing has taken place and the boundary of the elastic zone coincides with the boundary of the opening, p_{i_B} = zero, and there is no fractured material which requires stabilization above the opening.

Effect of Natural (Primitive) Earth Stresses

In applying the foregoing equations to horizontal openings we are no longer dealing, as we did in the example of the vertical circular shaft, with a material initially stressed equally in every direction in a horizontal plane. The initial vertical stress, σ_z, in materials in unconsolidated sedimentary deposits may be two to three times the magnitude of the initial horizontal stress, σ_x. In intact igneous and metamorphic rocks, on the other hand, the natural horizontal stress, σ_x, may be several times the natural vertical stress σ_z.

Shape of the Fractured Zone

Since the stresses may be unequal in the vertical and in the horizontal directions fracturing will extend for different distances in the vertical and in the horizontal directions. In addition, the weight of material loosened overhead exerts additional pressure on support and tends to cause shearing to extend further above the opening.

For purposes of numerical examples one of two idealized conditions may be assumed. We may assume that either:

(1) the sheared zone maintains a circular cross-section and compute the corresponding pressures on that basis; or

(2) we may assume that the initial stress field is isotropic and determine the extent of shearing and the resulting shape of the sheared zone.

The pressure on support is a function of the tangential stress at the boundary, which in turn is a function of the shape of the opening. The relationship between shape of an opening and the boundary stresses is illustrated in Fig. 11.

Example Calculations — Equilibrium in Pulverulent Materials

For a numerical example we will examine stress equilibrium around a horizontal circular opening in cohesionless granular material. Since stress relationships in such material are not isotropic (that is $\sigma_z > \sigma_x$) we shall make the simplifying assumption that the clastic zone maintains a circular shape, and therefore the boundary of the elastic zone is circular in cross-section.

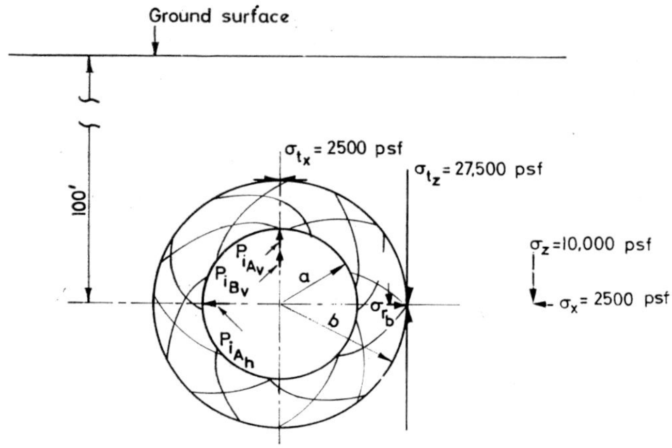

FIG. 20. Example showing stresses in the ground around an opening in cohesionless sand.

For purposes of an example calculation the following conditions are assumed (see Fig. 20):

γ = 100 pcf;
a = 5 ft;
b = 10 ft;
z = 100 ft;
$\sigma_z = 100 \times 100 = 10{,}000$ psf;
$\Phi = 37°$;

and from Fig. 10, chapter 4,

$$j = 4, \quad j-1 = 3, \quad j-2 = 2 \quad \text{and} \quad i = \frac{1}{4}$$

$$\sigma_x = \sigma_y = i\sigma_z = 10{,}000 \times \frac{1}{4} = 2500 \text{ psf}$$

c = cohesion = 0

At a circular elastic boundary the tangential stresses at ends of axes are (from equations (3) and (4)):

$$\sigma_{t_z} = 36_z - \sigma_x = (3)\,(10{,}000) - 2500 = 27{,}500 \text{ psf}$$
$$\sigma_{t_x} = 3\sigma_x - \sigma_z = (3)\,(2500) - 10{,}000 = -2500 \text{ psf}$$

The computed value of σ_{t_x} is tensile but tensile stress cannot exist in a cohesionless material. The lateral stress in this material cannot be less than that given by the equation

$$\sigma_x = i\sigma_z = \frac{1}{4}\sigma_z = +2500 \text{ psf}$$

Computed Support Pressures (Assuming $a = 5$ ft and $b = 10$ ft)

Vertical Pressure on Support

The vertical pressure on supports is partially due to the radial pressure required to inhibit further extension of the clastic zone (equation (16a)) and partially due to weight of sheared material in the clastic zone (equation 16b)

$$p_{i_{A,v}} = \sigma_{r_{b,\text{ vert.}}}\left(\frac{a}{b}\right)^3 \tag{16a}$$

$$\sigma_{r_{b,\text{ vert.}}} = \sigma_{t_x}\left(\frac{1}{2}\right)(1-\sin\Phi) \tag{10}$$

$$\sigma_{r_{b,\text{ vert.}}} = (2500)\,(0.20) = 500 \text{ psf}$$

$$\text{If } b = 10 \text{ ft}$$

$$p_{i_{A,v}} = 500\left(\frac{5}{10}\right)^3 = 63 \text{ psf}$$

This is the support pressure required to stabilize the clastic–elastic boundary.

Designating the vertical support pressure required to stabilize the material within the clastic zone by $P_{i_{B,v}}$ then

$$p_{i_{B,v}} = \frac{3}{2}\frac{\gamma a}{(j-2)}\left[1-\left(\frac{a}{b}\right)^{j-2}\right] \tag{16b}$$

$$= \frac{3}{2}(100)\,(5)\left(\frac{1}{2}\right)\left[1-\left(\frac{5}{10}\right)^2\right]$$

$$= (375)\left(\frac{3}{4}\right) = 280 \text{ psf}$$

Total unit vertical pressure on support at center of the arch is:

$$p_{i_v} = p_{i_{A,v}} + p_{i_{B,v}} = 63+280 = 343 \text{ psf}$$

Horizontal Pressure on Support

$$\sigma_{r_{b,\text{ horiz.}}} = \sigma_{t_z}\left(\frac{1}{2}\right)(1-\sin\Phi) = \frac{1}{5}\sigma_{t_z} = 27{,}500\left(\frac{1}{5}\right) = 5500 \text{ psf}$$

$$p_{i_{A\,h}} = 5500\left(\frac{a}{b}\right)^3 \qquad \text{(from equation (16a))}$$

If $b = 10$ ft

$$p_{i_{A,h}} = 5500\left(\frac{5}{10}\right)^3 = 690 \text{ psf}$$

This is the horizontal pressure on support at the horizontal axis of opening.

Computed Support Pressure assuming that the Clastic Zone Extends to the Ground Surface

The foregoing example assumed that the clastic zone extended only to a distance of one radius from the perimeter of the opening. For an opening at shallow depth it is possible that the clastic zone will extend to the ground surface.

We shall assume that this occurs and, using the conditions of the previous problem, compute the resulting pressure on supports.

Horizontal Pressure

Then $b = 100'$ and $a/b = 5/100 = 1/20$

$$p_{i_{A,h}} = \sigma_{r_b}\left(\frac{a}{b}\right)^{j-1} \tag{16a}$$

$$= 5500\left(\frac{1}{20}\right)^3 = \frac{5500}{8000} = 0.7 \text{ psf}$$

Thus the horizontal pressure on support is negligible.

Vertical Pressure

The maximum vertical pressure is found by computing pressures due to both (16a) and (16b) and summing them.

When the clastic zone extends to the ground surface, and to a considerable distance laterally, the vertical support pressure, $p_{i_{A,v}}$, as computed from equation (16a) becomes so small that it may be neglected.

Vertical suppport pressure is then that required to stabilize the loosened material in the clastic zone above the opening and may be computed from equation (16b).

Taking $a = 5'$ and $b = 100'$

$$p_{i_{B,v}} = \frac{3}{2}\frac{\gamma a}{(j-2)}\left[1-\left(\frac{a}{b}\right)^{j-2}\right]$$

$$= \frac{3}{2}(100)(5)\left(\frac{1}{2}\right)\left[1-\left(\frac{1}{20}\right)^2\right] = 374 \text{ psf} \tag{16b}$$

For the given conditions, and assuming that the clastic zone extends to the ground surface, the vertical pressure on support will approach a maximum limiting value of 375 psf regardless of the depth of the opening below the ground surface.

For ratios of b/a greater than 3 the term $\left[1-\left(\frac{a}{b}\right)^{j-2}\right]$ may be neglected and equation (16b) becomes

$$p_{i_{c,v}} = \left(\frac{3}{2}\right)\frac{\gamma a}{(j-2)} \tag{16c}$$

The pressure computed from (16c) is the vertical pressure at the "keystone" or top of the arch, which is required to stabilize the sheared material above the opening. This pressure decreases away from the center of the arch until it becomes zero at the ends of the horizontal axis.

Equilibrium in Cohesive Materials

If the material in the foregoing example calculation possesses some initial cohesion then the tangential stresses σ_t are reduced and σ_{r_b} required for equilibrium is reduced proportionately.

If we assume that the material contained enough clay to give it a cohesion of about 500 psf, this corresponds to an unconfined compressive strength of about 2000 psf (for $\Phi = 37°$). Taking the same conditions as in the previous example, $\sigma_{t_z} = 27{,}500$ psf

$$\sigma_{r_b} = (\sigma_t - Q)\left(\frac{1}{2}\right)(1-\sin\Phi) = (27{,}500-2000)\left(\frac{1}{5}\right) = 5100 \text{ psf}$$

Thus σ_{r_b} is reduced from 5500 to 5100 psf and p_i is reduced about 7 per cent, which is a relatively unimportant reduction in pressure.

Under the conditions of the previous problem the material would require a compressive strength $p_c = 27{,}500$ psf, or 190 psi, in order to stand unsupported. Such strength might be found in a hard clay, or in a very soft shale, or in a very slightly consolidated sand.

Pressures on Supports for Openings Located at Depth

The foregoing examples have considered pressures on supports for openings at shallow depths in sedimentary deposits. Next we shall consider the structural behavior of rock around openings located at greater depths, that is, at depths of several thousand feet, or more.

In previous examples we found that a pulverulent material will exert pressure on the lining of any sub-surface opening regardless of the depth at which it is located. Materials possessing small cohesive strength will stand unsupported at shallow depths but require support at greater depths.

Sound rocks, even of the weaker varieties, possess sufficient cohesion to support openings, even at considerable depths. For example, nearly all sound igneous and metamorphic rocks have compressive strengths in excess of 7000 or 8000 psi and many of them exceed 15,0000–20,000 psi. Since the tangential compressive stresses at boundary of openings in elastic rocks are equal to $2\sigma_z$ (under natural isotropic stress conditions), it is apparent that shear fracturing of rock at the boundary of an ideal circular opening will occur when $2\sigma_z$ exceeds Q. Thus for example in an igneous or metamorphic rock with a compressive strength, Q, of 12,000 psi the required magnitude of $\sigma_z \geqslant 6000$ psi and $z \geqslant 6000/1.2 = 5000$ ft. That is, the tangential stress at the boundary of a circular opening in this particular rock will exceed the strength of the rock when the opening is located at a depth of 5000 ft or more.

Example Calculations

For a numerical example we shall assume an opening excavated at a depth of 6000 ft in a massive unfractured rock which possesses the following physical characteristics:

Compressive strength $Q = 12{,}000$ psi

Angle of internal friction, $\Phi = 55°$ (The relatively high value for the internal friction is assumed to compensate for the interlocking characteristics of the slabs formed when a massive rock fractures by shearing.)

From Fig. 10, Chapter 4,

$$j = 10$$

At fracture the behavior of the rock under triaxial stress is defined by the following equation:

$$\sigma_1 = Q + j\sigma_3 \qquad \text{(Eq. 7, Ch. 4)}$$

At a depth of 6000 ft (assuming isotropic stress)

$$\sigma_z = (6000)\,(1.2) = 7200 \text{ psi}$$
$$\sigma_t = 2\sigma_z = 14{,}400 \text{ psi}$$

At the clastic–elastic boundary the following stress equilibrium conditions exist:

$$\sigma_{r_b} = \frac{1}{2}(1 - \sin \Phi)\sigma_{t_e} \tag{10}$$

and

$$\sigma_{r_b} = \frac{1}{2}(1 - 0.82)\sigma_{t_e} = 0.09\; t_e$$

but t_e represents only that portion of the tangential boundary stress which exceeds the compressive strength, Q, of the rock. Hence

$$t_e = 14{,}400 - 12{,}000 = 2400 \text{ psi}$$
$$\sigma_{r_b} = 0.09\; t_e = (0.09)\,(2400) = 218 \text{ psi}$$

This indicates that a pressure of 218 psi would have to applied to the rock surface to prevent shear fracturing. However, it is not practicable to install of support sufficiently strong to substain a load of 218 psi, even if it could be installed quickly enough to prevent fracturing. Therefore if we install support capable of resisting moderate side pressure of, say, 1 psi or 144 psf, the shearing must extend far enough into the rock to reduce p_i to this value.

$$p_{i_A} = \sigma_{r_b}\left(\frac{a}{b}\right)^{j-1} \tag{16a}$$

Using $\Phi = 55°$ for the fractured material

$$p_{i_A} = 218\left(\frac{a}{b}\right)^9 = 1 \text{ psi}$$

$$\frac{b}{a} = (218)^{\frac{1}{9}} = 1.82$$

$$b = 1.82a$$

Thus because of the high coefficient of internal friction, tan Φ, of the fractured material, shearing at the ribs extends only about 0.82 a or about 4.1 ft from the boundary in order to reduce pressure on supports to acceptable values.

Pressure additional to the 144 psf would be exerted on overhead support due to weight of rock in the clastic zone above the opening. The value of this additional pressure at the top of the arch is given by equation (16b).

Assuming an opening with a radius, a, of 5 ft then

$$p_{i_{B,v}} = \frac{3}{2}\frac{\gamma a}{(j-2)}\left[1-\left(\frac{a}{b}\right)^{j-2}\right] \tag{16b}$$

$$= \frac{3}{2}(173)(5)(0.125)\left[1-\left(\frac{1}{1.82}\right)^8\right]$$

$$= 163\left[1-\frac{1}{130}\right] = 163\,[0.99] = 161 \text{ psf}$$

The additional vertical pressure, 161 psf, must be added to the pressure, $p_{i_A} =$ $= 144$ psf, to obtain total unit pressure at the top of the arch,

$$p_{i_v} = 144+161 = 305 \text{ psf}$$

Thus if support is installed to inhibit shearing so that a circular fracture zone is formed the pressure on supports for conditions postulated will be as shown in Fig. 21 a.

The horizontal pressure on supports may be reduced by allowing shearing to extend further into the rock. For example, if the clastic zone were allowed to extend to a distance $b = 3a = 15$ ft, then

$$p_{i_A} = 218\left(\frac{5}{15}\right)^9 = 218\left(\frac{1}{19{,}700}\right) = 0.011 \text{ psi} = 2 \text{ psf}$$

and at the crown the weight of sheared rock would require a stabilizing pressure:

$$p_{i_{B,v}} = \frac{3}{2}(173)(5)(0.125)[1] = 163 \text{ psf}$$

and

$$p_{i_v} = 2+163 = 165 \text{ psf}$$

This is shown in Fig. 21(b).

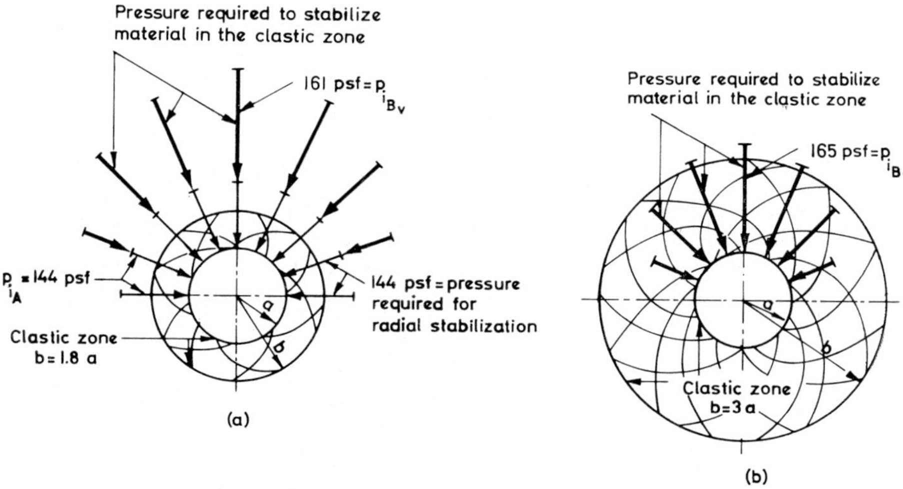

FIG. 21. (a) Pressures on supports for $b = 1.8a$. (b) Pressures on supports for $b = 3.0a$.

Configuration of The Elastic Boundary

Around a supported opening which is subjected to natural (primitive) isotropic stress the clastic zone extends further from the opening in the vertical direction than in the horizontal direction. This occurs because the radial pressure exerted by support is partially nullified in the vertical direction by the weight of the fractured material, as computed from equation (16b).

In the example used here the value of $p_{i_{B,v}}$ will always exceed the value of $p_{i,\text{hor}}$ for all practical values of $p_{i,\text{hor}}$. Thus support must be provided which is capable of resisting a greater load in the vertical direction than in the horizontal direction. It is to be noted that, for a given set of values for γ, a, and Φ, the value of equation (16b) approaches a maximum limiting value for relatively small extensions of the clastic zone. This maximum value is

$$p_{i_{B,v}} = \left(\frac{3}{2}\right)\frac{\gamma a}{j-2} \tag{16c}$$

It may also be noted that although the clastic zone is actually elliptical in shape, with the long axis vertical, it is desirable for the purpose of simplifying the calculations of support pressure to assume that the clastic zone is circular. This assumption causes no appreciable error in results because the value of $p_{i_{B,v}}$ so rapidly approaches its limiting value for minor extensions of the clastic zone.

Decrease of Opening Diameter due to Extension of the Clastic (or Plastic) Zone

In the foregoing examples we have spoken of allowing the shearing, and the clastic zone, to extend into the material around the opening in order to reduce the pressure on the supports. In most cases where the internal boundary of an opening is stressed beyond its elastic limit shearing will have extended some distance from the boundary before support can be installed. Therefore fracturing could not be entirely prevented even if it were desirable to do so. At the other extreme it might seem that it would be desirable to allow the clastic zone to extend to the maximum distance possible in order to reduce support pressure to a minimum. However, certain associated phenomena make such a procedure impractical.

(1) In materials which possess initial cohesion the extension of the clastic zone is a relatively slow process which continues over a period of days, weeks, or even months, depending upon the "viscosity" of the material.

(2) Normally, if fracturing is extensive, support must be installed soon after excavation in order to prevent the opening from filling with fragments. In cohesionless materials support must obviously be installed simultaneously with excavation.

(3) The extension of the clastic zone into the material around the opening is accompanied by a volume increase; that is, the sheared material occupies more space than the material in its initial state. Therefore, it is necessary to leave sufficient room between the support and the boundary of the opening for the desired amount of expansion, or it is necessary to install support which will yield sufficiently to allow the desired amount of expansion.

Thus in order to obtain an opening which will have a given dimension after stabilization it is necessary to make the initial excavation somewhat larger than its final dimension. The most economical system must be determined by balancing the cost of extra excavation, which will allow squeeze in and pressure reduction, against the cost of heavier support which will prevent so much "squeeze" but will be required to support a greater load.

The theoretical reduction in diameter required to reduce the ground pressure around an opening to allowable values may be calculated as shown on the following pages (see Fig. 22).

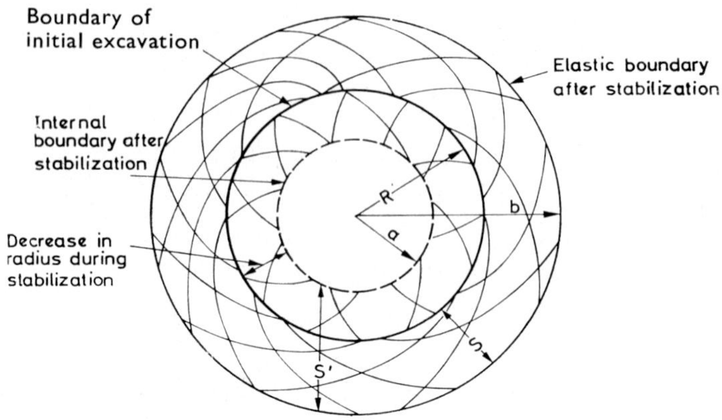

FIG. 22. Decrease in internal diameter of an opening with the extension of the clastic zone.

Equations for Calculating Decrease in Opening Diameter

The equation for decrease in radius of opening during extension of the clastic zone is derived as follows:

Vol. of $(b) = \pi b^2$
Vol. of $(S) = \pi(b^2 - R^2)$

After shearing, Vol. $S' = (\text{Vol. } S)(K) = \pi(b^2 - R^2)K$

$$\begin{aligned}\text{Vol. }(a) &= \text{Vol. }(b) - \text{Vol. }(S') \\ &= \pi b^2 - K\pi(b^2 - R^2) \\ &= \pi[b^2 - K(b^2 - R^2)]\end{aligned}$$

$$(a)^2 = \frac{\text{Vol. }(a)}{\pi} = b^2 - K(b^2 - R^2) \tag{17}$$

Labasse[(6)] gives values for K in weak shales as between 1.12 and 1.15. He suggests using a value of $K=1.1$ as a reasonable average coefficient of volume increase.

The following notation is used in these calculations (see also Fig. 22).

a = final radius of excavation;
R = initial radius of excavation;
b = radius of clastic zone;
K = coefficient of swelling.

Using a value of $K=1.1$[(6)] for the coefficient of volume increase, we shall apply equation (17) to determine decrease in diameter of openings used in previous examples.

In the first example the clastic zone was allowed to extend to a distance of $1.82a$ from the center of the opening in order to reduce lateral pressure on support to

144 psf. Initial radius of the opening was $R = 5$ ft. It is desired to find a, the radius after stabilization, and also the volume of material which will squeeze inside the circle marking the initial boundary of the opening.

$$(a)^2 = b^2 - K(b^2 - R^2) \tag{17}$$

$$(a)^2 = (9.1)^2 - 1.1(\overline{9.1^2 - 5^2})$$

$$= 83 - 64 = 19$$

$$a = (19)^{1/2} = 4.35 \text{ ft}$$

$$R - a = 5.00 - 4.35 = 0.65 \text{ ft}$$

The decrease in diameter of the initial opening would be 1.30 ft, and the diameter after stabilization would be 8.70 ft. This would still leave a reasonably large opening available as a travel-way and haulage-way.

The volume of material squeezed inside the original opening boundary would be 18.9 ft^3/ft or a volume equal to about 24 per cent of that originally excavated.

If the clastic zone had been allowed to extend to a distance, for example, of $3\,R = 15$ ft in order to further reduce the pressure on support then the reduction in diameter of opening would be:

$$(a)^2 = (15)^2 - 1.1(\overline{15^2} - 5^2)$$

$$= 225 - 220 = 5$$

$$a = (5)^{1/2} = 2.23 \text{ ft}$$

Thus the diameter decreases from 10 to 4.46 ft and the volume of material squeezed inside the original boundary is 62.8 ft^3/ft, or 80 per cent of the volume of the original excavation. Such conditions would, of course, be intolerable and the practical solution would be to install support sufficiently strong to stabilize the opening after the clastic zone had extended to a distance of about $2R$ from the center of the opening. Under such conditions the diameter of opening would become:

$$(a)^2 = (10)^2 - 1.1(\overline{10^2} - 5^2) = 100 - 82.5 = 17.5$$

$$a = (17.5)^{1/2} = 4.17'$$

The diameter of the stabilized opening would be 8.34 ft and the volume of material squeezed in would be 23.6 ft^3/ft, which is equal to about 30 per cent of the volume of the original excavation.

With the clastic zone allowed to extend to a distance of $2R$ the horizontal pressure on the support of the previous example would become:

$$p_{i_A} = 218\left(\frac{1}{2}\right)^9 = 218\left(\frac{1}{512}\right) = 0.42 \text{ psi} = 67 \text{ psf}$$

and vertical pressure at center of the arch would be

$$p_{i_v} = 163+61 = 224 \text{ psf}$$

This would probably represent the best compromise between the cost of required support and the cost of additional excavation.

Coefficient of Volume Increase

The foregoing examples have used a value of $K=1.1$. This figure was deduced by Labasse[6] from observations in shales. It is probable that this value may be somewhat less in other rocks, and in granular materials.

Coefficient of Internal Friction

In the preceding numerical examples it was assumed that the coefficient of internal friction, tan Φ, would be the same for the sheared material in the clastic zone as for the material prior to shearing.

For material which possesses considerable internal cohesion, such as a sound rock, the value assigned to tan Φ in the clastic zone should be larger than that for tan Φ in intact material. This is caused by the tendency of shear stresses to fracture the rock into interlocking slabs which possess some "beam action" as well as "arching" action. Even though it is fractured, this material still may exert considerable internal radial pressure, p_i, which acts against the boundary of the unfractured zone.Thus the actual pressure on support may be considerably less than the theoretical pressure computed in the examples. In many cases the rock around the periphery of such an opening, although closely sheared, may consist of slabs which are sufficiently interlocked so that no support is required.

Such conditions may be compensated for in computing pressure on supports by assigning higher values for tan Φ in the fractured material. Appropriate values or e ach case can only be determined by observations in each type of rock.

Application of Yielding Supports[29]

Figure 23 shows a hypothetical tunnel cross-section and the stresses on an element A in the ground outside the perimeter of the excavation. From the equilibrium conditions for such an element an equation is derived which expresses the relationship between diameter of the tunnel, the angle of internal friction of the fractured rock, and the radius of the fractured zone.

The amount by which the ground squeezes in is determined by the distance to which adjustments extend into the ground outside the tunnel, and also by the percentage of volume increase which accompanies stress readjustments in the ground.

In turn, the distance to which stress readjustments extend into the ground outside the tunnel is determined by the angle of internal friction of the ground, by the external natural ground stress, and by the pressure of tunnel support on the ground.

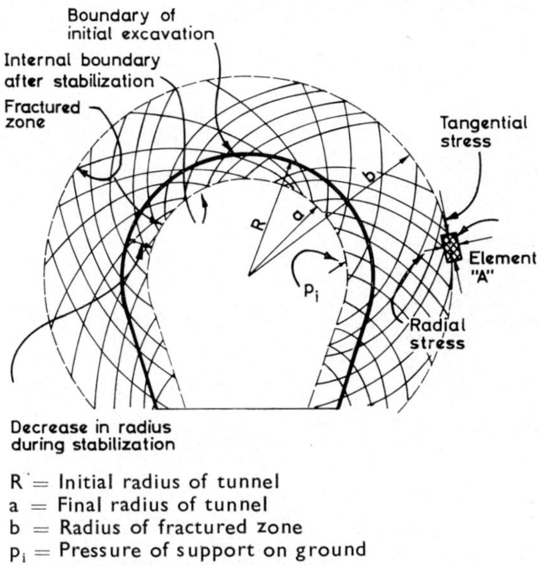

FIG. 23. Squeeze-in of the ground and corresponding equilibrium conditions around a tunnel in overstressed ground.[29]

Figure 24 is a graph of the equilibrium equation for various values of the angle of internal friction of the ground. For an example of its use, assume that the tunnel is driven through a zone of fault breccia at a depth of 1000 ft. The natural ground pressure at this depth may reasonably be assumed to be about 1000 psi, and the angle of internal friction of the fault breccia to be about 40°. Starting at a natural ground stress of 1000 psi in Fig. 24 (Point 1) we proceed horizontally on the graph to the line representing the angle of internal friction ($\Phi = 40°$) at (Point 2). Thence we proceed downward to intersect the R/a curve (Point 3) and thence horizontally to the R/a scale (Point 4) where we see that $R/a = 1.15$. Thus the initial radius of the excavation must be made 1.15 times larger than the required final radius of the excavation in order to allow for squeeze-in of the yielding supports.

Thus if we need a tunnel with a final clear radius of 5 ft, that is, a diameter of 10 ft, we will have to excavate it to an initial diameter of $10 \times 1.15 = 11.5$ ft to allow the ground to squeeze in while equilibrium is being established.

The graphs of Fig. 24 were drawn on the assumption that the size and spacing of the yielding steel arches would be such as to enable them to support a uniform external radial ground pressure of 20 psi (about $1\frac{1}{2}$ tons per ft^2) all around the periphery of the tunnel.

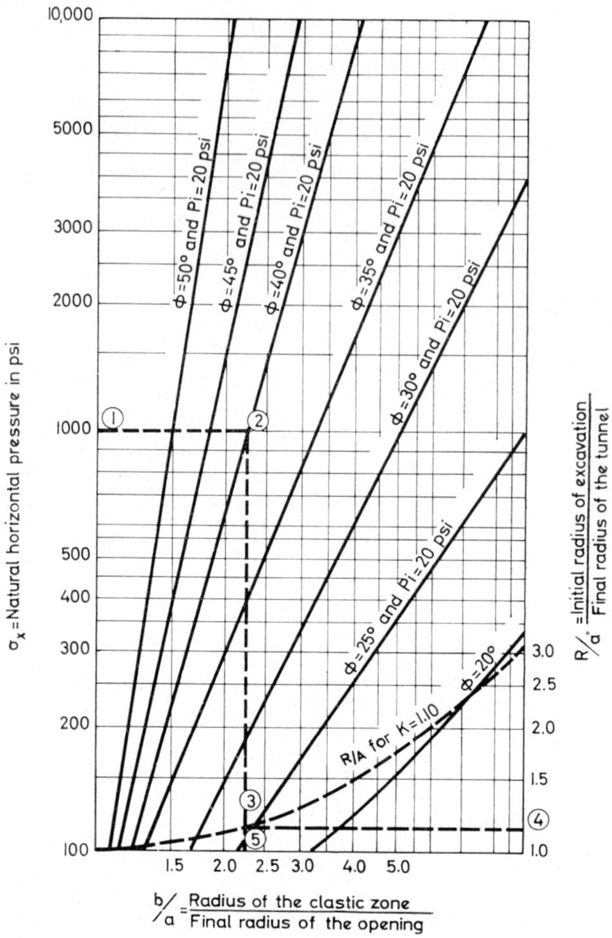

FIG. 24. Graph of the equilibrium equation for various values of the internal friction. (See the text for explanation of its use)[9, 29].

Model Tests[30]

Figure 25 shows the amount of fracturing and the closure produced in models constructed to represent circular openings in stratified rock. Support was simulated by a rubber cylinder which could be slipped into the hole and filled with water under pressure to simulate the desired support resistance. Results of the tests are summarized as follows:

Figure 25(a): The model was subjected to vertical pressure of 312 psi and horizontal pressure of 213 psi. No support was provided in the circular opening. The "rock" fractured and squeezed in until the cross-sectional area of the opening was reduced to 39 per cent of its original size.

Figure 25(b): The model was subjected to a vertical pressure of 312 psi and a horizontal pressure of 305 psi. Support resistance equivalent to a mine support pressure of about 11.3 psi was applied to the inside surface of the opening. The rock fractured and squeezed but the opening retained 66 per cent of its original cross-section area.

Figure 25(c): The model was subjected to a vertical pressure of 312 psi and to a horizontal pressure of 220 psi. Support resistance equivalent to a mine support pressure of about 50 psi was applied to the inside surface of the opening. Very little fracture occurred and the opening retained 82 per cent of its original cross-section area.

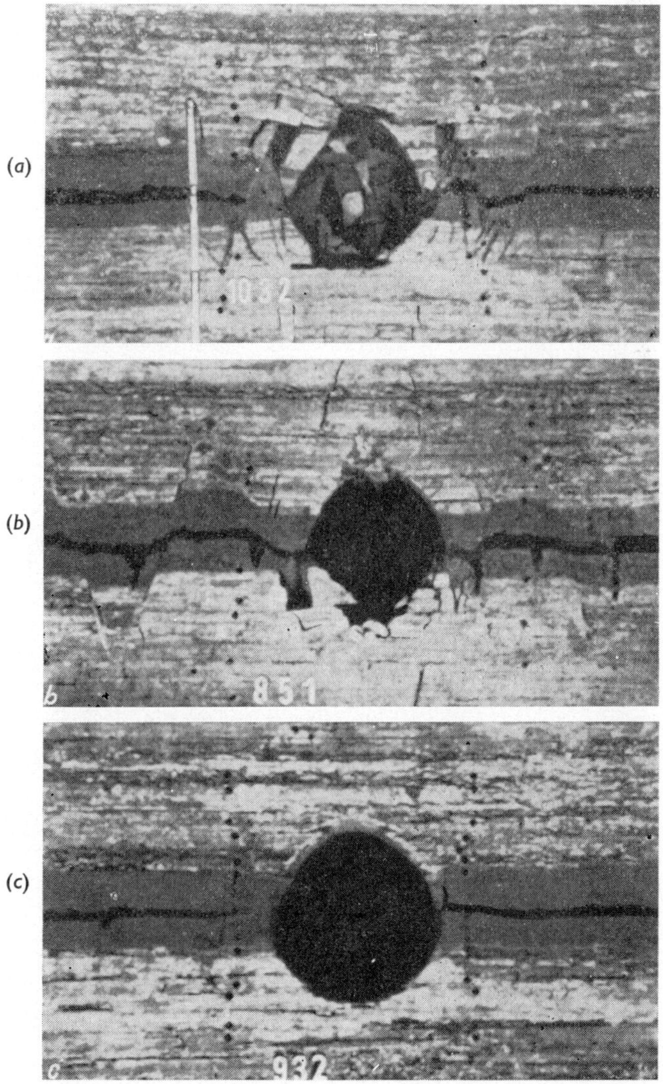

FIG 25. Model tests showing the deformations of circular openings as affected by the resistance of supports installed in the openings.(30)

Pressures on Lining — Horizontal Openings in Plastic Clay

The foregoing examples have considered the pressure on linings of openings in rock involving elastic equilibrium and in pulverulent materials involving clastic–elastic equilibrium.

The third class to be considered are the plastic materials, such as the saturated or partially saturated clays. Whereas the other types of materials considered previously herein possessed coefficient of internal friction generally in excess of tan 30°, the *partially saturated* clays ordinarily posses values of tan Φ ranging from about tan 10 to 25° with c values in the range from 1000 to 2000 psf, although these limits are only rough and there are many exceptions to them.

The *fully saturated* clays are generally considered to possess no internal friction; that is, tan Φ = zero. The softer of the fully saturated clays have c values of about 200–1000 psf, while the stiffer varieties may possess c values in the range from 1000 to 2000 psf.

The softer clays have the ability to retain their cohesion even after they have been subjected to shearing in the plastic zone, and it is necessary to take this property into account in equations for computing pressure on supports.

The clays without internal friction derive their self-supporting ability only from cohesion and such materials have the property of yielding slowly; that is, they somewhat resemble a thick viscous liquid which flows slowly when subjected to unbalanced stresses. This has been made apparent in tunnels where measurements of pressure on lining have been made over a period of years. Housel[13] describes the pressures measured on the lining of the Central Avenue Tunnel in Detroit, which was driven in saturated clay with a cohesion of about 200–300 psf and tan Φ = zero.

In this case readjustment of stresses occurred immediately after release of the compressed air and this was followed by an increase of vertical pressure which continued for a period of 5 years. This increase was more rapid at first, the rate of increase decreasing as the pressure on lining approached the depth pressure. Measurements were continued for another 5 years without detecting any further significant changes in pressure on lining.

In the case of the Chicago Subway Tunnels test section[14], vertical pressure on the lining began to increase immediately after the installation of the lining and within a period of from 1 to 3 months had reached essentially the full depth pressure.

The property of yielding with time has been noted in the softer rocks, and has been measured in the laboratory by Phillips[15]. It is especially noticeable in those rocks which consist of consolidated or partially consolidated clays, or which contain large proportions of clay. One might deduce that the rate of yield depends upon whether the rock derives its strength principally from internal friction, or from cohesion, since it is most apparent in those rocks with the least internal friction.

Equations for Calculation of Pressure on Lining

The equation for unit vertical pressure on the arch of a circular opening in plastic clay has been derived by Caquot and Kerisel[8].

$$p_{i_v} = \frac{\gamma z}{j-2}\left[\frac{a}{z}-\left(\frac{a}{z}\right)^{j-T}-T\left[1-\left(\frac{a}{z}\right)^{j-1}\right]\right] \tag{18}$$

where
$$T = \frac{c}{\tan \Phi}$$

Inspection of equation (18) shows that for $\Phi = 20°$, $(j-2) = 0$ and the value of the first term of the equation becomes indeterminate. When $\Phi = 20°$ it can be shown that the term approaches[9]

$$a\gamma \ln_e \frac{z}{a} \quad \text{or} \quad a\gamma\, 2.3 \log_{10} \frac{z}{a}$$

thus equation (18) becomes (when $\Phi = 20°$)

$$p_i = a\gamma\, 2.3 \log_{10} \frac{z}{a} - T\left[1-\left(\frac{a}{z}\right)^{j-1}\right] \tag{19}$$

Another discontinuity in equation (18) appears when $\Phi = 0°$. In this instance $(j-1) = 0$ and the second term becomes indeterminate. It can be shown that the term approaches (Caquot and Kerisel)[8]

$$4.6\, c \log_{10} \frac{a}{z}$$

thus equation (18) becomes (when $\Phi = 0°$)

$$p_i = a\gamma\left[\frac{z}{a}-1\right] - 4.6\, c \log_{10} \frac{z}{a} \tag{20}$$

Measured Pressures on Lining — Tunnels in Clay ($\Phi > 0°$)

Garrison Dam Tunnel — Test Station. The results of measurements, by personnel of the U.S. Army, Corps of Engineers, on loads on supports, and on lining, of a large diameter tunnel test section at Garrison Dam have recently been reported by Lane.[10]

This tunnel test section was 36 ft in diameter (excavated section). Temporary support was by means of steel ribs and lagging and permanent support was by concrete lining, 3 ft thick.

The tunnel was excavated in Fort Union clay–shale. The physical characteristics of this material showed considerable variation between individual samples but for the design of excavated slopes a cohesion, c, of 1400 psf was used, and a coefficient of internal friction of tan 20°. These values were determined as average from laboratory tests, as well as from an inspection of natural slopes occurring in the formation.[(11)]

It is of interest to compute the theoretical value of vertical pressure on support by means of equation (19). The distance from surface to test tunnel center was about 120 ft.[(12)] Taking the density as 110 pcf, we have for use in the equation:

$$c = 1400 \text{ psf}, \qquad \Phi = 20°, \qquad j = 2$$

$$T = \frac{1400}{0.36} = 3900 \text{ psf}, \qquad \gamma = 110 \text{ pcf}$$

$$a = 18 \text{ ft}, \qquad z = 120 \text{ ft}$$

$$p_i = a\gamma 2.3 \log \frac{z}{a} - T\left[1 - \left(\frac{a}{z}\right)^{j-1}\right] \tag{19}$$

$$= (18)(110)(2.3)\left(\log \frac{120}{18}\right) - 3900\,[1 - 0.15]$$

$$= 3780 - 3300 = 480 \text{ psf}$$

The pressure computed by this equation is somewhat less than the measured values of the average vertical pressure which ranged from about 600 to 2000 psf for the 36-ft diameter tunnel test section.

For adjacent tunnels, 35 ft in diameter, the measured average vertical pressure on ribs varied from about 900 to 3100 psf.

Maximum Measured Pressure. For each tunnel the *maximum* measured vertical pressure on supports was equal to about 15 per cent of the depth pressure due to the weight of the overburden. The natural depth pressure, σ_z, due to weight of overburden ranged from about 12,000 to 22,000 psf, varying with the depth of the cover over the tunnels.

The most difficult part of predicting pressure on tunnel supports lies in determining the proper values to be employed for cohesion and for internal friction. Generally, the strength characteristics as determined from tests on specimens in the laboratory will be more favorable than those existing in the soil mass because many pre-existing natural planes of weakness are so widely spaced that they reduce the strength of a soil mass around a large opening, but their effects are not apparent when relatively small specimens are tested in the laboratory.

In a subsequent section another method will be presented for computing pressure on supports of tunnels located relatively near the ground surface; that is, located at depths such that the z/a ratio is less than about 8 or 10.

Pressure on Lining. Several instances are cited in the literature of pressure measurements on the lining of tunnels constructed in saturated clay. These clays have little or no internal friction; that is, $\tan \Phi = 0$, approximately.

Housel[13] gives results of measurements of pressure on the concrete lining of a tunnel with an inside diameter of 9.5 ft and an outside diameter of 13 ft, situated at a depth of 60 ft below the surface.

The top 12 ft of overburden consisted of about 12 ft of sand while the remaining 48 ft of the material overlying the tunnel consisted of a plastic blue lacustrine clay with a shear strength which varied from about 400 psf at the top of the clay to an average of about 150 psf for the 40 ft immediately overlying the tunnel. Shear strength is shown in Fig. 26.

Pressure cells were installed, and measurements carried on, over a period of 10 years (beginning in 1931) after the completion of the tunnel. The initial vertical pressure on the lining was equal to about two-thirds of the "depth pressure". This vertical pressure increased slowly until at the end of 5 years (1936) the top and the bottom vertical pressures were approximately equal to the depth pressure. Readings of the pressure measuring cells were continued for another 5 years (until 1941), but no further significant changes in vertical or horizontal pressures were noted.

Figure 27 shows variation of pressure with time. Prior to December 4 and 10, 1930, an air pressure of 27 psi (3900 psf) was being maintained in the tunnel. After these dates the air pressure in the tunnel was atmospheric only.

It is of interest to compute the theoretical initial pressure on the lining using equation (20). The following physical data will be used as average:

$$\Phi = 0°, \qquad c = 200 \text{ psf}$$

$$\gamma = 120 \text{ pcf}, \qquad z = 66.5 \text{ ft}, \qquad a = 6.5 \text{ ft}$$

$$p_i = a\gamma\left[\frac{z}{a}-1\right]-4.6c\log\frac{z}{a} \tag{20}$$

$$p_i = (6.5)(120)\left[\frac{66.5}{6.5}-1\right]-(4.6)(200)\left(\log\frac{66.5}{6.5}\right)$$

$$= 780[9.2]-(920)(1.01)$$

$$= 7200-930 = 6270 \text{ psf}$$

The result corresponds approximately to the vertical pressure which was observed immediately after release of the compressed air but exceeds the subsequent vertical pressures, which decreased to about 5200 psf about 2 months after air release, and then began a slow increase over a period of 5 years, when it stabilized at about 7300 psf.

Since the average value of cohesion, 200 psf, can be determined only approximately, the results of the computations will be approximate to the extent that the values for physical constants used in the computations are approximate.

The theoretical support pressures around openings in plastic clay as found by computations are initial pressures. The cohesion found by laboratory tests is that of the clay on a short time basis, and, as has been noted herein previously, the value of c on a long term basis will be considerably less. This suggests that clay specimens should be tested for yield, and over a period of several weeks, or months, depending upon the length of time that temporary tunnel support may be in place prior to placement of the permanent lining.

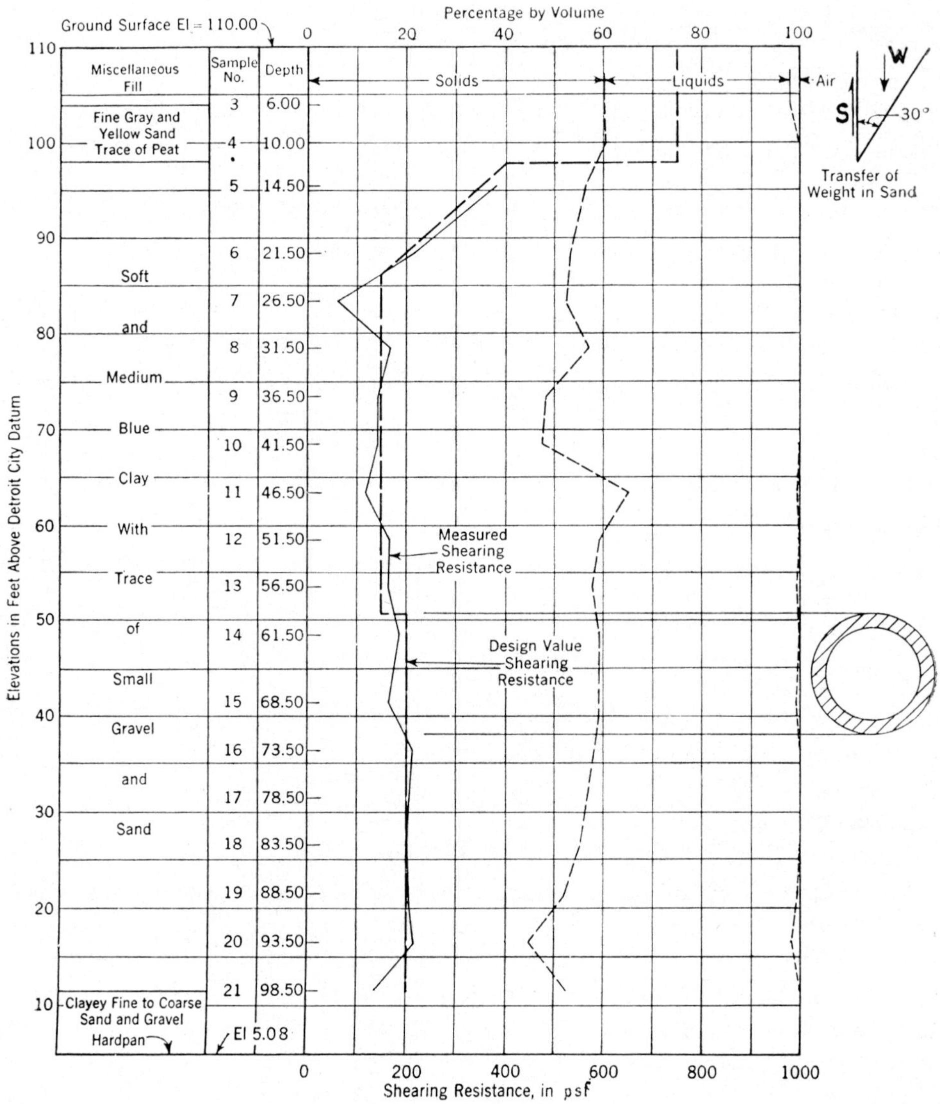

FIG. 26. Record of soils conditions at one location along the tunnel alignment.[13]

Materials with little or no internal friction derive their resistance to deformation only from their shear resistance. Under long term application of stress a slow yield takes place to relieve the shear stress. Such materials have the ability to partially support themselves, by arching action (which is derived from shear resistance), so that loads on tunnel support or tunnel lining are less than the depth pressure for a period of weeks, or months. However, over a longer period of time the clay yields in shear, somewhat in the manner of a thick fluid, and the vertical pressure

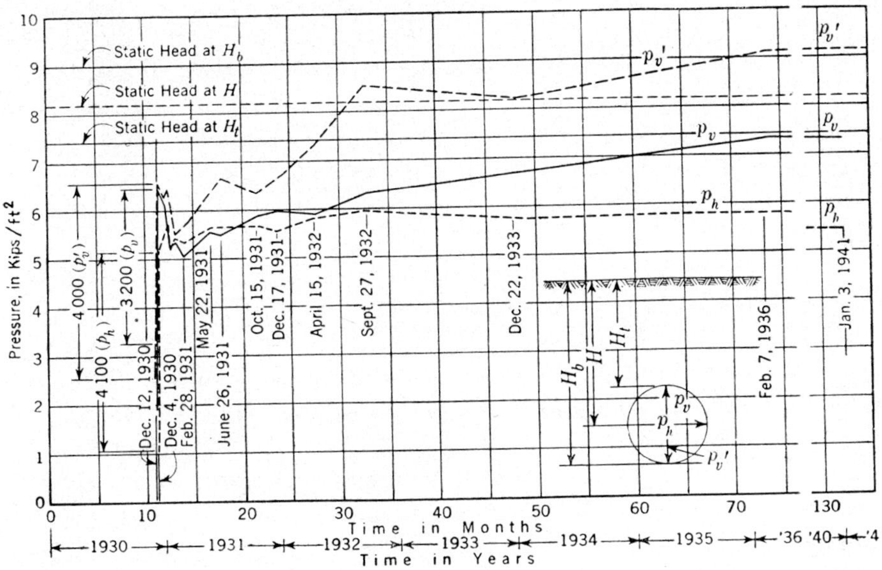

FIG. 27. Variation of the support pressure with time at one location on the tunnel.(13)

increases until it approaches the full depth pressure. In the instance just cited the vertical pressure increased from about $\frac{2}{3}\gamma z$ at the completion of the tunnel to γz after 5 years.

Chicago Subway Tunnels. Terzaghi(14) gives the results of measurements of pressure on the lining of a test section of subway tunnel constructed in Chicago. The tunnel test section was 19.3 ft wide and 21.3 ft high, with an arch of 9.6 ft radius. Temporary support consisted of steel ribs of 6-in. 20 lb. H-section lagged with steel linear plate. A test section 21 ft long was equipped with earth pressure measuring cells. The top of this section was located about 24 ft below the ground surface and the overlying material for 14 ft above the crown consisted of clay possessing an unconfined compressive strength ranging from 800 to 3200 psf. Taking the value of Φ as 0°, the value of cohesion would range from about 400 to 1600 psf. Sand and fill material overlay the clay.

When the section was first mined the initial load readings indicated that the arching effect in the clay was carrying almost 50 per cent of the weight of the overburden.

However, the vertical pressure increased steadily, in a little more than 1 month after mining the support was carrying almost the full depth pressure (when allowance was made for the supporting value of the compressed air (12 psi or 1728 psf).

After the compressed air was released the arching again became evident, with arching effect initially carrying about 40 per cent of the depth pressure. Pressure on supports again increased steadily and in about 6 weeks after release of the air the pressure on support had again nearly attained the value of the depth pressure.

Time Yield Effects. From the foregoing examples it is apparent that for an opening located at a shallow depth in purely cohesive material, arching action is most effective in reducing support immediately after excavation and pressure increases with time until the full depth pressure eventually is exerted on rigid support. Thus, the strength required for temporary support depends upon the time it will be in place before the permanent lining is placed, as well as upon the initial ground pressures which will be encountered.

The time required for full depth pressure to develop on support in plastic clay depends upon several factors. Among them are: the cohesive strength of the clay; the depth of the opening; the size of the opening.

For example, if a 10 ft diameter is located only 10 ft below the surface any adjustments in vertical stress and strain will penetrate the maximum distance possible (to the ground surface) within a short space of time. If the same opening is located at a depth of 100 ft, the readjustments in stress and strain can continue for months, or even years, before penetrating to the maximum distance.

This relationship between the depth of an opening and the length of time required for maximum support pressure to develop and stabilization to occur is a phenomenon commonly noted in the argillaceous rocks overlying coal seams.

Tunnels in London Clay. Ward and Chaplin[16] give results of measurements of stresses in cast-iron lining of several locations in London subway tunnels which were constructed from 50 to 60 years ago.

One of the tunnels had an external diameter of 12.5 ft and the other an external diameter of 22.5 ft.

In the smaller tunnel stresses were measured at sites where the depths below the surface were 50, 60, 71 and 86 ft. The larger tunnel measurements were made where the depth was 110 ft.

The measurements indicated that existing stresses in the cast-iron lining appear to be the result of a "hydrostatic" stress distribution in the surrounding ground. It appears that, over a period of 50 years, any initial differences between vertical stress and horizontal stress have been adjusted by yield of the clay.

Cooling and Ward[17] reported on the stresses measured (in 1952) on one cast-iron ring in each of two adjacent 25-ft diameter tunnels under construction at a depth of about 100 ft below street level in London clay.

Stresses measured on the one ring in tunnel No. 2 increased from those corresponding to zero ground pressure at installation to those corresponding to about 60 per cent of the depth pressure in 3 weeks, and further increased to those corresponding approximately to full depth pressure after about 2 months.

Measured stresses in the ring in tunnel No. 1 South were unevenly distributed and increased from those corresponding to zero soil pressure at installation to those corresponding to about 45 per cent of depth pressure at the end of $2\frac{1}{2}$ months, after which no further increase was detected.

A more extensive campaign of measurements was conducted in another tunnel (The Ashford Common Tunnel) of 9.3 ft outside diameter, located at a depth of 90 ft in London clay. The pressure on tunnel lining appeared to increase to attain a constant value about 2 months after construction. This pressure had been maintained for about 1 year (at the time the paper was written) and was equal to about 60 per cent of the load corresponding to full depth pressure.

No co-ordinated information on the physical properties of the soil column overlying this tunnel is available to the author. However, Tatersall *et al.*[18] give the average compressive strength as 300 psi (under undrained triaxial testing with lateral confining pressure of 90 psi) and the average inclination of the rupture plane to the principal stress plane as 59°. These results were obtained on clay samples cut and removed from the tunnel during driving and, according to the authors, the strength values obtained by testing samples from the tunnel were about three times the values obtained on samples taken from 4-in. cores bored from the surface to depths between 80 and 100 ft. There was a very wide spread between results obtained with various samples from the tunnel and the compressive strength varied from 145 to 655 psi.

Wakeling[19] summarizes the results of measurements obtained in a pressure tunnel located in London clay, at depths of 60–180 ft below the surface. This tunnel was 9 ft in inside diameter, and lined with 6-in. thick tapered concrete segments jacked into place so that they were prestressed against the clay periphery. The initial prestress varied from 0.06 to 0.55 of the depth pressure and at about the end of 2 years the pressure on the lining had increased to values between 0.33 and 0.76 of the depth pressure. The smaller lining stress generally corresponded to smaller initial prestress. However, the pressure was increasing more rapidly on those rings which had initially carried the least prestress.

MISCELLANEOUS METHODS FOR COMPUTING PRESSURE ON SUPPORTS FOR SHALLOW WORKINGS

For those tunnels which are constructed in cohesive materials, and are situated near the ground surface the equations given previously herein do not always give satisfactory results. For a tunnel located at a distance of one diameter or less beneath the surface, for example, the overlying soil resembles a beam and bending and tensile stresses may be important in determining the stability of such a beam.

When the depth of the opening exceeds one diameter, the "beam" becomes sufficiently thick so that bending stresses are no longer important and instead the stability is determined by the shearing strength of the material.

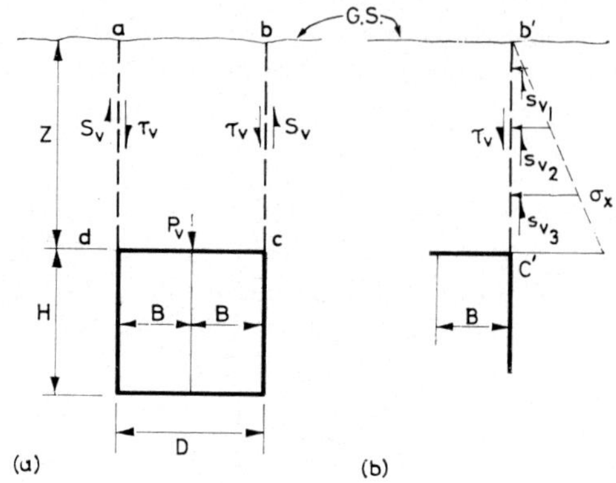

FIG. 28. "Subsiding Block" method for computing vertical pressure on supports for a near-surface tunnel in cohesive soil. (Shear stress, T_v, exceeds shear resistance, S_v.)

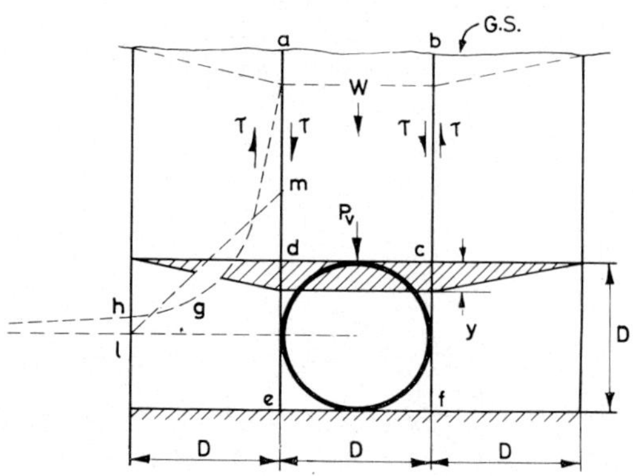

FIG. 29. "Relative-Yield" method for computing vertical pressure on supports for a near-surface tunnel in cohesive soil. (Shear strength exceeds shear stress.)[10]

For openings located at depths up to about six diameters below the surface Terzaghi[14], Housel[13] and Lane[10] have employed the "subsiding block" method which assumes that "blocks" of earth overlying the opening tend to shear and to subside along vertical planes, and that the pressure on support is equal to the difference between the downward force due to the weight of a "block" and the upward force resulting from the shearing resistance of the soil.

A modification of this method (the "relative yield" method) has recently been developed by Smith[11]. This modification takes into account the difference in compressibility, or rigidity, between the soil mass and the tunnel supports. If the soil mass at the sides of the tunnel is more rigid than the tunnel support, then the soil will support relatively more, and the tunnel will support less. On the other hand, if the support is very rigid, settlement of the overlying soil mass results in relatively more weight being taken by the support. This is the principle on which yielding steel arches and flexible support methods function. The support yields when the ground load reaches its carrying capacity to allow the adjoining ground to carry more of the load.

The general theory of each of the above methods is shown in Figs. 28 and 29.

The "Subsiding Block" Method for Computing Pressure on Supports

As a basis for investigating the stability of near-surface openings and for calculating the magnitude of earth pressure on supports the simplest case will be presented first. This involves an opening located near the ground surface in cohesive material, as shown in Fig. 28. The block of ground *abcd* may be considered as attempting to shear along vertical planes *ad* and *bc* and to subside on to the support. However, this subsidence is resisted by shearing resistance along the potential shear planes *ad* and *bc*.

Taking a unit length of opening, the total pressure on support will be:

$$P_v = 2B\gamma z - 2S_v$$

where S_v is the total shearing resistance along vertical planes.

In a material, such as a soft clay, which possesses cohesion and no significant internal friction ($\tan \Phi = 0°$) the unit shearing resistance on plane *ad* or *bc* will be equal to c and total shear resistance on a plane will be $S_v = cz$, and the equation becomes:

$$P_v = 2B\gamma z - 2cz \tag{21}$$

where P_v is total initial vertical pressure on support per unit length of tunnel.

For $P_v = 0$, equation (21) becomes:

$$B\gamma = c \tag{22}$$

Thus when $c \geqslant B\gamma$ the ground exerts no pressure on the support.

Taking as an example a tunnel 10 ft wide, situated at 20 ft below the ground surface in saturated clay with $c = 500$ psf we get

$$P_v = (2)(5)(120)(20) - (2)(500)(20) = 4000 \text{ lb}$$

In this case the initial vertical load on the support would be 4000 lb, which would be $4000/10 = 400$ psf unit vertical pressure.

If width of the tunnel were reduced to 8 ft, then:

$$B\gamma = (4)(120) = 480 \quad \text{and} \quad c > B\gamma,$$

Therefore, no initial ground pressure would be exerted on the support.

It should be noted that, in this example, the clay is presumed to have no internal friction, and that such a material yields plastically under long time stress until after a period of 1–3 months appreciable ground pressure develops on supports, although the full depth pressure may not develop until as much as 5 years after construction of the facility.

The fact that such soil temporarily supports most, or all, of its own weight may be taken advantage of in construction operations by employing lighter temporary supports during excavation operations, provided that the heavier permanent lining is placed before the full ground pressure develops.

Effect of Internal Friction

The foregoing example has considered material without any internal friction. If the material is partially consolidated clay, or if it is clay containing considerable granular material, then it will derive strength from internal friction as well as from cohesion.

The additional shear resistance may be thought of as manifested along vertical shear planes such as $b'c'$ in Fig. 28(b). The unit shear resistance at any point on plane $b'c'$ is equal to

$$S_v = c + \sigma_x \tan \Phi \tag{23}$$

where σ_x is the unit lateral stress, at any point, and is a function of the depth, z. The average shear resistance on plane $b'c'$ is equal to

$$S_{v_{\text{avg.}}} = c + \frac{1}{2} \sigma_{x_z} \tan \Phi \tag{24}$$

and the total shear resistance on the plane $b'c'$ is then

$$S_v = z\left(c + \frac{1}{2} \sigma_{x_z} \tan \Phi\right) \tag{25}$$

The lateral stress, σ_x, varies with the history and the origin of a soil, between $\frac{1}{3}z$ and z.

An equation of the form of (25) (which assumed transfer of load by "bridging") was used for computing the theoretical loads on supports at the Garrison Dam Tunnel test section.[(10)] In this case vertical movement above the tunnel section was observed to extend to a distance of about $2D$ above the top of the tunnel and shearing resistance was presumed to be effective only for this distance rather than completely to the surface. This method gave results which were reasonably close to the measured loads on supports for the initial shear strength of the ground. However, as work progressed and ground consolidated as the water table was lowered, the measured loads on supports became much less than those computed from the equations which assumed "bridging" action, and another method of computing theoretical vertical loads on supports was devised. This was the "relative yield" method developed by Smith.[11] This method gave results which were in much closer agreement with the measured loads on supports.

The "Relative Yield" Method for Computing Pressure on Supports

This method involves the determination of the modulus of elasticity of the tunnel support and of the "soil block." The vertical load due to the weight of overlying ground is then distributed in proportion to the relative rigidity of tunnel block and of adjacent soil blocks.

In equation (27) and Fig. 29 the width of adjacent ground subject to deformation on each side of the tunnel block has been taken as equal to the width of the tunnel block. This is in conformity with elastic theory, which shows that for a circular hole in an elastic medium subjected to compressive stress the significant increase of stress does not extend more than one diameter from the hole. The actual stress on a horizontal plane passing through the center line of the hole would be as shown by curve *hga* in Fig. 29, while the stress distribution assumed for equation (27) corresponds to that shown by line *lm*. The approximation is within the limits of that accuracy posed by other uncertainties in the equation.

The equations which follow were developed for use, and were employed in computing theoretical pressures on steel rib tunnel supports, and later on the concrete lining in the test tunnel section at Garrison Dam. The method was found to give results which checked closely with the measured pressures on support, and on lining.

The equations for pressure on support by the "relative yield" method are derived as follows:

E_t = modulus of elasticity of the tunnel block;

E_g = modulus of elasticity of the ground;

$$b = \frac{E_t}{E_g}; \qquad E_t = bE_g; \qquad W = \text{wt. of block } abcd. \tag{26}$$

$$\text{Unit strain} = \frac{\text{unit stress}}{\text{modulus of elasticity}}$$

In the tunnel block:

$$\text{Unit strain} = \frac{y}{D} = \frac{\dfrac{W-2\tau}{D}}{bE_g} \quad \text{or} \quad y = \frac{W-2\tau}{bE_g}$$

In the adjacent soil blocks:

The transferred load τ causes an average deflection $\frac{y}{2}$

$$\text{Unit strain} = \frac{\frac{y}{2}}{D} = \frac{\frac{2\tau}{2D}}{E_g} \quad \text{or} \quad y = \frac{2\tau}{E_g}$$

Since no shear failure takes place the deflection is the same for both blocks at the boundary

$$\tau = \frac{W}{2(b+1)}$$

And the load on tunnel supports is:

$$P_v = W - 2\tau = W\left[1 - \frac{1}{b+1}\right] \tag{27}$$

The foregoing derivation is adapted from Lane.[10]

Stiffness of the steel rib support, E_t, was about one-fifth to one-tenth the stiffness, E_g, of the adjacent ground. Stiffness of the concrete lining in the test tunnel was about $3\frac{1}{2}$–4 times the stiffness of the ground.

These figures, substituted in equation (27) give the loads on tunnel supports, and on lining:

$$\text{For } b = \frac{E_t}{E_g} = \frac{1}{10} \text{ (steel rib support)}$$

$$P_v = W\left[1 - \frac{1}{\frac{1}{10}+1}\right] = W\left[1 - \frac{1}{1.1}\right] = 0.09\ W$$

$$\text{For } b = \frac{E_t}{E_g} = \frac{1}{5} \text{ (steel rib support)}$$

$$P_v = W\left[1 - \frac{1}{\frac{1}{5}+1}\right] = W\left[1 - \frac{1}{1.2}\right] = 0.17\ W$$

$$\text{For } b = \frac{E_t}{E_g} = \frac{4}{1} \text{ (concrete lining on test section)}$$

$$P_v = W\left[1 - \frac{1}{\frac{4}{1}+1}\right] = 0.80\ W$$

For the ground a value of E_g = 2000 tons/ft^2 = 4,000,000 psf = 28,000 psi was established from laboratory tests and confirmed by field measurements of elastic rebound in large excavations. The advantages of employing yielding, or flexible, supports for temporary support was evident.

In retrospect computations to determine the tunnel modulii, E_t and E_c, plots of the measured vertical load vs. the vertical diameter change were employed. The following equation was derived for use in computing the modulus of elasticity of the tunnel support:

$$E_t = \frac{12EI}{R^3 S\left(I - \dfrac{H}{V}\right)} \tag{28}$$

where H/V = horizontal/vertical load ratio;
R = radius of the ribs;
S = spacing of the ribs;
I = cross-sectional moment of inertia of a rib

All of the quantities in the equation (28) can be determined except the H/V ratio, which may have to be estimated.

In the case of a relatively flexible support the H/V ratio will approach unity because vertical deformation of the support under load causes it to expand laterally against the ground and to build up the passive soil resistance.

For the Garrison Dam Tunnel test section (36 ft diameter) the H/V ratio for the steel-rib-supported section ranged from 0.8 to 1.1. In the stiff slotted concrete section the ratio was lower, ranging from 0.4 to 0.6, and averaging 0.5.

The value of I (moment of inertia for individual ribs) as used in equation (28) will be somewhat lower for the case of a sectional rib with bolted joints than for the same rib with welded joints.

Equation (28) is most useful for those tunnels located within a few diameters of the ground surface, in firm cohesive soils, or soft argillaceous rocks which behave elastically, under compressive stresses. This method of analysis gives results in those soils in which the shear strength of the soil exceeds the shear stress.

For shallow tunnels in firm rocks, which have modulii of elasticity of the order of several hundred thousand pounds per square inch, or more, the elastic deflection of the rock will be so small as to have negligible effect in increasing pressure on support. In such firm rocks the only purpose that temporary support serves is to hold in place loose pieces, blocks, and fragments which might otherwise fall into the tunnel.

Jannsen's Formula (The "Bin Formula")

Terzaghi[7] has modified the "bin formula" for use in computing vertical pressure on lining for near-surface tunnels in granular material.

The "bin formula" was originally derived by Jannsen for computing pressure of granular materials on the bottoms of storage bins. Such materials derive support from friction with the sides of the bin, and the resulting upward reaction carries a portion of the weight of the material. The proportion carried is a function of the width of the bin and the internal friction of the material.

Terzaghi's analogy considers the vertical planes of *gh* and *ij* in Fig. 30 as sides of a bin and plane *hj* as the bottom of a bin. Using this modified bin formula, the average unit pressure, σ_v, on plane *hj* is computed. σ_v is then the average pressure on top of the support. The pressure at the sides is found by considering the tunnel sides as retaining walls for the wedge-shaped blocks of ground, *hmn* and *ljk* at the sides. These blocks have applied to their top surfaces the pressure, σ_v, which is equivalent to that on top of the tunnel lining.

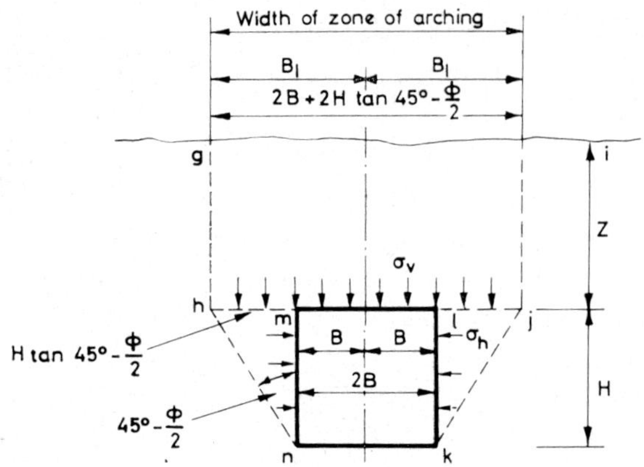

FIG. 30. "Bin Formula" for computing pressure on supports for a near-surface tunnel in cohesionless soil. (After Terzaghi.[7])

The bin formula as modified by Terzaghi is:

$$\sigma_v = (B_1)\,\frac{\gamma - c/B_1}{K\tan\Phi}\left[1-\exp\left(-K\tan\Phi\,\frac{Z}{B_1}\right)\right] \tag{29}$$

in which

$$B_1 = B + H\tan\left(45^\circ - \frac{\Phi}{2}\right)$$

and c is the cohesion and K is an empirical coefficient which varies in value from about 1 at the top of the support to about 1.5 at a distance of $2B$ above the top of the support.

When Z/B is large, the formula becomes:

$$\sigma_v = (B_1)\,\frac{\gamma - c/B_1}{K\tan\Phi} \tag{30}$$

For practical purposes, equation (29) may be reduced to equation (30) when the ratio Z/B reaches a value of about 4–6; that is, when the depth of the opening below the surface is at least equal to 2 or 3 times its width.

When $c \geqq B_1\gamma$ the vertical pressure on the lining will be equal to zero; that is, the opening will be self-supporting.

The unit lateral pressure, σ_h, on the supports at the sides of the tunnel may be computed in a manner similar to that used for pressure on retaining walls.

The average unit lateral pressure due to the weight of the wedges of earth, *hmn* and *ljk*, is given by the formula:

$$\frac{1}{2}\gamma H(i)$$

To this must be added the unit lateral pressure which is created by the vertical pressure, σ_v, on the tops of the aforementioned soil wedges (that is, the unit pressure, σ_v, on planes *hm* and *ij*), which is: $\sigma_v i$
then:

$$\sigma_h = \left(\frac{1}{2}\gamma H + \sigma_v\right)(i) \tag{31}$$

Support Loads vs. Tunneling Methods

The pressure which builds up on lining in tunnels through pulverulent materials varies according to the amount of ground which is "lost" from irregular cavities outside the support. Careless mining, or inadequate lagging or lining, allows runs of ground through the support with the formation of irregularly shaped cavities

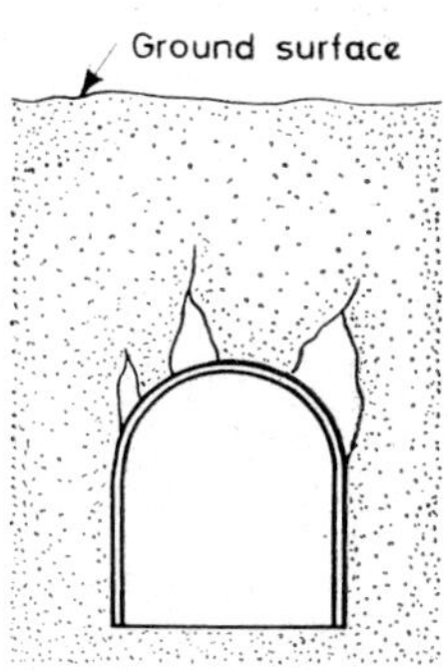

FIG. 31. Development of localized pressures on supports due to cavities formed in the ground outside the support.

outside the support. Such irregular contact between the ground and the tunnel supports results in concentration of ground pressures at contact points. (See Fig. 31.)

Under such unfavorable conditions pressure on supports will be maximum and may approach the pressures which are given by equation (29).

Examples of Sand Tunneling

The nature of ground pressures which developed during tunneling by different methods was observed during the driving of a 4 miles of sewer tunnel.

Nominal diameter of the tunnel, outside the temporary support, was about 12.5 ft. In cross-section it had a modified horseshoe shape, designed to be lined with concrete to a finished circular section of 9.5 ft.

A large portion of this tunnel was driven through relatively clean, poorly graded sand which possessed only that cohesion which was imparted to it by retained moisture. The tunnel was situated above the water table and was driven at an average depth of about 65 ft below the surface, although maximum depth was about 130 ft.

Tunneling by Driving Spiling

The first 1500 ft of this tunnel was driven by using 3 in. $\times$ 8 in. timber spiling which was driven over steel ribs. These steel arch ribs were fabricated from 6-in., 20-lb. WF sections and were spaced at 4 ft apart. Progress was about 45 ft per day (24 hr).

Difficulty was experienced in preventing sand from running in through cracks in the spiling and irregular voids formed outside the supporting timber. These caused the development of pressures which were sufficient to bend and distort the upper portions of the steel arch supports, although none of the steel arch supports were damanged sufficiently to require removal and replacement. It was apparent that high localized pressures were developing.

Tunneling with a Steel Shield

To eliminate the tedious spiling operation a steel canopy or "shield" was constructed and put into service. With this device daily advances of up to 153 ft were obtained (although the average daily advance was about 120 ft) and less support was required.

This device consisted of an arch-shaped steel shield, open at the bottom, about 22 ft long, covered with a $\frac{3}{4}$-in. thick steel skin plate, and equipped with hydraulic jacks to push the shield into the sand. The shield was supported on a yoke or "bent". Under the protection of the fore part of the shield mucking was carried out, and under the protection of the rear part the support was installed.

The shield created an opening in the sand which was about 2 in. larger radially than the outside radius of the temporary support. As the shield progressed forward its tail cleared the support which had been installed under its protection, and the sand settled onto the support.

Support Pressures

It was found that the ground could be supported with steel-arch ribs of 4-in. 13 lb. H section spaced at 4 ft center to center and lagged tightly with 2×8 inch lagging. Thus, both the steel quantity and the timber quantity were reduced more than 33 per cent from that required for the spiling or forepoling method used previously.

No noticeable deflection of the 2-in. lagging occurred and it appeared that the pressure of the sand on the lagging did not exceed 100 psf. Thus, the uniform annular space left by the shield appears to have been sufficient to allow extension of the clastic–elastic boundary so that pressure on the support was reduced to a negligible amount.

Actual Support Loads vs. Computed Support Loads

In most cases equation (30) gives values for pressure on lining which greatly exceed those observed in the field. For example, in the tunnel previously mentioned, which was driven through a relatively clean sand (dune and beach sand) at depths up to 130 ft, the magnitudes of pressures on support as estimated from deflection of the tunnel lagging did not exceed about 100 psf in that portion of the tunnel in which uniform support was provided by means of a tunneling shield.

Properties of the sand were: $\Phi = 37°$; $\gamma = 100$ pcf. The only cohesion was that due to contained moisture. For the tunnel, $B = 6$ ft; $H = 12$ ft; $z = 100$ ft.

Substituting these values in equation (30) we obtain

$$\sigma_v = (B_1)\frac{\gamma - c/B_1}{K \tan \Phi} = (12)\frac{\left(100 - \dfrac{0}{12}\right)}{(1)(0.75)} = 1600 \text{ psf}$$

Thus the pressure on lining indicated by this equation has a magnitude several times that of the pressure actually observed during shield tunneling but would probably be in line with localized pressures during spiling operations. In order to reduce the value of σ_v to correspond with values observed during shield tunneling it would be necessary to assign c a value of about 1100 psf. In a clean sand such cohesion does not exist.

If, under the same conditions, we compute the theoretical pressure on lining for a tunnel in clean sand above the water table by means of equation (16 b) we obtain values which are more nearly in accord with field observations during shield tunneling operations. Thus:

$$P_{i_{B,v}} = \left(\frac{3}{2}\right)\frac{\gamma a}{(j-2)}\left[1 - \left(\frac{a}{b}\right)^{j-2}\right] \tag{16b}$$

and in the present case the quantity $\left[1-\left(\frac{a}{b}\right)^{j-2}\right]$ may be neglected; therefore:

$$P_{i_{B,v}} = \left(\frac{3}{2}\right)\frac{(100)(6)}{2} = 450 \text{ psf}$$

which, although it still greatly exceeds actual observed pressures, is a much more reasonable figure and represents the maximum vertical pressure on support. This occurs at the crown (top of the arch, or "keystone").

A SIMPLIFIED METHOD FOR PREDICTING PRESSURES ON SUPPORTS IN WORKINGS AT SHALLOW OR MODERATE DEPTHS

A large proportion of mining operations are located at shallow to moderate depths beneath the Earth's surface and are excavated in rock which is essentially hard and intact except that it is cut by intersecting fracture planes spaced from a few inches to several feet apart, which divide the rock into angular fragments, or blocks.

The most intact of the massive rocks are generally cut by joint planes from 20 to 30 ft apart, and in the average massive rock joint and fault planes are probably spaced at an average about 15 ft apart.

Fracture planes which are so spaced that they are more than 2–3 ft apart have little effect on the stability of the rock around average mine tunnel, drift, or crosscut. Generally, the effect of natural fractures on the stability of an opening is limited if the fracture surfaces are in tight contact and if the fracture planes do not contain clay.

In most underground excavations in rock we are dealing with rocks which are cut by systems of intersecting fractures which are spaced from a few inches to a few feet apart.

Equation for Pressure on Supports

Equation (16c) is useful for computing the pressure on supports in granular material or in fractured rock:

$$P_{i_v} = \left(\frac{3}{2}\right)\frac{\gamma a}{j-2} \tag{16c}$$

where

P_{i_v} = the magnitude of the vertical pressure which must be exerted by support at the crown of the arch (psf);

γ = the density of the rock (pcf);

a = radius of the opening (ft).

j = $\dfrac{1+\sin\Phi}{1-\sin\Phi}$;

Φ = the angle of internal friction of the soil or rock mass.

The Clastic Zone

Equation (16c) is a simplification of equation (16), and in using equation (16c) we must assume that the zone of stress readjustment (the clastic zone) has extended to a distance of at least several diameters from the perimeter of the opening so that the natural (primitive) rock stresses no longer have any appreciable influence on the total support pressure required to stabilize the rock around the opening.

In a fractured material with a high coefficient of friction at shallow to moderate depths the pressure on supports will be that due to the rock in the clastic zone above the support, and the vertical pressure required at the top of the arch to stabilize the rock will be a function of the density (weight per cubic foot) of the rock and the radius of the opening.

The Angle of Internal Friction

The quantity $(j-2)$ in the equation is a function of the coefficient of internal friction of the material (which for convenience in making computations is expressed as the angle of internal friction). Based upon the condition of the rock we may estimate the magnitude of internal friction and compute the value of j; or the value of j may be read directly from Fig. 10, Chapter 4.

Some values for the coefficient of internal friction, the angle of internal friction, and the corresponding numerical values of j are given in Table 1.

TABLE 1

Coefficient of internal friction	Angle of internal friction	j
0.36	20°	2
0.47	26°	2.5
0.57	30°	3
0.75	37°	4
0.90	42°	5
1.03	46°	6
1.15	49°	7
1.23	51°	8
1.43	55°	10

Computation of Support Pressures

A hard, brittle rock which is fractured into large pieces, but all of whose surfaces are in tight contact, should be assigned an angle of internal friction Φ of about 46–49°. Thus, equation (16c) reduces to (taking Φ as 49°):

$$P_{i_v} = \left(\frac{3}{2}\right)\frac{\gamma a}{j-2} = \left(\frac{3}{2}\right)\frac{\gamma a}{5} = 0.3\gamma a$$

If the rock fragments are rough and the surfaces are in tight contact and well interlocked, then the rock can be assigned a Φ value of 51–55°. (The apparent high value for internal friction is due to the interlocking of irregular surfaces rather than to actual "rubbing" resistance.)

A fragmented rock whose fragments are coated with moist clay should be assigned an angle of internal friction of not more than 30°. Thus, equation (16c) would reduce to:

$$P_{i_v} = \left(\frac{3}{2}\right)\frac{\gamma a}{1} = 1.5\gamma a$$

and the required support pressure would be about five times that in the preceding example.

TABLE 2

	Angle of internal friction, Φ, (to be used in equation (16c))
STRATIFIED OR BEDDED ROCK (sedimentary rocks)	
Thin Bedded: Most of the beds are less than 12 in. thick with a large proportion being less than 6 in. thick	*42°–49°
Medium to Thick Bedded: Most of the beds in a sequence are over 12 in. thick and some are at least 24 in. thick	*49–55°
(It is assumed that there are no fractures across the bedding; if there are transverse fracrures then the stratified rock must be placed in one of the fractured classifications used for the non-stratified rocks below.)	
NON-STRATIFIED ROCK (igneous and metamorphic rocks)	
Massive Rock: Fractures and/or joints are spaced several feet or more apart.	>55° (No pressure on supports)
Moderately Fractured or Jointed Rock: Fractures or joints are generally 2–4 ft apart and are tightly closed.	49–55°
Blocky and Seamy Rock: Fractures are generally 1 to 2 ft. apart and are tightly closed.	46–49°
Brecciated Rock: Pieces are generally less than 1 ft. and most pieces are less than 6 ins. in the largest dimension.	
No clay on Fracture Planes	42–46°
Fault Clay, "Gouge", or Slickensides on Fracture Planes	30°
Granular, Pulverized, or Disintegrated Rock	
Containing No Clay	37–42°
Containing Clay	25°
Fault Clay or "Gouge" Containing Breccia	25°
Sedimentary Clays	(see page 230)

* These figures apply to an arched tunnel driven through stratified rock. For openings with flat roots in flat-lying strata the roof strength may be analyzed by methods described in Chapter 1, page 73 and Chapter 6, page 276.

A Classification of Structural Conditions of Various Rock Types

Presented in Table 2 is a classification which gives an approximate indication of the structural competence of rocks which are fractured in various degrees.

Effects of Clay

The massive rock, of course, exerts no pressure on supports, and neither do the thick bedded sandstones and mudstones or siltstones. Fractured rock generally may be considered to have an angle of internal friction of at least 46° if all fragments are in tight contact.

If fragments are loose or separated, but are rough and angular, the angle of internal friction will probably be from 37 to 42°. If fragments are coated with moist clay then the angle of internal friction should be taken as not more than 30°. If fragments are separated by layers of clay then the angle of internal friction should be taken as about 25°.

Those rocks which have Φ values of less than 25° are composed of the clays in various degrees of consolidation, and support pressures for such materials should be computed by means of equations (18), (19), or (20).

In general, there will be no pressure on supports for moderately sized openings in rocks with Φ values of 49–55° or more.

Effects of Water

The foregoing values for Φ were based on the assumption that the only water present in the rock would be the normal surface moisture on the rock fragments. If free water runs from cracks and seams, the Φ values for rocks in the breccia classification, and in the finer sizes, should be reduced by about 4 or 5°.

The Effect of Shape of Fragments

A granular or fractured material derives most of its strength from "internal friction". Therefore, other things being equal, the larger the coefficient of internal friction the less will be the pressure on the supports.

The internal friction of a fractured brittle material is derived partially from the "interlocking" effects of irregularities in the surfaces of the fragments. The importance of the shape of the fragments in determining the internal friction of the mass may be appreciated if we visualize the difference in behavior between a mass of loose spheres and a mass of cubical blocks. The latter can be held only in a tight container while the former will form a free-standing pile with considerable resistance to compressive stress.

The fragments composing the average rock mass are generally angular and rough, and the internal coefficient of friction of the mass (as determined by its resistance to compressive stress) is in excess of 1.0 and may exceed 1.5. (This assumes that the rock mass is subjected to stresses sufficient to cause movement along pre-existing fracture planes but not sufficient to create new fractures across the fragments.)

In sands, gravels, and alluvial materials where there are voids between fragments and some fragments are rounded, the coefficient of internal friction of the mass usually ranges from about 37 to 42°. (See Fig. 16, Chapter 4; see also page 527, Chapter 12).

Pressure Distribution on the Arch

The pressure required for stabilization is maximum at the top of the arch and decreases to zero at each of the ribs. This pressure is not that required to support a mass of loose rock, but is the pressure required to hold the rock fragments in

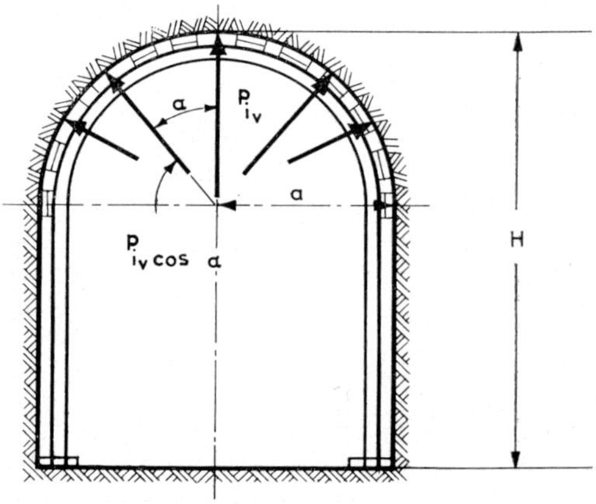

FIG. 32. Pressure on supports for an opening in granular material or in fractured rock.

contact so that they are "keyed" into place. For maximum efficiency the support should exert a uniform pressure against the surface of the rock, and the more blocking used the more closely will this ideal condition be approached.

The approximate radial pressure on the support at any point around the arch may be determined from the following equation:

$$P_i = \left(\frac{3}{2}\right)\frac{\gamma a}{j-2}(\cos \alpha) \tag{16d}$$

where α is an angle measured either clockwise or counter clockwise from the vertical around to the horizontal. (See Fig. 32.) Equation (16d) is obtained from equation (16b) by substituting $\cos \alpha$ for $(-\cos \omega)$ and by dropping the expression within the radicals (which becomes negligible when the value of b/a is more than 3 or 4).

Computation of Side Pressures

Generally there will be little side pressure on supports from any of the rocks which are classified structurally as "brecciated rock" or better (without clay on the fracture planes). For granular on pulverized rock without clay, the magnitude of the maximum unit side pressure may approach a value given by the following equation:

$$P_{i_h} = P_{i_v}(i) + H\gamma i$$

For those rocks which contain clay on fracture planes, or which consist largely of clay, the ultimate side pressure may approach 75–100 per cent of the vertical pressure on the tunnel arch. For further information on pressures in plastic clay refer to the section titled "Pressures on Lining–Horizontal Openings in Plastic Clay" (page 230).

Pressure Computations for Openings Other than Circular

Equations (16) through (16d) were derived for openings with circular cross-sections. Pressure on supports for openings with other cross-sections may be determined by drawing a circle which passes through the corners of the opening

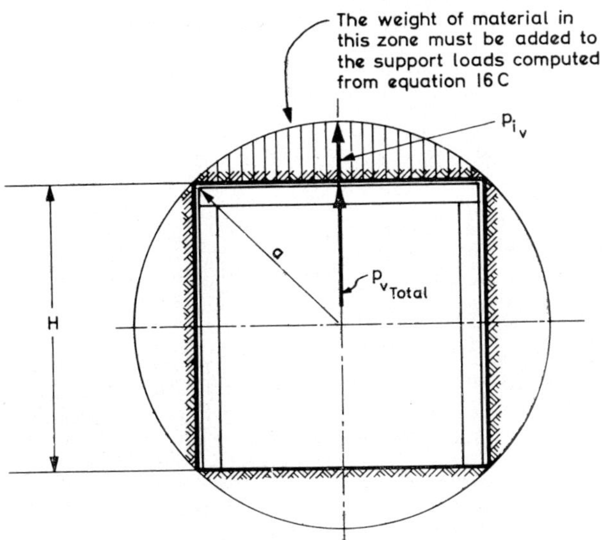

FIG. 33. Pressure on supports for a square opening in granular material or in fractured rock.

as shown in Fig. 33. That portion of the rock which then lies between the roof support and the upper boundary of the circle must then be considered to be a deadweight load to be added to the pressure which is computed from equation (16d).

The radius a to be used in equation (16c) is the radius of the circumscribed circle as shown in Fig. 33.

Estimation of Actual Loads on Existing Supports

In the absence of measuring devices installed on tunnel supports the equivalent rock load may be roughly estimated by observing the maximum deflection of lagging which bridge across from set to set.

The formula for calculating maximum deflection of a simple beam is

$$y = \frac{5wl^3}{384EI}$$

and taking the weight of rock (in place) as 170 pcf and $y = 0.1$ in.

$$\text{Depth of rock load} = \frac{0.54Ed^3}{l^4}$$

Taking the modulus of elasticity, E, as 1,500,000 psi (approximately that for wet Douglas fir) and substituting in the formula we find that the following loads would produce a deflection of 0.1 in.:

TABLE 3

Span	For 2-in. lagging equivalent rock load	For 3-in. lagging equivalent rock load
30-in. span	7.9 ft	26.7 ft
48-in. span	1.2 ft	4.0 ft
60-in. span	0.5 ft	1.7 ft

The figures above are the depths of uniform loads of solid rock which would be required to produce maximum deflection of 0.1 in. in the center of a lagging supported as a simple beam. The figures in Table 3 are computed for rough lagging (e.g. full dimension thickness). If 2-in. finished lagging ($1\ ^5/_8$-in. thick) is used (as on some tunnel projects) deflections will be 1.8 times as great as for 2-in. rough lagging.

The deflection is directly proportional to the load; therefore, a deflection of 1 in. would indicate 10 times the load of a 0.1 in. deflection.

The depths are indicated as loads of solid rock; actually, the rock would be fractured and an equivalent weight of fractured rock would, of course, occupy more space than that indicated in the table for solid rock.

The deflections shown are computed from formula for wet Douglas fir. It would be preferable to make a field check of the deflections, under various loads, with the type of lagging which is actually being installed for support.

It is probable also that the steel rib, being relatively rigid, will take a larger proportion of the rock load than would an equivalent length of lagging; this would tend to be the case in the fractured rock but, in the case of soft and plastic rock, the rock will tend to flow away from the points of high stress and toward the zones of lower stress, tending to equalize the pressure distribution. The load thus tends to be more uniformly distributed in soft rocks and less uniformly in those consisting of hard fragments, where point pressure may develop on ribs or even along the span of the lagging.

Another method of load estimation, in the absence of load measuring cells, is observation of the amount of squeeze of the blocks and wedges which are inserted between lagging and ribs, or between lagging and rock. With Douglas fir or similar timber, fairly heavy loads must develop before any squeeze on the blocking is apparent.

The deflection of the lagging is a much more sensitive indication of rock load than is the amount of squeeze of the blocking.

Examples of rock loads observed in a typical small tunnel are given in Chapter 2, "Structural Defects In Rock", Part A (page 17).

Terzaghi's Classification of Structural Condition of Rock

An approximate method for estimating pressure on support, devised by Terzaghi for use in connection with the design of tunnel support, is in extensive use.

He divided rock into various structural classifications by considering its degree of fracturing or crushing, presence of stratification, and the presence of water.

By referring to a table the approximate pressure on supports may be determined as a function of tunnel dimensions. The method, used in the design of tunnel support, is described in Ref. (31).

BIBLIOGRAPHY

1. JACOBI, OSKAR, Die Bewegungen zerbrochener Gesteinschichten um bergmannische Hohlraume, *Glückauf*, Nov. 9, 1957, pp. 1393–1417.
2. SCHOEMAKER, R. P., A review of rock pressure problems, *Mining Technology*, Nov. 1948.
3. FENNER, R., Untersuchungen zur Erkenntnis des Gebirgsdrucks (Study of ground pressures), *Glückauf*, Vol. 74, No. 32, Aug. 13, 1938, pp. 681–695; Vol. 74, No. 33, Aug. 20, 1938, pp. 705–715. Note: English translation prepared by *National Research Council of Canada, Division of Building Research*, Ottawa, Canada, 1955. Technical Translation TT–515.
4. LABASSE, H., Les pressions de terrains dans les mines de Houilles. Part II: Les pressions de terrains autour des puits, *Rev. Universelle Mines*, Vol. 5, No. 3, March, 1949, pp. 78–88.
5. MORRISON, R. G. K., and COATES, D. F., Soil mechanics applied to rock failure in mines, *Trans. Can. Int. Mining Met.* Vol. 58, 1955, pp. 405–415.
6. LABASSE, H., Les pressions de terrains dans les mines de Houilles. Part III. Les pressions de terrains autor des travers-bancs horizontaux, *Rev. Universelle Mines*, Vol. 6, No. 1, January, 1950, pp. 4–18.
7. TERZAGHI, K., *Theoretical Soil Mechanics*, John Wiley, New York, 1943, pp. 206–215; pp. 70–73, 194–196.

8. CAQUOT, A. and KERISEL, J., *Traite De Mecanique Des Sols*, Gauthier-Villars, Paris, 1956, pp. 474–484.
9. COEUILLET, M., Le soutenement en galeries et la mecanique des roches, *Rev. Ind. Minérale*, Vol. 39, No. 8, August, 1957, pp. 693–715.
10. LANE, K. S., Garrison dam — evaluation of results from tunnel test section, *Proc. Am. Soc. Civil. Engrs., J. Soil Mech. Found. Div.*, November 1957, Prod. Paper No. 1439.
11. SMITH, C. K. and REDLINGER, J. F., Soil properties of Fort Union clay shale, *Proc. Third International Conference on Soil Mechanics and Foundation Engineering*, Switzerland, 1953, Vol. I, pp. 62–66.
12. BURKE, H. H., Garrison dam tunnel test section investigation — A symposium, *Proc. Am. Soc. Civil. Engrs. J. Soil Mech. Found Div.*, Vol. 83, No. SM 4, November, 1957, Paper 1438.
13. HOUSEL, W. S., Earth pressures on tunnels, *Proc. Am. Soc. Civil. Engrs.*, Vol. 68, No. 6, June, 1942, pp. 929–950.
14. TERZAGHI, K., Liner plate tunnels in the Chicago (Ill.) Subway, *Proc Am. Soc. Civil. Engrs.* Vol. 68, No. 6, June, 1942, pp. 862–899.
15. PHILLIPS, D. W., Tectonics of mining, *Colliery Eng.* Vol. 25, June, August, September, October, November, 1948.
16. WARD, W. H. and CHAPLIN, T. K., Existing stresses in several old London underground tunnels, *Proceedings of the Fourth International Conference on Soil Mechanics and Foundation Engineering, London*, August, 1957, pp. 256–259.
17. COOLING, W. F. and WARD, W. H., Measurements of loads and stresses in earth supporting structures, *Proceedings of the Third International Conference on Soil Mechanics and Foundation Engineering, Switzerland*, 1953, Vol. II, pp. 162–166.
18. TATTERSALL, F., WAKELING, T. R. M., and WARD, W. H., Investigations into the design of pressure tunnels in London clay, *Proc. Inst. Civil. Engrs. London*, July, 1955, pp. 400–470 (Paper No. 6027).
19. WAKELING, T. R. M., Discussion — Earth pressures on structures and tunnels, *Proceedings of the Fourth International Conference on Soil Mechanics and Foundation Engineering, London*, 1957; Vol. III, pp. 230–231.
20. PECK, R. B. and BERMAN, S., Measurements of pressure against a deep shaft in plastic clay, *Proceedings of the Second International Conference on Soil Mechanics and Foundation Engineering, Rotterdam*, 1948, Vol. III, pp. 300–301.
21. DUVALL, W. I., Stress analysis applied to underground mining problems. Part I. Stress analysis applied to single openings, *U.S. Bur. Mines R.I. 4192*, 1948.
22. OTT, J. C., Quelques aspects du problème de la poussée sur les tunnels, *Bull. Tech. Suisse Romande*, 71me Année, No. 1, pp. 13–22, No. 4, pp. 41–50, 1945.
23. MOHR, F., Observations in shafts on rock movements due to mining. *International Strata Control Congress, Leipzig*, 1958, pp. 112–123 (sponsored by Deutsche Akademie des Wissenschaften zu Berlin, Sektion für Bergbau), Internationales Büro für Gebirgsmechanik.
24. YORK, B. and REED, J. J., An analysis of mine opening failure by means of models, *Mining Eng.*, July 1953, pp. 705–710.
25. SPACKELER, G., Observations on the distribution of stress around coal and salt workings *International Strata Control Congress, Leipzig*, 1958, pp. 305–319 (sponsored by Datsche Akademie der Wissenschaften zu Berlin, Sektion für Bergbau), Internationales Büro für Gebirgmechanik.
26. VENTER, J. and STASSEN, P., Pressure and movement in soft rocks, *International Strata Control Congress, Leipzig*, 1958, pp. 124–136 (sponsored by Deutsche Akademie Der Wissenschaften Zu Berlin. Sektion für Bergbau), Internationales Büro für Gebirgsmechanik.
27. ISAACSON, E. DE ST. Q., *Rock Pressure in Mines*, Mining Publications, London, 1958, pp. 30–31.
28. LABASSE, H., Les pressions de terrains dans les mines de Houilles; Les pressions de terrains autour des travers bancs, *Rev. Universelle Mines*, February, 1950.
29. WOODRUFF, S. D., Practical use of rock mechanics, *Mining Congr. J.* July, 1961, pp. 33–37, 55.
30. EVERLING, G., Modellversuche über das Zusammenwirken von Gebirge und Ausbau in Abbaustrecken, *Glückauf*, 5 December 1962, pp. 1471–1479.
31. TERZAGHI, KARL., Introduction To *Rock Tunneling With Steel Supports*, The Commercial Shearing and Stamping Co. Youngstown, Ohio, 1956.

CHAPTER 6

STABILITY OF WIDE OPENINGS

WIDE OPENINGS

A "wide opening" as the term is used herein is typified by an excavation created during the extraction of ore or coal. Stopes, rooms, and longwalls are all wide openings in that all have face lengths which are much greater than the thickness of the vein or seam which is being mined.

As such an opening is extended the immediate roof strata, or hanging-wall rocks, if intact, act as beams and eventually, if lateral extension of the opening continues, the stresses in the beam exceed its strength and it buckles and falls. The next strata above is exposed and it too eventually buckles and falls.

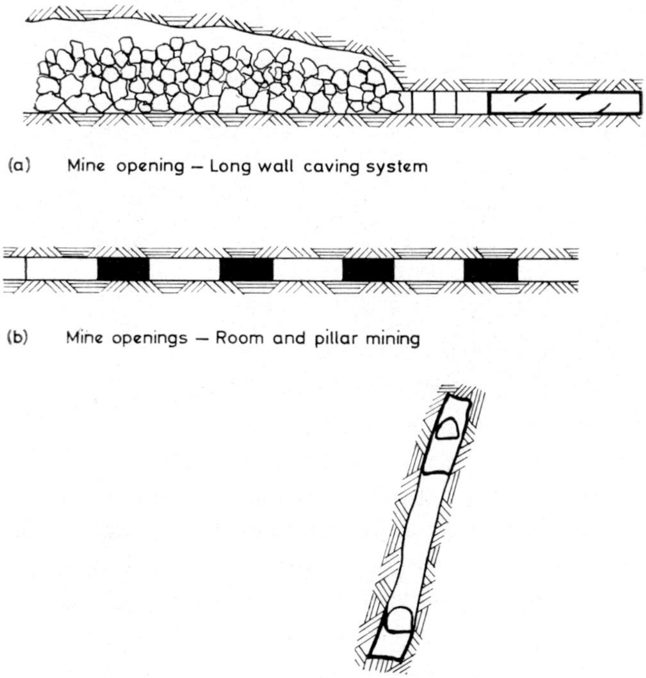

FIG. 1. Types of "wide" mine openings:
(a) Mine opening – longwall system. (b) Mine openings – room-and-pillar mining. (c) Mine opening – stoping system.

Even after the material in the roof (or hanging-wall) has been fractured it may still retain the ability to support itself through "arching action". This characteristic of pulverulent materials was illustrated in the Introduction to Part B, and the term "arching action" as used herein *does not* imply the formation of an opening with a vaulted arch but signifies, rather, the process by which fractured rocks may form partially self-supporting structures through the resolution of vertical forces into diagonal thrust.

The subsidence of beds overlying the mined-out area progresses upward toward the surface, but for small mined-out areas at great depth the subsidence may never reach the surface, stabilization being achieved by expansion of loosened or fractured material, or by the decrease in maximum thrust force as the rise/span ratio of "arching action" becomes more favorable.

For large mined-out areas subsidence penetrates eventually to the surface and total surface subsidence may vary from 30 to 100 per cent of the thickness of the mined-out bed. In the case of very deep workings, from 2 to 5 years may be required for surface subsidence to run its course and for stability to return.

Figure 1 shows typical stopes, rooms, and longwall systems, respectively.

CONTINUOUSLY ADVANCING FACES

Mine extraction operations involve the removal of a vein or seam over large areas. An individual area may comprise several square miles in the case of a coal mining operation. The gold mines of the Witwatersrand also extend over areas of many square miles. Other metalliferous deposits, although they are not so extensive as coal seams, frequently extend for thousands of feet both horizontally along the strike, and vertically down the dip.

During the mining of such mineral deposits large areas opened up and left unsupported so that the overlying rocks cannot bridge across from solid abutment to solid abutment. Thus the rear abutment must be located in the caved or filled area, behind the advancing face, that is the "goaf."

Thus the solid vein or seam just ahead of the face forms one abutment while the other end of the beam, or arch, must find its support in the filled or caved area at some distance behind the face. The long-wall mining systems used in European coal mining practice employ a continuously advancing face which may be several thousand feet in length. Extensive research has been carried on underground to determine magnitudes of ground pressures, locations of zones of "overpressure" (abutments), and the directions and magnitudes of ground movements in the vicinity of an advancing face. (See Volume III, Part B, Chapter 4, page 361, for an account of abutment pressures as determined by model tests in granular materials.)

INDUCED CLEAVAGE

The solid unmined vein or seam at the borders of a stope, room, or longwall supports the ends of the "roof beam" or "arch" and is therefore subjected to pressure greatly in excess of the depth pressure. These borders form the supporting "abutments" for the natural beam or arch and the overpressure at these locations will hereafter be referred to as "abutment pressure".

At depths of more than a few hundred feet the abutment pressures, resulting from concentration of depth pressure, σ_z, are of sufficient magnitude to induce shear fracturing of the softer roof rocks overlying coal seams, and fracturing of the coal in the seam. This condition is common to British and European coal mines, many of which operate at depths of from 1000 to 3000 ft, with a few approaching 4000 ft.

This type of rock failure, termed "induced cleavage" by Faulkner and Phillips[(1)] is important in most deep mines. The phenomenon was first described and explained by Faulkner and Phillips as observed in British coal mines, and this type of fracturing has been mentioned in a number of papers dealing with mining conditions on the Witwatersrand where induced cleavage cuts the strong quartzites of the roof and floor.

Because of the generally irregular nature of the wall and roof rocks associated with metalliferous vein deposits, induced cleavage is difficult to differentiate from natural cleavage. It is probable that a great deal of the transverse shearing in wall and roof rocks, which is generally assumed to be the result of geological forces, is in reality the result of forces induced by mining and extends only a short distance from the working faces.

Depth at which induced cleavage appears depends upon strengths of roof rocks, span of the opening, and physical properties of the vein or seam. Faulkner and Phillips have tabulated observations in a number of mines and have found that, in general, induced cleavage in shales first occurred at depths between 300 and 600 ft; the least depths at which cleavage was observed in sandy shales was 1800 ft, and in sandstone one instance of induced cleavage was observed at a depth of 2400 ft (maximum depth of the observations). It was noted that, in massive sandstones, the critical depths might be as great as 3000–3600 ft.

As to the effect of the width of a working on abutment stresses, Faulkner and Phillips [(1)] list the following observations at four longwall faces of varying widths:

Width (ft)	Depth (ft)	
75	585	No induced cleavage
150	555	Occasional induced cleavage
300	555	Induced cleavage prevalent
630	435	Induced cleavage prevalent; cleavage planes closer together and more regularly formed.

In long-wall workings with roofs of shale or mudstone the same authors found that stresses produced a flaggy roof near the ribs (shearing nearly parallel to bedding) to depths of about 900 ft. At greater depths induced cleavage was steeply inclined, cutting across the beds as previously described.

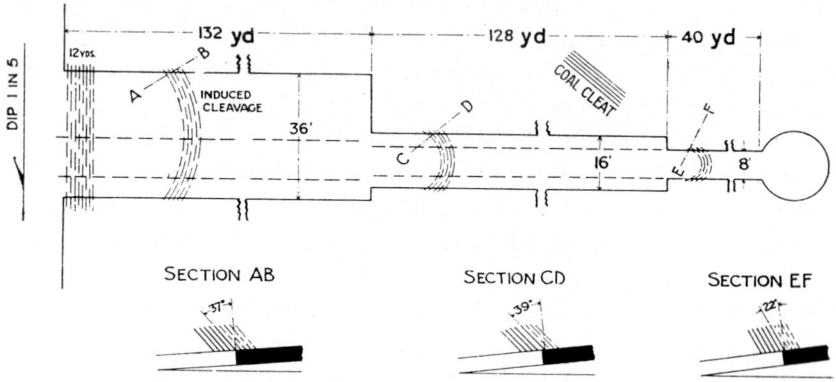

FIG. 2. The effect of changes in the width of a heading on the cleavage induced.[1]

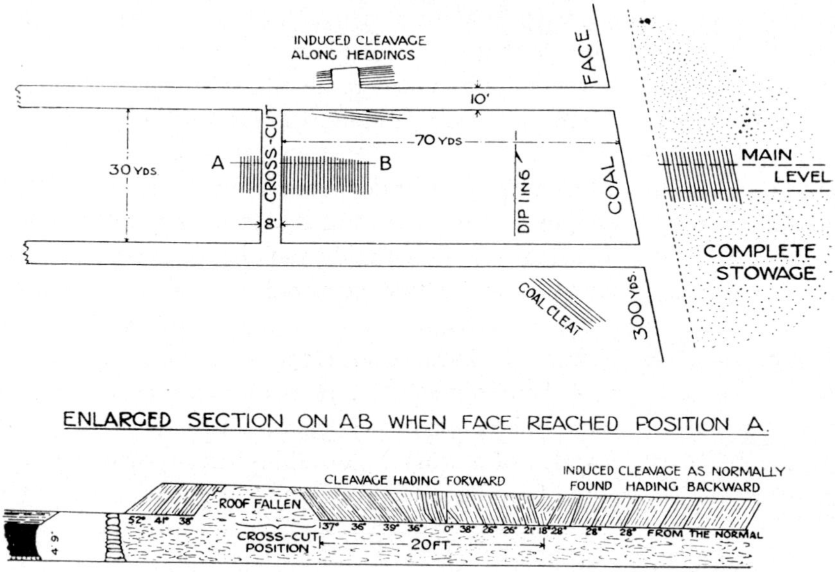

FIG. 3. Plan of headings in advance of a longwall face and section showing the effects of a cross-cut.[1]

Figures 2–9[1] show characteristics of cleavage induced ahead of long-wall faces.

As has been explained previously, such cleavage is induced when the major principal stress is sufficiently in excess of the minor principal stress to create effective shear stress exceeding the rock strength. The orientation of shear planes is the

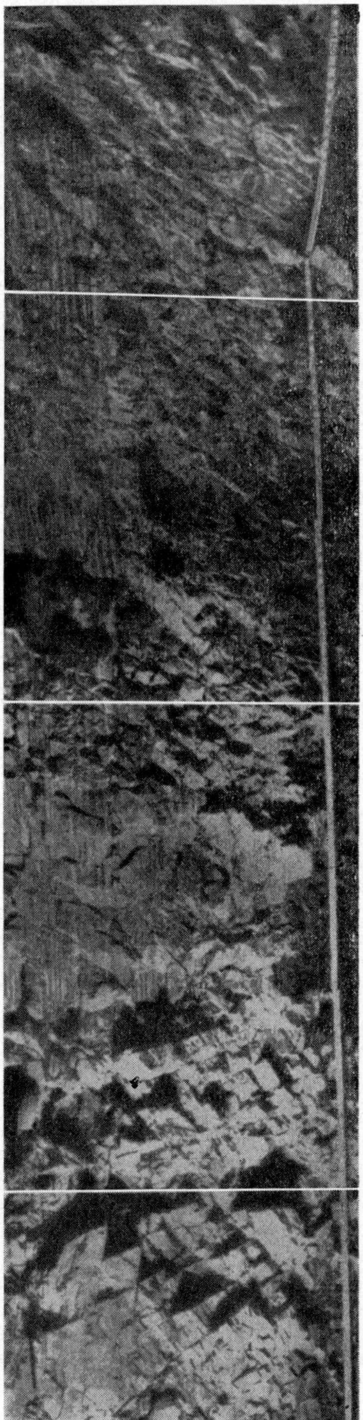

FIG. 4. Photograph showing the change in inclination of cleavage illustrated in the cross-section in Fig. 3.[1]

FIG. 5. Induced cleavage in shale and natural joints in sandstone.[1]

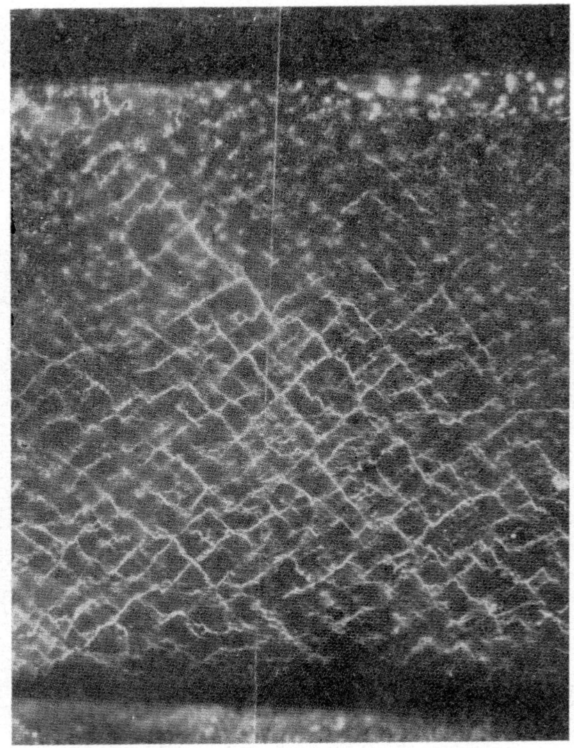

FIG. 6. Traces of planes of maximum shear stress on a face of a cube of paraffin wax under compression[1]

result of the directions of the principal stresses. The direction of the principal stresses may be rotated from vertical and horizontal by: (1) diagonal stresses resulting from the lateral thrust of a "voussoir beam" or "linear arch"; (2) "creep" or flow of coal or of roof or floor toward the free face. The effect of this creep has been analyzed mathematically and in some detail,[1] and it is evidently important in

FIG. 7. *Druckschlechten* (pressure cleavage) in the roof and floor of a copper-shale mine in Germany.[1]

determining the dips and directions of cleavage planes. However, the writer believes that diagonal stresses induced in abutments by "arching action" are at least as important in rotating the directions of the major principal stress away from a near vertical position. This reorientation produces shear planes inclined at angles generally about 25–35° from the vertical.

Where induced cleavage exists it obviously eliminates any possibility that the immediate roof can act as a beam. However, such roofs are observed, in many cases, to be self-supporting for appreciable spans and their strength can only be derived from "arching action" and the formation of a "linear arch".

"Induced cleavage" in the coal ahead of the face reduces the coal from an elastic solid to a fractured material and the position of the maximum abutment pressure moves to a distance ahead of the face such that lateral confining stress is sufficient to prevent further shearing.

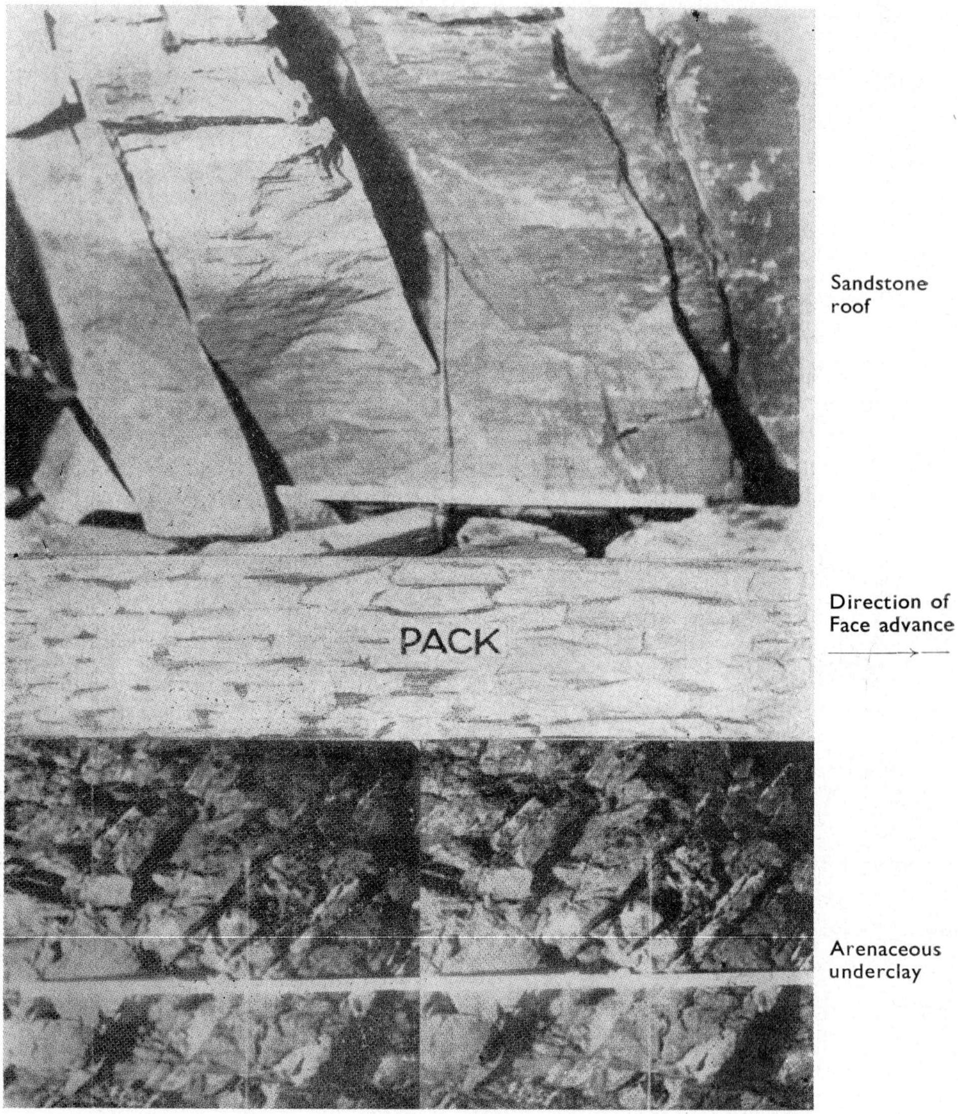

FIG. 8. Induced cleavage in a roof (sandstone) and a floor (arenaceous underclay).[1]

(a)

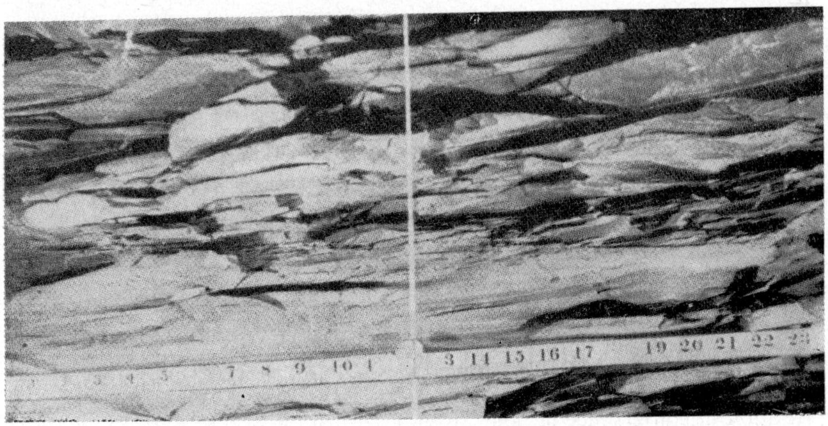

(b)

FIG. 9. (a) Induced cleavage. (b) "Flaggy" roof.[1]

EXTENT OF ABUTMENT ZONE

In a number of mines, measurements have been made to determine the distance which the abutment zone extends ahead of the face.

The relative magnitude of the overpressure has been estimated by measuring the amount of roof subsidence in workings already in existence ahead of a face.

Grond[2] found that roof subsidence ahead of the face in the Seam XIII of the Wilhelmina Mine, at a depth of 465 m (1520 ft) extended about 20–25 m (66–82 ft) ahead of the face although the rate of roof subsidence did not become significant until distance of 5–10 m (17–33 ft) ahead of the face. The immediate roof in this instance was about 10 m (33 ft) of sandy shale. This face had only been worked about 3–4 months and possibly the maximum extent of overpressures in abutments had not yet developed.

Other measurements made in a rock tunnel situated 10–24 m (33–79 ft) under the Coal Seam XII of the Queen Wilhelmina Mine showed an overpressure zone extending for about 80 m (262 ft) ahead of the longwall face. Depth was 320 m (1050 ft) and immediate roof was 20 m (66 ft) of sandy shale.

Faulkner and Phillips[1] have noted induced cleavage as far as 90 ft in advance of a long-wall face in British mines. This would indicate that significant abutment pressures extend more than a hundred feet ahead of a face in strong rocks.

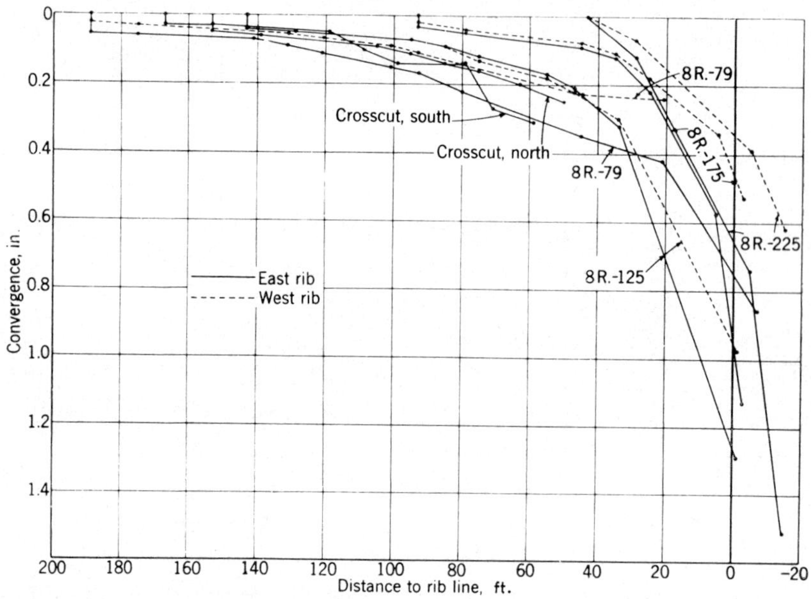

FIG. 10. Relation of convergence of points from 8R-79 to 8R-225, 15 Butt, to the distance from the rib line.[3]

Figure 10[3] shows results of measurements of convergence of roof and floor at various locations as a retreating rib line approached the measuring points. The coal bed was located at a depth of about 300 ft. The immediate roof consisted of about 10 ft of thin layers of coal and slate. These were overlain by the Pittsburgh Sandstone, at this location a strong sandstone about 50 ft thick. The remainder of the surface was limestone and shale.

Convergence was recorded as far as 190 ft from the rib line but important convergence was confined to a zone within about 100 ft from the rib line, and very rapid convergence did not commence until about 40 ft from the rib line. These curves are quite similar to those plotted from measurements ahead of longwall faces mentioned in preceding examples.

Horizontal Displacement

In addition to measurements of vertical subsidence of roof ahead of the face, Grond has measured horizontal displacements ahead of longwall faces in Dutch mines. Roof and floor rocks at 3–6 m (10–20 ft) ahead of a long-wall face were displaced several centimeters in a direction away from the advancing face. A point located 28 m (92 ft) ahead of the face was displaced 3 cm (1.2 in.) away from the face, while the face was advancing from 34 to 28 m (110–92 ft). The immediate roof consisted of about 10 m (33 ft) of shale.

It is apparent from this example that induced horizontal and vertical stresses exist ahead of an advancing face.

Witwatersrand

In the deep mines of the Witwatersrand high abutment pressures are prevalent but little information is available in the literature as to their extent. Joseph[(4)] says, regarding induced cleavage in mines of the Witwatersrand: "Usually they do not form more than a few feet in advance of the working face, but occasionally, if a stope has been standing, they may develop up to 20 ft or more ahead"; also, "cleavage cracks from early workings have been noted in later workings 80 ft above".

FRONT ABUTMENT PRESSURES

The foregoing examples have given some indication as to the lateral extent of the solid abutment zone on which significant overpressure acts. However, little exact information is available as to the magnitude of such overpressure. Van Interson[(5)] has deduced that, considering the ground as pulverulent material, theoretical maximum abutment pressure cannot exceed 1.7 σ_z.

Spruth[(6)] estimates normal maximum front abutment pressure to be about 1.20–1.25 σ_z but says that in solid rocks a "flash" overpressure may create a total stress of 2 σ_z.

In spite of the foregoing estimates there is much evidence pointing to the existence of front abutment pressures several times the magnitude of the depth pressure.

Jacobi[(10)] shows an estimated front abutment pressure equal to 5.5 σ_z as deduced from measurements at Neumuhl.

In some British mines induced cleavage in soft shales was noted at depths of 500 ft. At this depth σ_z would be about 500 psi but compressive strength of such rocks would be at least 3000–4000 psi. Thus abutment pressures would appear to be at least 6–8 σ_z in order to produce the observed cleavage.

In sandstone induced cleavage apparently does not occur until a depth of at least 2400 ft is reached. At this depth σ_z would be about 2400 psi and a sandstone with compressive strength of 10,000 psi would required an abutment pressure/σ_z

ratio of 10,000/2400 = about 4/1 = 4 σ_z to produce shearing. Stronger sandstones, for instance those with compressive strengths of 15,000 psi or more, would not shear until depths exceeded 3750 ft.

Consideration of pressures required to produce the phenomenon observed indicates that in intact rocks maximum front abutment pressures must attain values of at least four to six times the normal depth pressures.

Such values are still much less than those computed from theoretical considerations for stresses concentrated at the ends of wide openings in elastic materials (see Chapter 1).

Lateral Stress at Front Abutment

The principal compressive stress in the vicinity of an abutment is the result of the depth pressure, acting vertically, and a thrust acting laterally due to arching action. The direction of the resultant principal stress is inclined instead of vertical. At the abutment of a flat natural arch (linear arch) the lateral stress will be very high because the arch supports itself by resolving its weight into diagonal thrust. Even at relatively shallow depths the resultant diagonal compressive stress may be enough to cause induced cleavage.

It is thus possible for a diagonal compressive stress with magnitude several times that of the depth pressure to exist at an abutment. This can produce "induced cleavage" independently of the magnitude of the depth pressure and the vertical stress.

REAR ABUTMENT PRESSURES

There is some difference in the results of observations in British mines, and in German mines, as regards the location of the rear abutment, or zone of overpressure behind the advancing face. The British researchers have observed an increase of the span of the "pressure arch" (distance between abutments) with increase in depth of the workings, and in addition have observed a fairly sharply defined zone in which rear abutment pressure increases rapidly. This zone is identified by the rapid subsidence of the roof and the crushing of roadway supports. A rapid pressure rise has also been measured by pressure cells installed in the goaf.

British Observations

The approximate relationship shown in Table 1 between depth of the workings and distance between abutments has been observed in British Mines[7–9] and in German coal mines. Pressure measurements in British mines were made by means of pressure cells installed in relatively narrow "pack walls." Pressures measured on these pack walls were considerably in excess of the depth pressures but if we consider that a major portion of the measured overpressure on the pack wall was probably due to its relative width, the maximum unit load on the so-called abutment zone would not appear to exceed that due to depth pressure. In fact, a pressure equal to that of the depth pressure would in most cases suffice to induce the crushing of supports, and the compaction of fill which is observed in the mined-out areas at some distance behind the face and no overpressure (pressure in excess of depth pressure) would be needed to produce the observed phenomena.

TABLE 1

Depth (ft)	Width of maximum "pressure arch" (e.g. distance between abutments) (ft)
400	120
600	150
800	180
1000	210
1200	240
1400	270
1600	300
1800	330
2000	360

German Observations

An extensive program of measurements recently made in Germany in the mines at Neumuhl, Germany, has been reported by Jacobi[(10)]. These measurements failed to locate any sharply defined pressure increase in the goaf which could be identified as a "rear abutment," but found rather a gradual increase of pressure at increased distance from the face until the depth pressure was approached. These results differ from those of the British who have found rather a sharp pressure increase at a certain distance behind the face. The two conceptions are illustrated in Figs. 11 and 12.

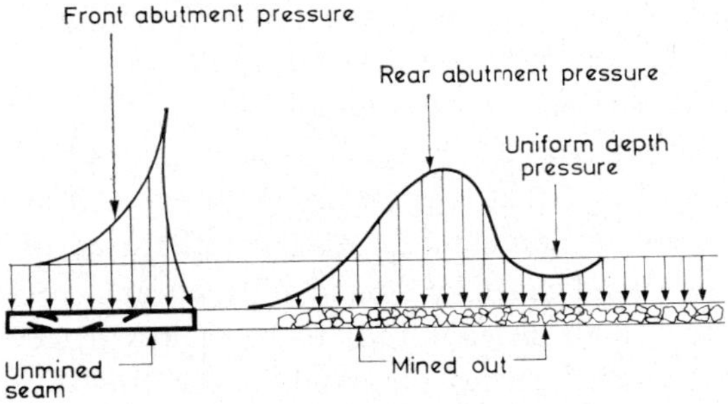

FIG. 11. Pattern of pressure distribution indicated by British observations.

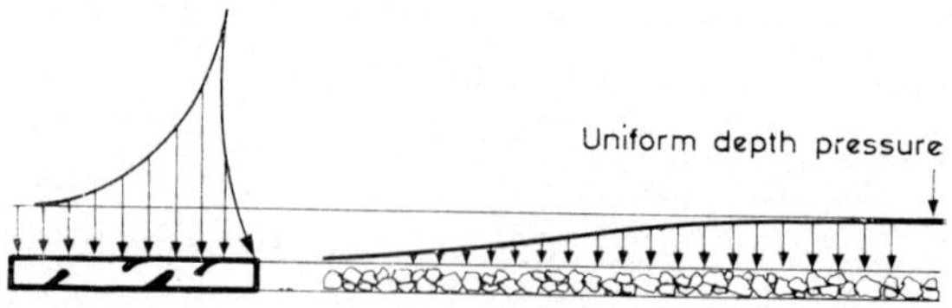

FIG. 12. Pattern of pressure distribution indicated by German observations.

It is evident, from the laws of statics, that the deficiency of vertical pressure in the zone behind the face must be compensated for by an excess of vertical pressure at some other location. All authorities agree, and measurements have demonstrated, that a zone of high "overpressure" exists immediately ahead of the face, and it would seem that a zone of overpressure, although of less intensity, would exist at some distance behind the face. Results of pressure measurements at Neumuhl have been summarized by Jacobi[(10)], and Fig. 13 shows the load distribution deduced from the measurements.

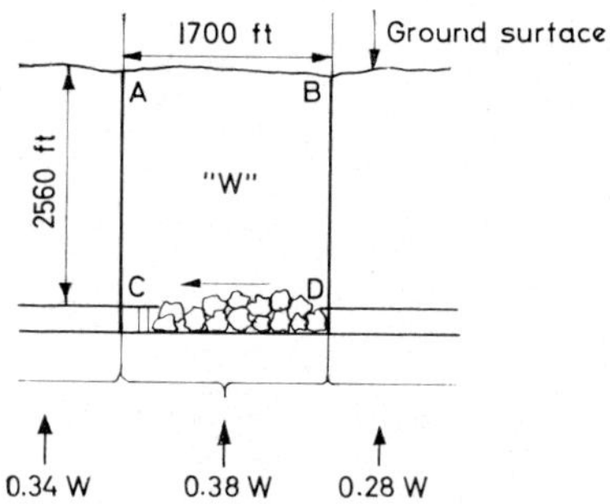

FIG. 13. General distribution of pressure due to the weight *W* of the undermined block *ABCD*. (Adapted from Jacobi.[(10)])

This seam was worked out over a length of 530 m (1700 ft) in $16\frac{1}{2}$ months. It represents the working out of a residual pillar with a width of about 250 m, and adjoining areas on both sides has been previously mined out.

As shown in Fig. 13, the toal excess vertical load measured on the front abutment was equal to 34 per cent of the weight of block *ABCD* and the total vertical load measured on the goaf was 38 per cent of the weight of block *ABCD*.

Thus it would seem either that the remaining 28 per cent of the weight of block *ABCD* must be borne by abutments outside of the area of the measurements, or that the remaining load must be spread over an area so large that increased unit pressures are too small in magnitude to be readily detected.

It is probable that the differences observed in the British and in the German mines are partially due to differences in the structural characteristics of the strata.

For instance, a badly fragmented or weak strata would tend to distribute pressure over a considerable area so that overpressure would be difficult to detect, while strong strata would tend to form a "linear arch" which would concentrate the overpressure within abutment zones of limited extent.

In addition, the presence of natural lateral stress may be an important factor in determining whether an abutment zone is marked by sharp increase in vertical pressure. The presence of natural lateral stress aids the strata over the seam and goaf to form a more rigid and cohesive structure, such as a "voussoir beam" or "linear arch" and such well-defined structures will tend to have well-defined abutments.

The increase in the distance between front and rear abutments which was observed with increased depth of workings may be partially explained as due to the increase in lateral stress with depth. This tends to stabilize the roof just as lateral stressing strengthens a pre-stressed concrete beam such as is used in civil engineering construction.

The relative dimensions of the area mined out during the Neumuhl pressure measurements[(10)] were such that the apparent lack of a "rear pressure abutment" may have been due to the supporting effect at the sides of the relatively narrow strip.

The pillar which was mined out was about 800 ft wide, 1700 ft long (along the direction of face advance) and situated at a depth of 2560 ft. Thus the width of the mined-out area was only about one-third of its depth, and it might be expected that a large portion of the weight of the undermined block would be transferred to the sides rather than to the front and rear. The existence of both front and rear abutments in French iron mines has been demonstrated by measurements, the results of which are presented below.

Abutment Pressures in French Iron Mines

Measurements made in French mines where bedded iron deposits are extracted by retreating-pillar methods have definitely proved that "pressure abutments" exist in the caved area of the goaf as well as ahead of the face. Deformation measurements were made in a gallery beneath the "Yellow Savage" iron ore bed. This bed was being mined by retreating pillar system with full caving of the roof behind the retreating pillar line. Depth beneath the surface was about 550–650 ft.

In the gallery measurements were made of the vertical deformation of the walls of the gallery. High pressure zones in the overlying workings caused vertical compression and vertical shortening of the walls, while zones of pressure release allowed vertical expansion of the walls of the gallery.

Figures (14(a)–14(e) illustrate the transition, with increasing width of the mined-out zone, from two abutments to three abutments, and finally to two separate sets of abutments.

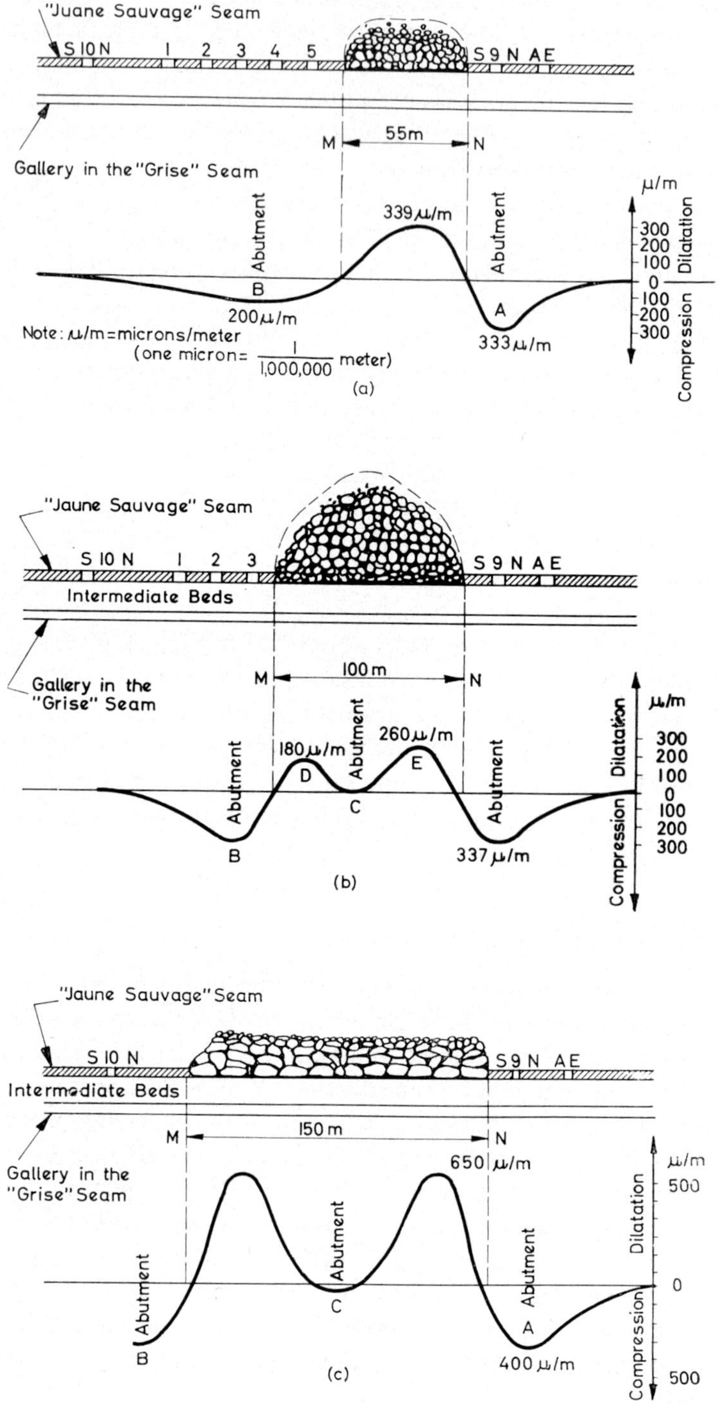
"Juane Sauvage" Seam
S 10 N
1 2 3 4 5
S 9 N AE
55m
M
N
Gallery in the "Grise" Seam
Abutment
339 μ/m
Abutment
μ/m
300
200
100
0
100
200
300
Dilatation
Compression
B
200 μ/m
A
333 μ/m
Note: μ/m = microns/meter
(one micron = 1/1,000,000 meter)
(a)
"Jaune Sauvage" Seam
S 10 N
1 2 3
S 9 N A E
Intermediate Beds
Gallery in the "Grise" Seam
M
100 m
N
180 μ/m
260 μ/m
D
E
C
B
337 μ/m
(b)
"Jaune Sauvage" Seam
S 10 N
S 9 N AE
Intermediate Beds
150 m
650 μ/m
500
0
400 μ/m
(c)

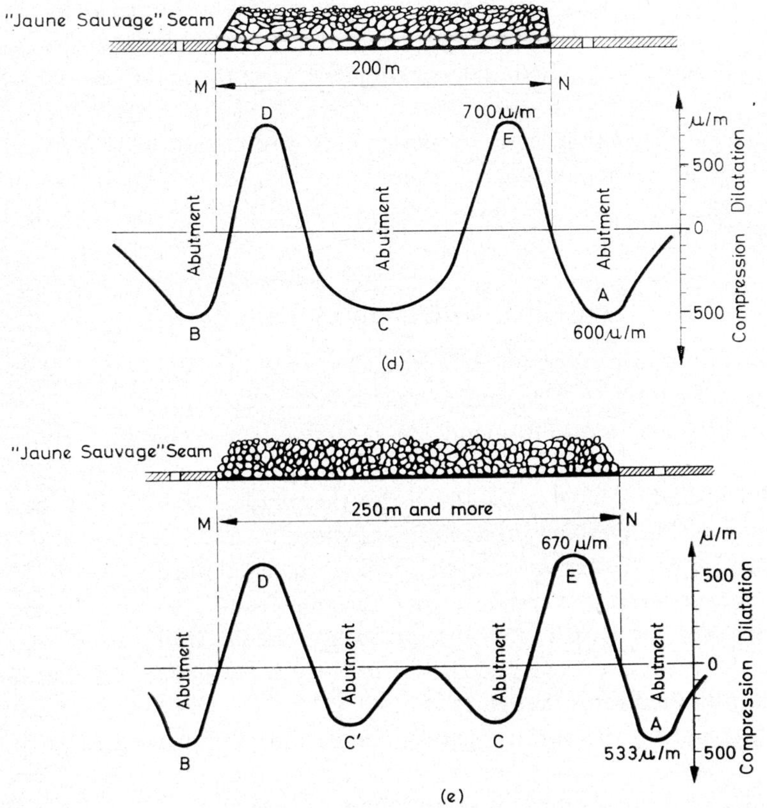

FIG. 14.

(a) Vertical deformations measured in the roadway X–X' which was situated in an unworked bed below the workings in the *Couche jaune sauvage*. (b) Vertical deformation measurements in roadway X–X' for a mined out width of 100 m. The pressure abutments are marked by the points A and B where the maximum vertical compression occurs. (c) Vertical deformation measurements for a worked-out width of 150 m. (d) Vertical deformation measurements for a mined-out width of 200 m. Deformations at the pressure abutments A and B have reached a maximum. (e) Vertical deformation measurements for a mined-out width of 250 m. The middle abutment at C has spit into two and pressures have decreased.[14, 19]

In Fig. 14(a), with width of the mined-out zone about 180 ft, compression measurements in the gallery indicate that pressure abutments exist at A and B at each edge of the mined-out area.

In Fig. 14(b) the width of the mined-out area has reached about 330 ft. The mined-out span has exceeded the arching capabilities of the rock and a third abutment is just beginning to form at C. Thus we now have two arches forming, each of which has a common abutment at C.

In Fig. 14(c) the mined-out span has increased to about 500 ft and pressures are increasing at all three abutments. In Fig. 14(d) the mined-out span has reached 650 ft and abutment pressures have reached a maximum. In Fig. 14(e) the mined-out span is about 800 ft and the rock above the caved area can no longer span to a center abutment. Thus two independent spans have formed, one with abutments BC'

and the other with abutments *AC*. Pressures on each abutment have decreased, and, other conditions remaining the same, the two independent systems of abutments will progress along with the working face, with the pressures remaining about the same.

It should be noted that these measurements give comparative vertical pressures only and the actual magnitude of abutment pressures in pounds per square inch cannot be derived from the measurements. (See Volume III, Part B, Chapter 4, page 361, for results of model tests.)

CRITICAL WIDTH OF AN EXCAVATION

At a certain width of span, *MN*, the abutment pressures reached a maximum as shown in Fig. 14(d). The single "rear abutment", *C*, common to both arches, then split into two "rear abutments" with consequent decrease in abutment pressures.

Tincelin and Sinou have called this width the "critical width". For the case shown it was about 650 ft. Designating the critical width by L, the relationship between the depth Z and L was determined, in these mines, to be about $L = 0.9Z$. This was not an exact relationship as variations in L ranging from -20 to $+10$ per cent were noted. (See Volume II, Chapter 3, page 88, for critical widths.)

Before the excavation reached the critical width subsidence of the surface progressed at a limited rate; that is, at about 0.5–1 cm per week. As soon as the critical width was reached, the surface subsidence became rapid, with rates of 10–20 cm per week.

The value of L is affected by the following factors:

(1) Rate of retreat of the pillar line.
(2) Thickness of the overlying strata.
(3) Structural characteristics of the overlying strata.
(4) Thickness of the mined-out seam.

COMPARISON WITH BRITISH OBSERVATIONS

Observations in British coal mines (Table 1) indicate that for comparable depths (600–800 ft) the average critical width L (note: L is equal to twice the width of the span of the pressure arch) is equal to, roughly, $L = 0.5\ Z\pm$. Thus the average critical width for this depth in British coal mines is about one-half of that observed in French iron mines. The differences are probably accounted for by the factors mentioned previously and by differences in mining methods. British coal mines generally employ long-wall systems while the French iron mines use a retreating-pillarline extraction system.

It is also of interest that the ratio of average critical width as found in British coal mines decreases with increased depth. At a depth of 400 ft it is about $L = 0.6\ Z\pm$, while at a depth of 2000 ft it is about $L = 0.4\ Z\pm$.

ABUTMENT PRESSURES IN STOPING SYSTEMS

The foregoing discussion has considered stress conditions around long-wall faces. The true long-wall mining system is rarely employed outside the European coal mines and the South African gold mines. However, the conditions associated with longwall mining, such as regularity of advance, absence of side abutments, uniformity of strata, etc., make such mines favorable places in which to measure rock movements and rock pressures. A great deal of this work has recently been done in Europe. In addition, stress conditions ahead of a long-wall face resemble somewhat the conditions behind a retreating pillar line during final total extraction.

The results of such measurements give us an insight into rock stresses, rock movements, and support pressures encountered along the borders of extensive mined-out areas which are being mined by other methods such as stoping systems, or room-and-pillar mining systems.

In the case of metal mine stopes, the lack of uniformity in wall rocks, the random fracturing, and the variations in dip and thickness of veins hinder attempts to measure those factors which can be measured under the more favorable conditions of long-wall mining.

It is the author's opinion that many of the pressure phenomena commonly observed in metalliferous mining can be explained on the same basis as those cited for coal mining.

For example, the "squeeze" on sill and raise timber which occurs after a stope back has advanced for a certain distance above the sill probably marks the rear abutment of a "pressure arch" which has its front abutment ahead of the advancing stope back. The author has observed this squeeze to begin after stope faces have advanced from 75 to 150 ft. above a sill, the distance depending upon a number of factors which include the depth of working, existence of natural lateral stress, speed of advance of the stope back, compactness of gob filling, and nature of adjacent workings.

Careful observation of the development of pressure on drifts, raises, and stope timber can give information as to the span of the "pressure arch" for a particular portion of a mine. The information can be usefully applied, as for instance in spacing development raises or levels so that a stope can be mined out before the rear abutment begins to crush supports and close access openings.

The access openings, such as drifts and raises leading to stoping systems, have many of the same characteristics and support problems which are found in the access roadways leading to long-wall faces. In a later section it will be pointed out that some of the support techniques employed in longwall roadways could profitably be applied to metal mine drifts and raises.

In the following sections the structural behavior of mine roofs is analyzed.

STRUCTURAL ANALYSES OF MINE ROOFS
ROOF ANALYZED AS A BEAM WITH FIXED ENDS

In stratified rocks when the strata are relatively sound and possess lateral continuity it is possible to make a theoretical investigation of the structural stability of a mine roof by considering the immediate roof as a series of contiguous beams of unit width with fixed ends. This method has been illustrated by Merrill in investigating the stability of the mine roof in the experimental oil-shale mine of the U.S. Bureau of Mines[(11)].

The most common case is that encountered in coal mine rooms where the roof is of intact stratified rock. Under such conditions the stability of the roof may be analyzed by considering the roof as a fixed-end beam with a thickness which is equal to that of those beds immediately overlying the opening which tend to form an integral unit, that is, those beds which do not separate easily along bedding planes. The strength of the beam is modified by the strength of the rock as it is subjected to long-time stresses. If the rock is subjected to a moderate stress applied over a long period of time it will eventually yield as much as if it had been subjected to high stress for a short period of time. Such behavior will eventually lead to the collapse of the beam.

Actually, an intact mine roof is a slab or plate (since it is fixed at all edges) rather than a beam. When the long dimension of a plate is twice (or more) the short dimension then the beam and plate theories give approximately the same values for the maximum tensile (flexural) stress. When this maximum flexural stress becomes equal to the modulus of rupture for the rock then failure of the beam occurs. (It should be noted that the tensile strength for brittle materials in flexure is much greater than the strength in pure tension.)

Beam Analysis

Methods for computing the stability of roof beams are covered in the section of Chapter 1 which is titled "Single Openings in Bedded Rock" (p. 73).

The equation of most interest is:

$$L = \sqrt{\left(\frac{2Rt}{\gamma F}\right)} \qquad \text{(Eq. 19, Ch. 1)}$$

where:

L = safe roof span;
R = modulus of rupture of the roof rock (tensile strength in flexure);
γ = density (weight per cubic foot taken as 156 pcf for Fig. 15);
t = thickness of the roof beam;
F = factor of safety (taken as 8 for Fig. 15).

Obert and Duvall[(11)] prepared a chart (Fig. 15) showing the computed relationship between safe roof spans, tensile strength of the rock, and roof thickness.

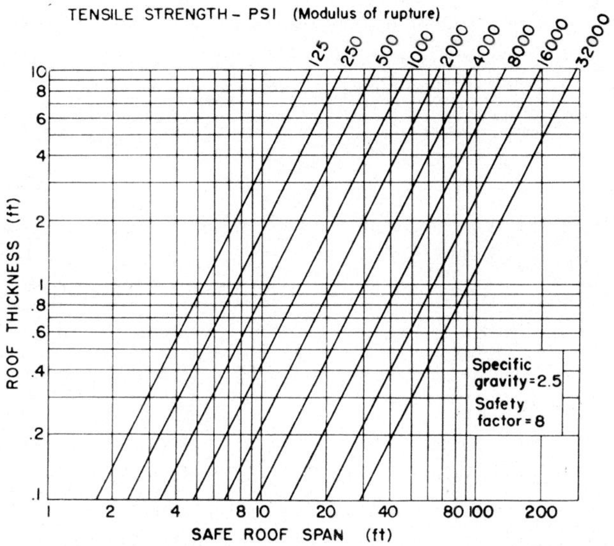

FIG. 15. Computed safe roof span vs. roof thickness and tensile strength of the roof rock.[11]

Application

The foregoing "fixed-end beam" simile is the simplest concept of the stability of wide openings. It is the type of behavior to be expected in near-surface consolidated sedimentary rocks in which the natural lateral stresses are relatively feeble in magnitude. This is the condition encountered in many U.S. Coal Mines where the roof consists of intact stratified rock and is located at a depth of not more than 400–500 ft.

Design of Oil-Shale Mine Roof[11]

Conditions in the oil-bearing shales of Colorado favor the application of theoretical methods for investigation of strength of mine roof. The strata are almost horizontal, are strong, contain few transverse fractures such as faults, fissures, or joints, possess lateral continuity, and are located at moderate depths. Also, the gradational nature of the oil content allows some latitude in choosing a structurally strong stratum for the roof. In addition, the laboratory tests for strength made on samples gave fairly consistent results from each stratum.

Average results of laboratory tests for physical properties of oil shale were approximately as follows:[11]

Compressive strength	16,000 psi
Modulus of rupture	4000 psi
Modulus of elasticity	1.1×10^6 to 3.5×10^6

Shear strength (parallel to bedding)	1200 psi
Shear strength (perpendicular to bedding)	3500 psi

Examination of drill cores for the area selected for the experimental room showed that the formation is a succession of relatively uniform flat-lying beds ranging in thickness from a few inches to 54 ft or more. The formation immediately above the proposed roof horizon was considered to consist of a bed approximately 8 ft thick with no thinner beds lying above it.

Model Tests

In addition to the theoretical investigations described previously, model tests were made on prototypes by Wright and Bucky. Mine models were constructed and tested to failure, using a centrifuge to provide adequate gravity loading. These tests were made with three-member models, and the bottom member, corresponding to the immediate mine roof, was assumed to be overlain by two thinner strata whose weight would be partially borne by the bottom member.

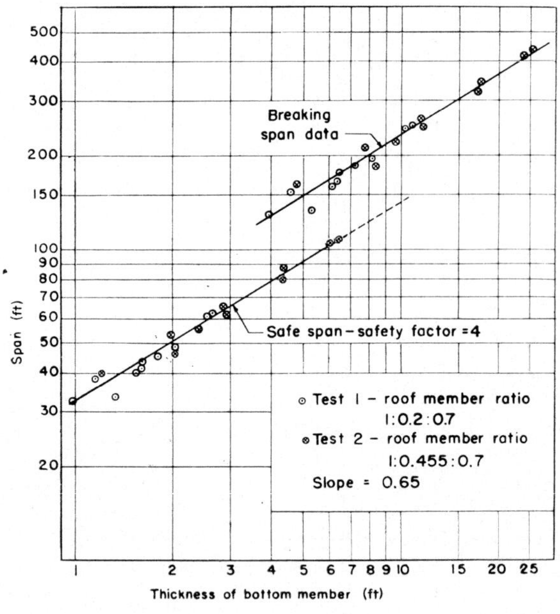

FIG. 16. Prototype breaking span and safe span vs thickness of the bottom member.[11]

The results of the tests are shown in Fig. 16. Wright and Bucky considered the length of span to be the diagonal distance between opposite corners of adjacent pillars; hence the measured safe roof spans are divided by $\sqrt{2}$ to determine recommended spans. This is shown in Fig. 17. The theory of multiple member roof beams is discussed in Chapter 1, page 73.

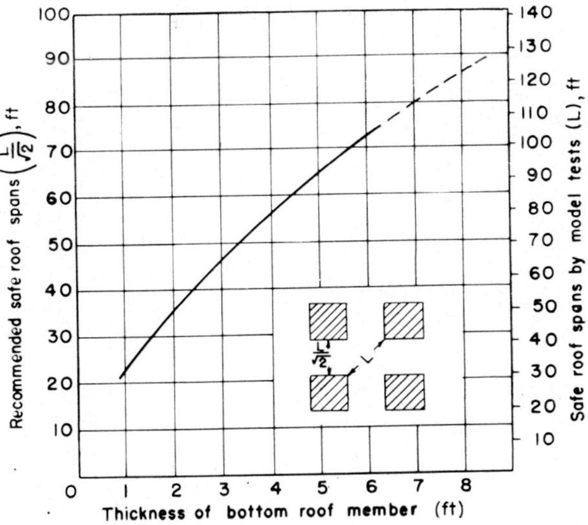

FIG. 17. Measured and recommended safe roof spans vs. roof thickness of bottom roof member.[11]

Experimental Room Results – Oil Shale Mine[11]

The test room was started in 1946 and widened, in a succession of operations, until it reached its final dimensions of 80 ft × 200 ft in June, 1951. In August, 1951, a slab 20 in. thick fell from each side of a transverse fracture across the roof, and 2 weeks later, a slab of the same thickness fell over the remaining area of the original 80 × 100 ft room. The remainder of the roof was still intact in January, 1954.

The breaking span of a 20 in. thick roof, computed from theory and using average physical property data, is 96 ft. Extrapolating the breaking span from model test data (see Fig. 16) the breaking span is 76 ft.

Roof Sag

Measurements of roof sag in the experimental room were made daily and data from Station *A* is shown in Fig. 18. In general, the sag rate was high during periods when the room was being widened and low when no mining was in progress. From elastic theory the sag rate during periods of inactivity always should be zero. The sag which occurred during periods of inactivity is referred to as "time sag" and probably corresponds to the "creep", "time yield", or "time strain" which is described in Chapter 4, page 175. During active periods the measured sag was greater than that which could be explained on the basis of elastic theory and part of such sag was attributed to "time sag".

For the final 80 ft × 200 ft room the total measured roof sag at Station A–0 was 0.605 in. Of this, 0.370 in. was attributed to sag during inactive periods (1510 days), and 0.235 in. to sag during widening periods (58 days).

An 8-ft thick roof behaving as an elastic beam should have sagged approximately 0.076 in. for an 80-ft span.

The difference between theoretical and actual roof sag is attributed by Merrill to two possible causes:

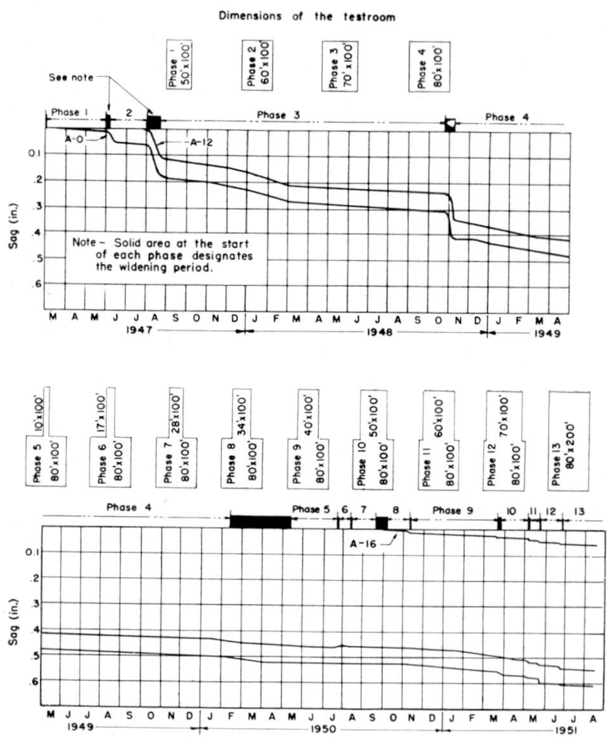

FIG. 18. Sag data, Station A. March 1947 to August 1951. (Letters indicate the months.)[11]

(1) Existence of "time sag" in addition to normal elastic deformation.

(2) Stratascope and drill core examinations show that the 8-ft roof contained fractures and separation of limited areal extent and therefore it may be assumed that the roof consisted of a number of partially bonded layers. Thus it would deflect more than a beam of homogeneous elastic material.

ROOF ANALYZED AS A PRE-STRESSED BEAM

The foregoing treatment has considered mine roof as a beam subjected only to gravity loading forces. Corlett and Emery[15] have pointed out that the natural (primitive) lateral stresses which are generally prevalent in geological materials will also have important effects on the stability of roof strata.

Figure 19(a) shows a simple beam composed of five separate pieces, each 4 ft long, butt jointed, but not coupled to each other.

(a) The beam would not support its own weight. The joints would open under tension and the beam would collapse. (Such a beam 6 in. wide and 12 inches deep would weigh about 80–90 lb. ft if composed of the average rock.)

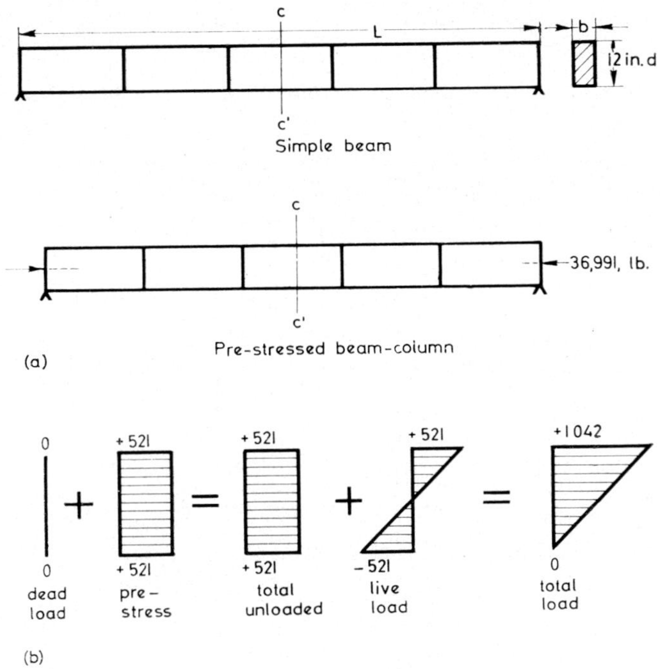

FIG. 19. (a) Pre-stressed beam. (b) Stress diagram for pre-stressed beam.[15]

(b) Neglecting the weight of the beam and assuming that it supports a live load of 125 lb./ft of beam length, the moment at midspan would produce stresses at $C-C'$ of 521 psi (compression in the top and tension in the bottom) if the beam could support them.

(c) If the beam is subjected to an axial compressive stress of 521 psi (an axial load of 36,991 lb.) tensile stress will be nullified and the beam will now be stable. (See stress diagram in Fig. 19(b).)

Natural lateral stresses of a magnitude of 500 psi or more might normally be encountered in sedimentary rocks at depths below 1000 ft or so and might be expected in igneous rocks and in metamorphic rocks at any depth more than a few hundred feet below the surface. (See Chapter 2.)

It appears that these stresses play an important part in stabilizing roof rocks above wide openings such as entries, rooms, flat stopes, and other openings which are excavated with wide flat roofs or backs.

Examples of Effect of Lateral Stress on Roof Stability

Emery[(16)] cites an example of a metal mine in a stratified roof in Canada where roofs of narrow openings failed by spalling rapidly and sometimes violently. Measurement by photoelastic coating techniques (Chapter 11) indicated the presence of high lateral compressive forces and the existence of sufficient prestress to allow the opening of rooms four times as wide as those normally driven. As a first step room widths were doubled and the results are being assessed before proceeding further.

Another example cited by Emery[(16)] is that of another metal mine where the ore occurs in two beds separated by 6–12 ft of barren quartzite. The beds dip at about 17° on the average. Initially, the upper bed was mined first by a room-and-pillar method and the roof was bolted and stood well over 50-ft spans.

Relaxation and high lateral prestress caused the floor to crack, and it generally collapsed when undermined in early attempts to remove the lower seam. It was found, however, that when the lower seam was mined first and the roof was bolted, relaxation and cracking were reduced to the point where the upper seam could be mined; the operation was adequately supported by the intervening slab through the pre-stress on its ends.

ROOF ANALYZED AS A "LINEAR ARCH"

Every miner is familiar with openings in moderately fractured rock which stand open without support, and with openings in badly fractured rock which stand with a minimum amount of support. Such conditions are common in the metalliferous mines and some concept other than "beam action" is required to explain the stability of the roof (or the hanging-wall) under such conditions.

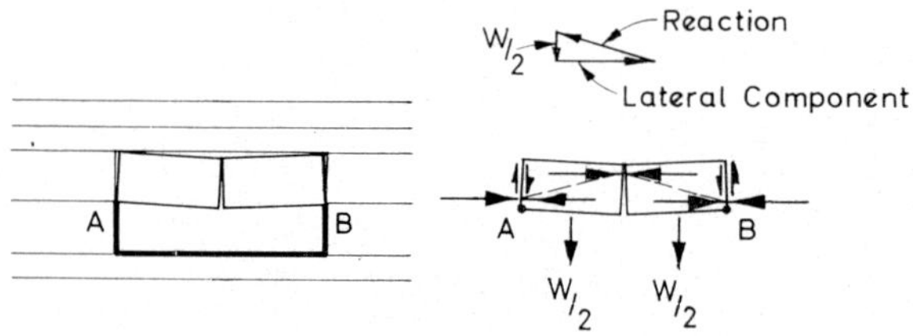

FIG. 20. Arching action in fractured strata.

The strength of such a roof is derived from "arching action". About 25 years ago Evans[(12)] demonstrated, both theoretically and experimentally, that fractured roofs may support themselves in much the same manner as do the voussoir arches used in civil engineering construction.

If we take the simple case of an opening overlain by strata which are fractured as shown in Fig. 20 it is apparent that the two blocks will be prevented from falling because they are prevented from rotating about points A and B. The restraining forces preventing rotation are simply the lateral reactions, as shown by the arrows. Friction forces at the ends of the "blocks" resist shear forces and prevent the blocks from moving vertically downward. This is the simplest example of "arching action". If this simple arch is cut by numerous steep fractures a structure similar to Evans's experimental brick beam (Fig. 21) will result.

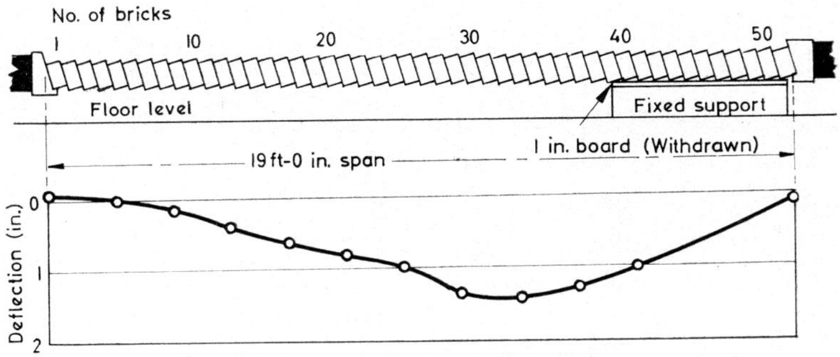

FIG. 21. Deflection of an experimental brick beam after removal of 1 in. of support.[12] (British Crown Copyright)

Evans termed the structure of Fig. 21 a "voussoir beam." The author prefers the term "linear arch" and will use that term hereinafter.

It is apparent that the decoupled rocks are supported entirely by compression and shear resistance and that the strength of the "linear arch" does not depend at all on the tensile, or flexural, strength of the rock, that is, resistance to bending moments, but on the compressive strength which is normally at least four to five times as great as the flexural tensile strength, in unfractured rock and is infinitely greater in fractured rock.

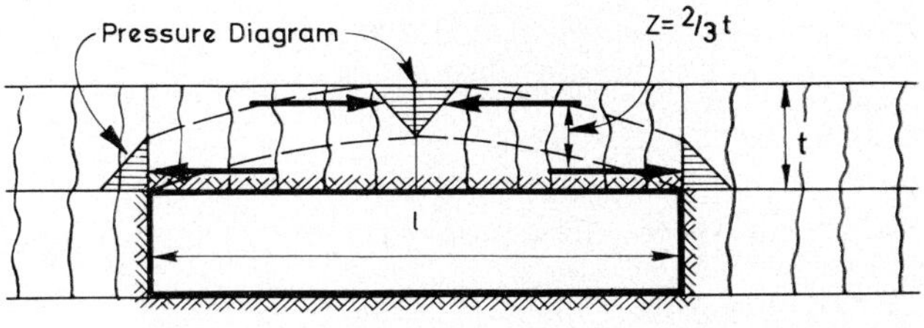

FIG. 22. Elements of the "linear arch" or "voussoir beam".[12]

The stability of the linear arch is a function of its thickness/length ratio as well as the strength and modulus of elasticity of the rock. The latter is an important property because it determines the amount of yield along the arch line, which in turn determines the "deflection" or "sag" of the linear arch. This in turn determines the length of the arm of the resisting moment which is shown as Z in Fig. 22.

Evans employed the compressive strength of the rock and the distance between centers of pressure to compute the approximate maximum resisting moment of a linear arch. He then equated the maximum resisting moment to the maximum bending moment of a gravity loaded arch to determine the maximum length of a self-supporting linear arch.

The basic equation derived by Evans is:[(12)]

$$l = \left(\frac{5Qt}{3\gamma}\right)^{1/2} \tag{1}$$

where (see Fig. 22)

l = maximum length of self-supporting linear arch;
Q = maximum compressive strength of the rock;
t = thickness, or vertical depth of the arch;
γ = density (weight in pcf).

The application of the equation (1) is explained in the Appendix at the end of this chapter. The equation gives only a first approximation of linear arch length and the final length must be determined by successive trial calculations which take into account the elastic yield of the arch and abutments.

The linear arch depends for its stability to a great extent on the rigidity of the rocks, that is, upon non-yielding arch line and abutments. For strata with high E values and with thicknesses in excess of 150–200 ft. the linear arch will be nearly as strong as a fixed-end beam in unfractured strata of the same thickness.

Maximum Spans for Self-Supporting Linear Arches

Figure 23 shows the calculated maximum spans for self-supporting gravity loaded voussoir beams, for various typical values of modulii of elasticity, E. These have been calculated by the same methods used by Evans except that, in the present case, abutment strain has been taken into account. Examples of calculations are given in the Appendix to this chapter.

As a compromise between the density of rocks common to coal fields and those common to metalliferous mines, γ is taken as 156 pcf (specific gravity 2.5); then equation (1) becomes:

$$l = \left(\frac{5Qt}{(3)(156)}\right)^{1/2}$$

and converting Q from pounds per square foot to pounds per square inch.

$$l = \left(\frac{(5)(144)Qt}{(3)(156)}\right)^{1/2} = 1.24(Qt)^{1/2} \tag{2}$$

where Q is expressed in pounds per square inch, t is in feet, and l is in feet.

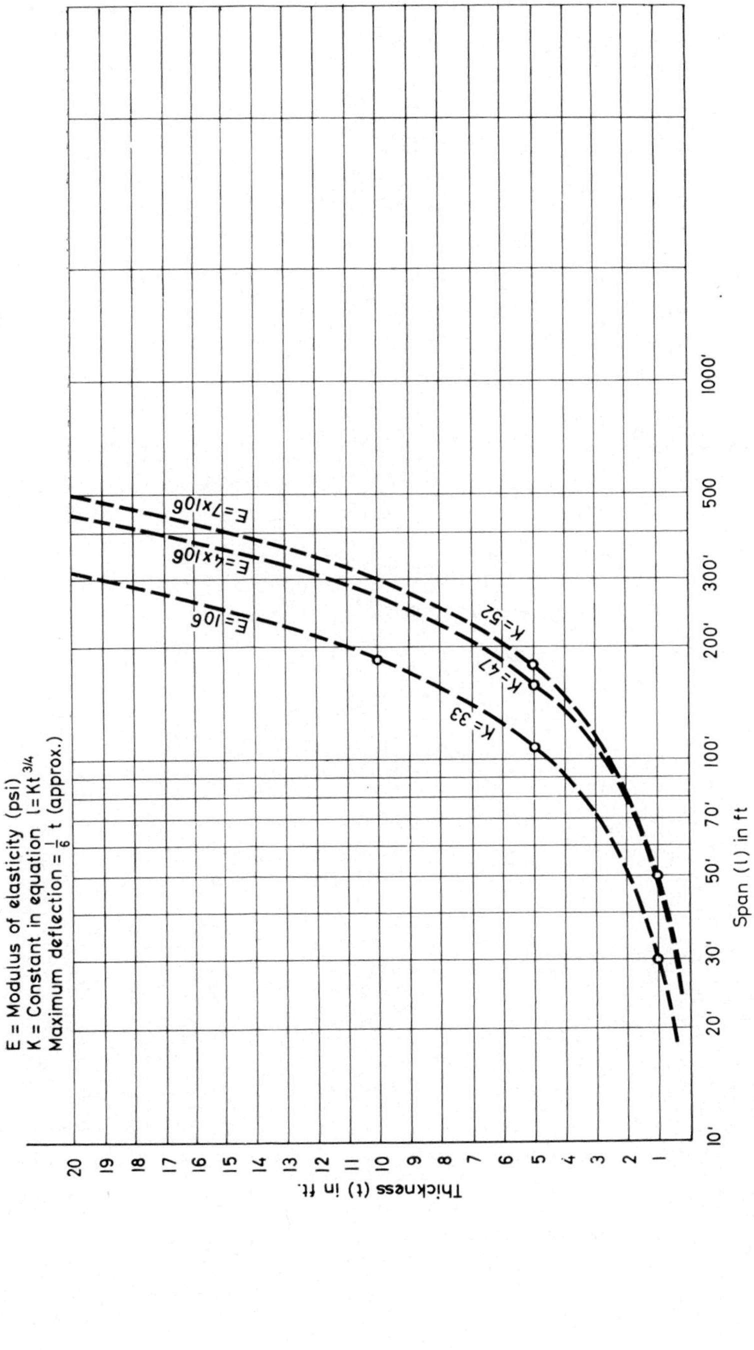

FIG. 23. Theoretical maximum spans for self-supporting "linear arches".

As a linear arch deflects the length of the effective arm Z (Fig. 22) of the couple producing the resisting moment, is reduced. The final deflection is determined by calculating deflection, determining increased stress due to deflection, calculating new deflection, new increased stress, and repeating the operations until the stress increases approach zero.

Deflection of the beam is due to bending moment and to compression along the arch reaction line which has the effect of shortening the arch and thereby reducing the value of Z, the arm of the couple producing the principal resisting moment. The bending moment is only a small fraction of the total moment of resistance (see Evans(12).)

The stability of linear arches less than about 40–50 ft thick is determined solely by the value of E. These values are indicated by the dashed lines in the graphs. For example, calculations show that an arch 10 ft thick (of strata with a modulus of elasticity $E = 10^6$ psi) would collapse when its length reaches 190 ft. However, even as it deflects to the failure point, the compressive stress, f_1, attains a maximum value less than 4000 psi because the arch and abutments yield sufficiently at this stress to permit buckling. For the arches shown on the graphs the dashed lines represent failure at stress less than the maximum compressive strength of the rock.

The dashed curves conform approximately to the equation $l = Kt^{3/4}$. K is shown in Fig. 23.

The maximum deflection of a linear arch less than 40–50 ft thick may be expressed as the ratio deflection/thickness. This value is a constant, that is, $\delta/t = 1/6$ approximately.

The foregoing discussion has related to gravity loaded linear arches, that is, arches which support only their own weight. If soft, or badly broken, strata overlie such an arch, then spans given in the charts will be greatly reduced. For such conditions maximum spans can be computed by adding the weight of the overlying strata to that of the supporting strata to obtain an "apparent weight". For example, a 10-ft thick arch overlain by 15 ft of soft strata: "apparent weight", γ_a, would be $156 + \frac{15}{10} \times 156 = 156 + 234 = 390$ pcf. This would be substituted for γ in equations (1) and (2) and would modify the constants in these equations.

The equations and the graph of Fig. 23 in this section are not intended to serve as design charts. They contain no factor of safety but show rather the theoretical maximum breaking spans and are presented herein as a basis for explaining phenomena which occur around wide openings.

Natural conditions will seldom be sufficiently favorable to permit the existence of such long spans. Factors tending to reduce lengths of self-supporting spans include diagonal fractures, clay seams, time yield of rock, fractures created by blasting, weight of overlying strata, etc.

Evans(12) mentions the case of a strong sandstone stratum 75 ft thick overlying a coal seam at a depth of 1200 ft. A deeper seam had been worked first and when the upper seam was opened up it was found that the coal had subsided about 4 in.

from its roof over an area 1200 ft long and 600 ft wide. Thus the roof was supporting itself over an area of about $16\frac{1}{2}$ acres. Such favorable natural conditions are seldom encountered but this example illustrates the natural strength of thick strata which have not been damaged by mining operations.

Deflection of a Linear Arch

The linear arch possesses the characteristic of flexibility; that is, it can deflect without collapsing as long as the center of pressure at the center of such a beam remains far enough above the centers of pressure at the abutments that compressive stresses do not exceed the compressive strength of the rock. This flexibility is

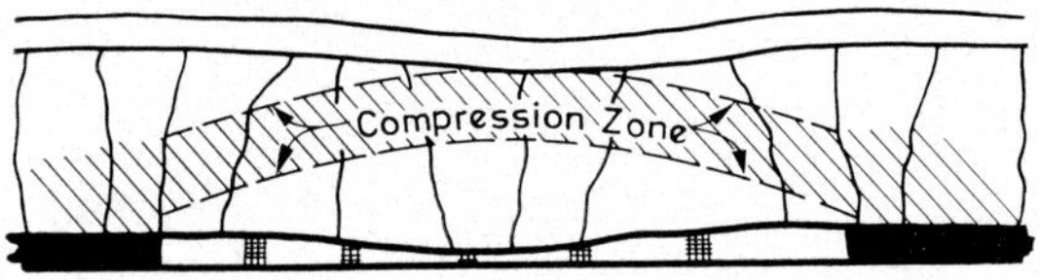

FIG. 24. Deflection of a thick "linear arch".

observed in the behavior of the roofs in the deeper mines; for example, in the mines of the Witwatersrand the roof, although cut by transverse induced cleavage planes, lowers slowly as an integral unit until it rests on the footwall. Figure 24 shows a simple illustration of how such a roof deflects, without collapsing. Flexibility is increased by the presence of transverse fractures, such as induced cleavage which have penetrated into the roof.

Effects of Time Strain

The foregoing discussion and calculations have not taken into account the effects of "time strain" or "rock creep". The behavior of rocks under long-time applied loads has been discussed in Chapter 4 and it has been noted that a rock subjected to a continuously applied load which exceeds about 60–75 per cent of its breaking strength may be expected to yield slowly under the applied stress and eventually to fracture within a finite time.

Thus, those linear arches which depend upon a high rock modulus of elasticity for their stability might be expected to yield slowly and eventually to collapse because the time yield has the same effect as would a slow reduction in the rock modulus of elasticity.

Effects of Natural Lateral Stresses

The presence of large lateral inherent compressive stress tends to stabilize a linear arch by reducing the amount of lateral yield of abutments and tends also to nullify any possible tensile stresses in the bottom portion of the linear arch. The effect of high lateral stresses is to create the equivalent of a "pre-stressed beam" described previously in this chapter.

Vertical or Steeply Dipping Openings

The foregoing illustrations have been typical of conditions associated with bedded deposits in stratified ground. However, the principle of the self-supporting linear arch applies just as well to tabular-shaped deposits which dip at steep angles,

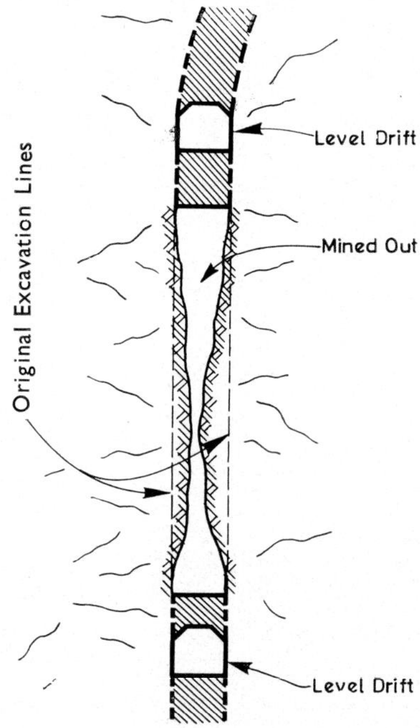

FIG. 25. Walls of a vertical stope showing deflection in the manner of linear arches (with the long dimensions of the linear arches vertical).

and which are located in stratified rock, or even in massive rock. For instance, many metalliferous vein deposits are enclosed in fractured rock. Taking as an example the vertically dipping vein of Fig. 25 enclosed in fractured rock we see that the linear arches in this case have their long dimensions vertical. The vertical stress (due to depth pressure) parallel to the length of the beam "pre-stresses" it, and gives it more rigidity than a horizontal linear arch.

The squeezing together of stope walls around large stoped areas is a phenomenon familiar to all metal miners. Usually, such squeezing is accompanied by exfoliation of large slabs (due to pre-existing fractures), but if the rock is initially intact, or if it is rock-bolted, the squeezing may take place as if the walls are intact elastic beams.

Stresses in a Simple Linear Arch Determined Experimentally

Wright and Mirza[17] have carried out experimental work to determine the magnitudes of stresses in a cracked beam similar to the "linear arch" which is shown in Fig. 20.

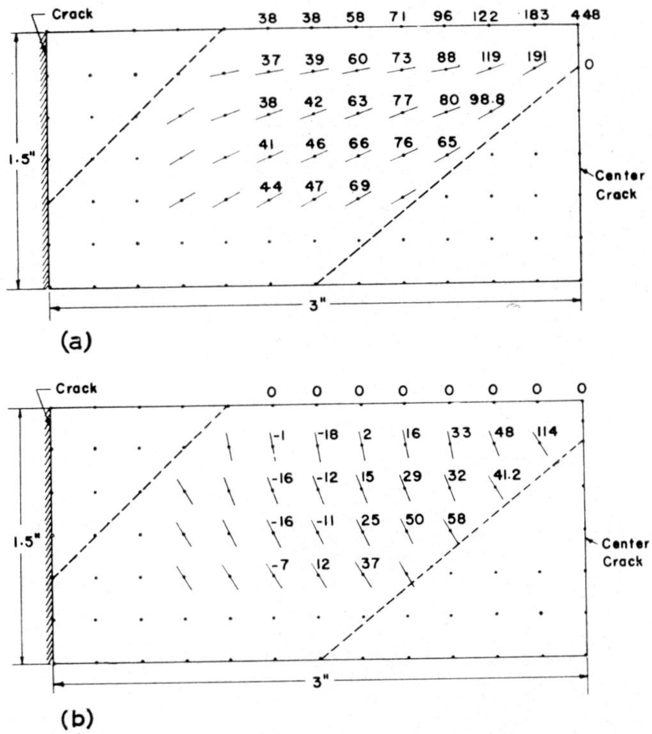

FIG. 26. (a) Maximum (compressive) principal stress in a cracked beam. (Stresses are in pounds per square inch (psi) and were determined by photoelastic methods.) (b) Minimum principal stress in pounds per square inch (psi).

Principal stresses in a beam 1.5 in. deep with a load of 2 lb/in. Diagonal lines indicate the directions of stress.[17]

Beams 6 in. long, $\frac{1}{4}$ in. thick, and of heights of 1 in., $1\frac{1}{2}$ in., and 2 in., were loaded uniformly and stresses were determined by photoelastic methods of analysis.

Figure 26 shows the stress distribution for one-half of a beam 6 in. long and $1\frac{1}{2}$ in. high which carried a total load of 12 lb. uniformly distributed. Maximum compressive stress occurred at the upper contact point and, in this case, was 448 psi.

It will be noted that the configuration of the stress distribution roughly conforms to the shape of an arch. The actual contact area at the top center of the beams tested was only about 18 per cent of the total beam height for all beams tested.

Stress Formula

It was determined that the maximum stress at the top of the center crack could be calculated directly by the following formula:

$$\sigma_{max} = 159.3w - 427.7w \frac{\text{depth}}{\text{span}}$$

where:

σ_{max} = compressive stress at top of center crack;
w = load (psi).

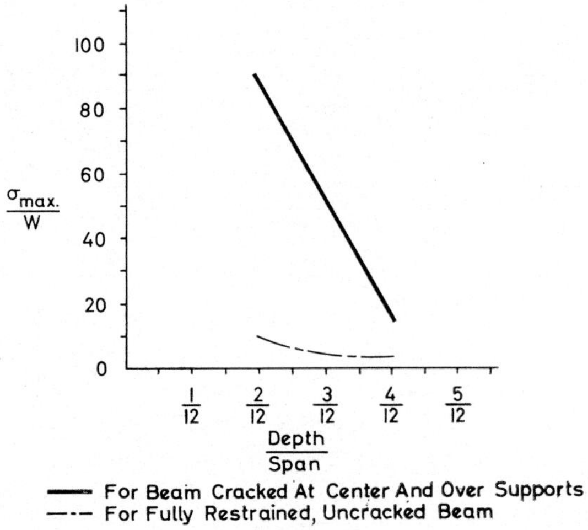

FIG. 27. $\frac{\sigma_{max}}{w}$ at center of span vs. depth of span ratio.[17]

Figure 27 is a plot of the maximum stress/load ratio against the depth of the beam, showing a straight-line relationship. Also shown is the maximum stress for a fully restrained, uncracked beam.

Implications of Test Results

In the formula which he derived for determining stresses in "voussoir beams" Evans[12] assumed that significant stress would be created across the contact surfaces for a distance equal to one-half the height of the beam. Under the conditions of the foregoing tests only the upper 18 per cent of the beam was stressed. If this were true also for rock *in situ* then maximum stresses created would be considerably in excess of those predicted by Evans' formula.

The material used for the beams in the tests conducted by Wright and Mirza was Kriston, a plastic with a modulus of elasticity of about 540,000 psi. The modulus of elasticity of the shales and mudstones commonly range from 1 to 3 or 4 million psi, and the moduli of the harder rocks, such as sandstones, and the igneous and metamorphic rocks may range from 4 to 10 million psi. Thus we might expect that a linear arch, *in situ*, would yield and sag less and that the contact surfaces at cracks would be proportionately greater than was the case with the Kriston model.

Natural (primitive or pre-existing) lateral stresses would also affect the stability of a linear arch *in situ*. Such stresses have the effect of converting the linear arch to a pre-stressed beam such as that shown in Fig. 19. Sag at the center of the linear arch is reduced by the natural lateral stresses and larger contact surfaces are maintained across the vertical cracks. Thus the compressive stress is spread over a larger area and the magnitude of the peak stress is reduced.

Stabilizing the Linear Arch

It is evident that in a linear arch which has a small vertical thickness the compress-ve stresses will be very high. This can be seen in Fig. 20 where the component, W, which is the weight of the beam, induces lateral reactions which have a magnitude several times the magnitude of the weight, W. These compressive stresses induce shear stresses which may cause fracturing and collapse of the arch.

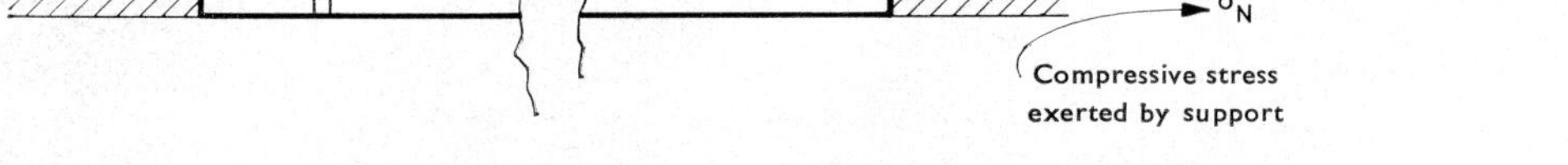

FIG. 28. Support methods for stabilizing the "linear arch".

It is apparent from stress relationships that a support installed immediately after excavation and tightened against the roof assists the roof in supporting itself.

Rock bolts, as well as structural types of support, create pressure normal to the potential shear planes and inhibit shearing due to compressive stresses in the arch. This is illustrated in Fig. 28 where shear stress, τ, which results from compressive stresses along the arch line, is inhibited by the stress, σ_N, exerted by the support.[13]

Rock bolts are particularly effective because they can be tensioned to exert high pressures on planes of potential shear fracture. Structural support is less effective because it cannot maintain continuous pressure on the rock but tends to relax with time, thus releasing the stabilizing pressures.

Rock bolts also suspend any decoupled rock in the zone underneath the arch line. In stratified rock containing some cross-fracturing the rock bolts clamp together the strata, preventing them from sliding along bedding planes, and create integral rock blocks which are rigid enough to form a linear arch.

The Voussoir Arch

Stress conditions in decoupled rocks over an opening may be compared to those in a voussoir arch. This structure is a masonry arch composed of segments, or voussoirs, which has been used in civil engineering works since Roman times. The concept of the voussoir arch applied to mining ground is shown in Figs. 29[12] and 30.

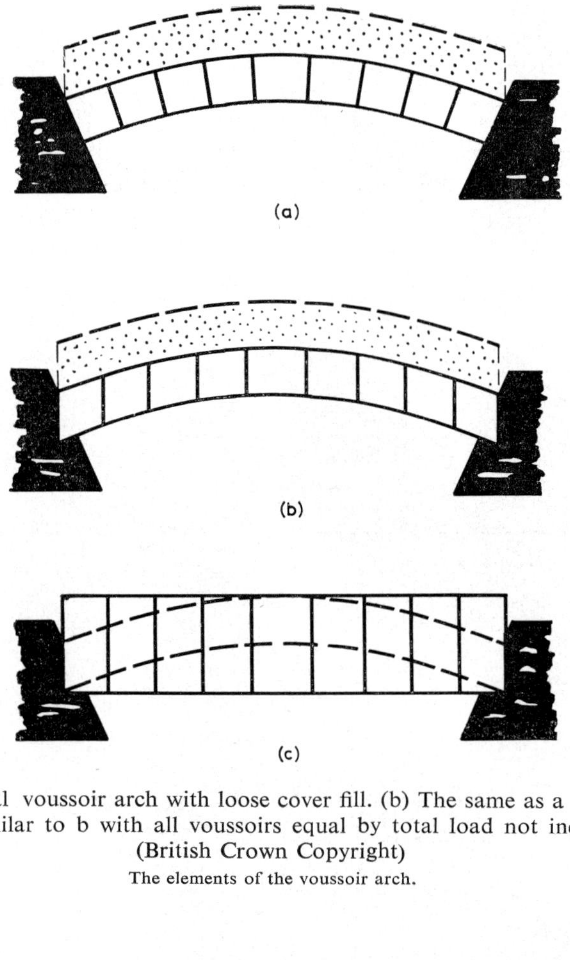

FIG. 29. (a) A normal voussoir arch with loose cover fill. (b) The same as a with all joints vertical. (c) Similar to b with all voussoirs equal by total load not increased.[12]
(British Crown Copyright)
The elements of the voussoir arch.

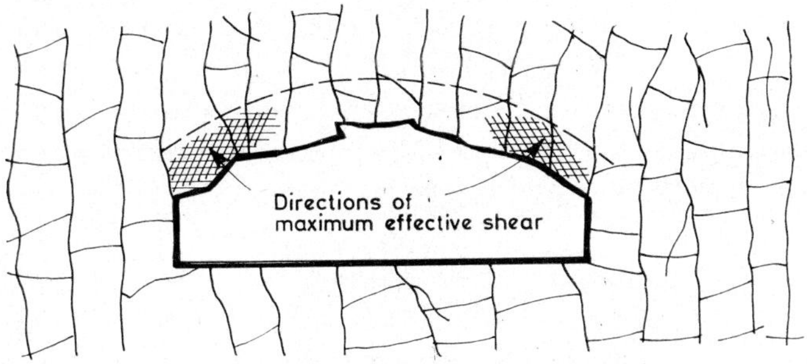

FIG. 30. The voussoir arch in massive fractured rock.

Arching Action in Fractured Massive Rock

A "voussoir arch" may be cut by random fracture planes without affecting its stability, provided that these planes do not approach, or coincide, with the direction of the maximum effective shear stresses created by the arch compression. Directions of these shear stresses are shown by the dotted lines in Fig. 30. If the directions of the pre-existing fractures, or planes of weakness, approach the directions of the maximum effective shear stresses then shear failure takes place along these fractures and rock will drop until the height of the arch increases enough to reduce compressive and shear stresses below the rock strength, and/or until the directions of maximum shear stresses are rotated away from the directions of the planes of weakness.

In stratified rocks, or in de-coupled rocks which are only moderately cross-fractured, the "voussoir arch" will be the "linear arch" discussed previously.

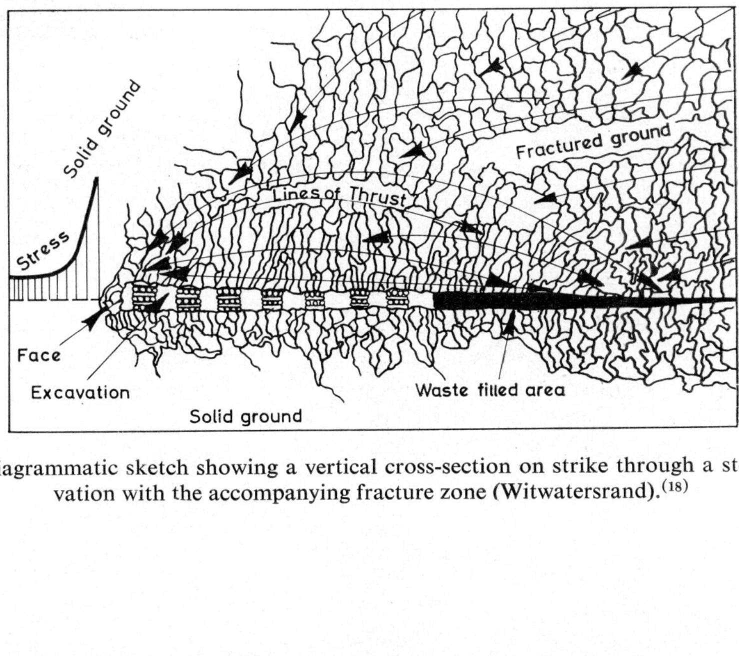

FIG. 31. Diagrammatic sketch showing a vertical cross-section on strike through a stoping excavation with the accompanying fracture zone (Witwatersrand).(18)

For a given span, the maximum compressive stress in an arch varies inversely as the ratio of rise to span. Therefore, in badly fractured rock above an unsupported opening the natural "arch line" (that is, the curved boundary above which no lateral tensile stress exists in the rock) moves upward until maximum compressive stress in the rock along the arch line is just equal to the compressive strength of the rock.

For a supported opening the arch line is much flatter because the confining effect of support strengthens the rock against compressive stress. Support installed creates a clastic zone extending inward from the rock surface to the elastic boundary. Maximum stress occurs along the clastic–elastic boundary. The theory of clastic–elastic equilibrium is discussed in Chapter 5.

In massive rocks cut by random fracture planes there is no plane of physical separation clearly defining a linear arch such as is found in stratified rocks. Figure 23, which shows length vs. thickness for various values of Q and E, indicates that, over those openings which are less than 30–50 ft wide there may exist a self-supporting linear arch with a thickness of less than 1 ft; that is, the "arch line" is less than 1 ft above the exposed rock surface. Such thin arches will be unstable since only a minor amount of sag causes a large increase in stress.

For rock in which no clearly defined separation, plane exists, the arch line adjusts itself upward until the compressive stress along the arch line is just equal to the compressive strength of the rock. Over an unsupported opening in badly fractured rock, therefore, the surface layer will be in a state of unstable equilibrium and anything which causes a slight increase in stress causes a layer of fragments to fall from the rock surface.

PLANNING WORKINGS TO MINIMIZE THE EFFECTS OF PRESSURE ABUTMENTS

The results of an extensive campaign of measurements and observations carried out by Tincelin and Sinou in the iron mines of the French Lorraine are presented here.[14, 19]

The bedded iron ores of this district are flat-lying sedimentary deposits which range from 6 to 30 ft thick and which are generally mined under 500–800 ft of cover. Ore is mined by a room-and-pillar retreating system with total caving behind the retreating pillar line. The system is illustrated in Fig. 32.

Abutment Pressures

The existence of a "pressure abutment" on the unmined ore at the edge of the caved goaf and another "pressure abutment" in the goaf were demonstrated by wall deformation measurements made in a gallery (drift or cross-cut) underneath a panel which was being mined. The results of these measurements are shown in Fig. 14 (a)–14 (e).

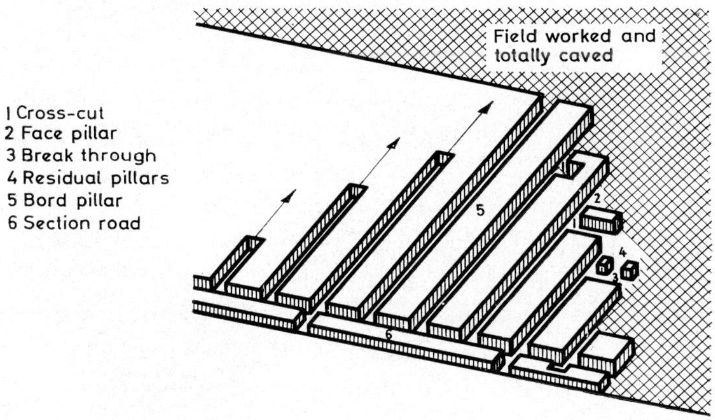

FIG. 32. Mining method used in the iron mines of the French Lorraine.[14]

Critical Width

Measurements and observations showed that the abutment pressures increased as the worked out and caved area was widened until a certain "critical width" was reached. This is the distance L (distance between the two pressure abutments resting on the unworked ore) which is shown in Fig. 14 (d).

In the Lorraine iron mines the "critical width" was related to the depth of overburden Z by the equation:

$$L = 0.9Z \begin{matrix} -20\% \\ +10\% \end{matrix}$$

At the moment when the width of the opening passes critical width the front abutment pressure reaches a maximum. At the same time the total thickness of the caved rock (up to the surface) rests on the floor in the worked out area (goaf) and the "rear abutment" pressure also reaches a maximum. (Fig. 14 (d)). The distance between the front abutment and the rear abutment is then $L/2$.

As the working width passes through critical width, and widening continues, the stresses acting on the front abutment and on the rear abutment decrease in magnitude and approach a nearly constant value (Fig. 14 (e)).

Between the front abutment (pressure on the unmined ore) and the rear abutment (pressure in the caved goaf) there is a zone where the rocks are unstressed. This zone has been referred to the "relaxation zone" or is sometimes referred to as the "Trompeter zone". Experience indicates that the distance, $L/2$, between the front abutment and the rear abutment remains approximately the same as the working is widened by the advancement of the working front (in this case the retreating pillar line).

Principles of Surface Subsidence

If the width of a working is less than the critical width, L, then there is little or no subsidence of the surface vertically above the worked out area. As soon as the critical width is exceeded rapid subsidence begins. The rate of subsidence before critical width is reached may be 0.5–1 cm per week whereas it will increase to 10–20 cm per week after the working exceeds critical width.

Abutment Widths

Figure 33 is a sketch showing an idealized cross-section taken through the zones affected by the working front. As shown, the total zone affected extends over a distance of about $3L/2$ and consists of the following zones:

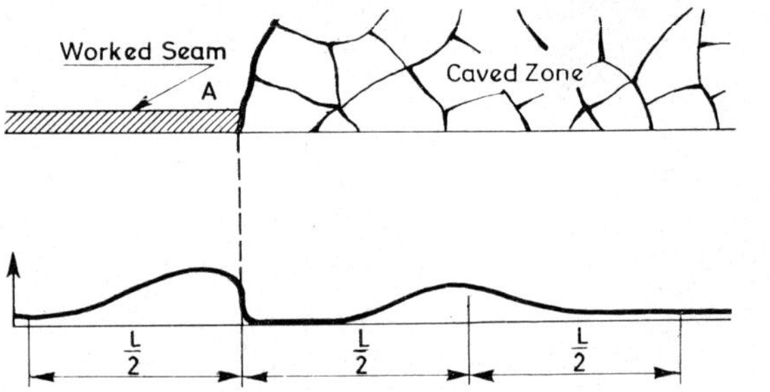

FIG. 33. Lateral extent of the zone which is affected by an advancing working front.[14]

(1) Pressure zone of the front abutment – this is a pressure zone with a width of about $L/2$ which bears on the unworked ore and, in plan, is a strip parallel to the break line.

(2) Relaxation zone situated between the front abutment and the rear abutment – this is a strip parallel to the break line within which the rock pressures are much less than the depth pressure. The width of the strip between the working front and the center of the rear abutment is approximately $L/2$ (Fig. 33).

(3) Restoration zone behind the rear abutment – in this zone the abutment pressure decreases from its peak value and the pressure on the floor of the goaf area decreases until it is equal to the depth pressure (pressure due to the weight of the overburden).

Dimensioning Workings

In the vicinity of a working front the pressures on the front abutment (that is the pressures on the unworked ore) vary with the width of the working and build up to a maximum at those points on the working front which are situated at a distance L (equal to the critical width) from other solid abutments. (Distance L is measured in a straight line across the goaf area.)

Case 1. Working Front Between Two Parallel Unworked Border Zones

Figure 34 shows a pillar line retreating between two parallel unworked border zones. The break line, AB, is advancing parallel to itself in the direction indicated

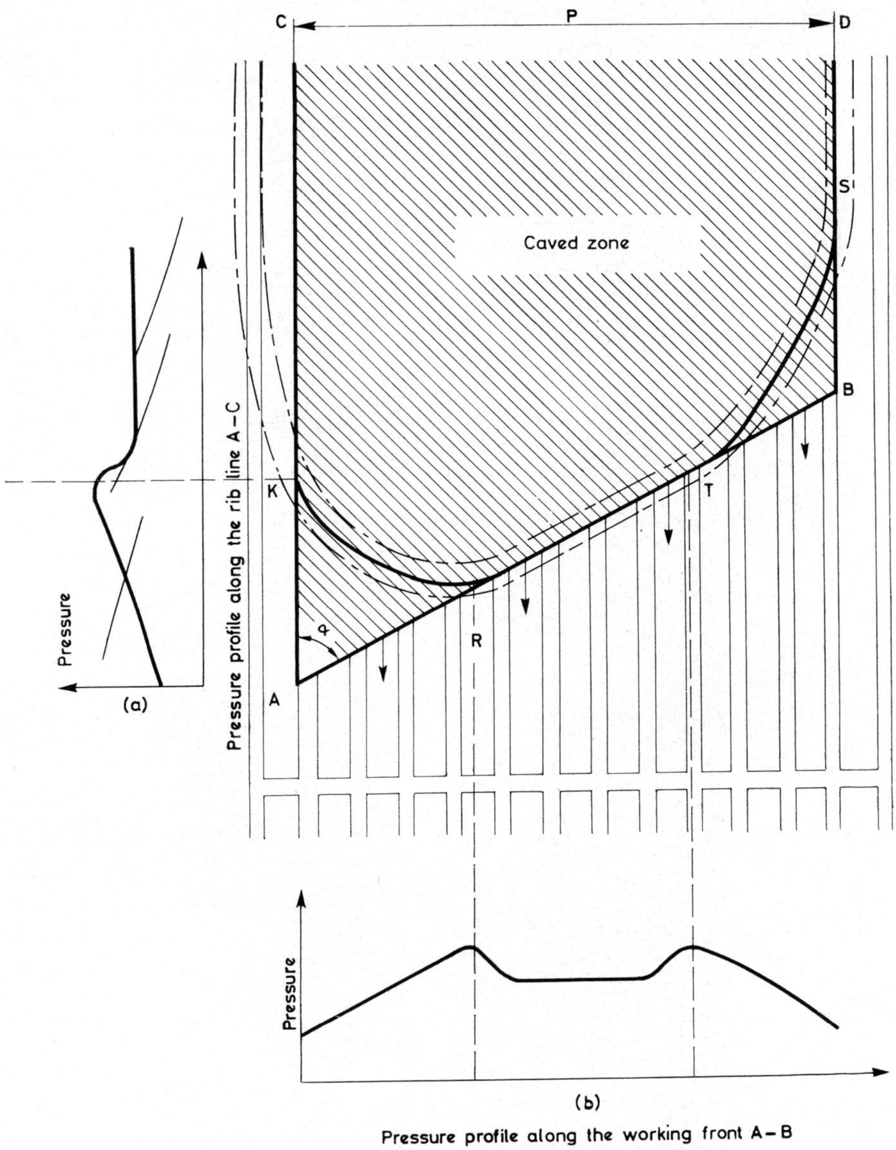

FIG. 34. Distribution along a working front which is advancing between two rib pillars.[14, 19]

by the arrows. All points on the break line between A and R are less than L distant from the face of the solid ore shown by the line AC. Similarily all points on the break line between T and B are less than L distant from the solid abutment indicated by the line BD.

The abutment stress along line AC is shown in Section (a) and the abutment stress along line AB is shown in Section (b) of Fig. 34. The peak pressures occur where the span L, measured across the goaf from the adjoining unworked ore, rests on the unworked ore at the break line.

The curvilinear triangles AKR and TBS each represent a structural plate, or a flat arch, which does not have a span width sufficient to break and thus must be completely supported on the solid ore of the abutments.

In determining the width of a working such as that in Fig. 34 it is important to size the working so that the two points of peak pressure, R and T, do not coincide since this would give excessive stresses in the abutment at one point.

In order to avoid such superposition of stresses it is necessary that the width, P, be less than L, or if wider workings are desired, then the distance AB should be made greater than L as was done in the layout shown in Fig. 34. In brief, the width and configuration of workings should be so selected that two points of maximum pressure are not superimposed for any long period of time.

Case 2. Working Stepped Faces

Figure 35 shows a method of working a retreating pillar line by "stepped faces". The working front consists of the faces AB, CD, and EF. It is very difficult to drive

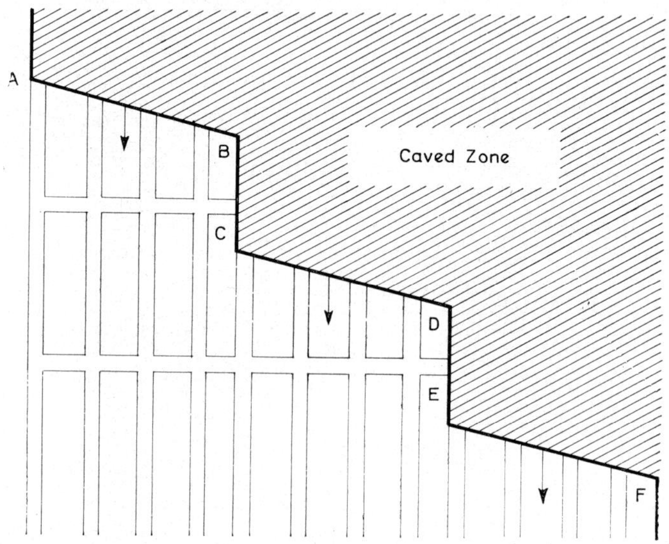

FIG. 35. A "stepped" working front.(14,19)

these fronts if the distance between the points *B* and *D*, and the points *D* and *F* (measured in a straight line across the goaf) is equal to *L*.

If such a situation exists then there will be two abutment pressures superimposed on each of the points *B*, *D*, and *F*.

One solution is to lengthen each of the working faces so that it is longer than *L* by 30–40 m. Similarily each of the offset distances *B–C* and *D–E* must be made longer than *L* by 30–40 m.

Another solution is to eliminate the steps from the face and work a single long face from *A* to *F*.

Case 3. Working Outward from an Acute Angle

Figure 36 illustrates the geometry and stress conditions involved when a working is started in one corner and worked outward from that corner with the width of face increasing as the working advances.

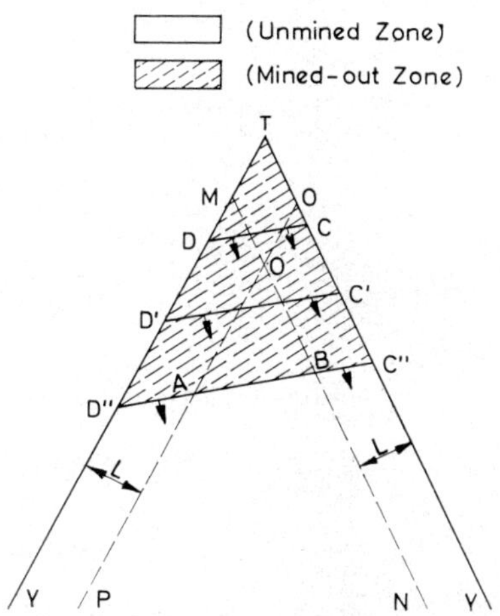

FIG. 36. Pressure zones for a working front which is progressing out of a corner. Maximum pressure will occur at *O* where the two abutment lines, *MN* and *OP* intersect.(14,19)

The working front is represented by line *DC* and advances parallel to itself to successive positions at *D′C′*, *D″C″*, etc. The worked and totally caved zone therefore forms a triangle corresponding to *TD″C″*.

The dashed lines at a distance *L* from the unworked ore represent the abutments of maximum pressure. When the working front reaches the point *O* the two points of maximum pressure coincide and pressure on the face will be maximum.

Case 4. Working Inward to a Corner

Figure 37 shows a working front, *DC*, advancing parallel to itself toward the peak of the corner, *T*.

The curve (1) has a length of $2L$ and curve (2) has a length of L. When the working face advances through the shaded area bounded by curves (1) and (2) points on the face will be subjected to superimposed abutment pressures which have their bases on the solid ore at the borders of the mined out area.

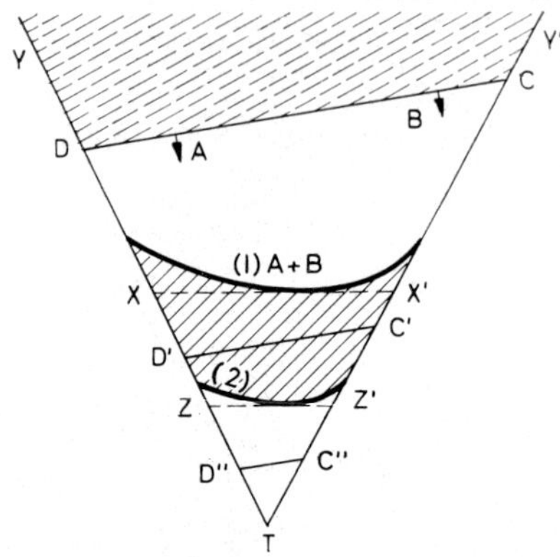

FIG. 37. Pressure zones for a working front which is progressing into a corner. High pressures will occur in the shaded area between lines (1) and (2). In this area the abutment lines will overlap.(14,19)

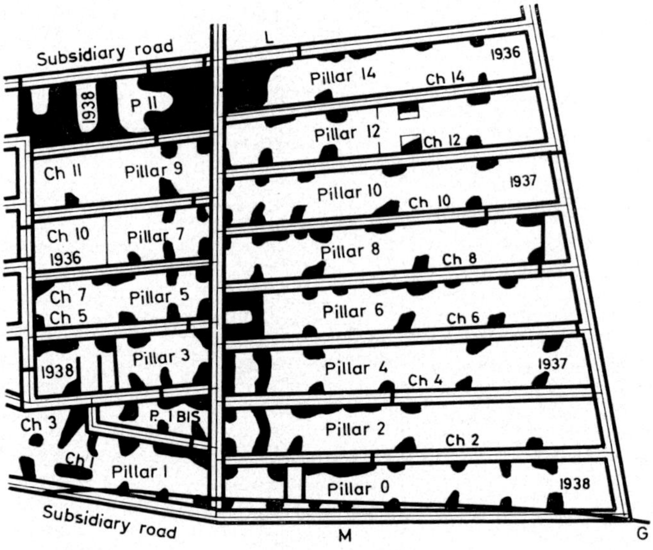

FIG. 38. This figure illustrates the high proportion of pillars which had to be abandoned because of excessive abutment pressures when the workings progressed from the lower right-hand corner into the corner at the upper left-hand.(14,19)

Figure 38 illustrates the extent of pillars which had to be abandoned due to excessive abutment stresses when an attempt was made to work into the corner shown at the upper left hand of the illustration.

APPENDIX

STABILITY CALCULATIONS FOR A LINEAR ARCH

The method for computing theoretical maximum lengths for self-supporting linear arches based on equations given by Evans[(12)] is presented on the following pages.

Figure 22 shows the elements of the linear arch used for calculation purposes, and Fig. 39 shows the basis for calculating the lateral yield of the arch abutments.

Let:

γ = density (taken as 156 pcf in these calculations);

l = $1.24\sqrt{(Qt)}$ = maximum possible length for a self-supporting "linear arch" assuming no elastic yield. (e.g. $E=\infty$);

Q = maximum compressive strength of rock;

t = thickness of the linear arch;

z = $^2/_3\, t$ = initial height of the arch line;

L = $l+\dfrac{8z^2}{3l}$ = length of arch line;

e = $\left(\dfrac{11}{24}\right)\left(\dfrac{f}{E}\right)$ = unit strain along arch line;

$\varDelta_{\text{arch}}$ = Le = total strain along arch line;

$\varDelta_{\text{abut.}}$ = $\left(\dfrac{f}{4}\times\dfrac{1}{2}\right)(5t)\left(\dfrac{1}{E}\right)(2)$ = total strain in abutments;

$\varDelta_t$ = $\varDelta_{\text{arch}}+\varDelta_{\text{abut.}}$ = total strain in arch and abutments;

L_1 = $L-\varDelta_t$ = revised length of arch line;

z_1 = $\sqrt{\left[\dfrac{3l}{8}(L_1-l)\right]}$;

δ = $z-z_1$ = deflection of arch line (max.);

f_1 = $f\times z/z_1$ = max. stress in deflected arch;

δ/t = ratio deflection/thickness;

f = $\dfrac{l^2}{1.54t}$ = max. compressive stress before deflection.

E = modulus of elasticity of the rock;

l = $Kt^{3/4}$ (when E controls arch stability);

K = $l/t^{3/4}$

Example:

Taking a "linear arch" with $t = 10$ ft, $Q = 5000$ psi and $E = 10^6$ psi. Then max. length of arch is (assuming no yield in arch or abutments)

$$l = 1.24\sqrt{(Qt)} = 1.24\sqrt{[(5000)(10)]} = (1.24)(223) = 277 \text{ ft}$$

Initial height of arch line:

$$z = \frac{2}{3}t = \frac{2}{3}(10) = 6.67 \text{ ft}$$

Length of arch line:

$$L = l + \frac{8z^2}{3l} = 277 + \frac{(8)(6.67)^2}{(3)(277)} = 277 + \frac{356}{832} = 277.43 \text{ ft}$$

Unit elastic strain along arch line:

$$e = \frac{11}{24}\left(\frac{Q}{E}\right) = \frac{11}{24}\left(\frac{5000}{10^6}\right) = \frac{(55)10^3}{(24)10^6} = 2.29 \times 10^{-3} = 0.00229$$

Total strain, along arch line:

$$\Delta = Le = (277.43)(0.00229) = 0.64 \text{ ft}$$

At this point in the calculations it is evident that the arch would collapse because the total strain along the arch line would reduce the length of arch line to less than the span of the arch, l.

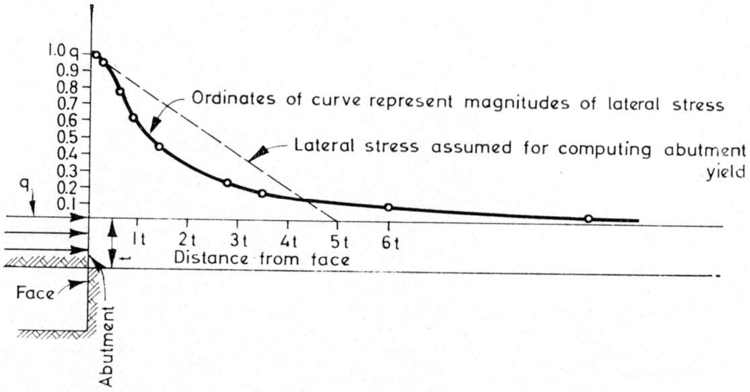

FIG. 39. Stress distribution in the abutments of a linear arch.
Computed from equations for stresses due to a flexible area load applied to part of the surface as a semi-infinite elastic solid. Terzaghi, *Theoretical Soil Mechanics*, John Wiley, New York, 1943, p 377.)

The equation does not take into account the deflection of the arch line due to elastic yield along the arch line and its abutments, although it gives reasonably close results for linear arches with $t > 150$–200 ft. In these cases, where L/l is large, the elastic yield along the arch line is less important in determining stability.

In order to determine the maximum span for a self-supporting linear arch of a given thickness, t, it is necessary to assume a length and compute the resulting stress and elastic yield, repeating the operation until the longest arch is found in which the compressive stress does not exceed the compressive strength, Q of the rock, and in which the elastic yield is not excessive.

For the arch, with $t = 10$ ft, of the previous example it is evident that 277 ft is too long. Next we shall assume a length of 160 ft. Then stress before deflection would be:

$$f = \frac{f^2}{1.54t} = \frac{25.600}{15.4} = 1660 \text{ psi}$$

Initial height of arch:

$$z = \frac{2}{3}t = 6.67 \text{ ft}$$

Length of arch line:

$$L = l + \frac{8z^2}{3l} = 160 + \frac{(8)(6.67)^2}{(3)(160)} = 160 + \frac{356}{480} = 160.74 \text{ ft}$$

Unit elastic strain along arch line:

$$e = \left(\frac{11}{24}\right)\left(\frac{f}{E}\right) = \left(\frac{11}{24}\right)\left(\frac{1660}{10^6}\right) = 0.00076$$

Total strain along arch line:

$$\Delta_{\text{arch}} = Le = (160.74)(.00076) = 0.11 \text{ ft}$$

Total strain in abutments:

An average stress of $(f/4)(\frac{1}{2})$ exists for a distance of $5t$ into each abutment.

$$\Delta_{\text{abut.}} = \left(\frac{1660}{4}\right)\left(\frac{1}{2}\right)\left(\frac{1}{E}\right)(5)(10)(2) = 0.02 \text{ ft}$$

Total strain — (arch + abutments):

$$\Delta_T = 0.11 + 0.02 = 0.13 \text{ ft}$$

Revised length of arch line:

$$L_1 = L - \Delta_T = 160.74 - 0.13 = 160.61 \text{ ft}$$

Revised height of arch:

$$z_1 = \sqrt{\left[\frac{3l}{8}(L_1 - l)\right]} = \sqrt{\left[\frac{(3)(160)}{8}(160.61 - 160)\right]} = \sqrt{\left[\left(\frac{480}{8}\right)(0.61)\right]}$$

$= \sqrt{[(60)(0.61)]} = \sqrt{36.6} = 6.06$ ft or the beam has deflected $6.67 - 6.06 = 0.61$ ft

Increase in stress:

The increase in stress is proportional to the shortening of the arm of the resisting couple from z to z_1, therefore:

$$\frac{z}{z_1} = \frac{f_1}{f} \quad \text{or} \quad \frac{6.67}{6.06} = \frac{f_1}{1660} \quad \text{or} \quad f_1 = 1830 \text{ psi}$$

Total yield of arch and abutments (revised):

Total yield of arch and abutments ($\Delta_{arch} + \Delta_{abut.} = \Delta_T$) is proportional to f_1/f.

$$\Delta_{T_1} = \Delta_T \left(\frac{f_1}{f}\right) = 0.13 \left(\frac{1830}{1660}\right) = 0.14 \text{ ft}$$

Revised length of arch line:

$$L_2 = L - \Delta_{T_1} = 160.74 - 0.14 = 160.60 \text{ ft}$$

Revised height of arch:

$$z_2 = \sqrt{[(60)(0.60)]} = \sqrt{36} = 6.00 \text{ ft}$$

Increase in stress:

$$f_2 = \frac{6.67}{6.00}(1660) = 1850 \text{ psi}$$

Total yield in arch and abutments (revised):

$$\Delta_{T_2} = \frac{6.67}{6.00}(0.13) = 0.14 \text{ ft}$$

Thus the stress and the total yield of arch and abutments approach limiting values for this set of values chosen for l, t, and E. That is a "voussoir beam" 10 ft thick, 160 ft long, with $E = 10^6$ psi will be self-supporting on a temporary basis; that is until it fails by "time yield" of the rock.

To find maximum length which will be self-supporting the foregoing calculations must be repeated for successively longer spans.

BIBLIOGRAPHY

1. FAULKNER, R. and PHILLIPS, D. W., Cleavage induced by mining, *Trans. Inst. Mining Engrs., London*, Vol. 89, Part 5, pp. 264–288; Part 6, pp. 289–297, 1935; also published as Excerpt No. 2786.
2. GROND, G. J. A., The precise topographical measurements in a coal mine underground works (Dutch Mines), *International Conference about Rock Pressure and Support in the Workings, Liège*, April, 1951, p. 52. (The foregoing conference was organized by *Institut National de l'Industrie Charbonnière*, Liège, Belgium. The name is ordinarily abbreviated thus: INICHAR.)
3. GREENWALD, H. P., MAIZE, E. R., HARTMANN, I., and RICE, G. S. Studies of roof movement in coal mines. 1. Montour 10 mine of the Pittsburgh Coal Co., *U.S. Bur. Mines R.I.* 3355, November, 1937.
4. JOSEPH, E. H. A., Rock bursts — The mechanism of crush bursts, *J. Chem. Met. Mining Soc. South Africa*, November, 1938.

5. VAN INTERSON, E. K. T., The marvel of mining, *International Conference about Rock Pressure and Support in the Workings*, Organized by INICHAR, Liège, April, 1951, p. 67.

6. SPRUTH, FRITZ, Distribution of pressure upon and near the coal face, *International Conference about Rock Pressure and Support in the Workings*, Organized by INICHAR, Liège, April, 1951, p. 36.

7. ALDER, H., POTTS, E. L. J., and WALKER, A., Research on strata control in the northern coalfields of Great Britain, *International Conference on Strata Control and Support in the Workings*, Organized by INICHAR, Liège, April 1951, p. 106.

8. EVANS, W. H., HOGAN, M. A., and VALLIS, E. H., An investigation of the loads on packs at moderate depths, *Trans. Inst. Mining Engrs., London*, Vol. 100, Part 12, pp. 340–361, 1941 (Paper 3003).

9. EVANS, W. H. and JONES, T. J., An investigation of the loads on packs at moderate depths. Part II, *Trans. Inst. Mining Engrs., London*, Vol. 105, Part 6, pp. 286–302, 1946.

10. JACOBI, O., The pressure on seam and goaf, Presented at *The International Strata Control Congress, Essen*, October 17 and 18, 1956; Published by Steinkohlenbergbauverein, Essen.

11. MERRILL, R. H., Design of underground mine openings, oil-shale mine, Rifle, Colorado, *U.S. Bur. Mines, R. I.* 5089, December, 1954.

12. EVANS, W. H., The strength of undermined strata, *Trans. Inst. Mining Met.* 1940–1941, pp. 475–500; Discussion 500–532; also Bulletin 439, April, 1941; Bulletin 440, May, 1941; Bulletin 444, September, 1941.

13. CORLETT, A. V., Rock bolts in the voussoir beam, *Trans. Can. Inst. Mining Met.*, Vol. 59, April, 1956, pp. 88–92.

14. TINCELIN, E. and SINOU, P., Summary of the results obtained from eight years research in strata control, *International Strata Control Congress, Leipzig*, 1958, pp. 282–304. Sponsored by Deutsche Akademie Der Wissenschaften Zu Berlin, Sektion Für Bergbau.

15. CORLETT, A. V. and EMERY, C. I., Prestress and stress redistribution in rocks around a mine opening, *Trans. Can. Inst. Mining Met.* Vol. 62, 1959, pp. 186–198.

16. EMERY, C. L., The strain in rocks in relation to mine openings, *Mining Eng.*, No. 1, October, 1960. The Institution of Mining Engineers, London, Paper No. 3834.

17. WRIGHT, F. D. and MIRZA, M. B., Stress distribution around a vertical grack in a mine roof beam, *Trans. Am. Inst. Mining Met. Engrs.*, June, 1963, pp. 174–179.

18. DENKHAUS, H. G., The significance of some properties of rock in relation to the problem of rockbursts in deep mining, *International Strata Control Congress, Leipzig*, 1958, pp. 29–48; Sponsored by the Deutsche Akademie Der Wissenschaften Zu Berlin, Sektion Für Bergbau.

19. DAWANCE, G. and TINCELIN, E., Application de l'auscultation dynamique a l'etude du developpement et de la repartition des contraintes autour des galeries, *Ann. Inst. Tech. Bâtiment et Trav. Publ.*, Supplement, March–April 1955, Nos. 87–8, Serie: Sols et Fondations (18).

CHAPTER 7

THE STABILITY OF MULTIPLE OPENINGS

THE equations for stress distribution around multiple openings in intact rock are given in the Section of Chapter 1 which is titled "Stresses Around Multiple Openings" (p. 65).

"Multiple openings" may be thought of as a series of parallel single openings, or wide openings, driven side by side; or they may be considered as an exceptionally wide opening with pillars located at regular intervals to stabilize the roof. Such systems are encountered in the "room-and-pillar" mining systems commonly used in American mines.

Figure 1 shows a typical room-and-pillar system using rib pillars and also a typical room-and-pillar system using square pillars.

In planning a mining operation it is desirable to be able to determine how much coal (or other mineral) may safely be removed during first mining without danger of extensive roof collapse; that is, what is the maximum amount of mineral which can be mined while still leaving enough in the pillars to safely support the roof, or hanging-wall.

PERCENTAGE RECOVERY AND PILLAR STRESSES

It is evident that the stresses in the pillars will be a function of the ratio of opening width to pillar width, and of the number of openings. It has been previously shown (Chapter 1) that, for a circular opening subjected to a unidirectional stress field of magnitude σ_z, the theoretical maximum tangential stress at the boundary will be $3\sigma_z$. If we create additional openings of the same size as the original and spaced at one diameter apart, the maximum stress at the perimeter of each will increase very little over that for a single opening. Therefore, an infinite series of openings of such size and spacing would be no more liable to collapse than would a single opening. However, in room-and-pillar mining where pillars are not recovered this arrangement would be equivalent to only a 50 per cent recovery. In order to secure a better recovery in those mines where pillars are not "robbed" it is necessary to make the ratio of W_o/W_p (width opening/width pillar) larger than unity.

As the ratio is increased two effects become evident:

(1) The maximum stress at the boundary of openings increases, although rather slowly.

(2) The stresses in the interior of the pillar increases rather rapidly.

Figure 34, Chapter 1, shows stress increase in interior of pillars with increase in W_o/W_p ratio. The five ovaloidal openings are used as an example because stress conditions for this number of openings are approximately the same as for an infinite number of openings. (See Fig. 31, Chapter 1.)

From the laws of engineering mechanics we can deduce that, for a mined area of large extent, the total load on the pillars must be equal to the total weight of the ground overlying the mined area. Thus we say that the average unit stress in pillars is:

$$\sigma_p = \frac{(\text{mined area})+(\text{pillar area})}{(\text{pillar area})}(\sigma_z) \tag{1}$$

This equation is shown graphically by the lower curve in Fig. 37, Chapter 1.

This equation will apply only in areas where pillars are of uniform size and the area is of such large extent that no significant support is obtained from the solid unmined area bordering the mined area. Also, the stress in a pillar may not be uniformly distributed unless the ratio W_o/W_p is relatively large. Duvall[(1)] has shown that for extraction percentages of 75 per cent or more the average stress in pillars approaches the magnitude of the maximum stress.

Figure 37, Chapter 1, shows the maximum stress concentration in pillars at the ribs of openings of various shapes as compared with the average stress concentration in pillars as computed from equation (1). The conditions shown apply to "rib" pillars (continuous pillars) and to the case of four or more pillars, that is, five or more openings.

Two idealized mine plans for 75 per cent recovery are shown in Fig. 1. In one case the area is supported by rib pillars and in the other case by square pillars.

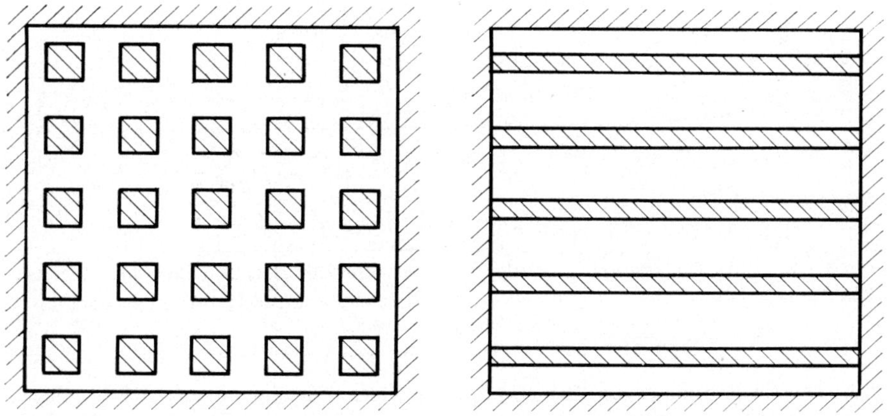

FIG. 1. Comparison of rib pillars with square pillars for 75 per cent recovery.[(1)]

For square pillar support the opening width is smaller and the pillar width is larger than for rib pillar support. The square pillar is more resistant to destruction by shear stresses than is an equal area of rib pillar because its dimensions allow the existence of larger lateral confining pressures in the interior of the pillar. These lateral pressures reduce the magnitude of the effective shear stress induced by vertical compressive stress.

Equation (1) may be re-written for square pillars:

$$\sigma_p = \left(1+\frac{W_o}{W_p}\right)^2 \sigma_z \tag{2}$$

$$\frac{W_o}{W_p} = \left(\frac{\sigma_p}{\sigma_z}\right)^{1/2} - 1 \tag{3}$$

Substituting the compressive strength, Q, of the coal (or rock) in the pillar we have:

$$\frac{W_o}{W_p} = \left(\frac{Q}{\sigma_z}\right)^{1/2} - 1 \tag{4}$$

which gives the maximum ratio of (width opening)/(width pillar) for square pillars. This fixes the maximum allowable percentage of extraction and by determining the maximum allowable roof span from Figs. 15 or 17 of Chapter 6, we can compute the pillar dimensions.

In the case of the Oil Shale Mine, Riflle, Colorado, tests by Wright and Bucky to study pillar loading and time effects[2] indicated that: (1) the safe cross-sectional area of the pillar could be determined by dividing the total load by the compressive

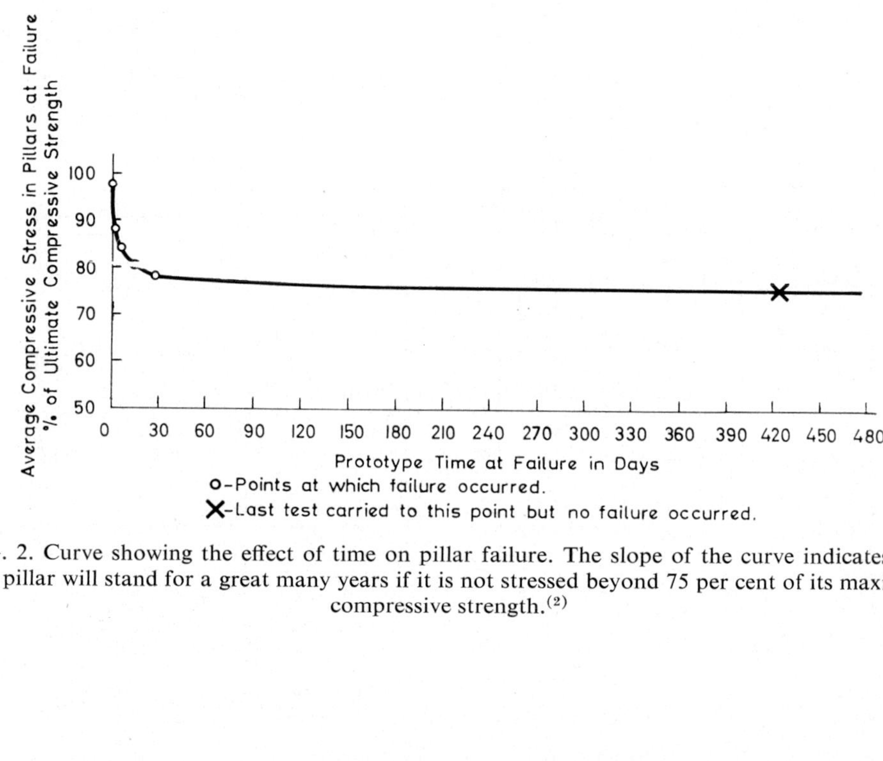

FIG. 2. Curve showing the effect of time on pillar failure. The slope of the curve indicates that the pillar will stand for a great many years if it is not stressed beyond 75 per cent of its maximum compressive strength.[2]

strength of the rock, and (2) the strength of the pillar was affected by time. However, pillars stressed below 75 per cent of compressive strength would stand for many years. (See Fig. 2.)

Figure 3 shows pillar dimensions based on a safety factor of 3, a compressive strength of 10,000 psi (which was the lowest value of compression strengths of thicker beds in the pillar) and a reduction in the compressive strength of pillar rock due to time, to 0.75 of measured strength.

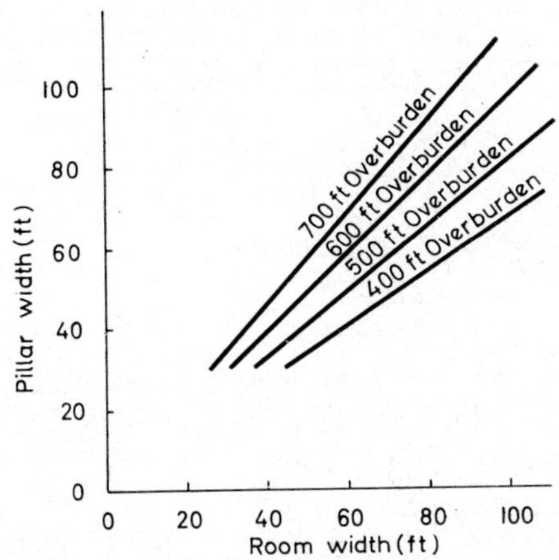

FIG. 3. Curves showing minimum safe pillar widths for square pillars as determined by room widths chosen for operation and by depth of overburden[(2)].

FRACTURE OF PILLARS

Figure 4 shows a pillar subjected to compressive stresses. These create shear stresses which are maximum at the edges of the pillar but decrease toward the center. The decrease of shear stress results partially from the increase of the lateral con-

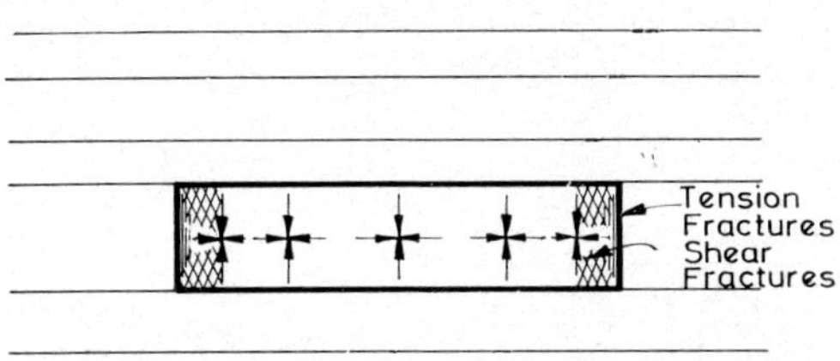

FIG. 4. Principal stresses in, and manner of failure of, an overstressed pillar.

fining stress, as well as a natural decrease of compressive stress away from the free edge. This explains why a square pillar will support more load than a rib pillar, or a rectangular pillar, of equal area. The latter types have one short dimension, and this limits the build-up of lateral confining stress.

Theoretical elastic analysis indicates that maximum stress occurs at the edge of a pillar. The physical properties of coal are such that the coal near the edges of a pillar yields, or fractures, so that maximum compressive stress actually occurs a short distance from the edge of the pillar. This is shown in Fig. 5.

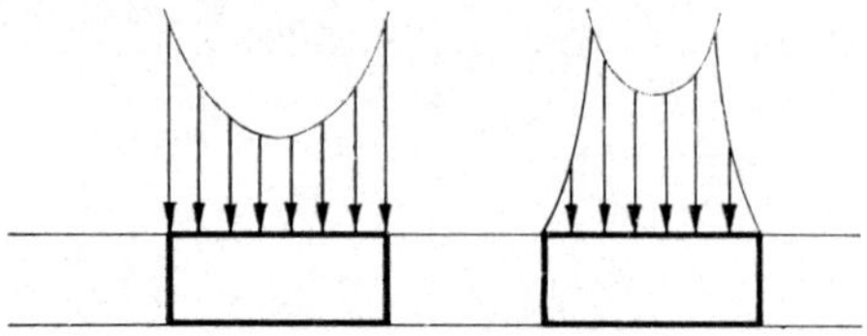

FIG. 5. Vertical stress distribution in pillars. (a) Unfractured pillar. (b) Fractured pillar.

Spalling or slabbing from the faces of a pillar may occur by "tension" fracturing induced by tensile stresses at the ends of natural planes of weakness in a seam or vein. This type of fracture was described in Chapter 3, p. 138, and it was pointed out that it is a surface phenomenon and does not extend far from the pillar face because of the restraining effects of the roof and floors.

Under excessive stress shear fracturing may extend deep into the pillar. If we consider the material of a seam or vein as being fragmented we may apply the laws of soil mechanics to obtain equations for the pressures on the material.

DISTRIBUTION OF VERTICAL STRESS IN A FRACTURED PILLAR

The ability of a pulverulent material to resist applied stress depends upon its angle of internal friction and upon the lateral confining stress. Thus the resistance of a fractured pillar increases rapidly inward from its edges. The equation for resistance to vertical pressure is[3] (see Fig. 6):

$$\sigma_p = \sigma_0 \exp\left[2\left(\frac{c}{t}\right)jx\right] \tag{5}$$

If $c = 0$, we have

$$\sigma_p = \sigma_0 \exp\left[2 \tan \Phi \left(\frac{x}{t}\right)j\right. \tag{6}$$

in which

σ_p = pressure on pillar at x;
x = distance from free face of pillar;
t = thickness of pillar;
$j = \dfrac{1+\sin\Phi}{1-\sin\Phi}$ (Values may be read from Fig. 10, Chapter 4.)
σ_0 = pressure at edge of pillar when $x = 0$;
c = cohesion which will be zero for pulverulent material.

Quoting Walker [3]:

The pressure on a fractured pillar increases rapidly from the edge inwards (Fig. 6). The advantages of lateral support at the edge are at once obvious and its carrying capacity is greatly improved by the support of a pack or filling. The pressure distribution across a pack or fill is very similar.

It is apparent that the carrying capacity of a big pillar will be much greater than that of a number of small pillars of the same area and also that the best shape of pillar is square. When a pillar is split reduction in the total support is unimportant, compared with the reduction in the effective support of the remaining pillars as a result of the removal of the lateral support.

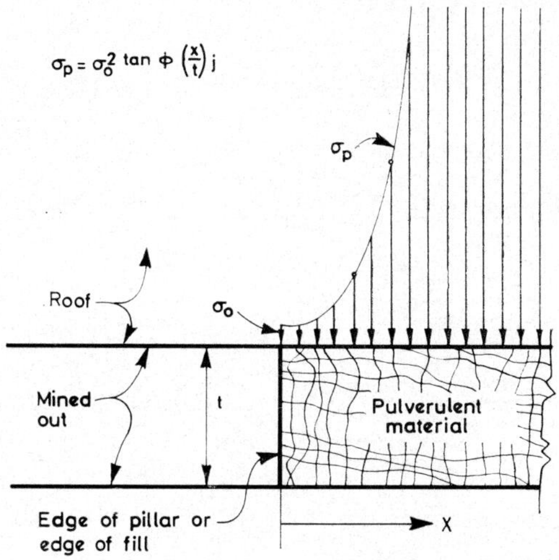

FIG. 6. Stress distribution on a fractured pillar or on fill material. (Adapted from Walker.[3])

SUPPORT DUTY OF COAL PILLARS

In practice coal pillars may be designed for different duties:

(1) For partial extraction without surface subsidence.
(2) For crushed pillar practice.
(3) Fot subsequent total extraction.

Partial Extraction without Surface Subsidence

The pillars are large enough to carry the limited abutment loads which are induced. These will be the loads induced at the ribs of each opening, and extending into the coal for a distance about equal to the width of the opening. Theory indicates that if a single opening will stand open without support, then multiple openings will also support themselves, provided that they are spaced at one diameter, or more, apart. That is, 50 per cent extraction is theoretically possible.

In practice this does not always work out because at great depths the stresses may fracture the coal in the pillars. In this case the pulverulent material may behave in accordance with the laws of soil mechanics and the supporting strength of the pillar depends upon its minimum width as well as upon its thickness.

Walker[3] mentions an instance where 40 per cent of the coal seam was extracted at a depth of 3900 ft. Average load on the pillars was 6500 psi, a value higher than the compressive strength of the coal obtained in laboratory tests.

Crushed Pillar Practice

A larger proportion of coal is extracted using this method than the one above, and limited subsidence of the surface results. The pillars, while yielding slightly, must be large enough to sustain the loads in the working area without crushing, although not large enough to withstand the back abutment load at some distance behind the working area.

Pillars small enough to crush are formed on retreat, generally by splitting larger pillars formed in advancing.

Subsequent Total Extraction

The pillars must be large enough to carry the maximum abutment loads which will be created along the retreating pillar line during the final extraction phase. During this final phase abutment loads along the pillar line will be similar to those which are found on the solid seam ahead of a longwall face, except that in the case of the retreating pillar line the width of the abutment zone will be greater and the maximum abutment stress will be somewhat less.

ROOM-AND-PILLAR AND BORD-AND-PILLAR WORKINGS

The "Bord-and-Pillar" method of working was gradually evolved from the room-and-pillar method as deeper seams were exploited in the North-of-England coal-field. Reduction of roof loads was effected by reducing the width of excavation and increasing the size of the coal pillars. As the depth of working increased, this principle became less economic and, as wider places were difficult to support and required frequent repair, except where strong and resilient roof strata allowed greater latitude, the tendency, especially after the introduction of coal face machinery, was to adopt the longwall method of working although it involved an increase in the amount of unproductive work.

The highest rate of production has been attained in the United States where the standard method of working is room-and-pillar. The seams worked in the U.S. are generally shallow and the workings have been intensively mechanized, whereas in Great Britain mechanized room-and-pillar working has largely been limited to depths of about 400 ft.[4]

The higher output obtained in the United States by room-and-pillar methods directed attention to the possibility of designing methods of working in which roof loads comparable with those experienced in the shallower seams could be obtained over a limited area at depth by utilizing the principle of transferring the main roof load. This led to experimentation with "yield pillars".

YIELD-PILLAR TECHNIQUE IN ROOM-AND-PILLAR WORK

The fact that the main roof load can be transferred to abutments with the maximum distance between abutments depending upon the prevailing conditions suggested that minute yield of coal pillars, over a limited area, would result in the transference of the main roof load beyond the narrow roadways on to solid coal barriers provided on the flanks. Thus, over a limited area vertical load is reduced and conditions are created which are similar to those in shallow workings.

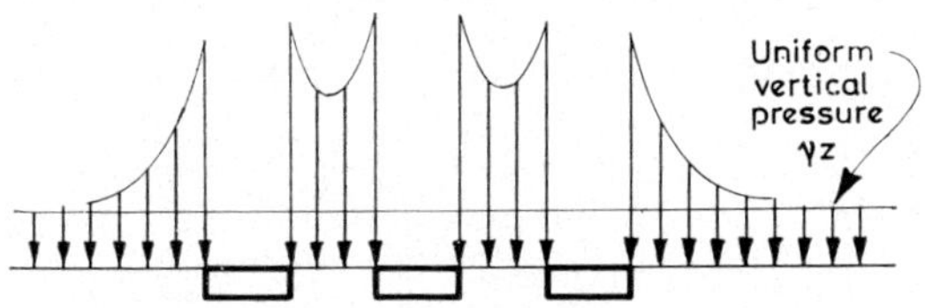

FIG. 7. Distribution of pressure on pillars in a homogeneous, elastic material.

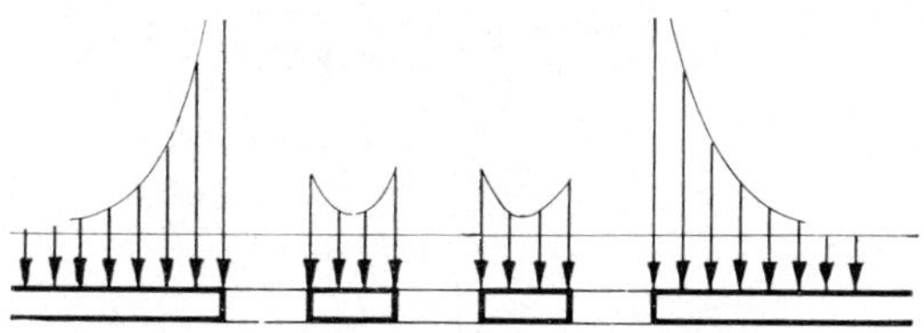

FIG. 8. Distribution of pressure when pillars yield elastically (without fracturing).

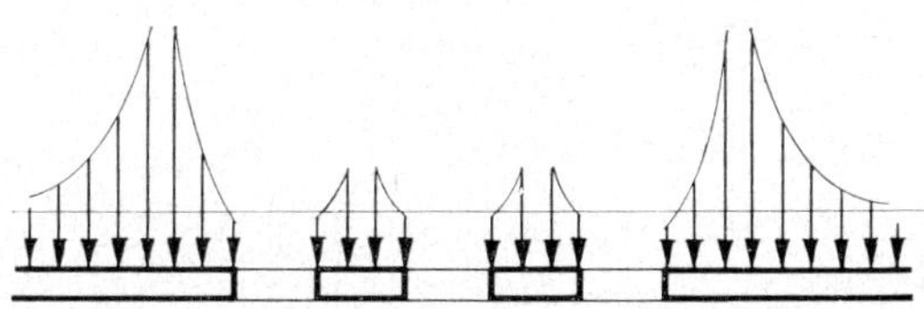

FIG. 9. Distribution of pressure when pillars yield by fracturing.

The principle of yield pillar technique involves the selection of such dimensions for pillar and roadways that the stress on pillar will be sufficient to create a minute amount of yield (vertical shortening). The shortening of the pillar results in slight downward deflection of the roof, which then transmits more vertical stress to the abutments.

(A crude simile to this would be the "continuous beam" used in structures. If the center supports of such a beam lower slightly the beam deflects enough to remain in contact with the supports, but a portion of its load has been transferred from the center supports to the side supports.)

Conversely, the introduction of more rigid support (e.g. larger pillars) within the excavation increases the intensity of stress in the area of the support. The increase in stress concentration at the edges of such support creates localized shear stresses which fracture the roofs of rooms, or "bords".

When pillars are made small enough to yield, the intensity of stresses at edges is reduced with the result that roofs of openings remain intact. However, if dimensions of pillars are made too small then the main span of the "beam" or "arch" may collapse with the result that pillars will be crushed by the depth pressure.

Figures 7, 8, and 9 illustrate the idealized pressure distribution for various types of pillar yield.

Application of "Yield Pillar" Methods to Coal and Potash Mining

According to Collardey[(5)], experiments with the usual room-and-pillar mining methods in French coal mines have shown that the limiting depth for successful application of room-and-pillar methods ranges from about 800–1600 ft. The limiting depth varies from district to district and from mine to mine, with the average being about 1300 ft. Exceptions to this are the potash mines where room-and-pillar methods are used at depths of 2000–2300 ft and the coal mines of the Provence Basin where coal is mined at depths of about 2000 ft.

It appears that the application of "yield pillar" or "compressible pillar" methods might allow room-and-pillar mining methods to be used at much greater depths than at present. Successful application of the method requires that the total width of the developed panel must be less than the span of the "pressure arch". (See Table 1, Chapter 6.) The method takes advantage of the fact that a large part of the weight of an undermined block can be transferred by "beam action" or by "arching action" to the "abutments". Pillars must be small enough to yield slightly and effect a transfer of load to the abutments. When workings are properly dimensioned pressure on the pillars will not be excessive, and roof and floor conditions will be good.

For a hypothetical case take a seam situated at a depth of 1800 ft. From Table 1, Chapter 6, we see that the average width of pressure arch, i.e. distance between "pressure abutments", is about 330 ft. If the pillared area consists of six rooms each 13 ft wide separated by pillars 50 ft wide as shown in Fig. 10, then the total width of developed zone will be about 380 ft, and the front abutment of the pressure arch will rest on pillars in the developed area as shown in the figure. This will result in heave of the floor and fracture of the roof adjacent to the pillars.

The pillars are too wide to yield, and thus they tend to carry load which would otherwise be transferred to the abutments. Excessive stresses may cause shearing of the roof adjacent to the pillar sides.

If, as shown in Fig. 11, pillar widths are reduced to 16 ft, then the same number of working places occupy a total width of only 180 ft, and the entire working area is well inside the pressure abutments. The narrow pillars can yield enough to avoid excessive stress build-up in the intermediate pillars and to transfer the pressures to a front abutment on the solid coal ahead of the developed area.

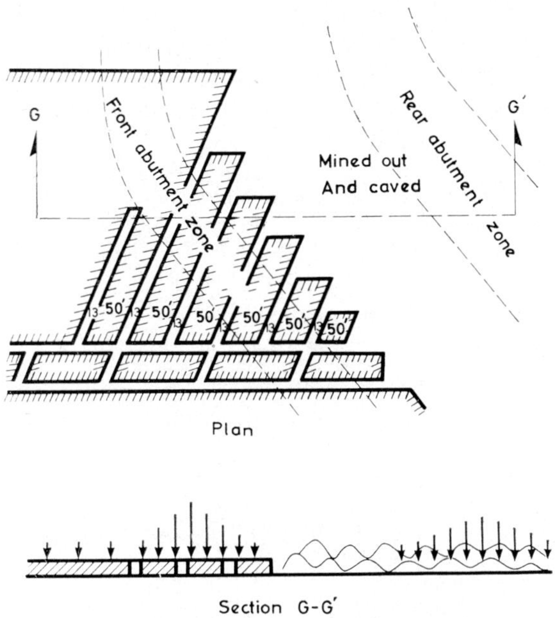

FIG. 10. Pillars are too wide to yield and too much area has been developed. The front abutment will damage the development openings.

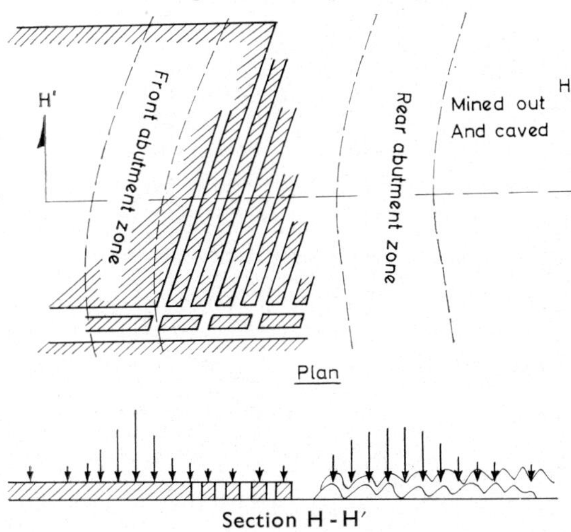

FIG. 11. Pillars are narrow enough to yield and to allow the "main roof load" to transfer to the abutments. The developed zone is narrow so that it lies entirely between the abutments.

In French potash mines pillars were at first made 25 ft in width (at depths of 2000–2300 ft), but much diffculty was experienced with fracturing of the roof adjacent to the pillars. When pillar widths were reduced to 10 ft the difficulties disappeared. In the Provence Basin where coal is mined at depth of about 2000 ft the widths of both rooms and pillars is 20 ft.

The determination of optimum widths for rooms and pillars in deep mining is obviously a subject for much experimentation and research. Any system adapted requires careful day-to-day supervision since the whole system is delicately balanced and any misalignment of workings could cause the loss of a section of workings.

Yield Pillars Between Multiple Roadways or Entries

Experimental work has been carried on in England with a view to determining the most favorable width of opening to pillar (W_o/W_p) ratio when driving development roadways. The total width of the development workings (two roadways plus a pillar) was small as compared to the possible span of the pressure arch which could be sustained; so these tests did not determine the optimum width of a panel of workings but only indicated the width of pillar which would yield under the existing depth pressures.

Underground Experimental Work with Yielding Roadway Pillars

Experimental roadways were driven in the Hutton Seam at Wearmouth Colliery[6]. The seam was from $2\frac{1}{2}$ to 4 ft thick at a depth of 1700 ft. The roof was a shale band 3 ft or more in thickness above which was coal 14–20 in. thick, overlain by more shale. The floor was a weak seggar clay.

Main Roadways

Two roadways were driven a distance of 650 yards. The roadway excavations were about 18 yards wide and the width of the separating coal pillar was varied from 11 to 36 yards.

These experiments showed that under the prevailing conditions a pillar 11 yards in width was too weak to carry the roof load, and yielded excessively, throwing excessive pressure on to the roadway supports. A pillar 33 yards wide was too strong to yield effectively and resulted in roof damage owing to concentrated loads at the sides of each excavation.

Room Entries

A number of experiments were made to find best dimensions for room entries. The following example is cited:[6]

Two other entries were driven in a westerly direction. The entries were 6 yards wide and the separating coal pillar was $6\frac{1}{2}$ yards wide. Trouble from both the roof and the floor was experienced and the width of the pillar was increased to $8\frac{1}{2}$ yards, then to $10\frac{1}{2}$ yards, and later to $13\frac{1}{2}$ yards without any improvement in conditions. After 128 yards had been driven it was concluded that the upper limit of yield pillar width had been exceeded and it was decided to reduce the width of the estimated lower limit.

The width of the entries was reduced to 5 yards and the width of the pillar to 3 yards. Roof and floor conditions were then very good and the pillar width was increased to 4 yards after which excellent conditions were maintained on both entries over a distance of 300 yards.

RELATIONSHIP BETWEEN WIDTH OF PILLAR AND EXCAVATIONS

In the above mentioned experiments, cited under "Main Roadways", the maximum ratio of W_o/W_p was about 1.64, which was found to be too large, and the minimum ratio was about 0.50, which was found to be too small.

In the second example cited, under "Room Entries", the initial ratio W_o/W_p which was 0.93 gave unsatisfactory results. The ratio was reduced to 0.45 without improving conditions.

In both instances the pillars were too wide, did not yield, and therefore the stress concentrations at sides fractures the roofs. When pillar widths were reduced to 3 or 4 yards excellent stability conditions were maintained, but it was necessary to reduce also the widths of the entries from 6 to 5 yards.

Consideration of the essential functions of the pillar in regard to (a) the required yield and (b) the support of the strata contained in the pressure arch which is formed will show the W_o/W_p ratio will decrease with depth, while at shallow depths the ratio may be increased.

DIMENSIONING "ABUTMENT" PILLARS FOR BORD-AND-PILLAR WORKINGS

Abutment loads developed by narrow workings were found to be comparatively small, but during the extraction of pillars on retreat those adjoining the goaf are subjected to the abutment load of a pressure arch which spans from goaf to pillars, across the face area. It had been deduced, therefore, that the widths of abutment pillars in the case of full extraction should not be less than half the width of the maximum pressure arch which exists under the prevailing conditions. Reference to Table 1, Chapter 6 shows that in accordance with this rule the width of abutment pillar will increase from 60 ft at a depth of 400 ft to 165 ft at a depth of 1800 ft. This width is the actual pillar width and not the distance between centers of roadways. It has been found that the most convenient shape for pillars is square, but if for any reason rectangular pillars are formed, the shortest length of such pillars should be one-half the width of the maximum pressure arch.

DESIGN OF PILLARS FOR PARTIAL EXTRACTION

Where only partial extraction is permissible, smaller pillars may be formed. About 50 per cent of the coal can be extracted in this way without appreciable subsidence of the surface at depths not greater than 400 ft, but the proportions which could be extracted at greater depth would generally be less, probably about 40 per cent at 1500 ft. Because the pillars formed by the extraction of such a proportion of coal are designed to carry only normal cover load, they will fall progressively if, for any reason, a concentrated load has to be carried, and therefore care must be taken not to form large areas of such pillars if it is possible that full extraction may be required later.

The extent of surface subsidence may be limited by "crushed pillar" practice in which pillars small enough to crush are formed on retreat, generally by splitting larger pillars formed in advancing.

Because a large coal-pillar is capable of carrying a much greater load then a number of small pillars of the same total area, the maximum proportion of extraction without appreciable surface subsidence can best be secured, where it is not necessary to consider the effect on other seams, by the working of wide strips of coal which alternate with solid pillars of similar width. The widths of these excavations and pillars should be about three-quarters of the width of the maximum pressure arch. By this means, 50 per cent extraction can be obtained at depths down to 2000 ft without appreciable effect on the surface.(4)

EFFECTS OF WORKINGS ON OTHER SEAMS

An abutment load in any seam induces areas of concentrated load in seams within a considerable vertical distance above and below. In designing workings of all types, provision must always be made for the accommodation of loads concentrated by workings in other seams, necessitating the provision of pillars larger than would otherwise be necessary in such areas. For this reason, dimensions can only be determined finally after full consideration of complete records of existing and projected workings.

STRESS DISTRIBUTION IN PILLARS AS DETERMINED BY UNDERGROUND MEASUREMENTS

Measurements in French Iron Mines

Figure 12 shows stress distribution in pillars 12 m square in the bedded iron ores of the Lorraine district of France as determined by sonic velocity measurements.

It will be noted that the outer edges of the pillars were fractured and largely destressed as evidenced by the low sonic velocities. Further inside the pillars the stresses increased to reach a peak at about 1 m in from the edge and then another peak at about 2–3 m in from the pillar edge.

Generally, the stresses on the center portions of the pillars were less than those at stress peaks except for pillar No. 5. The fact that stress values increased continuously from the edges to the center of pillar No. 5 may be an indication that

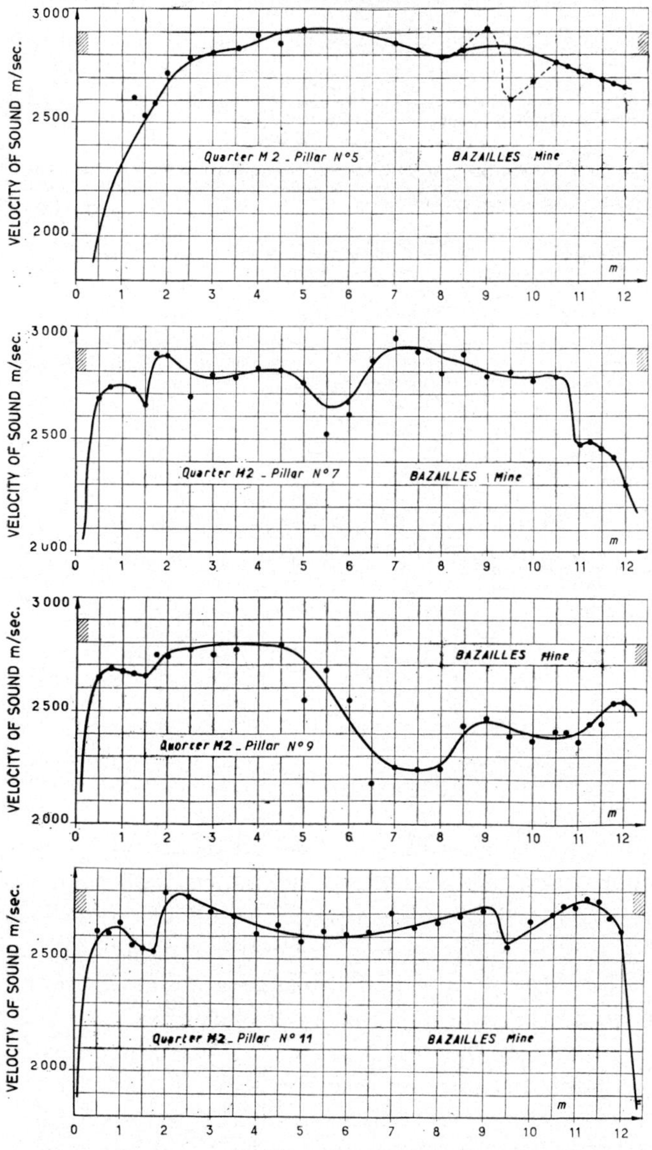

FIG. 12. Vertical stress distribution in square pillars in the Bazailles iron mine, France. (The general pattern of stress distribution was determined by measuring the velocity of sound waves through various portions of the pillars.[7])

fracturing had extended through the pillar, and that its resistance to compression was largely that of a fractured material laterally confined.

The stress measurements shown are only semi-quantitative because the velocity of sound does not vary directly with compressive stress applied to the rock. These measurements do serve to indicate the locations of peak stresses, however.

Procedures for determining stress distribution by means of sonic velocity measurements are described in Chapter 10.

Measurements in Swedish Mines

By means of a stress relief technique employing magnetostrictive gages placed in boreholes Hast measured pillar stresses in a lead mine at Laisvall, Sweden.[8,9] A plan of that portion of the Central Mine in which measurements were made is shown in Fig. 14, Chapter 2.

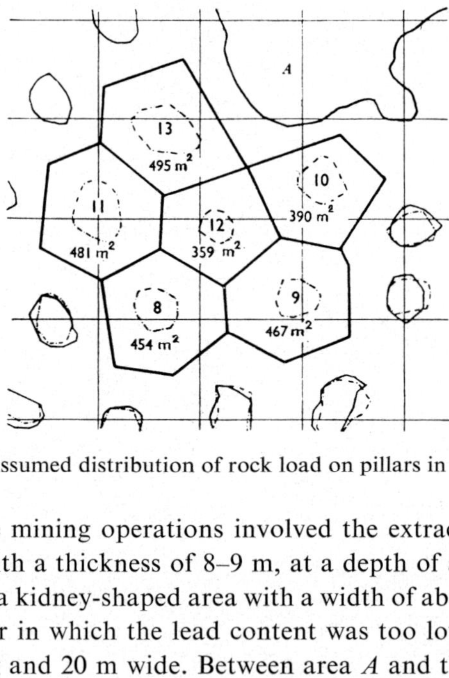

FIG. 13. Assumed distribution of rock load on pillars in Area II.[8,9]

In this portion the mining operations involved the extraction of a lead-bearing layer of sandstone with a thickness of 8–9 m, at a depth of about 100 m below the ground surface, over a kidney-shaped area with a width of about 200 m. *A* represents a portion of the layer in which the lead content was too low to justify extraction. It is about 70 m long and 20 m wide. Between area *A* and the outer boundaries of the mined-out space the roof supported by lead-bearing sandstone pillars which interrupt the continuity of the mined-out space. The cross-section of the pillars ranges between 43 and 107 m^2, and their spacing is about 22 m. Between pillars the roof is unsupported and practically horizontal. The free span between adjacent pillars is roughly 15 m. The age of the pillars ranges between 0 and 15 years.

Figure 13 is a plan of group II of pillars, representing one of four groups (I to IV) of pillars on which measurements were made. It will be designated as "area II" and its location is shown in Fig. 14, Chapter 2. Prior to excavation, the overburden pressure σ_z at roof level was about 25 kg/cm^2. If the influence of elastic deformations is disregarded each pillar would carry the load which acted, prior to excavation, on the polygonal area surrounding the pillar in Fig. 13. The cross-sectional area of the pillars, at an elevation of 1.2 m above floor level, ranges between 43 and 94 m^2. Depending on the size and spacing between pillars, the total load on the pillars computed on the basis of the aforementioned assumption ranges between 90,000 and 124,000 tons and the corresponding unit load, at an elevation of 1.2 m above floor level, between 132 and 209 kg/cm^2.

Stress measurements were made in all pillars of the group in Area II, in horizontal holes with a depth up to 2 m at an elevation of 1.2 m above floor level. Because the height H of the pillars is about 8.5 m, the measuring points are located about 0.14 H above the base of the pillars. The results of the measurements furnished information concerning the state of stress in vertical planes, roughly parallel to the adjacent outer face of the pillar. They permitted the computation from measurements of the vertical stress component σ_v and comparison of these stresses with the stresses computed on the basis of the load distribution illustrated by Fig. 13. Table 1 contains the data which have been obtained.

TABLE 1

Pillar no.	8	10	11	12	13	
σ_{vc} (computed)	206	135	137	209	132	
σ_v (measured)	204	144	125	180	111	
σ_v/σ_{vc} ratio	0.99	1.07	0.91	0.86	0.84	average 0.94

In pillar No. 9 measurements were made in three different holes, with the results shown in Table 2. Hast ascribed the erratic results obtained from the readings on this pillar to exceptionally intense fissuration of the rock contained in the pillar.

TABLE 2

σ_{vc} (computed)	= 205 kg/cm^2
σ_v (measured)	= 240, 168, 108 kg/cm^2
σ_v/σ_{vc} ratio	= 1.17, 0.82, 0.53

Data like those assembled in Table 1 were also secured for the pillars of areas I, III, and IV. The results were similar to those obtained for area II except for area IV in which the roof is supported by the oldest pillars in the Central Mine. At the time when the measurements were made, their age was about 15 years. The pillars are located in a bay-shaped recess of the mined-out space with a width of not more than about 60 m outside the lower right-hand corner of Fig. 14, Chapter 2. The stress readings on the pillars of this group were exceptionally erratic, very low; and on one of the pillars even equal to zero. From this fact and from what he observed in the mine Hast concluded that the pillars gradually break up with increasing age. This process is inevitably associated with a transfer of part of the load on to the rock surrounding the mined-out space.

The vertical normal stresses σ_v encountered in the rock at the outer end of the measuring holes in the plane at right angles to the centre-line of the hole were associated with horizontal normal stresses σ_h acting in the same plane. The ratio σ_h/σ_v ranged between 0.5 and 0.75.

BIBLIOGRAPHY

1. DUVALL, W. I., Stress analysis applied to underground mining problems. Part II. Stress analysis applied to multiple openings and pillars, *U.S. Bur. Mines R.I.* 4387, November, 1948.
2. WRIGHT, E. D. and BUCKY, P. B., Determination of room-and-pillar dimensions for the Oil-Shale Mine at Rifle, Colorado, *Trans. Am. Inst. Mining Met Engrs.* Vol. 181, 1949, pp. 352–359.
3. WALKER, L., The theory of strata control, *Mine & Quarry Eng.*, July, 1955, pp. 291–296.
4. Memorandum on the design of mine workings to secure effective strata control, *Trans. Inst. Mining Engrs., London*, Vol. 110, 1950–1951, pp. 251–278.
5. COLLARDEY, J., Critères d'application de la methode des chambers et piliers, *Ann. Mines Belg.* December, 1958, pp. 1110–1127.
6. ALDER, H., POTTS, E. L. J., and WALKER, A., Research on strata control in the northern coalfields of Great Britain, *International Conference on Strata Control and Support in the Workings;* organized by INICHAR, Liège, April, 1951, p. 106.
7. DAWANCE, G. and TINCELIN, E., Application de l'auscultation dynamique a l'étude du development et de la répartition des contraintes autor des galeries, *Ann. Inst. Tech. Bâtiment Trav. Publ.*, No. 87–88, March–April 1955, Serie: Sols et Fondations (18).
8. HAST, N., Bergtryckamatningar i gruvor (The measurement of rock pressure), *Jernkontorets Ann.*, Vol. 141, No. 10, 1957, pp. 633–744.
9. TERZAGHI, KARL, Measurement of stresses in rock, *Geotechnique*, Vol. 12, No. 2, June 1962, pp. 105–124.

CHAPTER 8

ROCK BURSTS

OPENINGS IN WEAK STRATA

The mode of failure of rocks overlying a wide opening depends upon the physical nature of the individual strata. Thin-bedded, soft, or weak rocks, such as shales and mudstones, may fail by buckling, in the manner of beams, when the span of an opening becomes too great for the strength of the beam. The immediate roof "beam" buckles and falls into the opening, and then successive strata above it follow suit until the loosening of the rock has progressed to such a height that "arching effect" becomes effective in supporting the remaining roof.

The fractured, or loosened, material occupies a slightly greater volume than the original rock in place. As the loosening progresses upward, a point will be reached where such material will exert considerable upward pressure on the surface of the unfractured ground. Thus a clastic–elastic boundary may be created.

The combination of the supporting effect of the loosened strata, and the reduction in diagonal thrust due to the increase in the rise/span ratio of the arch, that is, the clastic–elastic boundary, will eventually produce a stable arch, assuming that the opening is not widened further. The rise/span ratio of the clastic–elastic boundary depends upon the physical characteristics of the rock both before and after fracturing, as well as upon the depth pressure and the existence of natural lateral stresses.

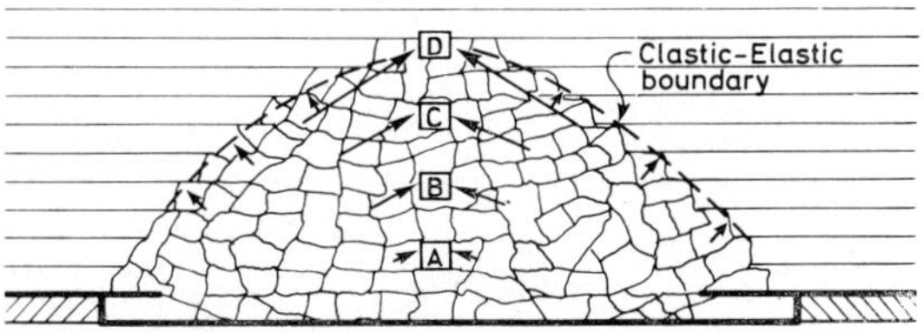

FIG. 1. Arching action; Relative diagonal thrust forces tending to support an element of rock at various distances above an opening.

The foregoing applies to the case of an opening which has a width less than "critical width". For such an opening the adjustments, and subsidence, do not penetrate to the surface. When an opening exceeds the "critical width" then adjustments extend to the surface and subsidence of the surface becomes evident.

Figure 1 shows the conditions above an opening which is of less than critical width. The ability of an element of fractured material to support itself varies with its height above the excavation. This is the result of two effects:

(1) The subsidence of element *C*, for example, involves a greater shortening of the diagonal thrust lines than does the subsidence of an element such as *A*. Thus the vertical force tending to support *C* is greater than that for *A*.

(2) The confining stress acting vertically on the element increases with increased distance above the opening.

Thus the diagonal thrust varies from a zero just above the excavation to a maximum just above the clastic–elastic boundary. It is apparent that the pressure on the abutments will be spread over a large area, and the peak pressure will be less, than is the case in a massive elastic medium where the loads are concentrated at the opening boundaries.

OPENINGS IN STRONG STRATA – ROOF STRESSES

The mode of stress failure of rock overlying openings in thick, strong strata may be visualized by recalling the stress conditions in the case of a self-supporting linear arch.

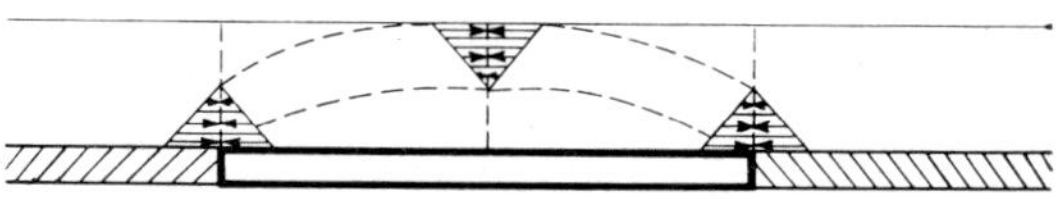

FIG. 2. Compressive stresses at the abutments of a linear arch.

Taking for an example the simplest case, that of the linear arch of Fig. 2, we recall that the maximum compressive stress occurs in the linear arch at the abutments. This compressive stress, σ_1, creates shear stresses inclined to the direction of maximum compression. For a rock with an angle of internal friction, $\Phi = 50°$ the shear planes will make an angle of 20° with the direction of the maximum compressive stress.

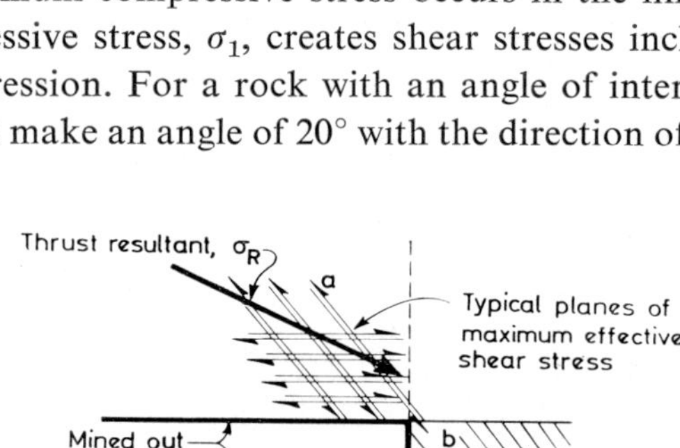

FIG. 3. Planes of maximum effective shear stress at the end of a linear arch.

There are two sets of shear planes, as shown in Fig. 3, but only those inclined in the direction of *ab* offer an opportunity for the rock to relieve itself of stress by fracturing. Therefore, fracturing under excessive stress will take place in the direction indicated by *ab*.

The foregoing analysis has not considered the effects of vertical shear, τ_v, on the stability of a linear arch because this stress in a gravity loaded beam is not sufficiently large to affect stability of the beam.

$$\text{Vertical shear (in psi)} = \tau_v = \left(\frac{l}{2}\right)\left(\frac{\gamma}{144}\right) \tag{1}$$

for a gravity loaded linear arch.

In rock with compressive strength of 10,000 psi and $E = 10^6$ psi a linear arch with thickness, $t = 200$ ft could theoretically support itself over a maximum span of 1600 ft (see Chapter 6). Then $\tau_v = 866$ psi.

Since the arch is subjected to a maximum lateral compressive stress, $\sigma_1 =$ 10,000 psi, the effect of $\tau_v = 866$ psi on its stability is negligible.

However, when we consider a linear arch loaded by a considerable thickness of overlying strata then τ_v may be large enough to influence stability.

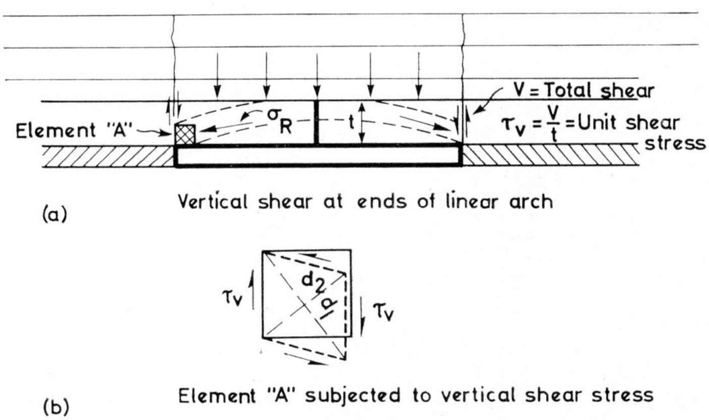

FIG. 4 (a) Vertical shear at the ends of a linear arch. (b) Element A subjected to vertical shear stress.

Figure 4 shows such an arch overlain by weak strata. In order to determine the effects of vertical shear and of lateral compressive stress, and effects of superimposing the two stress systems, an element will be removed from the end of the beam and its stability examined.

When subjected to vertical shear stress the element A is deformed as shown. It can be demonstrated (see any texbook on "Mechanics of Materials") that shear stress, τ_v, induces tensile stresses which are maximum along, and parallel to, diagonal d_1. These maximum tensile and compressive stresses are equal in value to the applied shear stress, that is,

Unit shear stress $= \tau_v = -\sigma_{d_1} = \sigma_{d_2}$

Since $-\sigma_{d_1} = \tau_v =$ maximum tensile stress, it is possible, under certain conditions, for a tension fracture to occur in an arch near an abutment. Such a fracture appears as an open crack at right angles to the direction of σ_{d_1}, that is, parallel to σ_{d_2}. Such tension fractures occur when vertical shear stress is large when compared to the lateral compressive stress.

However, it will now be demonstrated that, under the conditions normally encountered, tensile stress is neutralized by diagonal thrust in a gravity loaded arch and arch failure most commonly occurs by shear near an abutment, rather than by tension.

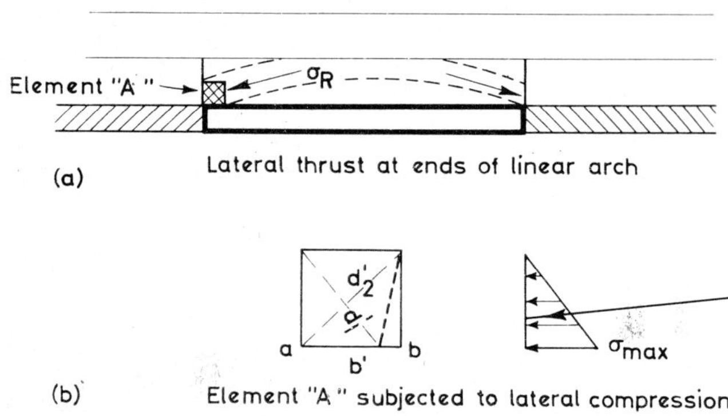

FIG. 5. (a) Lateral thrust at the ends of a linear arch. (b) Element A subjected to lateral compression.

Taking a linear arch (Fig. 5 (a)) which supports only its own weight so that the effect of τ_v is negligible we remove element A.

Figure 5 (b) shows element A and the resultant stress distribution. To simplify the example the maximum compressive stress is considered to be horizontal. It can be demonstrated by trigonometric relationships that for small displacements the unit strain along diagonal d_1' is compressive and is equal to half the unit strain along side ab. Since stress is proportional to strain,

$$\sigma_{d_1}' = \sigma_{\max}/2$$

Thus the diagonal compressive stresses created by arch thrust is equal to one-half the maximum compressive stress, $\sigma_{\max}$, at arch abutments. It can be seen that stress in direction of d_1' will be compressive.

The approximate maximum thickness of overlying strata which can be supported by a linear arch of any given dimensions may be computed by substituting the virtual weight, γ_v, for the actual weight, γ, in equation (1), Chapter 6:

$$l = \left(\frac{5Qt}{3\gamma}\right)^{1/2}$$

and converting from pounds per square inch to pounds per square foot:

$$l = \left(\frac{(5)(144)Qt}{3\gamma_v}\right)^{1/2}$$

and solving for maximum virtual weight

$$\gamma_v = \frac{240Qt}{l^2} \tag{2}$$

As an example, take a linear arch 100 ft thick and 400 ft long with $Q = 5000$ psi:

$$\gamma_v = \frac{(240)(5000)(100)}{(400)^2} = 750 \text{ pcf}$$

and total thickness of arch plus load is:

$$t\gamma_v/\gamma = \frac{750}{156}(100) = 480 \text{ ft}$$

that is, this linear arch, 100 ft thick, will support up to 380 ft of overlying strata without collapsing.

Maximum vertical shear stress would be (equation 1):

$$\tau_v = \left(\frac{l}{2}\right)\left(\frac{\gamma_v}{144}\right) = \frac{1}{2}(400)\left(\frac{750}{144}\right) = 1040 \text{ psi}$$

τ_v is only 1040 psi whereas it must exceed $\sigma_{max}/2 = Q/2 = 2500$ psi to produce tensile stress. Thus, if additional load is placed on the linear arch it will fail by shear rather than tension.

Tension fractures may be produced in those cases where the lateral component of the diagonal thrust is small, and where vertical shear stress is large. Such cases may occur at boundaries of wide openings in massive unfractured rock, or in the case of a strong, fractured rock in which the "compression arch" has a high rise/span ratio.

Failure is usually by shear fracturing induced by arch thrust near the abutments with the direction of shearing influenced by magnitude of vertical shear stress. The effect of increasing vertical shear, τ_v, is to steepen the inclination of principal compressive stresses and the resultant shear planes.

For a linear arch with no surcharge (e.g. small τ_v) the principal stress is as shown in Fig. 6 (a), and maximum effective shear stress acts at angles of $45° - \Phi/2$, or about 20° from the direction of maximum compressive stress. The angle of internal friction, Φ, for the rock is taken as 50°.

Figure 6 (b) shows the steepening of the arch-thrust resultant and the orientation of planes of maximum effective shear for a linear arch loaded by overlying strata.

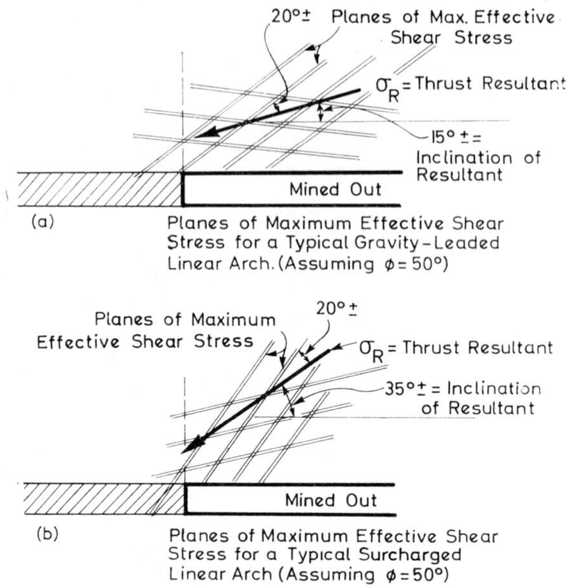

FIG. 6. (a) Planes of maximum effective shear stress for a typical gravity-loaded linear arch. (Assuming $\Phi = 50°$.) (b) Planes of maximum effective shear stress for a typical surcharged linear arch. (Assuming $\Phi = 50°$.)

VIOLENT STRESS FRACTURING
THEORY OF FRACTURE PROCESSES WHICH PRODUCE BURSTS

Rock Bursts and Coal Bursts ("Bumps" or "Bounces")

A rock burst is typified by violent fracturing of portions of rock around an opening, accompanied by audible noise, usually with fragments or slabs projected from walls, roof, or floor.

The process is known to the miners by a variety of names. Minor rock bursts, such as the flaking or spalling off of small fragments or slabs from walls or roof of an excavation accompanied by snapping or popping sounds are a fairly common occurrence in brittle rocks at depths of more than a few hundred feet. This mild form of rock burst is known to the miners by a veriety of names, including "popping", "spitting", "flaking", or "spalling".

A medium-sized rock burst may involve many cubic feet or cubic yards of rock. This may be loosened instantaneously from the walls, or roof, of an opening, accompanied by explosive sounds.

The largest rock bursts occur in connection with wide openings (and multiple openings) and may involve movement of mine roof or floor (or hanging-wall and foot-wall) and the loosening of hundreds or thousands of tons of rock. These occurrences are sometimes termed "pressure bursts".

Rock bursts and "coal bursts" in coal mines are generally designated by the term "bumps" or "bounces." A large proportion of these "bumps" or "bounces" are actually "coal bursts" due to fracturing of pillars and remnants and need not involve any actual fracturing of roof or floor rock.

The largest rock bursts create vibrations in the earth similar to those of an earthquake. For example, a large rock burst in the Kirkland Lake District of Ontario was recorded on a seismograph in Ottawa, about 300 miles away.

In former years rock bursts were sometimes referred to as "air blasts." This name originated from the fact that large quantities of rock loosened by some of the larger rock bursts displaced quantities of air, with the result that a blast of air passed through the mine workings. This gave rise to the theory that pockets of natural compressed air existed in the rock and caused the violent dislocation of the rock. This theory has now been discarded and most authorities agree that rock bursts are caused by fracturing of rock which is under excessive stress; the fracturing releases the rock to expand (or contract) to its normal dimensions, thus releasing the energy which was stored in the stressed rock.

The storage of this elastic energy in rock is similar to the action of a spring. If a spring is compressed it stores energy, and if released slowly the energy is recovered slowly. However, if the spring is compressed beyond its strength it fractures and energy is released suddenly, the snapping sound accompanies the fracture, and the flying pieces of the steel spring correspond to the noises and the projected rock fragments of a rock burst.

The quantities of energy released by rock bursts make it obvious that the rock stores large quantities of elastic energy, and, since rock is not strong enough in tension to store any appreciable energy by "stretching" then the stored energy must be accumulated under compressive stress. In order for the energy to be released suddenly a fracture must occur which will allow a portion of the rock to expand.

Rock, under excessive compressive stress, may fracture either in shear or in tension. The restraints present around wide openings tend to favor shear fracture rather than tension fracture.

Therefore, it can be deduced that a rock burst occurs when a mass of brittle rock, be it one ton or be it ten thousand tons, is subjected to compressive stress of sufficient magnitude to cause sudden failure along a shear plane.[48]

The rocks most subject to bursting are the strong, brittle, homogeneous rocks which contain few pre-existing fractures. Around wide openings bursts occur most frequently in hanging-wall rock, except in coal mines where they most frequently occur in the coal seam itself and technically should be called "coal bursts". The capacity of a rock for storing energy is inversely proportional to its modulus of elasticity, E. Coal has a very low E value and therefore a higher capacity for storing energy than do the rocks enclosing it.

Mechanics of Rock Bursts

The nature of shear failure in brittle, elastic material is shown diagrammatically in Fig. 7.

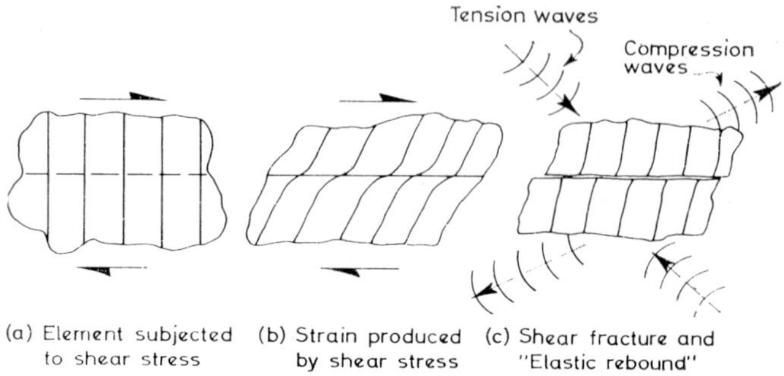

FIG. 7. (a) Rock element subjected to shear stress. (b) Strain produced by shear stress. (c) Shear fracture and elastic rebound.

Elastic Rebound

Relative movement of rock on opposite sides of the plane of maximum effective shear stress strains the rock to the point where fracture is produced. Upon fracturing, the rock "snaps" back to its unstrained condition. This action is known as "elastic rebound" and releases the elastic energy stored in the rock. This has the same effect on the rock as would an explosion or a blow with a heavy hammer, and produces vibrations or elastic waves which correspond to the seismic waves of an earthquake.(49)

On a tremendously large scale this process of shear fracturing and elastic rebound produces earthquakes and on a smaller scale the shear fracturing of rock around mine openings subjected to excessive compressive stresses produces rock bursts.

An earthquake, since it originates at depths in the earth where the rock is tightly confined, produces no loose rock fragments, whereas a large rock burst which occurs near a free surface may project hundreds of tons of rock, or coal, into an excavation.

Storage of Elastic Energy

The formula for elastic energy stored per unit volume of an unconfined elastic body is

$$U = \frac{144\sigma_1^2}{2E} \tag{3}$$

where U is energy stored expressed in foot-pounds per cubic foot, σ_1 is stress in pounds per square inch, and E is the modulus of elasticity in pounds per square inch.

When confined rock is subjected to compressive stress its capacity for storing energy is reduced, and is a function of m (Poisson's number).

Designating U as the energy stored by laterally unconfined rock and U_1 as the energy stored by a laterally confined rock, then:[(1)]

$$U_1 = \frac{(m+1)(m-2)}{(m)(m-1)} \cdot U \tag{4}$$

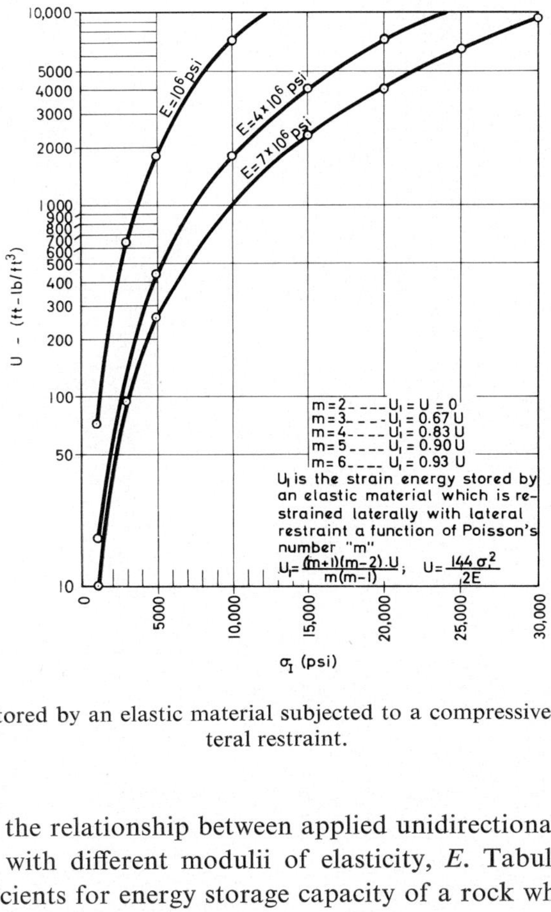

FIG. 8. Energy (U) stored by an elastic material subjected to a compressive stress, σ, with no lateral restraint.

Figure 8 shows the relationship between applied unidirectional stress and stored energy for rocks with different modulii of elasticity, E. Tabulated in the figure also are the coefficients for energy storage capacity of a rock which is confined on all sides, that is, subjected to triaxial stress. The energy storage capacity under confinement decreases as m decreases.

Poisson's number for most rocks ranges from about 4 to 6, with some of the harder rocks having m values up to 20. For coal, m ranges generally from 2 to 3.

Figure 9 shows stress–strain curves for selected coal samples. Values of E calculated from these curves vary from 500,000 to 600,000 psi for the Phelan and

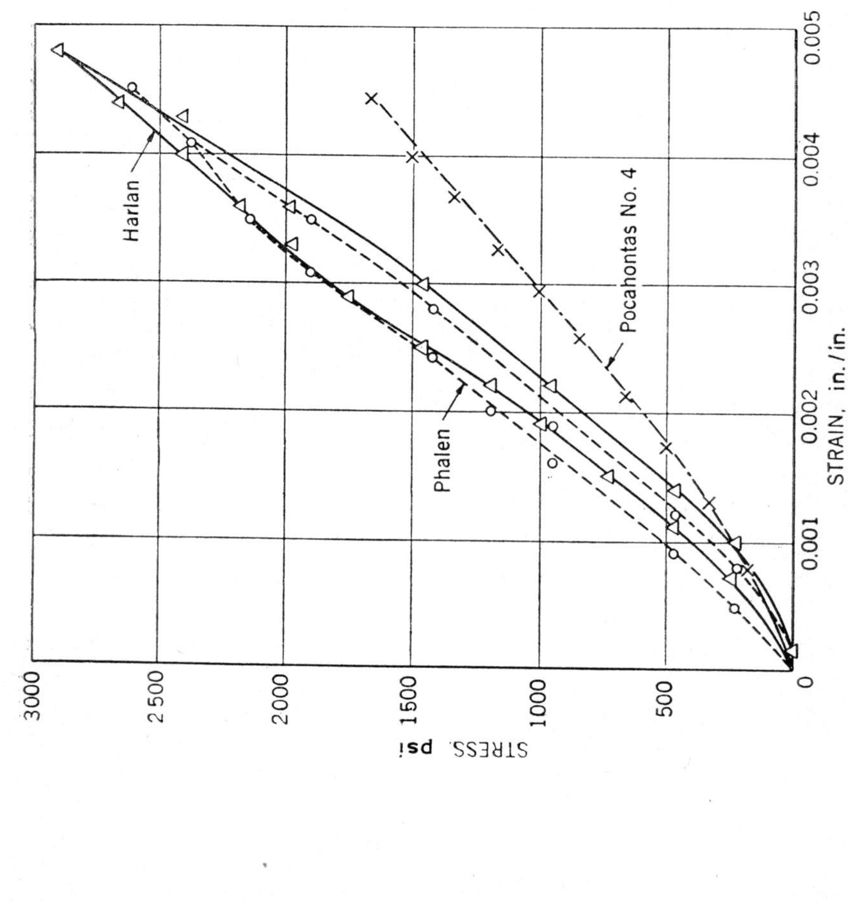

FIG. 9. Typical stress–strain curves for coal. (Very severe bumps have occurred in the Harlan, Phalen, and Pocahontas No. 4 coal beds.[2])

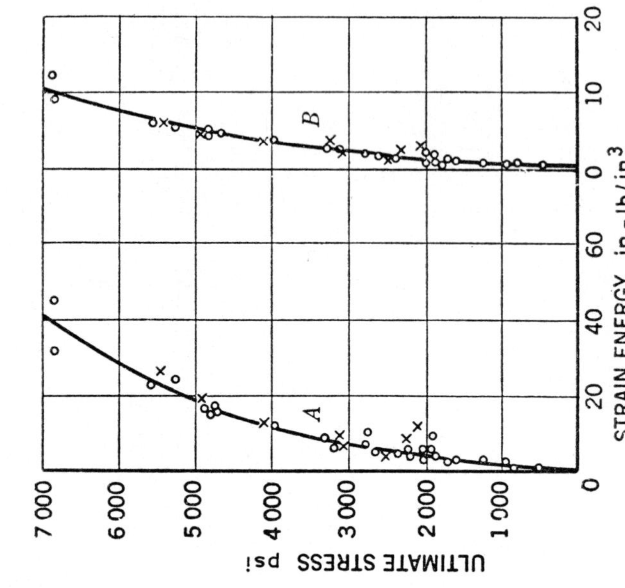

FIG. 10. Strain energy stored by coal when there is no lateral restraint.

(A) Energy estimated to be stored when the coal is stressed to its ultimate strength; (B) Energy stored when coal is stressed to one-half of its ultimate strength. (The circles represent coals from mines which bumps have not occurred and the crosses represent coals from mines in which bumps have occurred.)

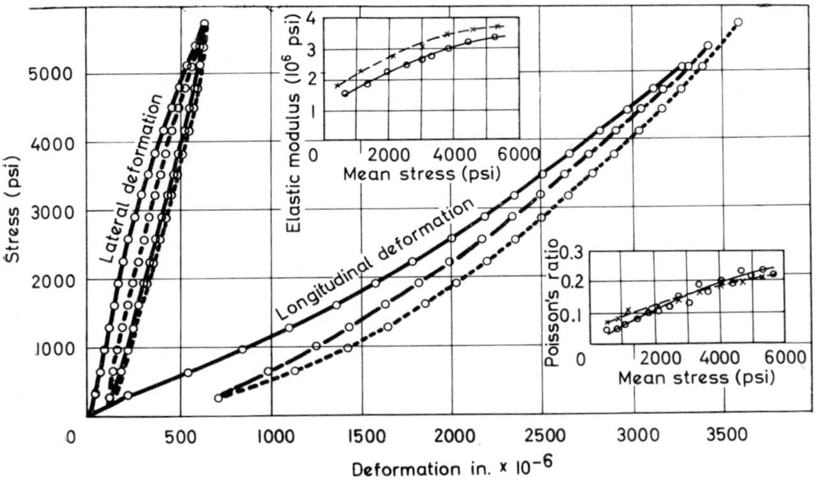

FIG. 11. Deformation tests on siltstone.[3]

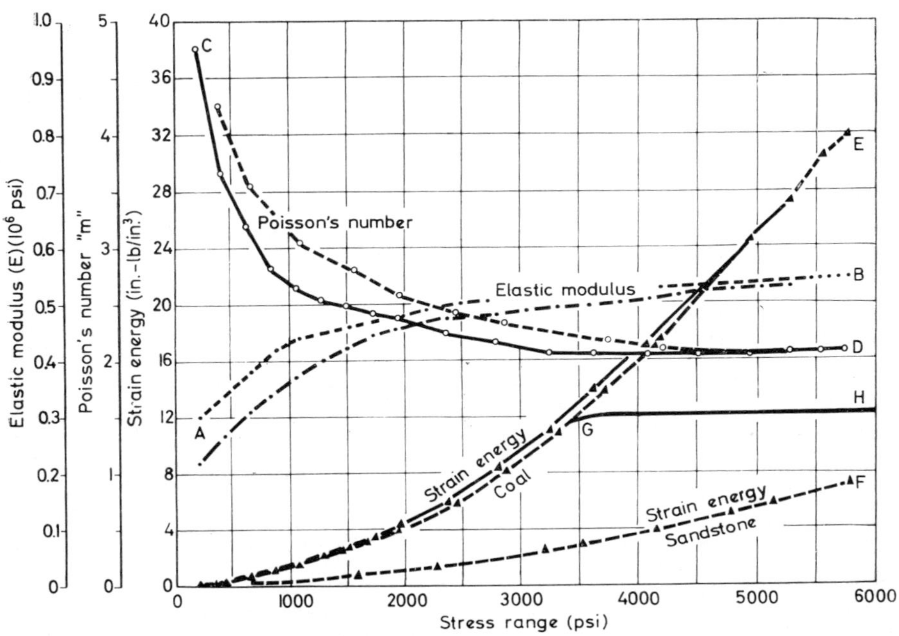

FIG. 12. Compression tests on coal showing the elastic modulus, Poisson's number, and energy storage capacity.[3]

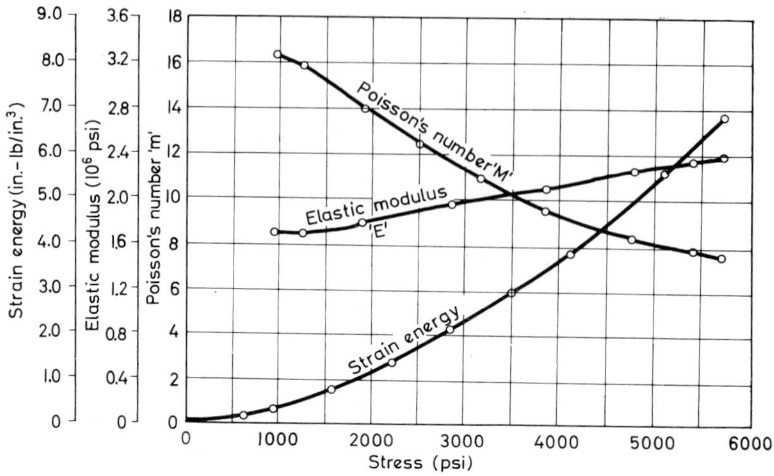

FIG. 13. Compression tests on sandstone showing the elastic modulus, Poisson's number, and energy storage capacity.(3)

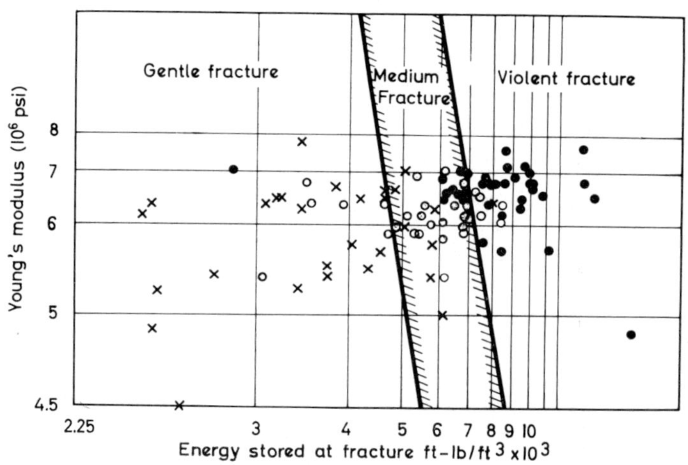

FIG. 14. Representation of violence of rock fracture in terms of rigidity and energy stored.(4)

Harlan coals to between 300,000–400,000 psi for the Pocahontas No. 4 coal. Fig. 10 shows strain energy stored by coal subjected to varying compressive stress.

The modulus of elasticity of coal is only about one-third to one-eighth that of the average rock enclosing the coal. This allows it to store as much, or more, energy than the rock in spite of its low *m* value, and its low compressive strength.

Figure 12(3) shows variation of physical constants of coal with increasing pressure and comparison of its energy storage capacity with that of a sandstone.

Figures 11 and 13 show how the physical constants of siltstone and of sandstone vary with compressive stress.

Figure 14 shows relative violence of fracture of test specimens of Witwatersrand quartzite as a function of U and E. Compressive strength of these rocks averages from 20,000 to 25,000 psi. The crosses indicate "gentle fracture", the circles "medium fracture", and the black dots "violent fracture". It is apparent that violence of fracture depends upon the amount of energy stored at the time of fracture.

Quantities of Energy Involved in a Burst

The amount of energy released by a burst ranges from the minute amount released when a chip of rock flakes off the face up to millions of foot-tons of energy released by a large "pressure burst" in a deep mine.

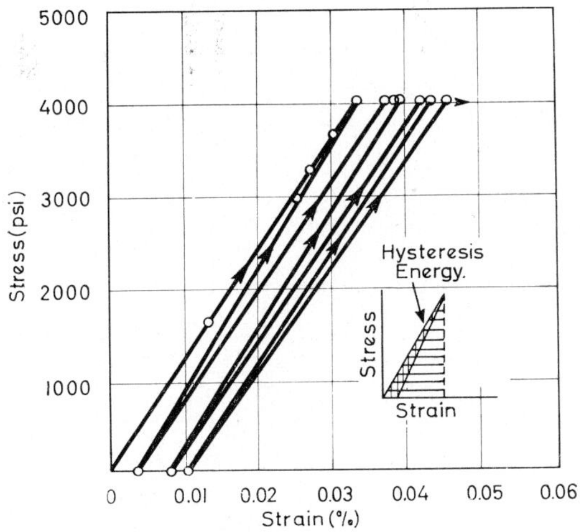

FIG. 15. Hysteresis in quartzite.[4]

One large rock burst in the Kirkland Lake District of Ontario, Canada, was recorded on a seismograph in Ottawa, about 300 miles away.

J. S. Jones[5] estimated, on the basis of seismographic measurements, that the quantities of energy released by a typical medium-heavy rock burst in the Kolar Gold Field of India must have involved the liberation of at least 15 million foot-tons of energy which could not be stored in less than 7.5 million cubic feet of rock. Such a rock "block" could have, for example, dimensions of about 500 ft $\times$ 500 ft $\times$ $\times$ 50 ft. Jones, in his computations, apparently assumed that modulus of elasticity and applied stresses were such that the rock involved stored an average of at leas 4000 ft-lb./ft^3.

A news item[6] tells of a recent serious rock burst on the Witwatersrand which involved the sudden upward movement of the foot-wall over an area 240 ft $\times$ 50 ft.

It was estimated that the energy involved in the burst was equivalent to $1050 \times \times 10^6$ ft-lb.

In the case of "bumps" or "coal bursts", no estimates of energy involved appear to be available in the literature. Most coal mines are relatively shallow in comparison to the 7000–10,000 ft depths of the deeper metal mining districts afflicted with heavy rock bursts. Thus it appears that because of the pressures associated with the greater depths, the quantities of energy involved in coal bursts would not approach those involved in rock bursts of the aforementioned metal mining districts.

Amount of Elastic Strain

It should be noted that, although a large "block" of rock may be involved in a burst, very little change in the overall dimensions of such a block is required, either to store, or to release, large quantities of elastic energy. For example, a unidirectional compressive stress of 25,000 psi applied to a laterally unconstrained rock with a modulus of elasticity $E = 4 \times 10^6$ produces a dimension change of 1/160 and for $E = 10 \times 10^6$, a dimension change of 1/400. The maximum amount of strain before rupture is determined by the ratio:[48]

$$\frac{\text{Compressive strength}}{\text{Modulus of elasticity}} = \delta$$

For most rocks which are subject to bursting this ratio appears to range generally between 1/160 and 1/300.

The foregoing examples, although they present the simplest cases, serve to illustrate the point that very little evidence of rock movement may be visible in the workings despite enormous quantities of energy released in a major rock burst. This energy can be stored and released by elastic compression and sudden release of rock involving dimension changes of rock "blocks" which may be as little as 1 ft in 300 ft.

TYPES OF ROCK BURSTS

Minor Burst — Single Opening

As excavation of an opening proceeds tangential compressive stresses are induced around the perimeter by concentration of pre-existing natural stresses. Thus any element, such as the one shown in Fig. 16, is subjected to tangential compressive stress. Tensile stress may be induced at the ends of those pre-existing fractures which are approximately parallel to the direction of the applied compressive stress. Thus the surface may "flake off" or "slab off" by "extension fracturing", that is, by the extension (by tensile fracture) of natural fractures, incipient fractures created by blasting, or other planes of weakness.

If no properly oriented pre-existing fractures are present near the rock surface then the rock may fail by shearing along inclined planes as shown in Fig. 16.

Prior to its detachment from the rock surface each element is under compression and is therefore deformed and shortened. At the instant that fracturing occurs the

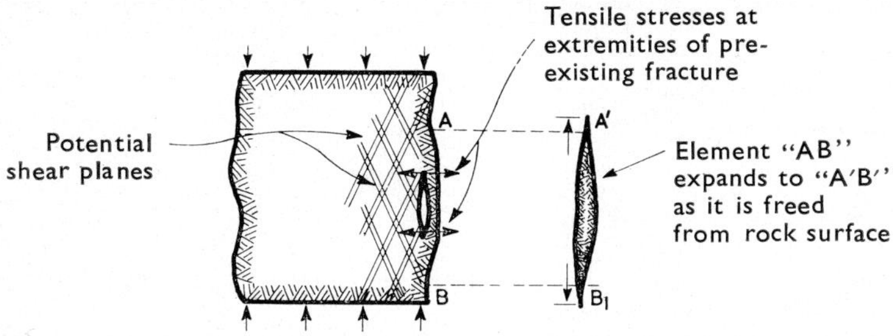

FIG. 16. A minor rock burst. Mechanics of the process of "spitting", "popping", or "spalling". (Rock failure either in tension or by shear.)

element expands and this action results in its projecting itself from the wall, sometimes quite violently. This process is known by miners by a variety of names, among them "flaking", "spitting", "popping", "spalling", and "slabbing".

Moderate Bursts — Single Opening

A typical burst in a single opening subjected to excessive tangential compressive stress is shown in Fig. 17. Here the stresses are higher and extend farther into the rock than in the case of spitting and spalling, and failure is produced along planes

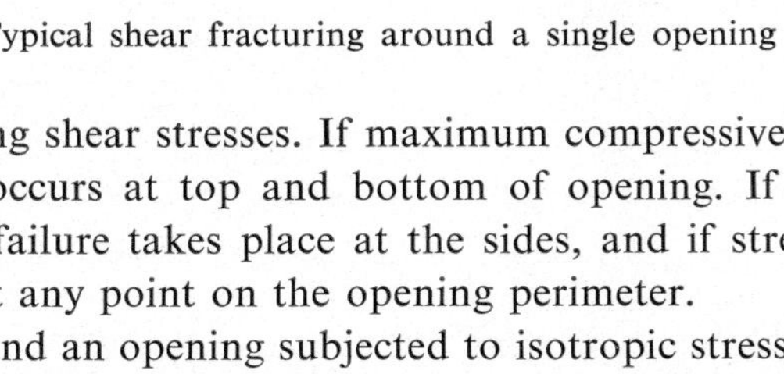

FIG. 17. Rock burst. Typical shear fracturing around a single opening under isotropic stress.

of maximum effecting shear stresses. If maximum compressive stress is horizontal then shear failure occurs at top and bottom of opening. If maximum stress is vertical then shear failure takes place at the sides, and if stress is isotropic then failure may occur at any point on the opening perimeter.

Shear failure around an opening subjected to isotropic stress has been produced in the laboratory by York and Reed[7], who subjected thick-walled cylinders made of pottery plaster to external pressures sufficient to produce failure. In these experiments the failure took place along shear planes as defined by theory and as illustrated in Fig. 18. An infinite number of these planes of maximum effective shear stress exist around the boundary of such an opening. As a result triangular slabs sheared out along intersecting planes as shown. In these experiments failure began when the applied external stress reached a value of from 300 to 500 psi, which would give a compressive stress at the inside boundary equivalent to 600–1000 psi. This figure compares with 300 psi, which was the ultimate compressive strength of the pottery plaster used.

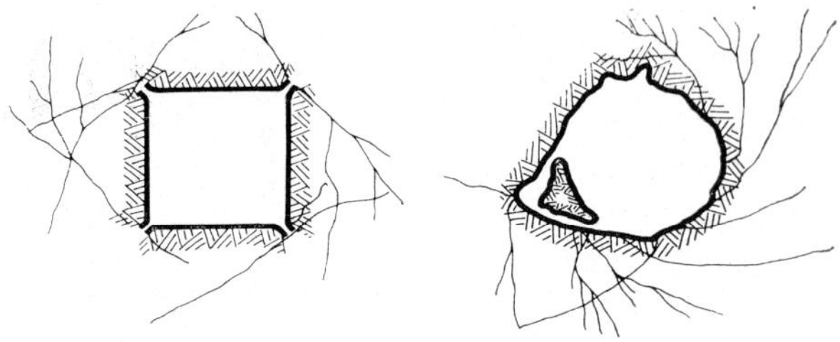

FIG. 18. Typical fracture under isotropic stress. Experimental results on models subjected to hy drostatic pressure (by York and Reed[7].)

Bridgman[8] has experimented with thick-walled hollow cylinders of various brittle rocks subjected to high hydrostatic external pressures. Failure of the rock ft the internal boundary of the cylinder took the form of violent spalling of minute aragments from the inside surface, with the interior gradually enlarging as pressure was increased until it was packed with a fine impalpable "sand". No flowage of the rock was noted, even at external pressures of 90,000 psi applied to cylinders of porphyry, andesite, and granite.

Theory of the Fracture Processes Involved in Major Rock Bursts or "Pressure Bursts"

Because of large abutment pressures and large volumes of rock involved those rock bursts which occur in connection with multiple openings or wide openings involve the release of larger quantities of elastic energy than those associated with single openings.

On the Witwatersrand bursts associated with abutment pressures around wide openings are generally termed "pressure bursts", and this term will be used generally herein to designate any burst occurring in or around a wide opening or a multiple opening.

The author employs the term "pressure burst" merely because it is a convenient term for designating rock bursts of large magnitude. It will be demonstrated that the mechanics of a "pressure burst" are the same as those for a moderate burst around a single opening; that is, the burst occurs when a mass of rock which is subjected to excessive compressive stress suddenly ruptures along a shear plane.

One possible method of classifying pressure bursts is by the location of the main fracture which produces the burst. On this basis we would have two classes:

(1) Roof or floor (or hanging-wall and foot-wall) bursts:
  (a) by rupture of the linear arch in roof or floor;
  (b) by rupture of a massive roof or floor.

(2) Seam (or vein) bursts:
  (a) in coal seams;
  (b) in metalliferous veins.

Rupture of the Linear Arch

Shear stresses near the abutment of a linear arch are shown in Fig. 19. The direction of maximum effective shear varies with the type of arch loading, being flatter for a gravity loaded arch and steeper for an arch with surcharge. (See also Fig. 6.)

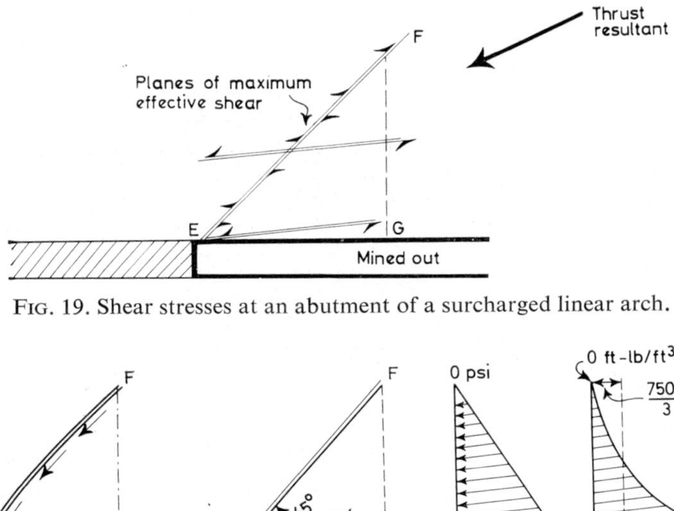

FIG. 19. Shear stresses at an abutment of a surcharged linear arch.

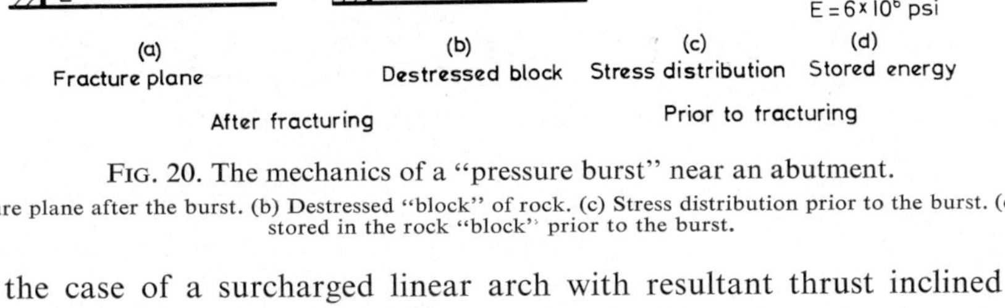

FIG. 20. The mechanics of a "pressure burst" near an abutment.
(a) Fracture plane after the burst. (b) Destressed "block" of rock. (c) Stress distribution prior to the burst. (d) Energy stored in the rock "block" prior to the burst.

For the case of a surcharged linear arch with resultant thrust inclined about 20–25° to the horizontal, fracturing takes place along planes having directions similar to *E–F*. Thus a fracture, progressing upward into the rock from point *O* releases a large volume of rock to expand suddenly. This is shown in Fig. 20.

Rupture takes place suddenly along a plane such as *E–F*, relieving the block of rock *EFG* of compressive stress and, within a few milliseconds, it expands in the direction indicated by the arrows.

We can compute the amount of elastic energy stored in the portion of the rock *EFG* prior to rupture. This will be the energy released by the rupture. For convenience in calculation the dip of the rupture plane will be taken as 45° and dimensions of the triangular rock "slice" as 50 ft × 50 ft × 1 ft thick as shown in Fig. 20 (b). These dimensions would be reasonable for a linear arch 100 ft, or more, in thickness.

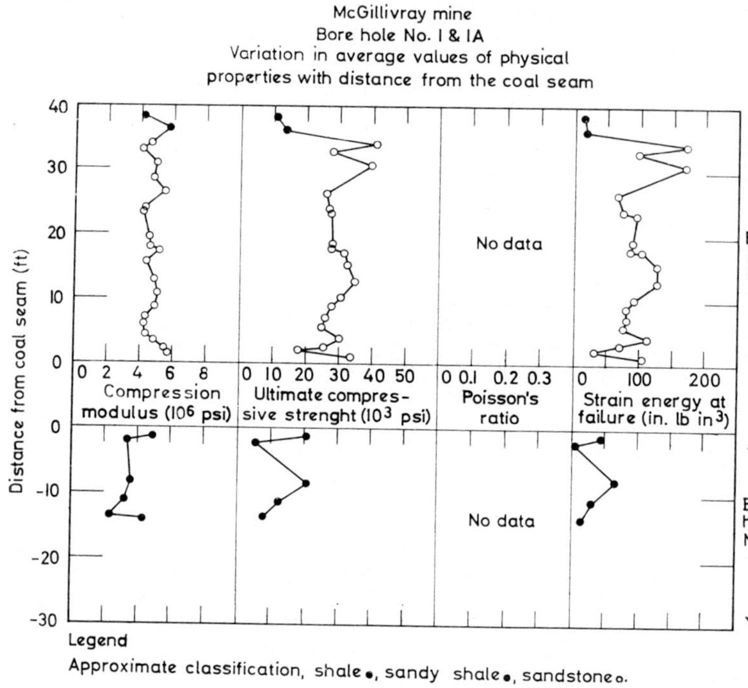

FIG. 21. Physical properties of the rock strata above and below the coal seam, McGillivray mine, Canada.(9)

Rocks with compressive strengths of about 20,000–30,000 psi and $E = 4\times10^6$ to 7×10^6 are typical of bursting rock in the Canadian coal mines at Coleman, Alberta (see Fig. 21)(9) and Springhill, Nova Scotia(10) as well as the gold mines of the Witwatersrand, South Africa.(4)

Using for this example a rock with compressive strength, $Q = 25{,}000$ psi and $E = 6\times10^6$, we find, from Fig. 8, that a cubic foot of rock will store about 7500 ft-lb of elastic energy when stressed to the breaking point.

The distribution of compressive stress in the rock block prior to rupture is presented approximately by Fig. 20 (c) which shows stress varying from 25,000 psi to 0. Energy stored per unit volume, U, varies as the square of the stress ($U = K\sigma_1^2$);

therefore the line representing stored energy is a parabola, and the area under the parabola represents the total energy stored in the block (Fig. 20 (d)). Since the area under a parabola is equal to $\frac{1}{3}$ (base $\times$ height), the average value of U in the example is $7500/3 = 2500$ ft-lb/ft^3 for the block EFG.

The volume of the block is

$$\frac{50 \times 50 \times 1}{2} = 1250 \text{ ft}^3$$

Total energy stored in the block is $1250 \times 2500 = 3{,}125{,}000$ ft-lb/linear ft of face. If fracture extends for 200 ft parallel to the face the total energy released would be 625×10^6 ft-lb.

Thus, in strong rock, a large quantity of energy may be released by a burst even though the rock block involved in a burst is relatively small in comparison to the overall dimensions of workings and depth of cover.

In the example the expansion along dimension GE will be:

$$50 \frac{(25{,}000)}{(6 \times 10^6)} = 0.21 \text{ ft},$$

snd in direction FE expansion would be 0.30 ft.

After fracture the rock in block EFG, which formerly was a part of the self-aupporting linear arch, is no longer an effective part of the arch and the compressive stresses must realign themselves to form an arch with a higher rise/span ratio, thus reducing maximum compressive stress to within the strength of the rock. This is shown in Fig. 22.

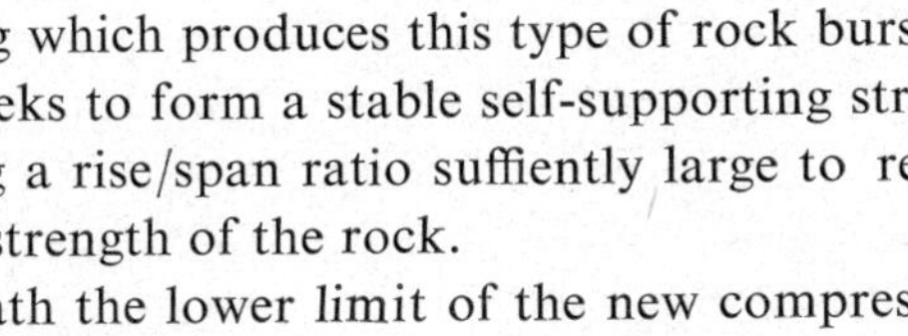

FIG. 22. Re-alignment of the arch line after a "pressure burst".

Thus the fracturing which produces this type of rock burst is part of the process by which the rock seeks to form a stable self-supporting structure, with the compression arch having a rise/span ratio suffiently large to reduce maximum stress to values within the strength of the rock.

The rock underneath the lower limit of the new compression arch, OIJ, is now partially suspended from the arch and partially resting on supports in the goaf area. A strong rock has sufficient tensile strength so that its entire weight is suspended from the new compression arch, while a weak rock will add its weight to any already on the supports, packing, or filling in the goaf.

The example given was for a burst near an abutment (vicinity of the face). Examination of Fig. 23 reminds us that high stresses exist also at other places in a linear arch and, if the arch has sagged away from overlying strata, then a burst can occur at a location other than near an abutment. In the case shown the result would be buckling of the arch at its center with additional weight thrown on support in the goaf, particularly near the location of the center of the collapsed arch. Tension cracks would open on the underside of the lowering arch, particularly near the point of the failure.

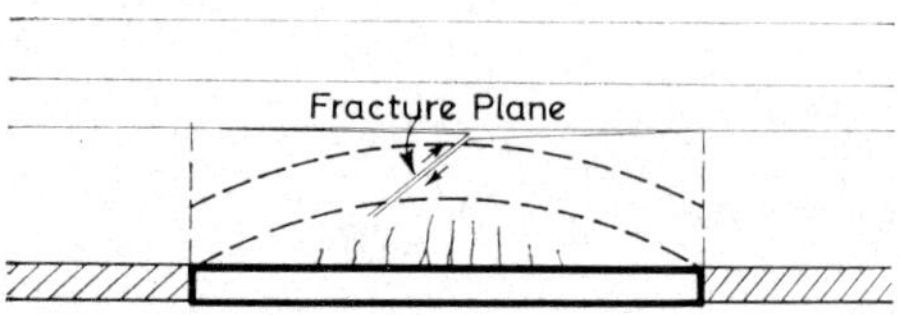

FIG. 23. "Pressure burst" by fracturing near the middle of a linear arch.

Roof Fracture Ahead of The Face

If the seam or vein is relatively soft and yields for a few feet ahead of the face so that the maximum pressure "peak" is transferred ahead, then a rock burst may occur with the plane of fracture being located over the unmined coal. In such case the actual rupture plane will not be visible until the face is mined ahead to expose it.

Rupture of Massive Roof or Floor

The foregoing discussion has considered conditions in lightly fractured rock. In a massive homogeneous rock similar rupture may take place. In this case the horizontal stress component at the abutment is small and the vertical component is large as compared to the linear arch. Thus the resultant maximum compressive stress acts at a steeper angle and shear planes are correspondingly rotated, so that major fracture planes dip steeply.

"COAL BURSTS" OR "BUMPS"

In contrast to those rock bursts which involve rupture of the roof or floor, the majority of the "bumps" which occur in coal mining operations involve only the violent fracture of coal and do not involve any rupture of rock in either the roof or the floor. It would contribute much to an understanding of the phenomena if such bursts were designated in the literature as "coal bursts" to indicate that no fracture of the enclosing rock is involved.

Strengths of the roof and floor rocks in mines where "coal bursts" occur may be much less than strengths of rocks associated with rock bursts. Holland[(11)] lists the physical characteristics of roof and floor rocks from a number of *U.S.* mines where "coal bursts" have occurred. Compressive strengths ranged from 7000 to

19,000 psi (one was as low as 4000 psi) for roof rocks and generally from 3500 to 9000 psi for floor rocks. Modulii of elasticity of roof rocks generally ranged from 3×10^6 to 7×10^6.

Energy Storage

Coal has a relatively low compressive strength, usually from 2000 to 5000 psi, and a modulus of elasticity usually ranging from 300,000 to 600,000 psi. The formula for elastic energy stored per unit volume of material is (See Fig. 8):

$$U = \frac{144\sigma_1^2}{2E} \qquad (3)$$

Thus, in spite of its low compressive strength, Q, its low E value still allows it to store large quantities of elastic energy, before fracturing.

Taking, for example, a coal with $Q = 5000$ and $E = 500{,}000$ we find that it can store 3600 ft-lb/ft³ before fracturing if unconfined. When completely confined its energy storage capacity varies inversely with Poisson's number, m, in accordance with equation (4).

Mechanics of a "Bump"

Nearly all "coal bursts" occur in seams overlain by strong, thick strata such as sandstones or conglomerate. The principal reason for this seems to be that such strata support themselves over long spans without caving; therefore the pillars which support the long spans are subjected to excessive vertical stresses. Holland(11) mentions roof spans of extending as much as 225 ft, under 800 ft of cover, before roof failure occurs.

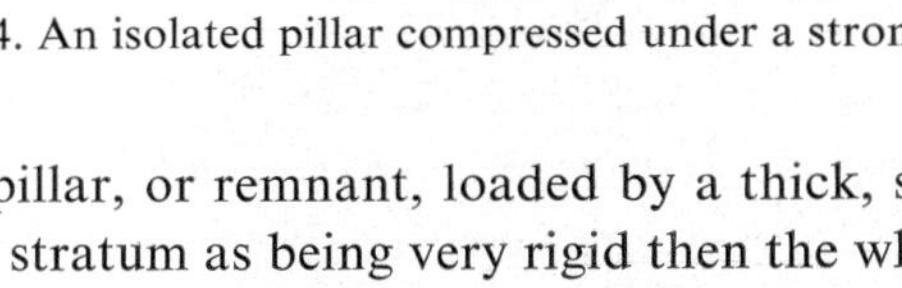

FIG. 24. An isolated pillar compressed under a strong roof.

Figure 24 shows a pillar, or remnant, loaded by a thick, strong roof stratum. If we take the sandstone stratum as being very rigid then the whole pillar will be compressed vertically by the same increment. At the edges of the pillar the coal relieves itself of stress by fracturing, but the rigid stratum cannot sag around the edges to obtain support from the fractured material. Therefore, more weight has to be supported by the solid interior. This interior, even though it may be stressed beyond its unconfined compressive strength, cannot fracture because the lateral confining stresses prevent shear failure.

If some incident relieves the lateral confining stress suddenly, or if stress in the pillar reaches the point where vertical stress is greatly in excess of lateral confining stress, then fracturing may extend almost instantaneously through the overstressed unfractured portion of the pillar. The elastic energy stored in the coal causes fragments to expand suddenly as fracturing frees them from the seam and to project themselves into the opening (Fig. 25). Broken coal accumulates in the opening until it piles up sufficiently to create some lateral stress which inhibits further fracturing.

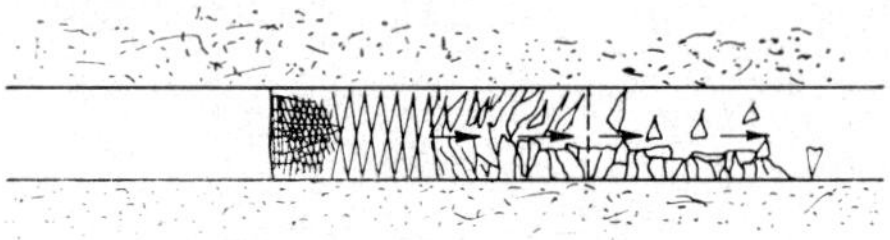

FIG. 25. Bursting of an overstressed isolated pillar.

Release of Gas

In addition to the elastic energy in the coal, transportation of the broken coal into the workings may be aided by the quantities of gas released as the coal is fragmented.

Coeuillet[(1)] has computed that 1 m^3 of gas contained in a ton of coal at 2 atm pressure will, if released by the sudden pulverization of the coal, be able to do 7 ton/m of work. Similarly, 10 m^3 of gas contained at 2 atm will do 70 ton-m of work if suddenly released.

Bursts in gassy seams frequently release large quantities of gas and it is probable that much of the coal transported during such a burst is "blown", or carried by the suddenly released gas. Most authorities do not believe that the gas itself contributes directly to the "burst" or to the fragmentation of the coal. Such fragmentation is considered to be the effect of excessive stress, with the resulting fragmentation allowing the gas to escape.

A "coal burst" in the Crowsnest Pass Coal Field, British Columbia, released an estimated 1.8 million ft^3 of gas.[(11)]

Khodot[(12)] says that research has demonstrated that during an outburst in a gassy seam a unit volume of coal may release 20 vols. of methane, sometimes more. He notes that about 40 m^3 of gas is required to transport 1 m^3 of coal a distance of 10–15 m (1 m^3 = 35.31 ft^3 = 1.308 yd^3. 1 m = 3.28 ft).

Ettinger[(13)] notes that experiments indicate that there is no difference between those coals susceptible to "instantaneous outbursts" and those not so inclined as far as their capacities for total absorption of gas. However, he has noted a measurable difference in their speed of absorption or release of gas.

Coal bursts may take place without any fracture of the rock in roof of floor, all the fracturing taking place in the coal seam. In some instances, less common, fracturing of roof and/or floor may accompany bursting of the coal, one incident probably triggering the other.

Conditions which Produce "Bumps"

Holland and Thomas[(2)] list the following natural conditions which favor the presence of "bumps" in coal mines:

(1) The cover over the bed is approximately 500 ft or more thick.

(2) A strong, overlying stratum, usually a massive sandstone or a conglomerate, occurs immediately above or close to the coal bed.

(3) The floor is strong and does not heave readily.

(4) The thickness and physical and chemical characteristics of the coal bed apparently are not critical factors, as bursts have occurred in coal beds as thin as 34 in. and as thick as 23 ft and as weak as the Pocahontas No. 4 (crushing strength 2140 psi in a 3-in. cube) and as strong as the Island Creek (crushing strength, 4700 psi in a 3-in. cube).

(5) Faults have been observed in several areas where bumps have occurred in coal beds. However, their complete absence in other areas where such phenomena have occurred demonstrates that faults do not necessarily contribute to bumps.

(6) Bumps in coal mines have been recorded as occurring in beds dipping as much as 35°. In the United States the greater number of bursts have occurred in beds dipping less than 10–12 per cent, probably because most of the coal is produced from such beds.

Stress Conditions in a Typical Coal Pillar

Figure 26 shows the probable stress distribution in a typical coal pillar in narrow workings.

The following conditions apply to the pillar of Fig. 26:

(1) The least lateral dimension of the pillar is 4–6 times the width of the rather narrow passageways that cut it out.

(2) The pillar is surrounded by other pillars that are the same size and are composed of coal having uniform physical properties.

(3) The pillars are loaded by a thick massive sandstone roof and are supported by a floor that offers great resistance to yielding. The stresses are comparatively low at the edges of pillar but quickly build up to maximum a short distance in the pillar. They then gradually drop to a uniform stress, which continues to within a few feet of the opposite side, when the stress conditions described are repeated. For a passageway of given height the intensity of the stress developed and the distance over which the higher stress acts increases with its width.

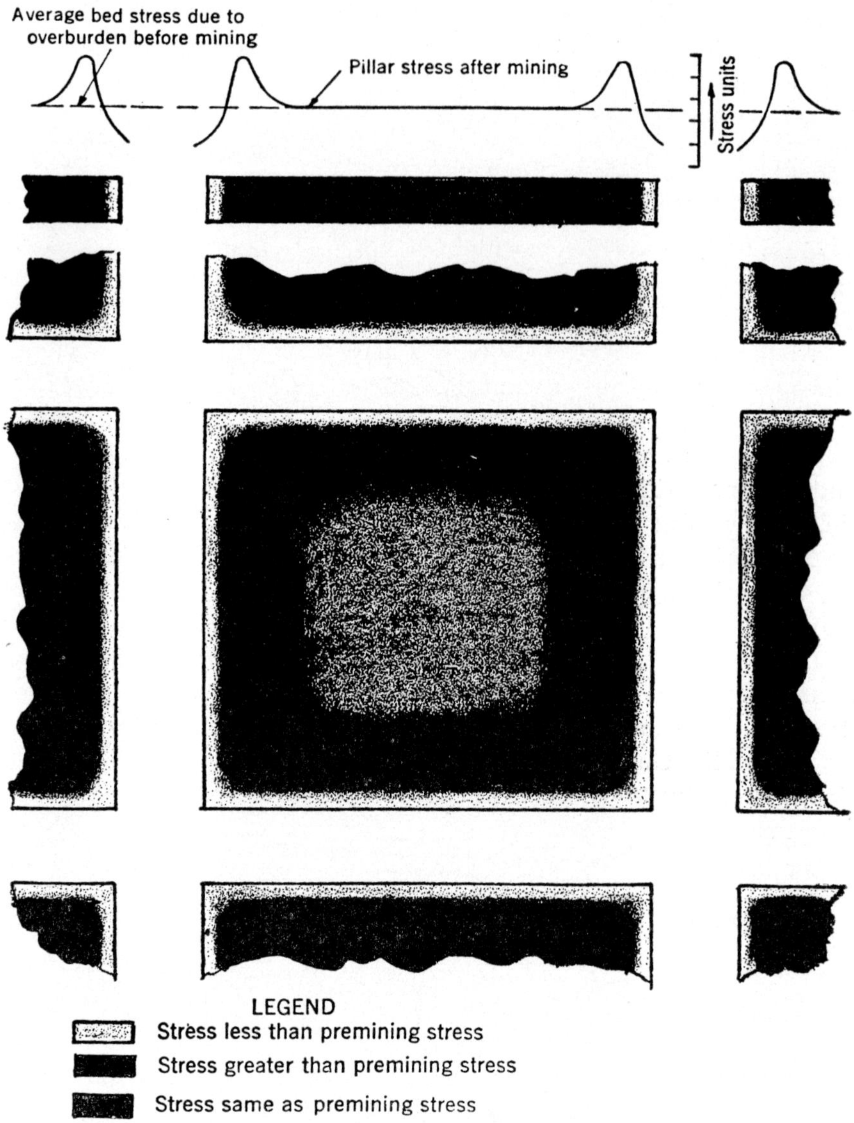

FIG. 26. Stress concentrations in individual pillars beyond abutment area. High stresses induced by an extensive goaf area.[2]

Stresses are maximum near the ribs of passageways, and if the passageways are widened, the zone of high stress extends further into the pillars.

If abutment stresses, such as would occur along a pillar line adjacent to an extensive mined-out area, are superimposed on the pillar system described above it is obvious that the stresses on pillars shown would be increased by several hundred percent.

Localities in the abutment area near a pillar line where very high stresses may be anticipated are:

(1) Areas at, or close to, the extraction line itself.
(2) Pillar areas adjacent to passageways (especially wide passageways).
(3) Pillars that are larger (longer and wider) than surrounding pillars.
(4) Points formed on active pillar lines or where pillar lines advance parallel to old goaves.
(5) Areas in which the coal in the pillars and the top and bottom rocks have appreciably different physical properties to the rocks in adjacent areas.

Analysis of Coal Mine "Bumps" and Prevention Measures

Holland and Thomas[(2)] investigated and analyzed 117 "bumps" which occurred in coal mines and found that almost three-quarters of them occurred within pillar-line points. (A pillar-line point is that area lying within 200–300 ft of converging pillar lines.)

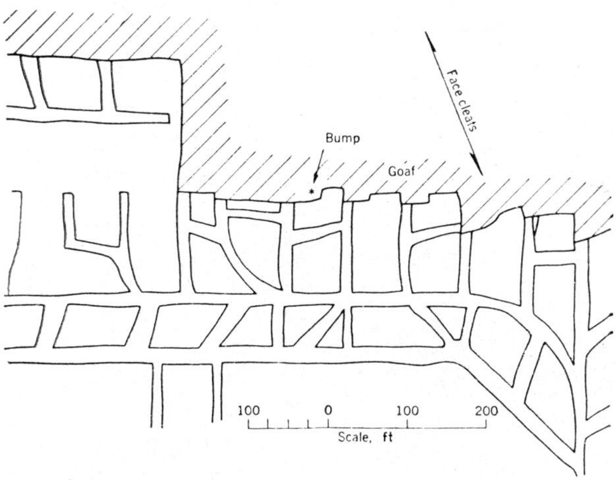

FIG. 27. Bump on a pillar line in a notch about where one would expect the maximum stress from a "roof bridge". (Cover, 1650 ft; coal thickness 56 in.; roof and floor strong.[(2)])

Figures 27–29 show actual conditions under which some bumps have occurred.

Holland and Thomas[(2)] recommend the following methods for preventing bumps, or minimizing their effects:

(1) In pillaring operations all coal should be recovered. The practice of leaving sprags, pillar remnants, or complete pillars in the goaf should be avoided. If it is impossible to recover a pillar remnant, then its load-carrying capacity should be destroyed.

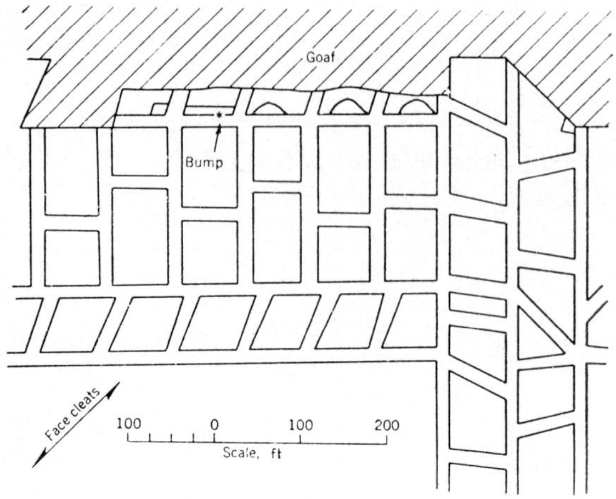

FIG. 28. Bump on a pillar line in which the back ends of the pillar were being slabbed, resulting in a wide passageway.

(The increased abutment stress caused by widening the passageway was superimposed upon the abutment stresses from the goaf.) (Cover 1350 ft; coal thickness, 53 in.; roof and floor, strong).[2]

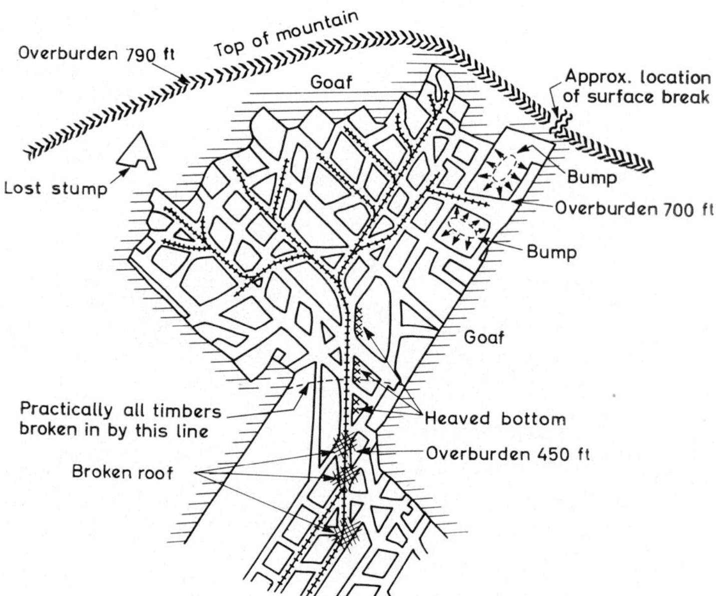

FIG. 29. Pillars that burst on a pillar–line extraction point; a slab started at the back end of the pillar was being advanced parallel to an old goaf, which had been standing for many years, toward a new extraction line.

(The pillars that bumped were substantially larger than the neighboring pillars. Cover 700 ft; coal thickness, approximately 7 ft; roof and floor, strong.)[2]

(2) Pillar-line points should be avoided in so far as is possible. While all points cannot be eliminated, the number can be greatly reduced by using long pillar lines and by careful planning. It is particularly important to plan mining operations to avoid long entries protected by barrier pillars 200–500 ft wide on each side. Such conditions can be eliminated by a retreat system of mining in which entry barrier pillars are brought back when general robbing operations are carried out.

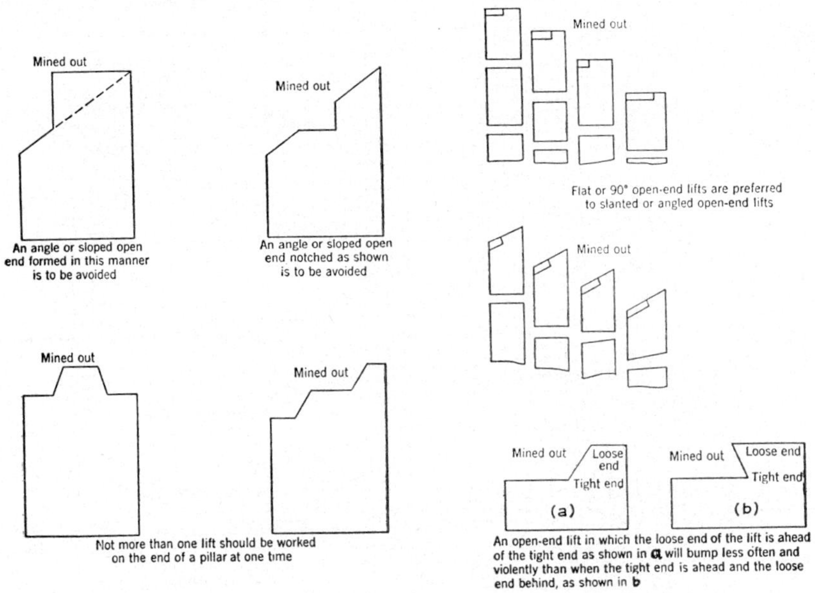

FIG. 30. Open-end pillar extraction practices which should be carefully considered in areas where rock bursts or bumps are likely to occur.[2]

Or, one side can be robbed on the advance and the other side recovered in the retreat; such a procedure eliminates one point that occurs when barrier pillars between two old goaves are recovered.

(3) Roof spans projecting over the goaves should be kept as short as possible or else provided with such support that the roof beds do not fracture. The lengths of roof spans can be reduced by orienting the direction in which the roof rocks break most readily. Long pillar lines also help to reduce long spans between unmined areas and the pillars on the extraction line.

(4) Development work should not be done in abutment areas. Under conditions that conduce to bumps, development work, blocking out pillars, etc., should be so planned that such work is not less than 3 or 4 pillars ahead of the pillar line. Development places should not be advanced toward the pillar line in an abutment zone because of the probability of encountering a high stress area.

(5) Pillars on or near the extraction line should not be split; nor should they be slabbed, other than on the goaf side, except when absolutely necessary. If they must

be slabbed, other than on the goaf side, the slab should be kept as narrow as possible.

(6) Open-end pillar recovery is believed to be the most desirable practice under conditions that conduce to bumps. Certain practices in open-end pillar recovery are believed to decrease the probability of a bump occurring, while others are

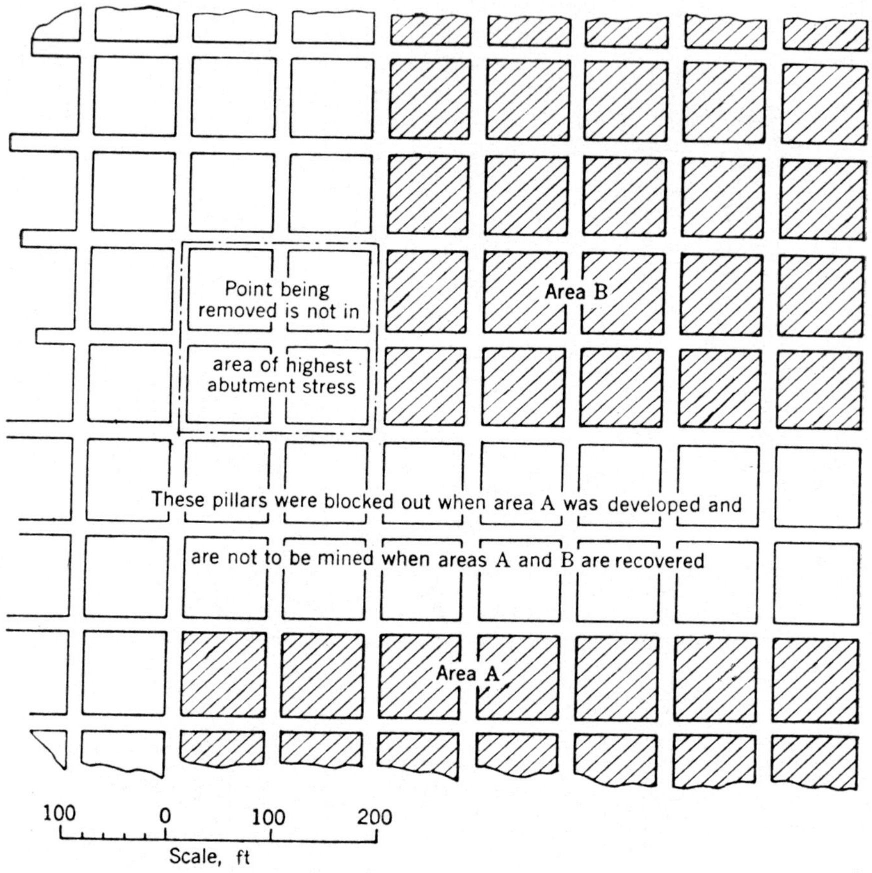

FIG. 31. Suggested method for keeping the area being mined away from the area which is subjected to the highest abutment stress.(2)

thought to increase that probability. These are illustrated in Fig. 30. The lifts taken in open-end work should not exceed 14 ft in width.

(7) When a pillar line is worked so that one end abuts on an old goaf and excessive pressure causes bumps on the point, the trouble can be controlled to some extent by leaving one or two rows of pillars adjacent to the goaf unmined. (See Fig. 31.) If the pillar lines are long, the coal lost from this practice will not be excessive. Likewise, to prevent development close to an old goaf, one or two rows

of pillars may be left unmined. (See Fig. 32.) To recover these pillars when the next lift is extracted, the roof should be strengthened, at the time of development, by roof bolting. Timbers and cribs may also be used as additional support if needed. Such practices, incidentally, are consistent with approved ventilation practices.

(8) Pillars blocked out should be approximately the same size and shape, preferably square and large enough to insure adequate support of the overburden.

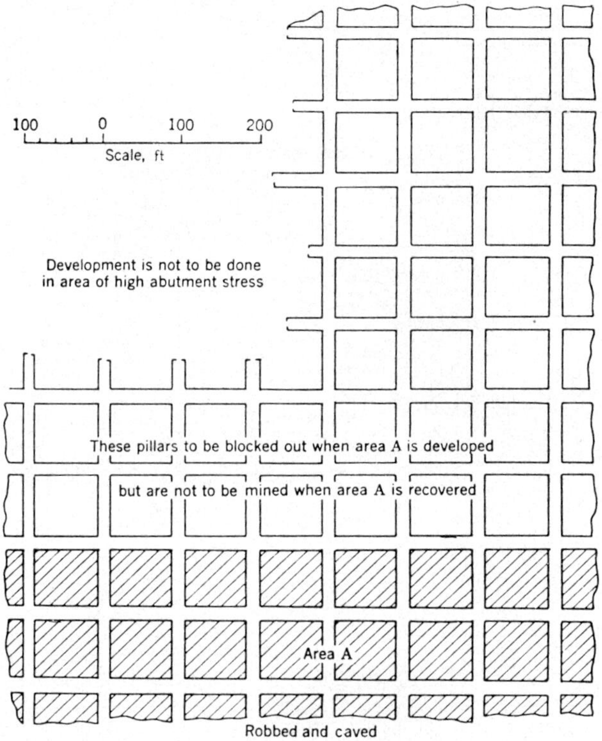

FIG. 32. Development in high-stress areas adjoining old goaves can be avoided by leaving one or two rows of pillars blocked out and unmined when the first section developed (*A*) is recovered.

(9) Passageways in development work should be kept as narrow as practicable, preferably not wider than 14 ft.

(10) Most bumps are the result of improper mining methods and practices. Doubtless, some are caused by the physical condition of the coal and adjacent rocks. Therefore, in areas where rolls, change in dip, change in coal thickness, and change in texture or hardness of the coal occur, particular attention should be given to selection of mining practices and methods that will give the miner as much protection as possible.

Long-wall Methods

Long-wall mining has been advocated as a means of preventing bumps in coal mines. While extensive mining has been done under heavy cover by this method without bumps, yet serious bursts have occurred under certain conditions with its use. (See Figs. 33 and 34.)

As may be noted, the coal had been mined on two or more sides of the area where the bumps occurred, and the pillars left were so small that the abutment

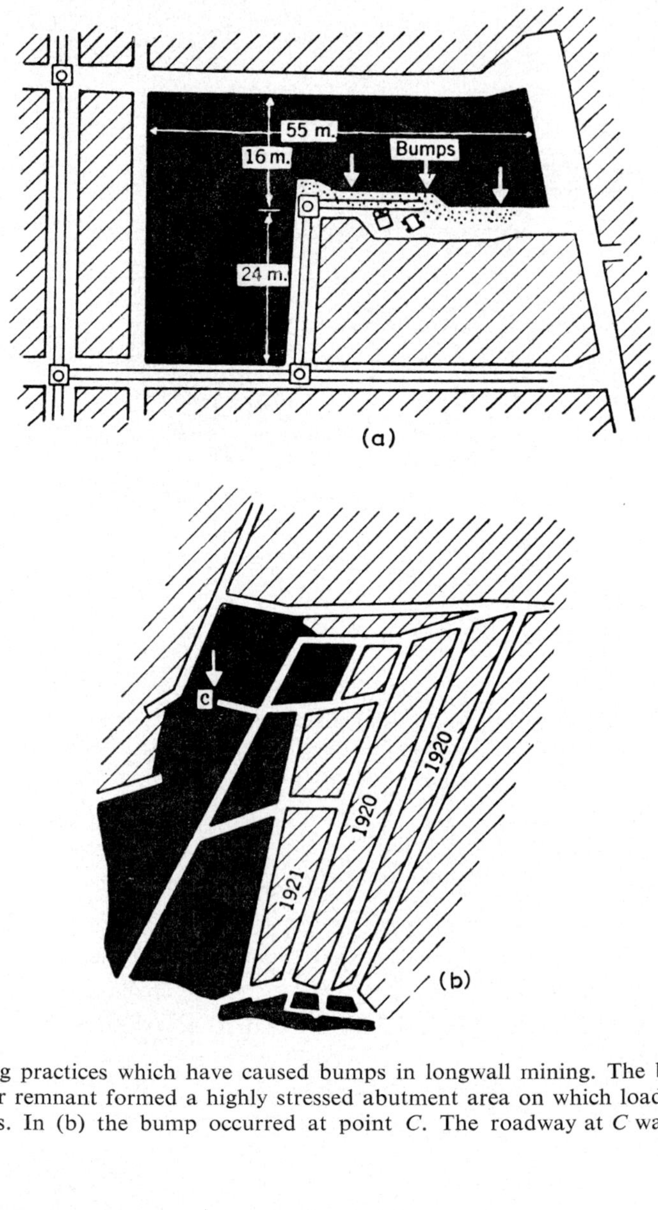

FIG. 33. Mining practices which have caused bumps in longwall mining. The bumps occurred where the pillar remnant formed a highly stressed abutment area on which loads were imposed from all sides. In (b) the bump occurred at point C. The roadway at C was 35 m long.[2]

load from more than one undermined section of overburden overlapped in the areas being mined. The bumps illustrated all occurred in the coal.

Figure 34 shows the typical conditions that conduce to bumps in retreating long-wall. Here, the pillar being mined forms a gigantic point where abutment loads of two undermined areas can overlap.

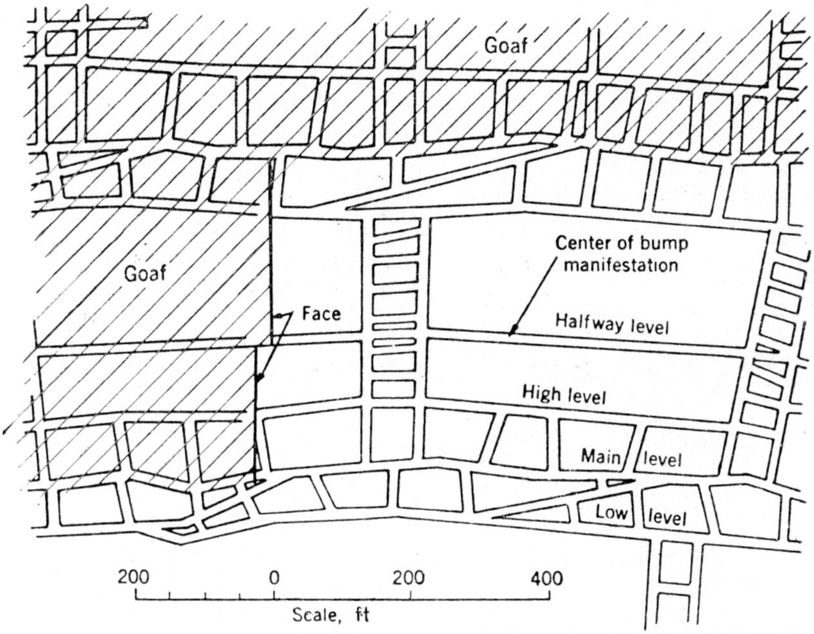

FIG. 34. Conditions under which bumps have occurred at Springhill, Nova Scotia.

The system of mining was changed from room-and-pillar to retreating longwall in the hope of preventing bumps. In certain parts of the mine bumps have continued to occur. Note that the coal being mined forms a gigantic point. Cover, 2500–2800 ft; roof and floor, strong.

ROCK BURSTS ASSOCIATED WITH METALLIFEROUS VEINS

Since coal has a high energy storage capacity at low stress values, coal-mine bumps may occur at relatively shallow depths, provided that roof rocks are strong. In the case of metalliferous veins the vein material is frequently as hard as its enclosing rocks and difficulties from bursting of the vein itself are normally encountered only at considerable depths. Bursting may occur by fracture of the roof or floor near the face, rather than by bursting of the face itself. However, isolated pillars, remnants, or pillars left for level support, may be stressed to the point where they burst.

Generally, if difficulty is encountered with bursting, the same rules which apply to coal mining, as outlined in the preceding section, will help to prevent bursts.

The gold mines of the Witwatersrand, some of which operate at depths approaching 10,000 ft (deepest development opening is 11,000 ft), experience a great deal of difficulty, and damage from the incidence of bursts at, or near, the stope faces. A long-wall mining system is used but, because of the high depth pressure stresses at these depths, bursts seem to be unavoidable.

DE-STRESSING TO PREVENT BURSTS

Recently on the Witwatersrand a de-stressing technique has been employed at stope faces. This consists of drilling holes, 10 ft deep and 3–4 ft apart, directly into the face and loading and blasting them. The purpose is to create fractures in the rock extending several feet ahead of the face. This allows rock at, and immediately ahead of, the face to yield slightly, thus relieving itself of maximum stress. The stress "peak" is transferred several feet ahead on to the unfractured vein. This technique reduces the probability of bursts of the exposed face and also tends to spread the "peak" stress over a wider base, thus reducing it slightly. It also moves the zone of "peak" stress, that is, the zone most likely to burst, far enough forward so that any bursts will be cushioned by several feet of partially destressed rock. Although this method has not completely eliminated rock bursts it has reduced the incidence of severe bursts by 77 per cent and has reduced the number of bursts occurring on day shift (when men are at the working faces) by 92 per cent.[21]

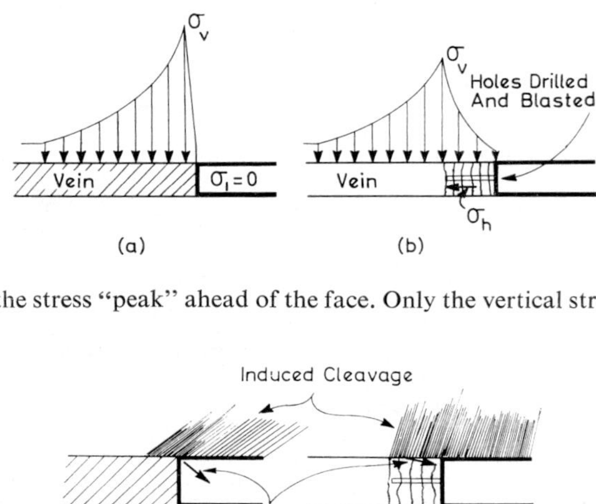

FIG. 35. Transfer of the stress "peak" ahead of the face. Only the vertical stress is represented here.

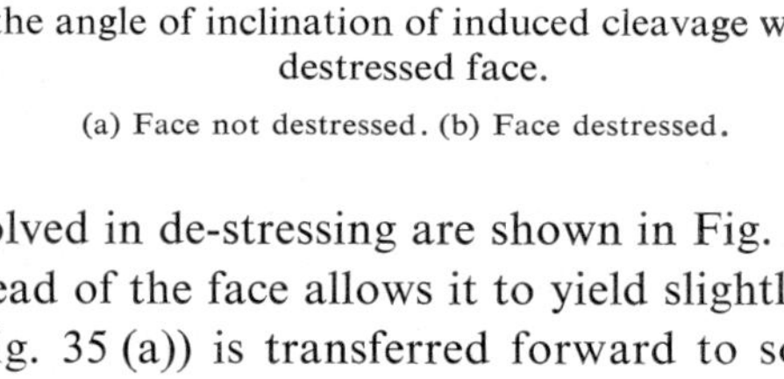

FIG. 36. Increase in the angle of inclination of induced cleavage which is observed over a destressed face.

(a) Face not destressed. (b) Face destressed.

The principles involved in de-stressing are shown in Fig. 35 and 36. The fracturing of the ground ahead of the face allows it to yield slightly so that the maximum pressure "peak" (Fig. 35 (a)) is transferred forward to some point as shown in

Fig. 35 (b). In effect the pressure peak occurs at the clastic–elastic boundary. When the pressure peak occurs at or near the face the lack of any lateral restraint allows the face to fail violently under vertical stress. When the pressure peak is transferred ahead of the face (by de-stressing) into the zone where lateral restraining stress exists such violent shear failure is inhibited. In addition, the violent shear failure of roof or floor is also inhibited, probably because the zone of maximum shear stress in roof and floor is shifted ahead to where appreciable lateral restraining stress exists.

Many bursts occur by rupture of the roof ahead of the face. The plane of shear failure which produces the burst is exposed after the face has been advanced 5–7 ft subsequent to the burst. In these cases the rock has evidently yielded enough to transfer the peak stress that distance ahead of the face.

Effect of De-stressing on Roof Conditions

The process of de-stressing has produced notable improvement in roof conditions. The principal effect is to cause the planes of induced cleavage in the roof to form at a steeper dip.

In one instance[14] in a stope at a depth of 9100 ft the induced cleavage planes which formed in the hanging-wall dipped toward the face at 30–40° and were spaced from 6 to 9 in. apart. These planes, together with natural planes of weakness, caused a thickness of from 3 to 6 ft of hanging-wall to fall continuously.

TABLE 1[14]

	Before de-stressing	After de-stressing	Percentage improvement
Number of stopes	17	17	—
Fathoms stoped	18,800	19,300	2.7
Incidence:			
number of bursts	48	32	33
incidence of bursts per 1000 fathoms stoped	2.55	1.66	35
Severity:	26	6	77
number of severe bursts			
percentage of severe bursts to total bursts	54	19	—
incidence of severe bursts per 1000 fathoms stoped	1.38	0.31	77
Time:			
number of bursts during day shift	13	1	92
percentage bursts during day shift to total bursts	27	3	—
number of severe bursts during day shift	6	0	100
percentage severe bursts during day shift to total bursts	$12\frac{1}{2}$	0	—

After de-stressing was started, the induced cleavage in the roof steepened to a dip of 70–75°. Cleavage planes were spaced at 1–4 in. apart, and were less prominent than before. The hanging-wall was easily controlled and the fathomage stoped increased from 72 to 138 fathoms per month.

The improvement in self-supporting ability of the hanging-wall apparently resulted from the steeper dip angle of induced cleavage planes. It will be recalled from Chapter 4 that a linear arch is strongest when planes of weakness approach a direction normal to the thrust line.

TABLE 2[14]

	Before de-stressing	During de-stressing
Number killed	6	–
Number injured	38	1
Total casualties	44	1

The dips of the induced cleavage planes are determined chiefly by the directions of the principal stresses. When induced cleavage forms at, or just ahead of the face, the direction of the least restraint (e.g. the smallest principal stress, σ_3) is approximately as shown in Fig. 8–36 (a). As the pressure peak is moved further ahead by de-stressing, the direction of least restraint is rotated to some position, such as shown in Fig. 8–36 (b). Thus the cleavage induced by the principal stresses is steepened.

De-stressing of Remnants

De-stressing has not been entirely effective when used on remnants. Stress conditions for remnants correspond to those for pillars, and if we refer to Fig. 37, which shows stress distribution in a heavily loaded pillar, it is apparent that when

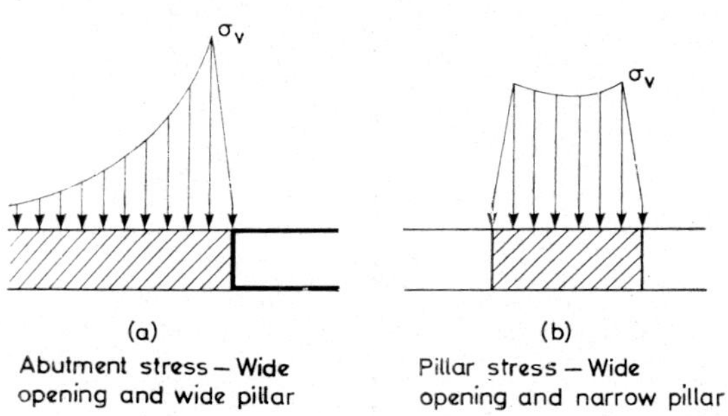

FIG. 37. Stress distribution, abutments and pillars.

the width of an opening becomes very great as compared with the width of pillar, the "abutment stress" or zone of high stress near the pillar boundary extends completely across the pillar. Thus the "peak" stress cannot be transferred further ahead by de-stressing because the whole pillar approaches peak stress.

In this connection it has been noted[(14)] in the mines of the Witwatersrand that the frequency of rock bursts is greatest when the remnant area is between 11,000 and 14,000 ft^2 and that the frequency decreases suddenly when the area becomes less than about 7000 ft^2. From this it is deduced that a pillar with less than this area is so highly stressed that fracturing extends completely through it.

De-stressing of Pillars to Prevent Bumps

Talman and Schroder[(15)] describe the methods used to de-stress pillars in a coal seam 4–8 ft thick under 700–1400 ft of cover. Immediate roof is strong sandstone about 100 ft thick. It was observed that coal blocks less than 45 ft wide had never been known to "bump" in the Pocahontas seams. Therefore, an attempt was made to form pillars less than 45 ft wide in the abutment zone on retreat. However, when pillar work backed up to a surrounded large block the block became highly stressed and subject to bursting.

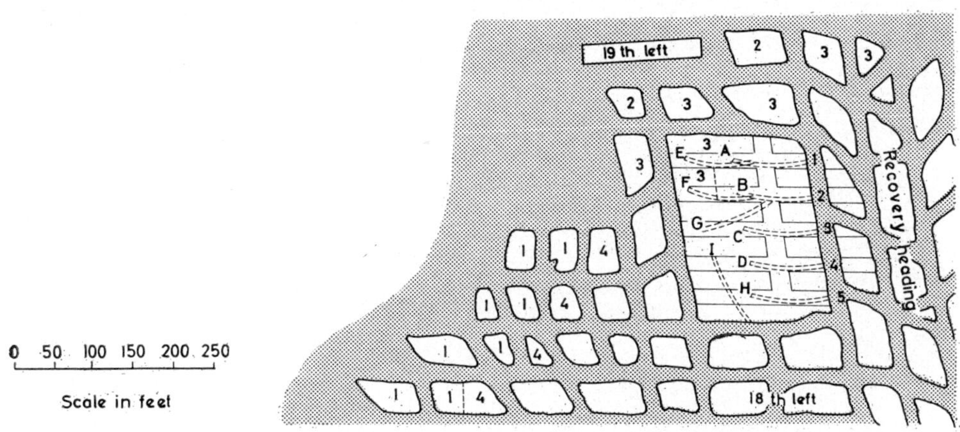

FIG. 38. Original plan for drilling and developing the block:
Drill holes *A*, *B*, and *C* while extracting No. 1 pillars and then develop rooms 1 and 2 up to and including the first break-through. Hole *D* is to be drilled along with holes *F* and *G* in room. 2. On completion of hole *G*, room 1 is to be tut through into the airway and No. 2 pillars extracted. Holes *H* and *I* are then to be drilled, room 2 cut through into che airway, and No. 3 pillars mined. On completion of hole 1, No. 4 pillars are to be extracted and rooms 3 and 4 developed through, at which point the block will be considered sufficiently developed to permit orderly extraction of pillars to a line to be established by indicated conditions as the pillar line moves outby. Holes are to be drilled in alphabetical sequence, with drilling and mining conducted on opposite shifts.

It was deduced that auger holes bored into the pillar might cause it to fracture and thus relieve the accumulated elastic energy. A method was accordingly developed which involved the boring of 24-in. diameter auger holes into the block at intervals. Principles of the method are shown in Fig. 38 and 39. Holes are drilled along the lines of proposed rooms; but due to rotation of the auger the holes "drift" to the right and do not follow a straight line.

The method has been quite successful in reducing bumps but the following precautions must be observed (from Talman and Schroder[15]):

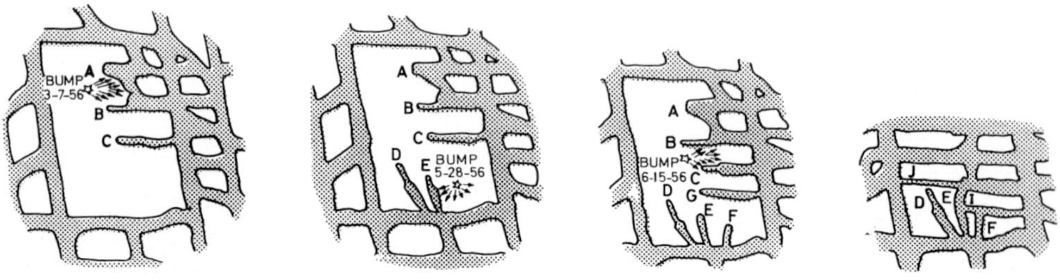

FIG. 39. *First:* Shows progress on the large block with a revision to the original plan dictated by initial results with drilling and mining. *A*, *B*, and *C* were drilled to their projected depths only after light medium bumps had continuously interrupted drilling operations by fouling and damaging the augers. Most of this trouble was attributed to light bumps and right-hand rotation of the drill, which caused sections to operate on curve alignment. *Second:* It was evident that the major stress in the block was not relieved by inby drilling and that further activity on that end of the block might cause a major bump, with injury to workmen. It was decided to try to activate a major bump by drilling from the outby end of the block. Hole *D* was drilled to the projected depth, and hole *E* set off a bump at 55 ft. *Third:* After hole *E* had induced a major bump, hole *F* was drilled only to reduce the size of the corner block. Hole *G* was drilled to ascertain remaining pressure. Since no pressure was found in hole *G*, loading operations were activated in the room following hole *C*, which had been drilled to a depth of 75 ft. Beginning at the mouth, loading operations had advanced 52 ft, or within 23 ft of the end of the hole, when a medium bump occurred. *Fourth:* Here the vulnerable block is shown completely relieved of stress. Holes *I* and *J* were drilled in proper sequence as protection holes from room development.

(1) Pre-weakening holes must be projected on room centers.

(2) After a hole is drilled, room development is to follow the auger hole regardless of the direction it takes.

(3) The drill rig shall not be positioned less than 35 ft from the mouth of the hole being drilled.

(4) Drilling and section mining operations shall not be conducted on the same shift.

(5) A section loading operation shall not be conducted less than two blocks away from a vulnerable block while augering is in progress.

(6) Mine development shall stop 35 ft outby end of auger holes in cases where all stresses have not been released. With proper protection and specified distance from face, the hole may be extended.

(7) No rooms or openings are to be driven without a protection hole on the vulnerable side.

(8) When there is any doubt as to whether a hole has been closed by subsidence and/or bumps, it must be re-drilled before it is used as a protection hole.

(9) It must be verified that auger holes have cut through bottom bone and coal to the mine floor.

(10) There must be as few men as possible at the face at all times.

A hole drilled into a pillar which is already stressed nearly to the breaking point will have two effects:

(1) Tangential stress at periphery of the auger hole will be two to three times the stress prevalent in the pillar and extension and shear fracturing extend from the hole boundary into the pillar. (See Chapter 1.) In some cases this fracturing widens the hole to 6 or 7 ft.

(2) The reduction in area of the pillar transfers additional stress to the remaining portions. This additional stress, plus the removal of lateral restraint frequently produces a series of light to medium bumps as shear fracturing extends into adjacent portions of the pillar.

EFFECTS OF FAULTS IN PRODUCING BURSTS

Faults may influence the occurrence of bursts in underground workings in one of three ways:

(1) A fault may form a plane of weakness which destroys or reduces the ability of a portion of the rock to support itself by "beam" or "arching action". As a result excessive loads may be imposed on the seam at the face, or on pillars or remnants adjacent to the fault, with "coal bursts" or vein bursts produced.

Other discontinuities, such as dikes, or clay seams, may produce similar effects.

(2) A fault is a plane along which fracturing has been produced by excessive shear stress. In most cases fracturing, together with subsequent "time yield" of the rock, has relieved practically all the shear stress. However, in the case of younger faults, and those still active, high stress may still exist in the vicinity of the fault. "Tight faults" (little or no gouge) and faults in the brittle elastic rocks are most likely to retain some stress; that is, the stress may not have been completely relieved during "elastic rebound". Figure 16, Chapter 2, illustrates a method for determining the direction of resultant principal stresses adjacent to such a fault. By resolving shear stress into the principal stresses it is possible to predict the points where maximum stress may occur at boundaries of openings in the vicinity of the fault.

(3) Movement along active faults may produce, or "trigger", bursts. In Chapter 2 it was noted that in the Sunnyside Coal Mines, Utah, disturbances have been noted simultaneously along the same fault zone at points from 7000 to 13,000 ft apart. It is possible that these bursts were produced, or encouraged, by movement along an active fault. It is even possible that extensive workings along such a fault may encourage local fault movement by removing the rock adjacent to the fault plane and thereby removing some of the restraining frictional force.

Earthquaes which occur along active faults at some distance from mine workings may "trigger" bursts in zones which are already highly stressed. Budryck[(16)] tells of bursts which occurred simultaneously with earthquakes at the Centrum Mine in Upper Silesia. On February 2, 1950, an earthquake with epicenter located at a depth of 6.2 miles occurred in this mining district and caused bursting to occur along a length of 235 m of mine galleries. On May 28, 1950, another quake with epicenter located at a depth of 5.6 miles triggered a pressure burst which affected 320 m of mine galleries. It is unlikely that such an earthquake would seriously affect any workings which were not already stressed nearly to the breaking point.

LINEAR ARCH VS. BEAM ACTION

It has been suggested that a rock burst occurs when a rock "beam" fractures under excessive stress.

A true beam depends for its strength upon the tensile strength of the rock composing it. If any transverse natural fractures exist in the rock then such a beam cannot form because the tensile strength of the rock is zero.

If we assume an intact stratified rock, then such a beam may form, but its energy storage capacity is very limited because the modulus of rupture of rock in tension is generally limited to a low value.

Most rocks which are subject to bursting have modulii of rupture of 4000–5000 psi and compressive strengths of 20,000–25,000 psi.

Assuming that such a beam does form and that the modulus of rupture of the rock is 5000 psi (and the modulus of elasticity is 4×10^6 psi), then the maximum amount of energy which can be stored in the highly stressed portions is (from Fig. 8) about 450 ft-1b./ft^3 before fracture occurs and the energy is released.

On the other hand, if we assume that the rock is stressed in compression, as in a linear arch, then the rock can be stressed to about 20,000–25,000 psi before fracture occurs. From Fig. 8 we find that it will have stored about 7000 ft-1b. of elastic energy per cubic foot before fracturing.

The low tensile strengths of rocks limit their energy storage capacities in tension. Most rock bursts which involve rupture of a roof or floor would seem to be the result of rock fracture under excessive compressive stresses existing near the abutments of natural arches.

LOCALITIES IN WHICH ROCK BURSTS AND COAL BURSTS OCCUR

The two major districts in which rock bursts are a most serious problem are the Kolar Gold Fields of India and the Witwatersrand gold mining district of South Africa.

The Witwatersrand is the world's most extensive metalliferous mining district, and also the deepest. Development openings at one mine extend to a depth of more than 11,000 ft. The depth pressure, together with mined-out areas which extend over many square miles, and very strong quartzite roof and floor rocks, have made rock bursts a serious problem in that district. The mines are worked by long-wall systems.[20–23]

The Kolar Gold mines extend to depths of almost 10,000 ft. The steeply dipping veins are enclosed in hard, strong metamorphic rocks which are subjected to intense natural lateral stresses. Rock bursts result from this combination.[17–19]

Canada

Rock bursts have occurred at depths as shallow as 125 ft in the fluorspar mine at St. Lawrence, Newfoundland, where the depth pressure could not be sufficient to fracture the granite. It is assumed that such bursts are the result either of high lateral natural (primitive) rock stresses, or of the rupture of long-span linear arches.

At the Springhill No. 2 Mine (coal) at Springhill, Nova Scotia, working has progressed to a depth of 4200 ft.[28, 29, 30] The vertical pressure at this depth, together with the thick and exceptionally strong sandstone and shale roof rocks and large mined-out spans, causes exceptionally violent bursts. A major "bump" (rock burst) in October, 1958, resulted in the deaths of seventy-five miners in the No. 2 Colliery of the Cumberland Coal and Railway Co. and resulted in permanent closure of this mine. The mines are worked by a retreating long-wall system.

At the Crowsnest Pass coalfield, Western Canada, a retreating pillar extraction system is used.[24] Enclosing rocks are sandstones, shales, and conglomerates, which are heavily faulted. At the McGillivray Mine, Coleman, Alberta, which operates at a depth of about 2000 ft, bursts appear to be due to the vertical pressure combined with long mined-out spans and massive, strong roof and floor strata.

Kirkland Lake District, Canada:[25, 26] Mining extends to depths of more than 6000 ft in this gold mining district. Country rocks are metamorphic. Slabbing of drifts, raises, and crosscuts is indicative of the presence of high lateral natural stresses. The Sudbury District has also experienced a number of bursts.[26]

British Mines

Bursts have occurred in the South Staffordshire Coal seam[43] and the western portion of the South Wales Coalfield where anthracite is mined.[44]

United States

Before they were worked out at depth, the copper mines of the Keweenaw Peninsula, Michigan, extended to depths of about 6000 ft.[27] Bursts occurred from depths of 600 ft downwards; generally, the intensity increased with depth. Thick, strong basalt flows enclose the copperbearing vein. The relatively flat dips of the copper lodes exposed them to the depth pressure and this, together with the strong roof floor rocks and extensive mined-out area, made rock bursts a serious problem.

The Sunnyside coal mines at Sunnyside, Utah, are worked at depths from 500 to 2500 ft. Again, a combination of heavy vertical pressure, thick sandstone roof strata, and a strong sandstone floor makes rock bursts a serious problem. In addition to those bursts which are associated with large mined-out areas, some bursts have occurred in isolated development headings when they approached certain faults. This indicates that tectonic stresses probably exist in zones adjacent to certain of the faults.[31, 32]

Gary district of West Virginia: Coal mining is carried on at depths of 700–1400 ft. Bursts occur where the seam is overlain by massive sandstone beds up to 100 ft thick. Pressure concentration on pillars and abutments due to depth, long mined-out spans, and strong roof seems to be responsible. Mining is by room-and-pillar methods.[11, 15, 33]

Poland

The coal mines of Upper Silesia are located in a seismically active area and many quakes affect the area. The epicenters of these quakes are located generally at depths of 20,000–30,000 ft and along certain well-defined fault lines. Few of the quakes are violent enough to cause serious damage on the surface, but some of them trigger the heavier rock bursts in the coal mines. Most of the minor rock bursts occur in the absence of earthquakes.[16]

Germany

Rock bursts occur in the coal mines of Upper Bavaria.[35] These mines operate at depths of about 2000 ft. Bursts appear due to depth pressure, large mined-out areas, and strong roof rocks.

Neumuhl Colliery: This colliery is worked at a depth of about 2500 ft. Bursts here have occurred where roof rocks were thick and strong, and they are apparently due to these rock characteristics, depth pressure, and long spans. Working is by long-wall methods. Klein-Schiersted and Heringen Potash Mines have also experienced rock bursts.

Russia

Kizeliev Basin: Coal is mined under a massive sandstone roof, up to 80 ft thick. In some mines bursts occur at depths as little as 650 ft. In others bursts occur only at depths of 1300–2000 ft. Again, depth pressure, long spans and strong roof appear to be responsible. Working is by long-wall systems.[37, 38]

Donetz Basin: Coal is mined at depths of 1000–1500 ft. Bursts are caused by the same factors as above. Working is by longwall systems.

France

Fuveau District in the Marseilles region: Bursts are apparently due to depth pressure and a strong limestone roof. The Provence Basin, the Briey Basin, and the iron mines of the Lorraine have also experienced bursts.[39, 40, 41]

India

Coal is mined at depths of about 1200 ft under a strong sandstone roof. Bursts are due to usual causes.

Eastern Alps

This lead–zinc mining district extends into the corners of three nations, Austria, Italy, and Jugoslavia.[36, 45] About 2500 rock bursts had been recorded in the district during the period from about 1905 to 1956. The country rock is limestone and dolomite. Most bursts occur between depths of 1000 and 1300 ft, although bursts occur at intervals less frequently at both shallower and greater depths. Most bursts occur in the vicinity of north and northeast trending faults and appear to be caused by stresses in the vicinities of such faults. Some evidence exists that movement is still occurring along these faults.

Civil-Works Tunnels

Delaware Acqueduct: During construction of the Delaware Acqueduct in New York State, bursting of the roof of the tunnel was observed to occur at depths as little as 500 ft. Violent and continuous bursting was noted at depths of more than 1000 ft. The roof of the tunnel tended to slab off to form a peaked arch. This is the type of fracture caused by excessive natural lateral stresses.[46]

Duchesne Tunnel, Utah: Bursting occurred in this tunnel under relatively shallow cover. This indicates that the quartzite was under considerable lateral stress.

Alps tunnels: Bursting has caused difficulties in a number of civil-works tunnels constructed in the Alps. Most famous of these is the Simplon Tunnel, constructed in the early part of this century. Others in which bursting caused difficulty were the Tauern, Karawanken and Wocheim tunnels.

Miscellaneous Mines

Other mines which have experienced a moderate amount of difficulty from rock bursts are Morro Velho (gold mine), Brazil; the tin mines and the coal mines of Cornwall and Derbyshire, England; the Hillgrove District (gold mines) of New South Wales, Australia; Cassidy Colliery, Vancouver Island, B.C., Canada; and some coal mines in the State of Washington. Other districts are Pribram (lead–silver mines), Czechoslovakia, and coal mines of the Istria Basin, Jugoslavia.[36,42]

Coeur d'Alene mining district of Idaho: The Sunshine Mine has experienced occasional rock bursts during the past 25 years. Most have occurred in hard quartzite at depths of 3000 ft or more. It is not clear as to whether vertical stresses were principally responsible or whether lateral stresses induced by stoping played a part in these bursts.

The Galena Mine, developed to a depth of 3000 ft within the past few years, has experienced several severe rock bursts since 1959. Mine workings are adjacent to a large normal fault, and it is not clear as to whether residual stresses along the fault might have caused the bursts or whether it resulted from abutment stresses at the boundaries of stoped areas.

Granite Quarries

Rock bursts have been observed to occur at the surface in the granite quarries of Vermont and Georgia. Such bursts can occur only as a result of intense natural lateral stresses.[47]

Conclusion

The foregoing has listed many of the mining districts in which occurrence of rock bursts is so serious that accounts find their way into the literature. Doubtless, there are many more where bursts are so infrequent or so unimportant that no mention is made of them in the literature. For example, the author saw the results of one rock burst which occurred on the 2000 ft level of a lead–zinc mine in the Bingham Canyon, Utah, District. This burst occurred in a crosscut penetrating a tight fold in quartzite. It loosened several cars of rock and bent back the rails laid in the bottom. This was an isolated incident, such incidents being rare in this district.

It is probable that almost any mine which reaches more than 2000 ft in depth in brittle rocks will have a certain amount of bursting, although such incidents may be so infrequent as to present no hindrance to mining operations.

EFFECTS OF ROCK CREEP

The fact that rocks, when subjected to sufficient pressure, deform slowly and continuously with time, furnishes an explanation for the fact that rock bursts and falls of ground do not necessarily occur at blasting time, when blasting vibrations would be expected to trigger such disturbances. It appears that the most highly stressed rock is undergoing a continuous deformation which redistributes the maximum stresses. Readjustment of one portion may result in over-stressing another portion, with resultant sudden failure.

The nature of "rock creep" and its effects are discussed more fully in Chapter 4.

BIBLIOGRAPHY

1. Coeuillet, R., Pressions de terrain et incidents dynamiques — degagements instantanes et coups de toit, *Rev. Ind. Minérale*, July, 1954, pp. 770–806.
2. Holland, C. T. and Thomas, E., Coal mine bumps: some aspects of occurrence, cause and control, *U.S. Bur. Mines Bull.* 535, 1954.
3. Phillips, D. W., Tectonics of mining, *Colliery Eng.* Vol. 25, June, August, September, October, November, 1948.
4. Roux, A. J. A. and Denkhaus, H. G., An investigation into the problem of rock bursts. — An operational research project, *J. Chem. Met. Mining Soc. South Africa*, November, 1954, pp. 103–124.
5. Morrison, R. G. K., Report on rockburst situation in Ontario mines, *Trans. Can. Inst. Mining Met.* Vol. 45, 1942, pp. 225–272.
6. Pressure burst survey may increase African mine life, *Mining World*, January, 1956, p. 68.
7. York, B. and Reed, J. J., An analysis of mine opening failure by means of models, *Mining Eng.* July, 1953, pp. 705–710.
8. Bridgman, P. W., The failure of cavities in crystals and rocks under pressure, *Am. J. Sci.* Vol. 45, 1918, pp. 243–268.
9. Hardy, H. R., Jr., *Physical Properties of Mine Rock Under Short Period Uniaxial Compression.* Part 1. *Results of McGillivray Mine, Coleman, Alberta.* Canada Dept. of Mines and Technical Surveys, Mines Branch, Ottawa. Report No. F.R.L. – 243 of the Fuels Division, June, 1957.
10. Cameron, E. L. and Hardy, H. R., Jr., Rock pressure studies in the mines of Springhill, Nova Scotia. A progress report. Part III. Determination of rock strength characteristics, *Trans. Can. Inst. Mining Met.*, Vol. 59, 1956, pp. 242–251.
11. Holland, C. T., Cause and occurrence of coal mine bumps, *Mining Eng.*, September, 1958, pp. 994–1004B.
12. Khodot, V. V., The mechanism of instantaneous outbursts (in Russian), *Ougol*, December, 1951, pp. 6–11.
13. Ettinger, I. L., The characteristics of tendency of a coal to instantaneous outbursts (in Russian), *Ougol*, October, 1952, pp. 31–34.
14. Roux, A. J. A., Leeman, E. R., and Denkhaus, H. G., De-stressing: A means of ameliorating rock burst conditions. Part I. The conception of de-stressing and the results obtained from its application, *J. South African Inst. Mining Met.*, October, 1957, pp. 101–127; December, 1957, pp. 256–268; August, 1958, pp. 66–69.
15. Talman, W. G., and Schroder, J. L., Control of mountain bumps in the Pocahontas No. 4 Seam, *Mining Eng.* August, 1958, pp. 888–891.
16. Budryck, W., Pressure bursts related to research on earthquakes in Upper Silesia, *Congress on Ground Pressure, Leipzig*, October 14–15, 1958, pp. 261–269.
17. Spalding, J., Deep mining problems in the Kolar Gold Field, *Eng. Mining J.* August, 1947, pp. 70–75.
18. Cowlin, W. R., and Isaacson, E. de St. Q., Planning and research necessitated by rockbursts in an underground shaft, *Mine & Quarry Eng.*, February 1959, pp. 75–81.
19. Taylor, J. T. M., Research on ground control and rockbursts in the Kolar Gold Field, India, *Trans. Inst. Mining Met.* Vol. 72, 1962–1963, Part 5, pp. 317–396 (Bulletin No. 675, February 1963).
20. Irving, C. J., Ground control at depth, Transvaal and Orange Free State Goldfields, *Trans. Can. Inst. Mining Met.*, Vol. 59, 1956, pp. 27–31; also *Can. Mining Met. Bull.*, February 1956.

21. ROUX, A. J. A., LEEMAN, E. R., and DENKHAUS, H. G., Destressing: a means of ameliorating rockburst conditions. Part I. The conception of destressing and the results obtained from its application, *J. South African Inst. Mining Met.*, October 1957, pp. 101–127.
PART II. HILL., F. G. and PLEWMAN, R. P., Implementing de-stressing with a discussion on the results so far obtained, *Ibid.* December, 1957, pp. 256–268.
22. HILL, F. G., An investigation into the problem of rock bursts. An operational research project. Part I. The approach to the problem and analysis of the rock bursts that have occurred on the E.R.P.M. during the years 1948–1953, *J. Chem. Met. Mining Soc. South Africa*, October 1954, pp. 63–83. Part II. ROUX, A. J. A., and DENKHAUS, H. G., An analysis of the problem of rock bursts in deep level mining, *Ibid.* November, 1954, pp. 103–120.
23. WEISS, O., The theory of rockbursts and the possibilities of geophysical methods in predicting rockbursts on the producing mines of the Witwatersrand, *J. Chem. Met. Mining Soc. South Africa*, January 1938.
24. MILNE, W. G., and WHITE, W. R. H., A seismic investigation of mine bumps in the Crowsnest Pass coal field, *Trans. Can. Inst. Mining Met.* Vol. 61, 1958, pp. 356–363. Also *Can. Mining Met. Bull.* November, 1958, pp. 678–685.
25. ROBSON, W. T., Rock burst incidence, research and control measures at Lake Shore Mines Ltd., *Trans. Can. Inst. Mining Met.* Vol. 49, 1946, pp. 347–374.
26. MORRISON, R. G. K., Report on the rockburst situation in Ontario mines, *Trans. Can. Inst. Mining Met.*, Vol. 45, 1942, pp. 225–272.
27. CRANE, W. R., Rock bursts in the Lake Superior copper mines, Keweenaw Point, Mich., *U.S. Bur. Mines Bull.* 309, 1929.
28. BROWN, A., *et al.*, Rock pressure studies in the mines at Springhill, Nova Scotia; a Progress Report *Trans. Can. Inst. Mining Met.*, 1956, Vol. 59, pp. 242–251.
29. BROWN, A., Ground stress investigations in Canadian coal mines, *Mining Eng.*, August, 1958, pp. 879–887.
30. CAMPBELL, W. F., Deep coal mining in Springhill No. 2 Mine, *Mining Eng.*, September, 1958, pp. 987–992.
31. PEPERAKIS, J., Mountain bumps at the Sunnyside Mines, *Mining Eng.*, September, 1958, pp. 982–986.
32. HUNTSMAN, L. P., and ELKIN, H. H., Reducing the bump hazard at Sunnyside, Paper presented to the Annual Meeting of the A.I.M.E. at San Francisco, February, 1959 (Paper No. 59 F90).
33. NORRIS, W. JR., Pillaring operations with continuous mining machines under bumping conditions, *Mining Congr. J.* March, 1960, pp. 41–43.
34. TALMAN, W. G., Auger drilling to control mountain bumps — Pocahontas No. 4 Seam, *Coal Age*, January, 1955, pp. 68–71.
35. LANGECKER, F., Contribution to the question of bumps, *International Conference About Rock Pressure and Support in the Workings, Liège*, April, 1951, Paper G6, pp. 460–462. Organized by INICHAR.
36. HAGEN, W., Gebirgeschlage im Erzbergbau, *Glückauf*, 21 June 1952, pp. 630–634.
37. AVERCHINE, S. G. and GRIGORIEV, S. E., Roof bumps in the Kezeliev Basin (in Russian), *Ougol*, June, 1952, pp. 1–10.
38. CHATILOV, V. A., Instantaneous outburst in the Andreiev seam at the "Red October" Mine (in Russian), *Ougol*, November, 1952, pp. 15–18.
39. COEUILLET, R., Pressions de terrain et incidents dynamiques-degagements instantanes et coups de toit, *Rev. Ind. Minérale* July, 1954, pp. 770–808.
40. PETERSCHMITT, E., Les engegistrements seismiques des effondrements spontanes du Bassin de Briey, *Rev. Ind. Minerale*, April, 1962, pp. 262–281.
41. TINCELIN, E. and SINOU, P., Effondrements brutaux et generalises coups de toit, *Rev. Ind. Minérale*, April, 1962, pp. 24–262.
42. RICKARD, T. A., Air blasts in mines, *Eng. Mining J.*, November 24, 1923, pp. 900–901.
43. First progress report of the Safe Working of Mines Committee of the South Staffordshire and Warwickshire Institute of Mining Engineers on the problem of bumps in the South Staffordshire Coal Seam, *Trans. Inst. Mining Engrs., London*, Vol. 105, Part, 9, pp. 458–503, 1946.

44. Pescord, R. F., Rock bursts in the western portion of the South Wales coalfield, *Trans. Inst. Mining Engrs., London*, Vol. 107, Part 10, 1948 (Paper No. 3235).
45. Tschernig, E., Rock bursts in lead and zinc mining of the Eastern Alps, *International Strata Control Congress, Leipzig*, 1958, pp. 260–269.
46. *Rock Bursts — A Symposium*, A.I.M.E. Technical Paper No. 1468, *Mining Techn.*, May, 1942.
47. White, W. S. Rock bursts in the granite quarries of Vermont, *U.S. Geol. Survey* Circular 13, 1946.
48. Woodruff, S. D., The causes of structural deterioration of rock around underground openings. Unpublished Thesis, University of Idaho, College of Mines, 1956.
49. Woodruff, S. D., Practical use of rock mechanics, *Mining Congr. J.* July, 1961, pp. 33–37, 53.

CHAPTER 9

DESTRUCTION OF ROCK BY MOISTURE, ATMOSPHERE, AND TEMPERATURE CHANGES

Absorption of Moisture

This is an important cause of sloughing and "slaking" of wall and roof rocks when these contain appreciable quantities of clay minerals which are affected by free moisture or by changes in relative humidity. Shales, siltstone, and some sandstones are the most seriously affected. Clays may also be encountered at any depth in zones of hydrothermal alteration in the igneous and metamorphic rocks such as the granites, monzonites, diorites, gneises, schists, etc. Clays are also encountered in these rocks in zones of weathering near the surface.

Humidity Effects on Shales

U.S. Bureau of Mines R.I. 3588 [(1)] describes a series of experiments to determine the effects of changes in temperature, relative humidity, and presence of free moisture on typical coal mine "draw slate". Rocks tested were typical of those found in coal mine roofs.

Although commonly referred to as "slates" the roof rocks in bituminous coal mines are not generally well enough indurated to be classified as such but correspond more nearly to shales. The "draw slate" referred to hereafter was in reality an irregularly laminated bituminous shale consisting mostly of kaolinite and fragments of carbonized plant material. It also contained about 0.19 per cent pyrite and about 2–3 per cent moisture when excavated. The "second slate" was a fine grained gray shale with streaks of carbonaceous material. It contained some micaceous fragments and about 0.11 per cent pyrite. The "third slate" was a dark gray shale, highly carbonaceous, containing a fair amount of micaceous material and about 0.80 per cent pyrite.

The accompanying figures[(1)] illustrate results of the tests conducted on shale specimens approximately $1.5 \times 1.5 \times 5$ inches which were subjected to controlled variations in temperature and humidity. Sample numbers containing the letter *G* were coated with $\frac{1}{8}$ in. coating of cement–sand mortar to simulate gunited rock. Sample numbers containing the letter *E* were coated with a coal-tar-base paint to test effectiveness of such paint in protecting rock from effects of moisture.

Some of the test specimens were cut with the long axis parallel to the bedding and others were cut with the long axes normal to the bedding. Invariably the expansion of specimens with increase of humidity was three to five times as great in a direction normal to the bedding as in a direction parallel to bedding.

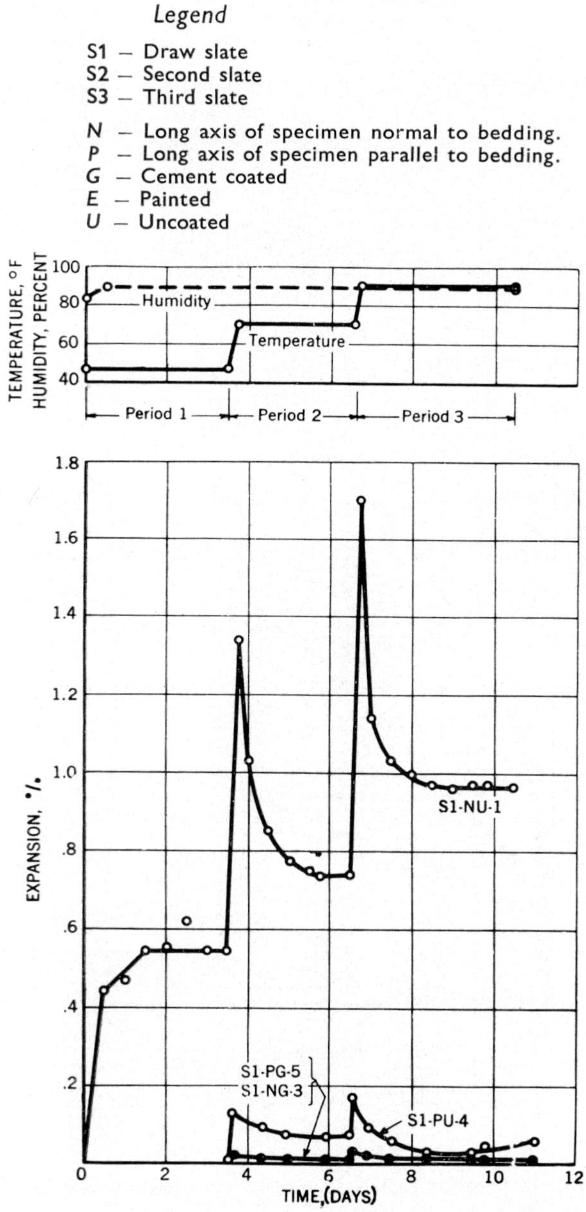

FIG. 1. Response of untreated and cement-coated draw slate to changes in temperature at constant relative humidity.[1]

Figure 1 shows results of subjecting a draw slate to a constant relative humidity of 90 per cent and varying the temperature. A sudden sharp increase in expansion following each temperature rise resulted from free moisture condensation on the surface of the specimen. This condensation took place because the rock specimen

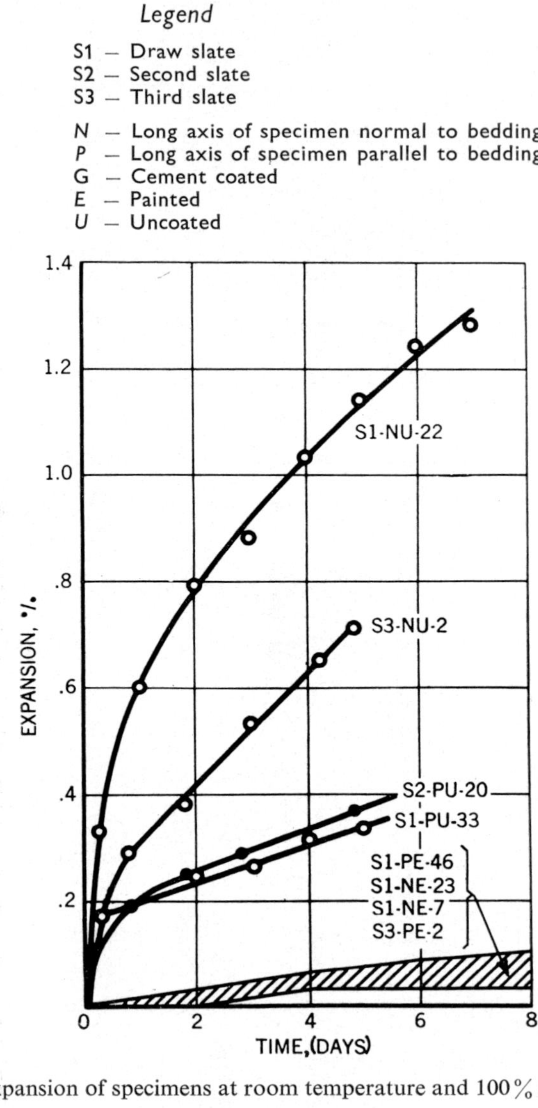

FIG. 2. Expansion of specimens at room temperature and 100% relative humidity.[1]

remained relatively cool for some time after air temperature was raised. As the temperature of the specimen attained the air temperature the free moisture evaporated and the specimen shrank until it reached stability for the particular humidity being used.

As shown, the expansion for *S*1-*NU*-1 at 43°F reached a balance at 0.55 per cent. At 70°F it reached a balance of 0.75 per cent expansion and at 90°F a balance of 0.95 per cent expansion. Expansion of *S*1-*PU*-4 which was cut with its long axis parallel to bedding, was only about one-tenth as great.

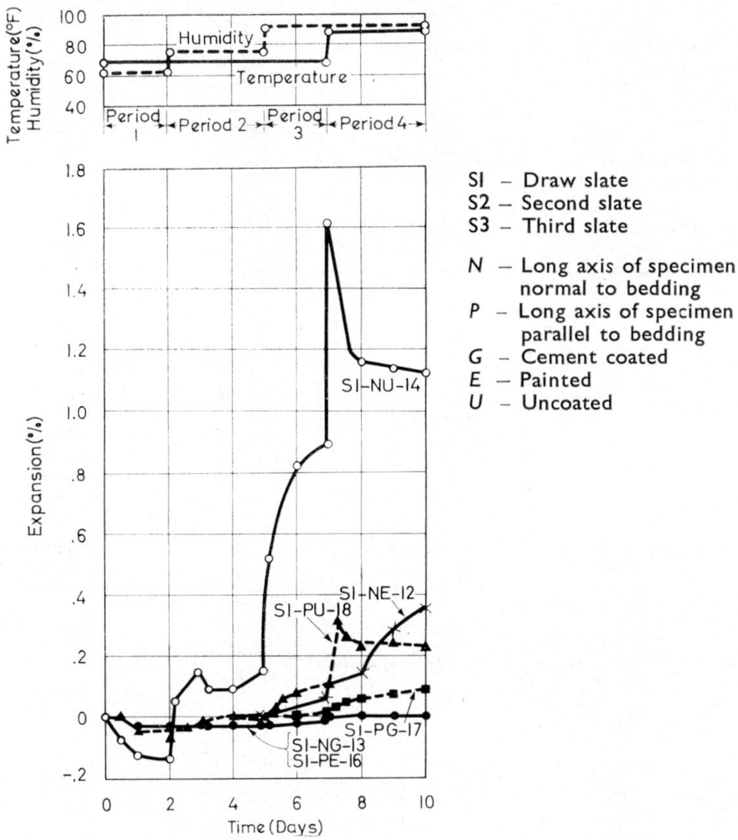

FIG. 3. Response of untreated, cement-coated, and painted draw slate to changes in humidity and temperature.[1]

Figure 3 illustrates expansion of draw slate with varying humidity and Fig. 4 shows action of the "third slate" with varying humidity. Action of the third slate follows the same pattern as the draw slate but at the end of the test the draw slate reached a balanced condition at about 1.1 per cent expansion while third slate had reached a balanced condition at only about 0.5 per cent expansion.

In another series of tests slates and specimens of coal were exposed to 60°F temperature and 97 per cent humidity for seven days. Untreated specimens reached equilibrium with the atmosphere in 3–4 days. Maximum gain in weight (and moisture content) of slates ranged from 1.24 to 1.44 per cent. For coal, moisture absorption was only 0.15–0.35 per cent. Painting of specimens retarded absorption of moisture so that, in general, painted specimens absorption in 7 days averaged only 20 per cent of the moisture absorption of untreated specimens.

Figure 2 shows expansion of specimens kept at 100 per cent relative humidity and at room temperature.

The changes of humidity in tests described heretofore were sufficient to start disintegration of all untreated slate specimens.

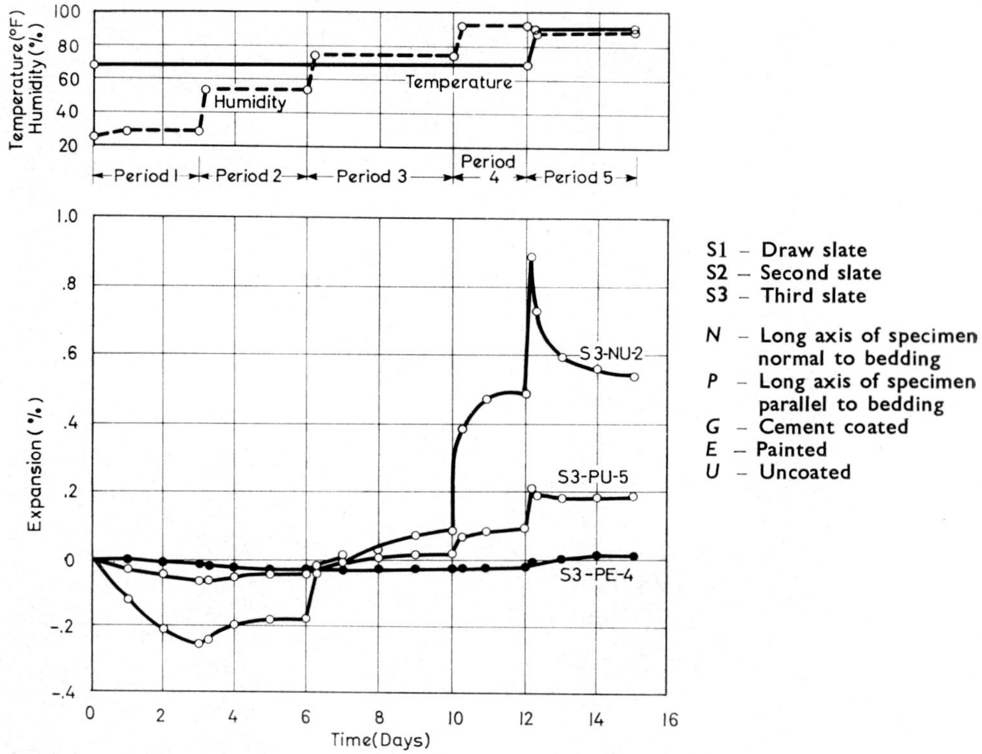

FIG. 4. Response of untreated and painted third slate to changes in humidity and temperature.[1]

Saturation of Shale Specimens

Upon being saturated by immersion in water, 90 per cent of the specimens reached equilibrium within 2 days with some samples showing expansion of as much as 6.5 per cent and all untreated samples disintegrated during immersion.

Effects of Drying

It was noted that specimens taken from the mine some time prior to testing had dried out appreciably by the time they were tested. Additional samples of fresh slate were then taken. Some were dried in air and then immersed whereupon they disintegrated within 15 min. For comparison, a fresh sample was immersed immediately upon removal from the roof and kept in water for $2\frac{1}{2}$ months without visible deterioration. It was then removed from the water, dried, and again immersed, whereupon it fell apart in 15 min.

During tunnel construction in sedimentary rocks the author has observed the peculiar behavior of argillaceous rocks under the influence of moisture. One typical

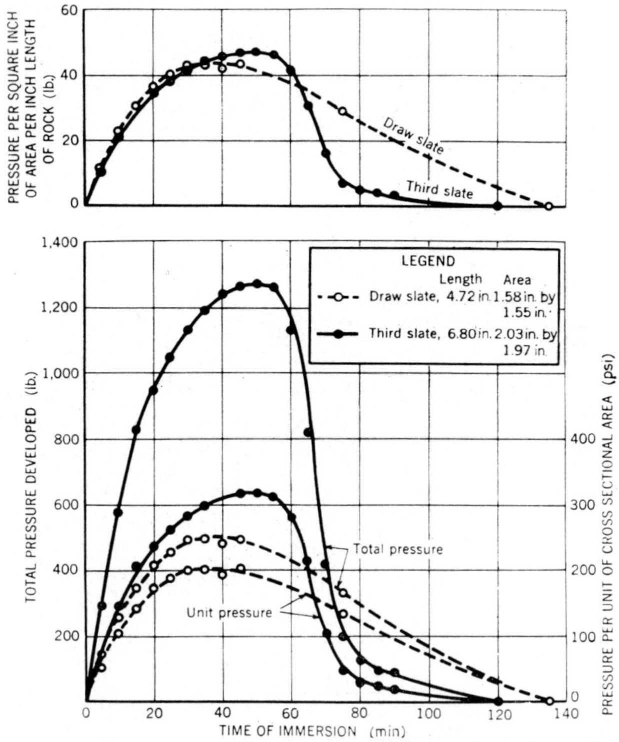

FIG. 5. Pressure developed by restraint of swelling of slates immersed in water.[(1)]

formation was a thin-bedded to laminated carbonaceous shale containing occasional seams of lignite, which matches closely the description of the coal-mine roof rocks tested by the U.S. Bureau of Mines. Temperature of the air in the tunnel varied from 68 to 72°F for the duration of the job, and although no tests were made for relative humidity, water flowing down the tunnel floor gave the ventilation air an opportunity to become saturated. Rock in the walls and roof showed little deterioration and fragments lay partially submerged along the tunnel floor for months without losing strength. Fresh rock fragments taken outside and allowed to dry remained firm although a few cracks formed, roughly parallel to the bedding. However, when these dried specimens were submerged in water they deteriorated into a mass of mud within a few hours.

Physical Characteristics of Slates Tested[1]

Compressive strengths of 2-in. cubes varied from 1900 to 3200 psi parallel to bedding and 2500–6300 psi normal to bedding for the various slates.

Two draw slate specimens were kept at 70–90°F and 90 per cent humidity for 7 days and then crushed. Crushing strengths were about 60 per cent of the strengths of fresh specimens.

Drying slate at 105°C increased its compressive strength parallel to the bedding about 30 per cent and normal to the bedding about 100 per cent.

Modulus of elasticity for the slates averaged about 2 million psi.

Stresses Induced by Humidity Changes

The tests described heretofore have indicated that partially dried roof shales, when subjected to 100 per cent humidity, will expand up to 1.3 per cent in a direction normal to the bedding. In a coal mine the strata ordinarily lie approximately flat and the roof of an "entry" consists of strata as shown in Fig. 6. Expansion in a direction normal to the bedding will not create excessive stresses in roof strata since they are free to expand vertically. However, the moisture will tend to weaken bonding materials along bedding planes and this may cause sloughing of strata from the roof.

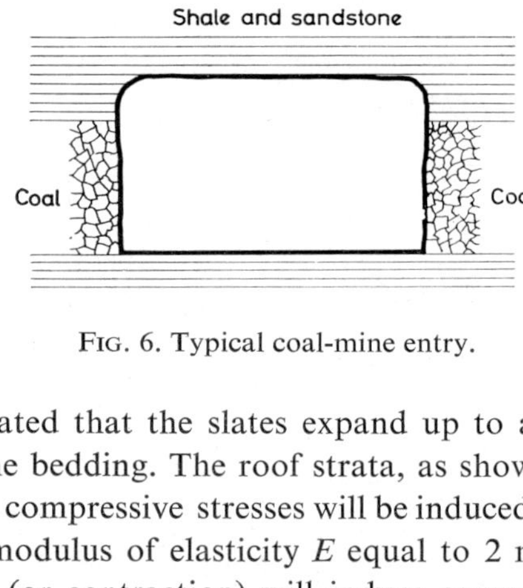

FIG. 6. Typical coal-mine entry.

The tests also indicated that the slates expand up to about 0.45 per cent in a direction parallel to the bedding. The roof strata, as shown in Fig. 6, are not free to expand laterally and compressive stresses will be induced parallel to the bedding. Since the rock has a modulus of elasticity E equal to 2 million psi then each 0.1 per cent of expansion (or contraction) will induce compressive (tensile) stress of 2000 psi. Thus, expansion (contraction) of only 0.2–0.3 per cent will create stresses exceeding strengths of most roof rocks.

Rapid changes in the air temperature causes free moisture to condense on rock surfaces and produce temporary expansion of much greater magnitude than do simple changes in humidity.

Summary

The experiments described herein have indicated that humidity changes cause spalling and slabbing of agrillaceous rocks by causing alternate expansion and contraction thus creating excessive stresses in rock which is under restraint.

The ideal solution would be to maintain such conditions of temperature and humidity that exposed rock would neither gain nor lose moisture throughout the year, and its moisture content would be maintained at that which it contained when it was first exposed. Another solution would be to dry the rock and keep it permanently dry. However, removing sufficient moisture from mine air would be a difficult and expensive process. The experiments also indicated that a policy of wetting the rock as soon as it is exposed and keeping it permanently wet would be preferable to allowing the alternate wetting and drying which ordinarily occurs.

The tests described above pertained to shaly rocks. It would seem that any rock containing appreciable quantities of clay minerals would follow similar behavior patterns under similar conditions. Sedimentary rocks which contain large percentages of clay, in addition to the shales, are the siltstones and some sandstones. Zones of hydrothermal alteration and zones of weathering in the metamorphic and igneous rocks also contain appreciable quantities of clay minerals. It is not likely that the clay minerals in these rocks are as well oriented as those in the shales, and expansion resulting from moisture absorption should be equal in all directions rather than varying with the bedding as it does in thin-bedded rocks.

The samples tested in the above experiments consisted principally of kaolinite. Another clay mineral which causes great difficulty when encountered in underground operations is bentonite whose principal constituent is the clay mineral montmorillonite. This mineral has a high absorptive capacity and expands to several times its original volume when given access to moisture. Fortunately, it is not of common occurrence. When bentonite is encountered the only control for swelling is to seal it from contact with moisture.

CHEMICAL EFFECTS OF THE ATMOSPHERE

Water, oxygen, and carbon dioxide are the principal substances which react with the rock forming minerals to cause chemical deterioration. Water may act as a reagent by itself or it may act as a carrier for dissolved oxygen or carbon dioxide.

Carbon Dioxide

Carbon dioxide dissolved in water forms carbonic acid which in turn reacts to dissolve calcareous rocks. This combination also reacts to dissolve and remove calcium minerals from the igneous and metamorphic rocks. The large natural caves, such as Mammoth Cave, were formed by the solution of calcium carbonate from limestone by the foregoing process. Inasmuch as the process is a slow one it is not likely to produce immediate effects in weakening rock around underground openings.

Oxygen

The sulfide minerals are probably the most susceptible to attack by oxygen, and even though they may be present in amounts smaller than 1 per cent, reaction with oxygen still exerts deleterious effects on rock strengths. Pyrite, FeS_2, is the most common and widely distributed of the sulfide minerals, occurring in significant amounts in shales, some sandstone, and in zones of hydrothermal alteration in igneous and metamorphic rocks. Bituminous shales, particularly those associated with coal beds, contain up to 1.5 per cent pyrite, and veins and wall rocks of most metalliferous mineral deposits are heavily mineralized with pyrite.

Pyrite oxidizes to form sulfates and oxides of iron. The sulfates are soluble and may leach out or react with other minerals in the rock. A typical reaction of this type caused considerable difficulty during construction of the Djenjen Cued Dam in Morocco.[(2)] The difficulty was experienced because of the decomposition of shales exposed by excavation. An upper strata consisted of marly calcareous shales and lower strata of siliceous shales. The principal constituent of the shales was determined to be the micaceous mineral illite, which resembles muscovite. Tests indicated that it did not swell in water. Analyses showed that the shales contained pyrite so finely disseminated that it was unidentifiable except by chemical tests. Calcareous shales contained up to 4.5 per cent pyrite and the siliceous shales up to 6.5 per cent pyrite. Disintegration of the shales was caused when pyrite reacted with air and moisture to form ferrous sulfate which then reacted on the calcium carbonate to form anhydrite ($CaSO_4$) which, in turn, hydrated to gypsum ($CaSO_4 \cdot 2H_2O$) with an increase of volume.

Iron minerals in the rock are also susceptible to oxidation. The iron in pyrite, if it is not converted to sulfate, will form the crumbly red–brown oxides as will also the iron in siderite ($FeCO_3$). Ferromagnesiam minerals are also susceptible to oxidation, with the dark colored basic rocks, such as basalt, being particularly vulnerable to weathering which converts the iron minerals to oxides. However, the process is much slower than the oxidation of sulfide minerals and probably is important only over a period of many years. Siderite breaks down much faster than the ferromagnesian minerals, but at a much slower rate than the sulfides.

Water

This acts as a carrier for carbon dioxide and also enters into chemical reactions itself. The most famous case of this kind occurred during the driving of the Simplon Tunnel through the Alps when beds of anhydrite were encountered. Moisture in the tunnel reacted with the exposed anhydrite beds to form gypsum. The increase in volume of gypsum over anhydrite caused the bottom and sides of the tunnel to swell and heave and no supports were strong enough to resist the pressure caused by this chemical reaction. The only means of controlling this disturbance was to seal off the anhydrite beds from contact with the moisture. Occurrences of this kind are rare but something which may be occasionally encountered in sedimentary rocks.

Difficulties are often encountered in supporting openings which are driven through basic dikes. It appears that the cause of the difficulty is the chemical alteration, and resulting volume increase, of the plagioclase feldspar albite when it is exposed to moisture and carbon dioxide of the atmosphere.

EFFECTS OF TEMPERATURE CHANGES

Figure 7 shows results of thermal expansion tests on shales.

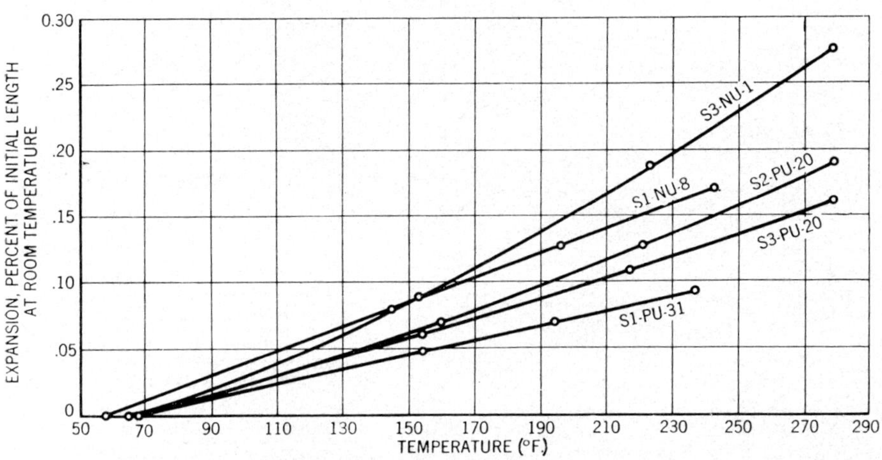

FIG. 7. Thermal expansion of shale specimens.[1]

Temperature Fluctuations

In mining operations the daily temperatures of mine air at intake may vary as much as 40–50°F and the seasonal variations may be even larger. However, such wide temperature fluctuations exist for only a short distance after entering the intake because air temperature rapidly adjusts to that of the rock. Thus, temperatures in deep mining operations do not vary more than a few degrees the year around. In shallower workings temperatures may vary as much as 10–20 °F from season to season, and by lesser amounts from day to day.

Stresses Induced

In the tests mentioned above the expansion of shale (per degree Fahrenheit) parallel to bedding was about 5×10^{-6} in./in. and 1×10^{-5} normal to the bedding. With rock spans restrained at the ends as shown in Fig. 6 the stress created by temperature changes can be computed from the formula:

$$\text{Unit stress} = \text{Modulus of Elasticity} \times \text{unit strain}$$

Modulus of elasticity for the rock tested was about 2 million psi.

Parallel to the bedding a temperature change of 1° causes expansion of 5/1,000,000 in./in. which, with rock restrained, creates stress of $5/1{,}000{,}000 \times 2{,}000{,}000 = 10$ psi. A 20° temperature change will create a stress of 200 psi in the rock. Thus, ordinary temperature changes will not cause excessive stresses parallel to the bedding.

Normal to the bedding a temperature change of one degree will create stresses of $1/1{,}000{,}000 \times 2{,}000{,}000 = 20$ psi in the rock, and temperature change of 20°F will create stress of 400 psi. Compressive stress of this magnitude will not cause failure since compressive strength of rock tested was about 2000 psi, or more. However, some cracking may occur as result of tensile stresses.

Values of thermal coefficient of expansion for some other rocks are listed below:[3]

	(in./in. per °F)
Limestone	0.6 to 4.7×10^{-6}
Granite	3.1 to 4.2×10^{-6}
Sandstone	3.2 to 6.9×10^{-6}
Marble	2.0 to 5.6×10^{-6}
Slate	5.1×10^{-6}

The maximum coefficient of each of the rocks listed is about the same as that for shale parallel to bedding, that is, 5×10^{-6} in./in. per °F. However, the stress created will be proportional to E (about 10 million psi for granite) and, for granite, the stress created by a 1° temperature change will be 50 psi. A 20°F change will create stress of about 1000 psi, which is still far below the compressive strength of granite although such stress might cause tension cracks.

A very rapid temperature change would cause surface layers to expand (or contract) quickly, before rock underneatch had changed temperature and resultant shear stresses between rock layers could cause surface to crumble and peel off. However, temperature changes in nature do not ordinarily take place rapidly enough to create such conditions.

Summary

In view of the small coefficient of expansion of rock and the limited range of temperature which exists in most underground operations it is not likely that expansion and contraction due to temperature changes are important factors in producing sloughing of rock from walls and roofs. However, some sloughing due to this cause may occur in the first few hundred feet of inby air intakes, since air temperatures in this reach have not had opportunity to adjust to rock temperatures and will fluctuate almost as widely as outside temperatures.

Temperature variations do produce indirect effects which are much more destructive to rock than thermal expansion. These effects are variations in humidity, and deposition of free moisture on rock surfaces, both of which have been discussed in this chapter.

CONCLUSIONS

This chapter has covered some of the minor causes of structural deterioration of rock around underground excavations. These are summarized below.

Absorption of Moisture

This is next in importance to pre-existing stresses in producing failure by the sloughing and "slaking" which result when argillaceous rocks are subjected to cycles of varying humidity, such as daily or seasonal variations occurring in mine ventilation circuits. Moisture softens the rock and causes expansion and contraction. These produce stresses which cause sloughing of the surface layers of rock.

Chemical Action of the Atmosphere

The most common destructive effect under this classification is the reaction of oxygen with pyrite to produce sulfates and iron oxides. Another effect is the alteration of albite by moisture and carbon dioxide. The accompanying weakening and expansion destroy the rock. Effects of carbonic acid on calcareous rocks and reaction of water with anhydrite are of less importance, the former because it is a slow process and the latter because it is rarely encountered.

Temperature Changes

Values of the thermal coefficient of expansion for common rocks indicate that temperature changes are not important causes of rock failure except near the portals of intake airways where temperature may fluctuate widely.

BIBLIOGRAPHY

1. HARTMANN, I. and GREENWALD, H. P., Effects of changes in moisture and temperature on mine roof. 1. First report on strata overlying the Pittsburgh Coal Bed, *U.S. Bur. Mines R.I.* 3588, October, 1941.
2. CAILLERE, S. and KRAUT, F., Le rôle de la minéralogie dans la construction des barrages, *Le Génie Civil*, Vol. 128, No. 18, September 15, 1951, pp. 345–347.
O'ROURKE, C. E., *General Engineering Handbook*, McGraw-Hill, New York, 1940, p. 181.

CHAPTER 10

STRESS AND STRAIN MEASUREMENTS IN SOILS, ROCKS, AND SUPPORTING STRUCTURES

FOR the development of information for the design of underground openings and their supporting structures, four principal classes of measurements are of interest. These are:

(1) Measurement of strains in the ground surrounding an opening.

(2) Measurements of the convergence movements of rock surfaces.

(3) Measurements of pressures on mine filling material.

(4) Measurement of loads on structures for supporting ground and stresses in the supporting structures.

No great difficulty is experienced in conducting measurements in ordinary soils. Pressure measuring instruments are relatively easy to place and soil displacements are large enough to be readily measured. In hard rock the situation is different. Even the simplest methods available for measurement of internal rock strains have been relatively costly and as a result such measurements and the development of the data therefrom have been left to large organizations which are engaged in mining operations or in the design and construction of underground facilities.

The following stresses are important in influencing the behavior of soil or rock around subsurface openings:

(1) The magnitude and directions of natural (pre-existing, primitive, or inherent) stresses in such soil or rock.

(2) The magnitudes and directions of induced (concentrated or re-aligned) stresses. These are induced by creation of an opening.

PHYSICAL CHARACTERISTICS WHICH MAY BE MEASURED

Listed below are some of the physical features of a rock which are modified when it is subjected to the stresses induced by creation of an opening:

(1) Closure of roof and floor or closure of sides.

(2) Tangential deformation of exposed surfaces.

(3) Changes in velocity of sound waves passed through the ground.

(4) Changes in the modulus of elasticity of the ground.

(5) Nature of sub-audible vibrations originating in rock.
(6) Deformation of boreholes.
(7) Deformation and restoration of "slots" in rock surface.

Measurements of strains and stresses include the following:

(1) Measuring strains in rock at exposed rock surfaces.
(2) Measuring strains in rock remote from a free surface.
(3) Measuring convergence of roof and floor (or hanging-wall and foot-wall).
(4) Measuring absolute movements of roof and floor (or hanging-wall and foot-wall).
(5) Measuring pressures on mine filling material.
(6) Measuring ground pressures on supporting structures.
(7) Measuring stresses in supporting structures.

MEASURING STRAIN (DEFORMATION) IN ROCK

The closure of roof and floor, or of walls and ribs, is the most conspicuous phenomenon associated with subsurface openings and the easiest to measure. However, such measurements do not yield information as to the stresses existing in the rock. Methods for determining the actual magnitudes of stresses within the rock involve measurement of deformation of rock blocks which are freed from the main mass and allowed to expand.

The determination of rock stresses depends upon the measurement of extremely small deformations. The modulii of elasticity of hard rocks may range from 2 to 7 million psi. Thus, large changes in stress values are produced by very small changes in dimensions (strain). In order to measure these minute changes in dimensions it is necessary to employ instruments capable of measuring to within a few ten-thousandths of an inch.

The accurate interpretation of strain is also complicated by the fact that any opening made in rock for the insertion of a measuring device causes a change in the initial state of stress. In an elastic material a stress concentration is created near the boundary of the opening.

In order to convert strain measurements to stress values it is necessary to determine the value of the modulus of elasticity of the rock. This can be determined accurately in laboratory specimens but the modulus of elasticity of rock *in situ* generally is somewhat less than the laboratory value; thus an uncertainty is introduced into the computed stress values.

It must be emphasized that rock *stresses* are not measured directly, but that measuring techniques are designed to measure *strains* and the *stresses* are then computed by using suitable values for the rock modulus of elasticity.

There are two general methods for determining absolute rock strain. These are: (1) The strain relief method and (2) the strain restoration method.

In the strain relief method strain gages are fixed to tunnel walls at selected locations. A groove is then cut around the location of the strain gages, freeing the rock surface to expand. The amount of the expansion is a function of the initial stress within the rock and of the modulus of elasticity of the rock.

In the strain restoration method strain gages are fixed to the rock surface and readings are taken. A deep slot is then cut into the rock above the gages and the rock is allowed to expand. A "flat jack" or "Fryssinet jack" (this is a flat, expandable, sheet metal pressure cell) is cemented into the slot and expanded by application of hydraulic pressure until the strain gages indicate that the rock has been restored to the state of strain existing prior to cutting of the slot. The pressure in the jack is then assumed to be equal to the original pressure in the rock tangential to the surface of the opening.

STRAIN RELIEF METHODS

The strain relief method was first employed in 1932 by Lieurance at Hoover (Boulder) Dam.[1-3]

Strain gage rosettes were set in the rock and the rock was then relieved by drilling, with a jackhammer, a series of holes 30 in. deep on a 4-ft diameter circle around the gage points, and then by chipping out between the holes. The method was expensive and cumbersome. However, the measurements showed that the initial horizontal stress in the rock surrounding the tunnel at a depth of about 150 ft below the canyon floor was equal to approximately three times the present overburden pressure.

Another instance of the early application of the strain-relief method was in the Gorge Penstock Extension Tunnel (power system of the City of Seattle). Results of these measurements were later analyzed by Terzaghi and Richart.[3]

The strain relief method was also applied in the Prospect Mountain Tunnel (Colorado-Big Thompson Project of the U.S. Bureau of Reclamation). On this project it was desired to determine the natural state of stress of the rock in order to know whether or not to use reinforcing steel in the concrete tunnel lining. Strain gage rosettes were located approximately at the crown, at the ends of the horizontal axis, and at 45° below the horizontal at each side, forming in effect a ring of five measuring points at each location along the $12\frac{1}{2}$-ft diameter tunnel.

After the strain gages were cemented in place a standard diamond drill was used to cut an annular slot in the rock around the gages, thus allowing that portion of rock to expand. The inside diameter of the drill was 6 in. and the width of the kerf was about $\frac{3}{8}$ in.

The surface on which the gages were mounted required careful selection and preparation. It was ground smooth and was free from pits, cracks, or large grains. The rock surface was thoroughly dried before gages were cemented to the rock and was dried with a heat lamp after gages were cemented in place. Strain gages were sealed with waterproof mastic to protect them against moisture.

Results of the tests cited above are described in Chapter 2, which deals with natural stresses in undisturbed soils and rocks.

Measuring Strains in Rock Remote from a Free Surface

Methods for measuring strains in rock at a distance from an accessible opening involve the drilling or boring of holes in which measuring instruments can be placed. After a measuring instrument is affixed in place it is "overcored" and the rock annulus containing the gage is allowed to expand until it is essentially stress free.

Stress Measurement by Means of Magnetostrictive Gages

A method employing magnetostrictive gages, as devised by Hast in Sweden[11] has proved to be a satisfactory method for determining absolute strains in rock at a distance from an accessible opening. The gage which Hast has designed works on the principle of magnetostriction; that is, certain alloys of cobalt, nickel and iron, when placed in a magnetic field, shorten or lengthen. Conversely, when such a substance is placed in a magnetic field and stress is applied its magnetic permeability changes.

Hast's Procedure

The procedure is described in great detail by Hast [11, 33]. A hole with a diameter of 26 mm is drilled into the rock, commonly at right angles to the exposed surface, in increments of 50–100 cm. After each increment is drilled, a gaging device is installed in it. Then the stresses in the rock surrounding the increment are reduced to zero by increasing the depth of an annular groove, with an inner diameter of 87 mm, concentric with the hole, until the gage readings indicate that the stresses have disappeared. Either a strain gage or a calibrated and pre-stressed pressure gage could be used. Hast selected the pressure gage because he believes that it permits more reliable location of defective spots in the rock.

The prestressed pressure gage, like a strain gage, registers a change in stress or strain only in the direction of its centerline. Changes in stresses in other directions remain undetected. Therefore the measurements are made in succession in three different directions, X_1, X_2 and X_3. These directions are commonly so chosen that they intersect at 60°.

Hast's gage is composed of three parts as shown in Fig. 1: a pressure cell (*a*), a set of wedges (*b*), and a well-bearing (*c*). The cell (Fig. 2) has a length of 10 mm and a diameter of 6 mm; it consists of a nickel alloy spool serving as a core for a copper coil (*e*) and a protective outer cover (*f*). It operates on the principle of magnetostriction. After the gage is installed in the drillhole, oriented in direction X_1, the cell is pre-stressed by rotating a rod which connects a crank located outside of the drillhole to the set of wedges (*b*) in the gage, shown in Fig. 1. The degree of

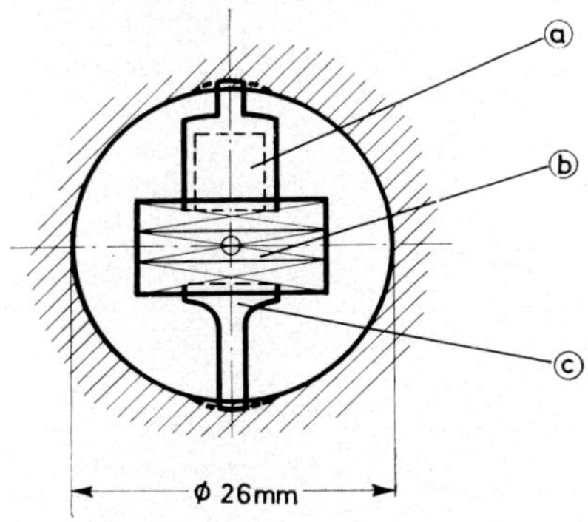

FIG. 1. Section through Hast's pressure gage.
(a) pressure cell, (b) set of wedges, (c) bearing.[11, 33]

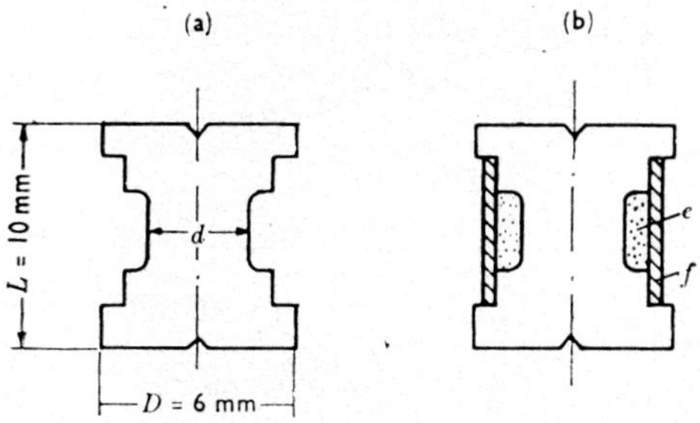

FIG. 2. (a) Nickel alloy spool forming the core of Hast's pressure cell. (b) Section through pressure cell; (e) is copper coil, (f) is permalloy mantle.[11,33]

prestressing is registered by the pressure cell. After the initial reading on the cell has been made, the depth of the annular groove is increased. During this process the pressure on the cell decreases until the bottom of the groove has advanced to a distance of about 5 cm beyond the cell. Then it assumes a constant value which indicates that the stresses in the rock surrounded by the groove have disappeared.

The next step is to rotate the cell about 60°, into position X_2, advance it to a point located beyond the zone of stress relief, prestress it and increase the depth of the groove until the readings on the cell have ceased to decrease.

The distance between two consecutive setups of the cell is at least 10 cm. As a consequence the readings in the three directions X_1 to X_3 are made at three different distances from the outer end of the hole. The readings are correlated by means of interpolation diagrams like that shown in Fig. 3. In this diagram each of the curves labelled X_1, X_2, X_3 connects the points obtained by the measurements in one of the three directions X_1 to X_3, plotted against the distance from the cavity wall. If a cell is located close to a joint, abnormal readings are obtained. The hole to which Fig. 3 refers crossed joints at distances of about 1.5 and 3.0 m from its outer end. The points representing the readings in the proximity of the joints are connected with the smooth curves by dash lines, to indicate their abnormal position.

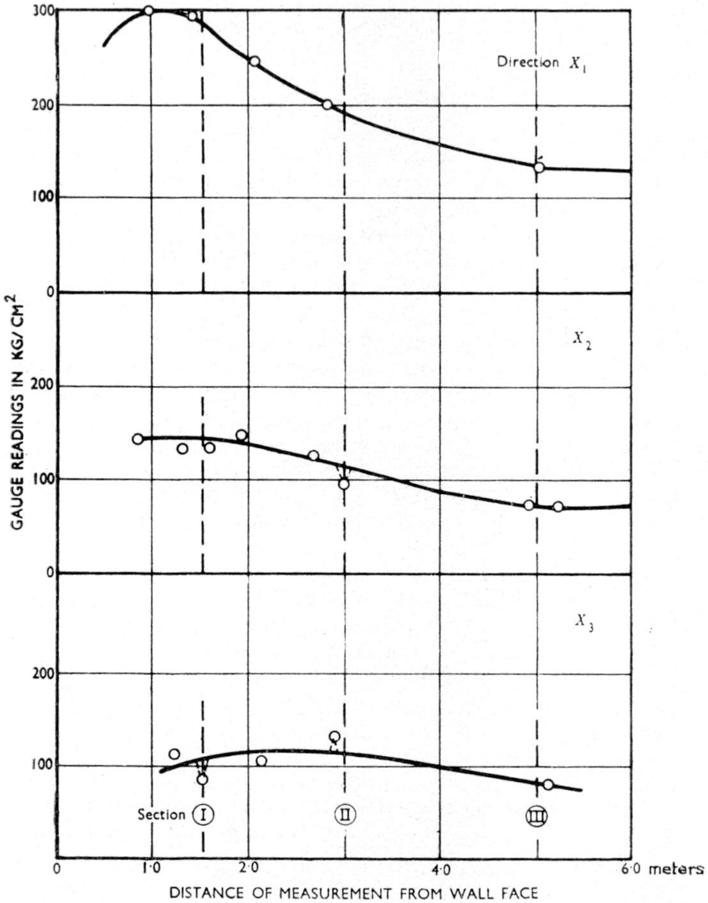

FIG. 3. Small circles represent results of individual gage readings in the directions X_1 to X_3. The smooth curves were traced for the purpose of computing magnitude and directions of maximum and minimum normal stresses in Sections I to III.[11,33]

In order to calibrate a cell, the cell is installed in a cylindrical hole with a diameter equal to that of the drillhole (26 mm), which was drilled across a prismatic block made out of a perfectly elastic material with a modulus of elasticity E of the same order of magnitude as that of the rock. Then the block is subjected to axial compression and the cell readings are plotted against the unit load on the block. In the drillhole in the field each cell is prestressed until the pressure on the cell is well in excess of the maximum stress which is anticipated in the rock. The prestressing of the cells produces a slight indentation in the walls of the drillhole on either side of the gage, as shown in Fig. 1 and during the test, while the diameter of the hole is increasing, the depth of the indentation decreases. Equations have been derived for computing the decrease as a function of the cell pressure and its effects on the cell readings are considered in the stress computations. After the field test is completed, the core surrounded by the annular groove is recovered, cut into 20-cm sections and for each one of the sections the E-value is determined by a laboratory test. In this manner the data are obtained which are required for a correlation between the stresses recorded by the gage and the stresses in the rock. If the hole penetrated rocks with different elastic properties, the conversion factor varies for different parts of the hole. Examination of the cores also makes it possible to find out after the tests whether the drillhole was located close enough to the centerline of the groove.

The results of the observations in one drillhole provide information only concerning the intensity and the distribution of the normal stresses in a plane at right angles to the centerline of the drillhole. Yet, the rock is in a three-dimensional state of stress and the orientation of the principal stresses is unknown. In order to get complete information concerning this state of stress, Hast performed the measurements in three drillholes oriented at right angles to each other.

Time Required for Measurements

Hast's detailed description of his instruments and procedures conveys the impression that the determination of the stresses in the rock surrounding small-diameter drillholes is still a major operation which must be carried out with utmost care and precision. The time required for obtaining the data shown in Fig. 3 (drilling, installation of gages and measuring the stresses) was about 1 week. The cutting of an annular groove in sections of 50 cm concentric with a drillhole requires very experienced and skillful personnel, and the difficulties increase with increasing depth of the hole. Yet between 1952, when the first measurements were performed, and 1961 the maximum depth to which stresses were successfully measured was increased from 2 to 20 m.

Both Hast's procedure and his pressure gage are patented in Sweden, England, South Africa, U.S.A., Canada, and various other countries.

Measurement of Diametral Borehole Deformation for Stress Determination

Another method for determining rock stresses is the accurate measurement of borehole horizontal and vertical axes to determine the relative deformation produced in the cross-section of the borehole by stresses in the rock. Maximum vertical diametral deformation of a horizontal borehole is produced by vertical stress when the is no horizontal stress acting. When the vertical stress and horizontal

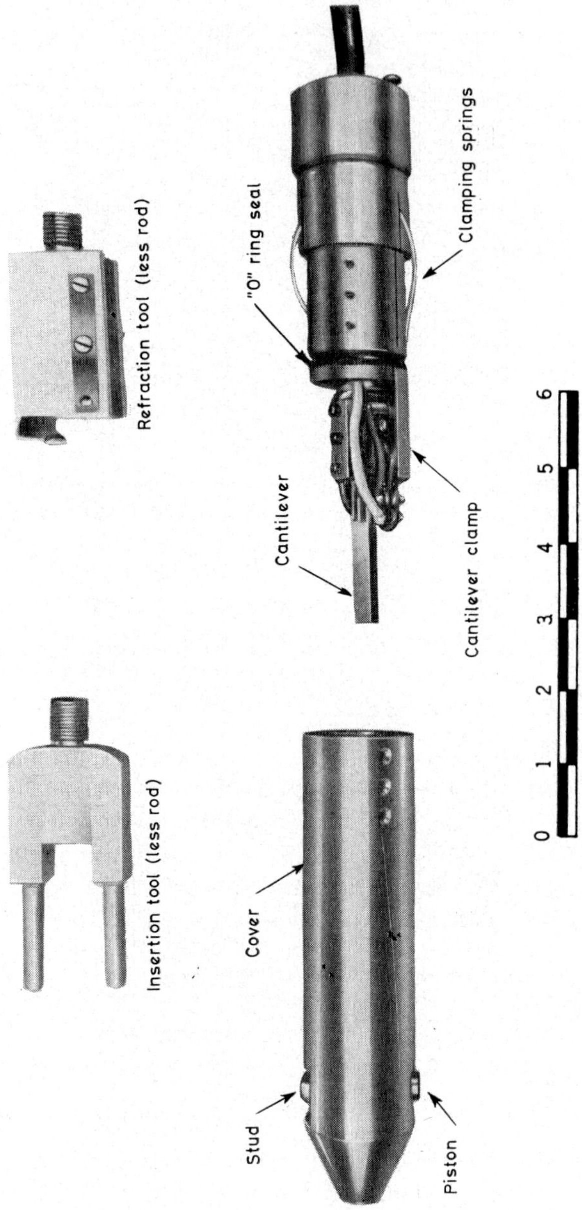

FIG. 4. Exploded photograph of borehole gage.[39]

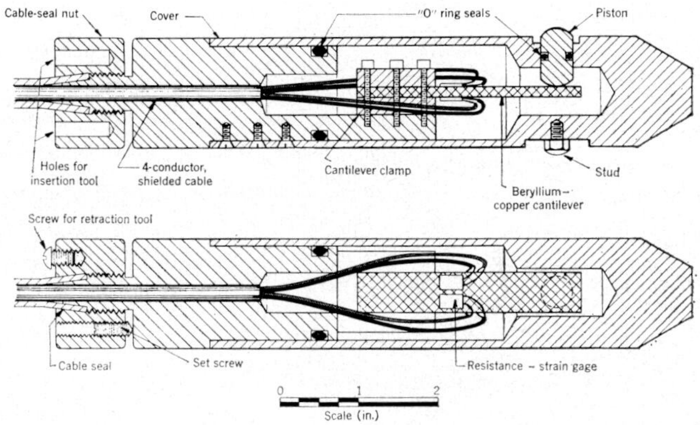

FIG. 5. Sectional views of borehole gage.(30)

stress in the rock are equal there will be no differential deformation along the two axes of the borehole.

An accurate instrument for measuring borehole diametral deformation has been devised in Germany(16). The formulas used for determining stress from the borehole measurements are derived from the formulas of Kirsch. Among other factors the rock modulus of elasticity, E, and Poisson's number, m, must be known.

The U.S. Bureau of Mines has recently devised and applied a gage for accurately measuring borehole deformation.(30, 31) A typical gage is shown in Figs. 4 and 5.

The gage is intended for use in comparatively competent rock, which will core in lengths averaging 1 ft or more and which is relatively elastic (stress proportional to strain). This gage has an outside diameter of 1.395 in. and is designed to fit into a standard EX diamond drill hole (EX bit diameter is 1.460 ins). The gage has an accuracy of 20 microinches for the overall system, which corresponds to a stress accuracy of 13 psi for a rock (in a uniaxial stress field) with a modulus of elasticity of 3.0×10^6 psi.

In operation the piston actuates the beryllium–copper cantilever, on which four bonded resistance strain-gage elements are mounted and connected to form a four-arm bridge. The bending strain produced in the cantilever is transduced by the resistance strain gage into an electrical analog, which is measured on a conventional resistance strain-gage indicator. The cantilever also produces the contact force for the piston. Thus, the mechanical system has only two moving parts.

Theory and Equations

The deformation of the hole in a uniaxial stress field and in plane stress is given by:[31]

$$U = \frac{dS}{E}(1+2\cos 2\theta) \tag{1}$$

where U = deformation of hole (change in length of a diameter);
a = radius of hole;
d = diameter of hole = $2a$;
S, T = perpendicularly applied stress (for a uniaxial stress field $T = 0$);
θ = angle (counterclockwise) from S to r (see Fig. 6);
E = modulus of elasticity.

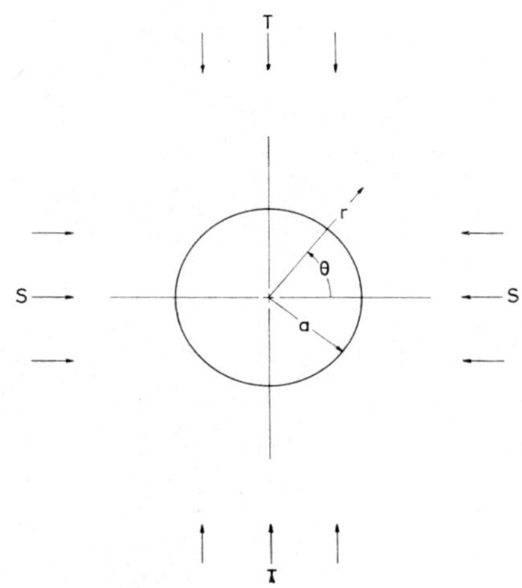

FIG. 6. Cross-section of hole in plate.[31]

When $\theta = 0°$, the deformation is in the direction of the applied uniaxial stress, and equation 1 reduces to

$$U = \frac{3dS}{E} \tag{2}$$

When $\theta = 90°$, the deformation is

$$U = -\frac{dS}{E} \tag{3}$$

and the minus sign signifies that, as the stress increases, the hole (at that point) is expanding.

The deformation versus the angle θ for one quadrant of the hole ($\theta = 0°$ to $\theta = 90°$) is plotted in Fig. 7.

For a biaxial stress field and plane stress, the deformation is related to the biaxial stresses S and T by

$$U = \frac{d}{E}\,[(S+T)+2(S-T)\cos 2\theta] \tag{4}$$

When $\theta = 0$,

$$U = \frac{d}{E}\,(3S-T) \tag{5}$$

and when $\theta = 90°$,

$$U = \frac{d}{E}\,(3T-S) \tag{6}$$

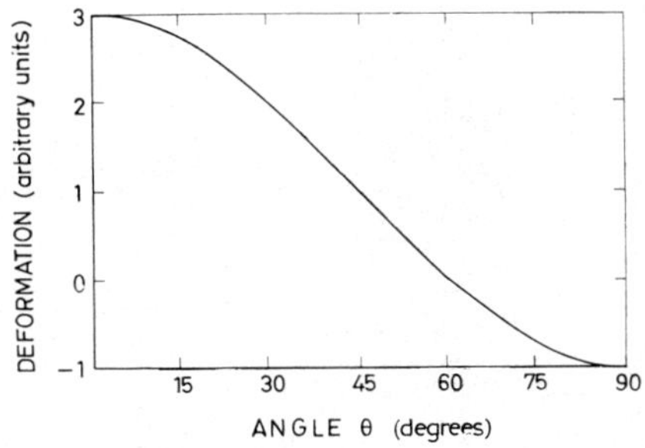

FIG. 7. Borehole deformation vs. angle θ[31]

If the deformation is measured across three different diameters and the modulus of elasticity and Poisson's ratio are known, the magnitude and direction of the stresses S and T can be computed.[31] The equations for these conditions will be found in the Appendix to Chapter 2, page 118.

STRAIN RESTORATION METHODS

Strain Restoration by Means of Flat Jacks

The strain restoration method has been used extensively by European engineers for determination of rock stresses in connection with civil works construction and mining operations.[8,9] The method has recently been used in the United States as a check on strain relief stress measurements at an underground powerhouse site by the California Division of Water Resources. (See page 518, Chapter 12.)

One application of the method involves the cutting of a slot in the rock just large enough to accommodate a "flat jack" or Fryssinet jack 70 cm square. (The Fryssinet jack is a flat, expandable, sheet-metal "flask" or pressure cell which is expanded by the application of hydraulic pressure.) Measuring points A and B, as shown in Fig. 8, are established prior to cutting of the slot and the distance between points is accurately determined. The flat jack is then placed in the slot and cemented tightly in place with quick-setting cement mortar. Hydraulic pressure is applied to the flat jack until measurements show that the distance between points *A* and *B* has been restored to its original dimension. The pressure in the flat jack is then a function of the original pressure in the rock before the slot was cut.

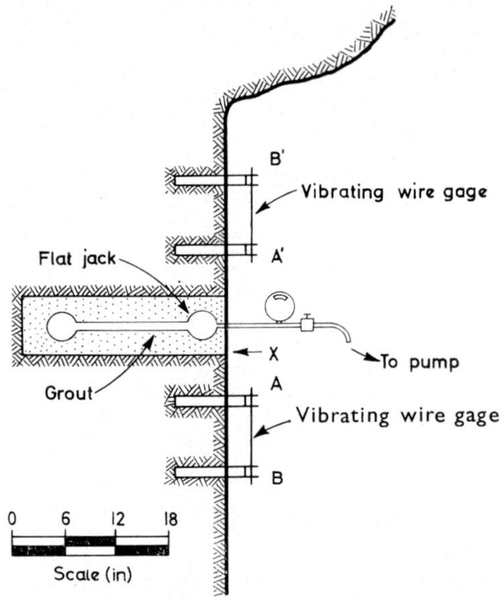

FIG. 8. Cross-section showing installation of a flat jack and instrumentation.[8,9]

Extensive experimentation with this system[8] has shown that the pressure required to restore the original strain varies with the locations of the measuring points relative to the slot.

It was found that best results were obtained by setting the measuring points along the vertical axis of symmetry and within a distance equal to about two-thirds the length of the flat jack. Thus, in Fig. 8, the distance *X–B* should be equal to or less than $\frac{2}{3}$ of the jack length. At the same time the point *A* must be set far enough from the slot so that its anchorage will not be loosened during cutting of the slot. When the measuring points are set in this location and hydraulic pressure is applied to the flat jack the original distance between the measuring points will be restored when the pressure in the jack reaches a value approximately equal to the original stress in the rock.

In practice when a flat jack 70 cm (about 28 in.) long and 70 cm wide was used the distance *A–B* was made about 30 cm (about 12 in.).

Among the experiments carried out in French iron mines where bedded deposits are mined by room-and-pillar methods were measurements of pressures at the sides of entries with rectangular cross-sections about 5 m high and 4 m wide situated at a depth of about 200 m (650 ft).[8] Hydraulic pressures required to neutralize rock deformation varied from about 80 to 95 kg/cm^2 (about 1130–1280 psi).

The stress readings obtained were not the original rock stresses but some multiples thereof. Stress concentrations are usually two to three times the initial stress, the exact values depending upon relative dimensions of the excavation and initial ratio of lateral and vertical stresses.

Limits of Jack Pressures

Hydraulic pressures up to about 200 kg/cm^2 were applied to some of the jacks used in the experiments above. The limiting capacity of such jacks is about 180–200 kg/cm^2 (2500–2800 psi). When higher pressures were applied, leaks developed in the jacks. Thus the ability of these jacks to measure pressures is probably confined to the measurement of rock stress of less than 3000 psi. The U.S. Bureau of Mines has recently developed flat jacks capable of carrying pressures up to 10,000 psi. These are described in Chapter 12.

Application of the Method in Australia

An adaptation of the flat jack method was used for measuring stresses in the rock around tunnels for the Snowy Mountains Hydroelectric Project in Australia.[10]

The usual procedure at each site was to make measurements in pairs at right angles, and if possible in three mutually perpendicular directions (for example, by making measurements in the walls and roof of a tunnel). One of the directions was always vertical and the other two were horizontal.

As the next step a two-dimensional photoelastic analysis was performed on model openings of the same shape as the tunnels where the measurements were made. From these analyses stress concentration factors were derived, which, when applied to the measured stresses, were assumed to give the magnitude of the original natural stresses in the directions of measurement. These of course were not necessarily the principal stresses. Results of stress measurements at three locations are summarized in Chapter 2, "Natural Stresses in Undisturbed Soil and Rock".

As used on the Snowy Mountains Hydroelectric Project, the flat jack methods was modified in that measuring points were set on opposite sides of the slot.

Strain Measurement Gages

Vibrating wire gages, described elsewhere in this chapter, provide convenient and sensitive means for measuring the small strains associated with this type of test. Wires may be very finely tuned by strumming them and comparing tones. Even a relatively insensitive ear can distinguish the difference in pitch when wires are not tuned to the same frequency.

Photoelastic Coatings for Measuring Surface Strains in Rock

The photoelastic coating method for measuring surface strains in rock involves the bonding of a coating of a material which is optically birefringent to the rock surface where strain is to be measured. If the bond is sufficiently satisfactory any surface strain in the rock is transferred to the coating and its effects can be observed through a reflection polariscope.

The photoelastic coating gives the same information as would a system of strain gages installed at the same location while the equipment required for its application and the observation and interpretation of results is much less complicated and expensive than that required for a strain gage installation.

Since the photoelastic coating is a strain recording device it is apparent that the material will record only the *changes* in rock surface strain which take place after it is bonded to the rock, and that absolute strains will not be measured unless that portion of the rock is cut free and allowed to expand until it has reached stability.

The measurement of rock strains by means of photoelastic coating is a recent development but the results achieved to date are important enough so that a detailed progress report has been published .[(29)]

This method appears to be a very important advance in the processes of measuring surface strains in rock and in encouraging the everyday application of the results of such measurements in mining operations.

The aforementioned report, which is titled *Photoelastic Coating Technique Applied to Research in Rock Mechanics*, has been reprinted in Chapter 11 of this Volume.

BOREHOLE PRESSURE CELLS

The borehole pressure cell is installed in tight contact (usually cemented in place) with the rock at the bottom of a borehole and registers changes of pressure on the cell. Such pressure changes may be created by advancing extraction faces or by other changes in dimensions of adjacent openings. If the cell is constructed so that it has the same modulus of elasticity, E, as the rock in which it is enclosed, then the deformation of the cell will be the same as that of the rock and the stress change will be equal to the strain change times the modulus of elasticity.

If the modulus of elasticity of the cell differs from that of the rock then the cell must be calibrated to determine the relationship between the actual rock pressure and the pressure recorded by the cell. Such a cell may be calibrated by enclosing it in a large block of rock of the appropriate type and subjecting the assembly to varying pressures in a large hydraulic press.

Blade-type Stressmeter Load Cells

The blade-type stressmeter is illustrated in Fig. 9. This stressmeter consists of two welded blades about $8 \times 1 \times \frac{3}{4}$ in. thick with a 0.01-in. thick glycerine film in between. A thin film is used with a view to reducing the temperature effect. Glycerine is used because of its being non-corrosive with steel and of low compressibility. The blades are provided with rims to increase the accuracy and response to the applied load. The closure of the plates is of the order of 0.001 in. per 1000 psi for a particular design. The glycerine film impinges against a circular metallic diaphragm, the periphery of which is clamped in place by the screw head. An active electric resistance wire strain gage is cemented on to the diaphragm while the compensating

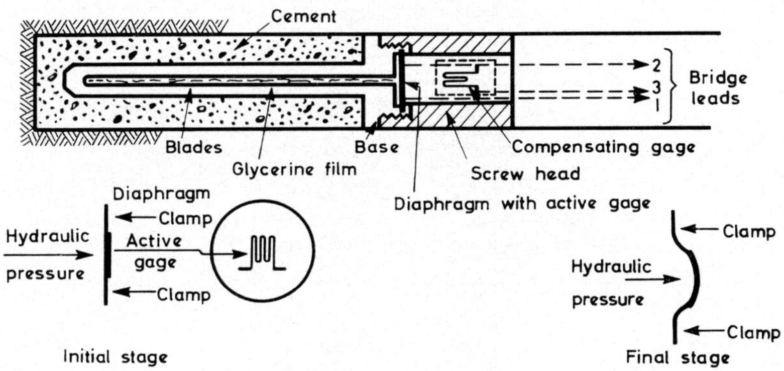

FIG. 9. Blade-type stressmeter.[12, 13, 34, 35]

gage with identical characteristics is fixed on a separate piece and is held in position by screw from the extension of the screw head. The compensating gage is provided to counteract the effect of temperature, etc., on the active strain gage, thus insuring that the measured resistance change in the active gage is due to pressure only. The gage leads are taken through stressmeter head to a three-core cable. The leads from the stressmeter are taken to a junction box of multimeter capacity and finally to a Wheatstone Bridge. The operation of a selector switch incorporated in the bridge enables connection to be made with any desired stressmeter for observation. This technique assists in taking observations from a remote position.

The working principle is as follows: The instrument is inserted at the bottom of a long hole in solid ground ahead of the working face. As the rock surrounding the stressmeter deforms, it exerts pressure on to the blades via a filling medium such as concrete made with expanding cement to counteract shrinkage. The blades of the stressmeter then tend to close in and the occluded glycerine film, being practically incompressible, transmits the resultant pressure to the diaphragm which, together with wire gage, then deforms. This amount is measurable by the amount of change in the resistance of the gage as determined by Wheatstone's Bridge method.

Calibration curves of known load against resistance change for each stressmeter, are obtained in the laboratory before installing underground. Later these curves may be used to translate observed resistance into the equivalent stresses existing in the underground rock. The changes in stress value, with respect to face advance, gives an idea of the location of front abutment zone.

Electrical Resistance Strain Gage Load Cells for Boreholes

A type of borehole cell which employs electrical resistance strain gages (SR-4) as sensitive elements for recording rock strain has been devised by the Fuels Division of the Mines Branch of the Department of Mines and Technical Surveys of Canada.[(14)]

Electrical resistance strain gages of the types illustrated in Fig. 16 are mounted on thin disks which in turn are installed as diaphragms in a thin cylindrical shell which forms the body of the load cell and also houses a temperature compensating gage, a cable clamp and a leveling device. The whole apparatus is made small enough to be inserted in a borehole $2\frac{1}{4}$ in. in diameter. It is cemented into place in a hole and records changes in rock stress due to advancing extraction faces.

In such a cell the load-bearing disc is stressed more than the surrounding material because it is a "strut" surrounded by voids. To determine the ratio between actual rock stress and stress recorded by this meter it is necessary to insert the apparatus into a hole in a block of rock and subject it to loading in a large hydraulic press.

Hole Cement for Pressure Cells[(12)]

Material for cementing a stressmeter in a hole must be non-shrinking. As observed, concrete normally shrinks on drying and, if it be used to seal the instrument in a borehole, there will be danger of losing contact. The transmission of load from the rock via the sealing material should then not only be imperfect but also uncertain. To avoid this possibility, the following expanding materials have been tried with partial success: rough plaster, pottery plaster, industrial fine plaster, and rapid hardening cement. Dental plaster of proper consistency has also been used with partial success in Canadian mines. The most suitable material, however, consists of:

112 lb. of Portland cement;

112 lb. of sharp (dry) sand to $\frac{3}{16}$ in. size;

Three quarters of a pint of Conplast, (a liquid used to increase fluidity and to ensure homogeneity and thereby strengthen the mixture);

Three pounds (1 bag) Cebex 113 (inclusive of $CaCl_2$) (a powdered expanding material used in conjunction with $CaCl_2$, the latter being to retard the chemical action); the quantity may be doubled if extra expansion is required;

$4\frac{1}{2}$ gal water.

When mixing, the following procedure is adopted:

(1) Add Cebex 113 to cement and sand and mix well in dry state.

(2) Add Conplast to water in a container large enough to hold the complete charge.

(3) Add the dry mixture gradually to the liquid portion, stirring continuously until all the lumps are disposed and the grout is in complete suspension.

Measurement of Rock Modulus to Determine Stresses

The modulus of elasticity, E, of a rock varies with changes in stress applied to the rock. An instrument called a "modulus meter" has been developed in England,[15] to be inserted in boreholes to determine the modulus at any given point. This instrument comprises a means of applying pressure to the interior of a borehole at any point at which it is required to measure the stress in the rock and a means of measuring at the same point the distortion produced in the hole by the applied pressure. From such observations it is possible to plot a graph of applied pressure vs distortion, the slope of which is the modulus of the rock. By suitable calibrations in the laboratory the relationship between the rock modulus and the stress may be determined.

The instrument consists essentially of a cylindrical shell which may be expanded radially by hydraulic pressure. Two "feeler" studs protrude through the side of the shell to make contact with the rock in the borehole. These feelers indicate the amount of expansion produced by any given hydraulic pressure applied to expand the shell.

Where rock is of such type that significant readings can be obtained the modulus meter is a very convenient instrument to use. Readings can be taken at close intervals along a borehole thus giving a complete record of stress distribution along the hole. This contrasts with other types of measuring devices which must be cemented, or wedged, in place and depend upon a changing stress field for readings.

SONIC METHODS FOR DETERMINATION OF ROCK STRESSES

The velocity of sound waves through rock varies with the pressure to which rock is subjected. A simple relationship exists between the modulus of elasticity and the velocity of sound in a perfectly elastic medium: that is,

$$V = \left(\frac{E}{\gamma}\right)^{1/2} \qquad (7)$$

where V is the velocity of sound, E is Young's modulus, and γ is the density.

For a perfectly elastic material E is a constant; hence the velocity of sound is independent of the pressure, and for such material the proposed method would fail. Thus the sonic velocity method for stress measurement depends upon the fact that rock modulus of elasticity increases with the pressure to which a rock is subjected up to the point where the pressure exceeds the strength of the rock and fracturing begins.[17]

Difficulty is caused by the high velocities of sonic waves in rocks (ranging from about 5000 to 20,000 fps) and by the smallness of the percentage of velocity change in some rocks, especially at high pressures. [13] Where the distances being traversed by sound waves between boreholes are only a few feet the apparatus used must be capable of measuring time in microseconds. Transducers are used in conjunction with cathode-ray oscillograph for measuring such time intervals.

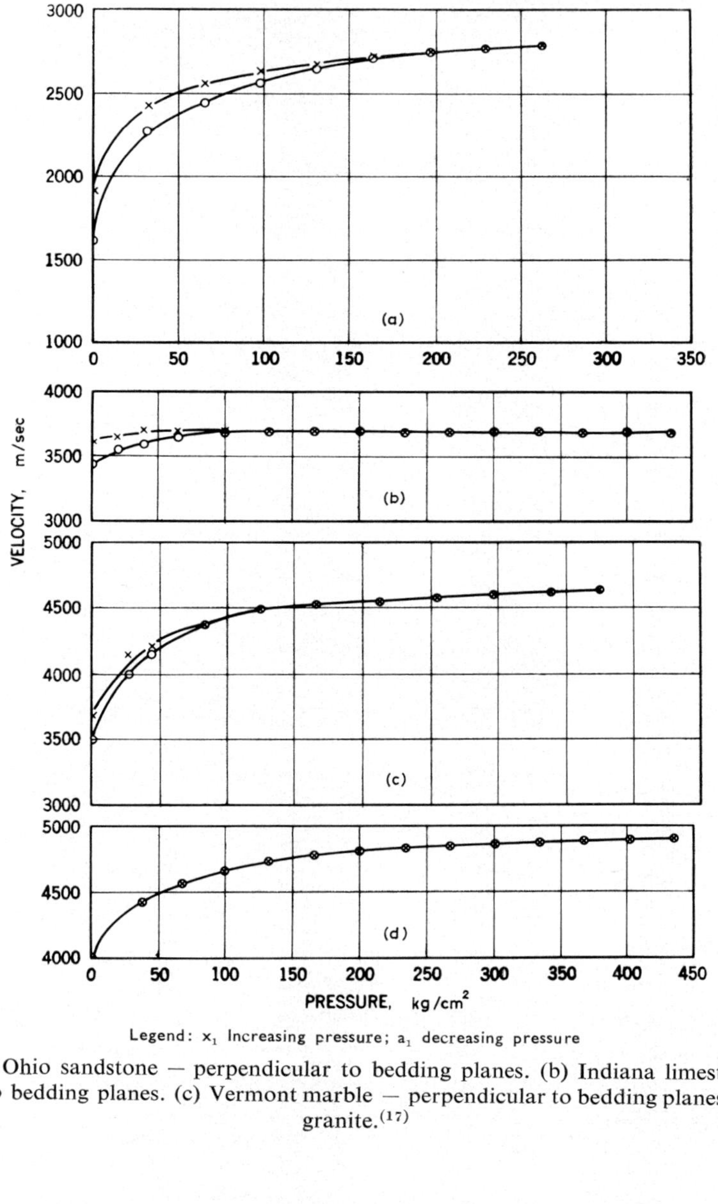

FIG. 10. (a) Ohio sandstone — perpendicular to bedding planes. (b) Indiana limestone — perpendicular to bedding planes. (c) Vermont marble — perpendicular to bedding planes. (d) Quincy granite.[17]

Determination of Stress Distribution

The method usually employs two boreholes with the sound-producing device bottomed in one borehole and the sound pickup apparatus placed in the other borehole and expanded into tight contact with the rock.[9, 19] This sound may be produced by the impact of a "hammer" or a "plunger" as it is shoved against the bottom of the hole. An impact switch on the impactor closes a circuit at the instant of impact and gives a time reference for the start of the sound travel which is then picked up by the transducer in the adjacent borehole. In order to obtain a series of measurements at different depths in the walls, or pillars, the boreholes are deepened after each measurement and a new measurement taken at the additional depth.

Measurements may also be made by setting the transducers at successively deeper depths in the borehole and initiating the sound at the collar of the hole. Another method employs three holes, with the sound initiated in the center hole and transducers placed in the two outer holes.

Laboratory Tests

Figures 10 and 11 show how the sound velocity varies with pressure for various types of rocks. It will be noted that the percentage increase in velocity is greatest in the lower stress ranges, that is, for stresses generally between zero and about

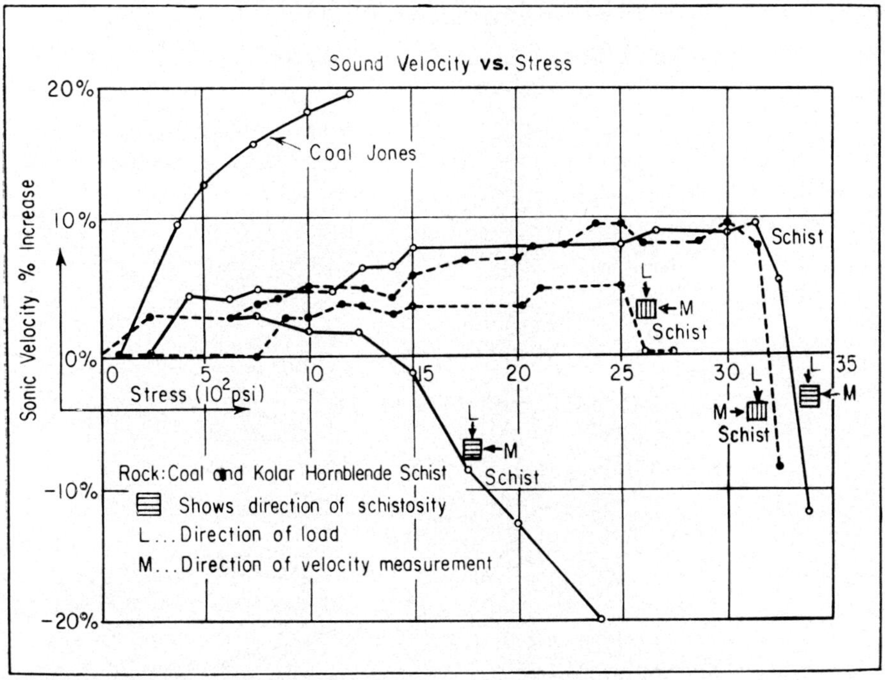

FIG. 11. Sound velocity vs. stress in hornblende schists of the Kolar Gold Field, and in coal.[12]

1000 psi. This indicates that such measurements may be most effective in measuring stresses around workings at shallow depths, and that the effectiveness of the method will decrease for deeper workings where initial rock stresses are on the order of several thousand pounds per square inch.

Listening Apparatus

Figure 13 shows a "geophone" for receiving subaudible vibrations. The geophone consists of a two-layered Rochell Salt crystal ($2\frac{1}{2} \times \frac{3}{4} \times \frac{1}{2}$ in. thick) housed in spring board fashion in $1\frac{1}{4}$ in. diameter by 8 in. long steel tube. The crystal is wrapped in silver foil. The vibrations received by the crystal compress it and generate a proportionate e.m.f. by piezoelectric effect. This is conducted through the silver foil to a lead for amplification. The electrical impulses on being amplified (about half a million times) pass through a coil within the influence of a magnet which results in the production of a proportionate break in the linear trace of the style used for recording.

To make the unit proof against very high-intensity shocks, a device is incorporat ed in the circuit which causes the record to vary logarithmically for abnormall- strong input signals.

Stress Distribution in Pillars

Figure 12 shows the results of sound velocity measurements across square pillars located in a French iron mine where the bedded deposit is mined by room-and-pillar methods. The pillars were 12.5 m square, and, as indicated by the black dots, readings were taken at 1-m intervals. As shown by the relative velocities a fractured zone extended into the pillar on each side for a distance of about 2 m. Adjoining each of these fractured zones was a zone of peak pressure, while still farther inside the pillar was a central interior zone of less pressure.

THE MICROSEISMIC METHOD FOR PREDICTING ROCK FAILURE

The counting and measurement of the magnitude of subaudible vibrations have been found to be a fairly reliable indicator as to the state of stability of rock surrounding an underground opening. The underlying idea is that the frequency and magnitude of subaudible vibrations increase when a rock is stressed to near its breaking point. Such vibrations are believed to be the result of the fracturing of the individual crystals, grains, or fragments composing the rock mass.

In practice a geophone (Fig. 13) installed in a hole drilled into the rock picks up the minor mechanical vibrations and converts them to electrical impulses. These may be transmitted to earphones for listening or they may be recorded for later inspection.

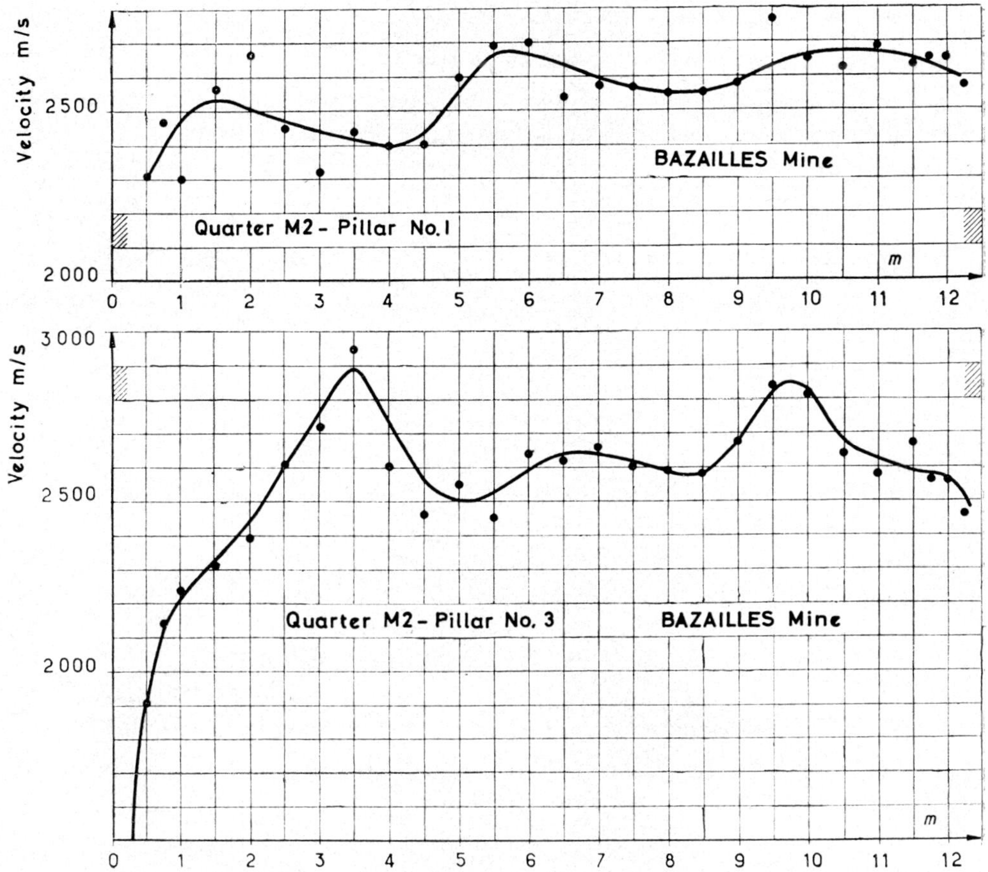

FIG. 12. Distribution of vertical stress on pillars as indicated by sound-velocity measurements in square pillars of the Basailles iron mine, France.[19]

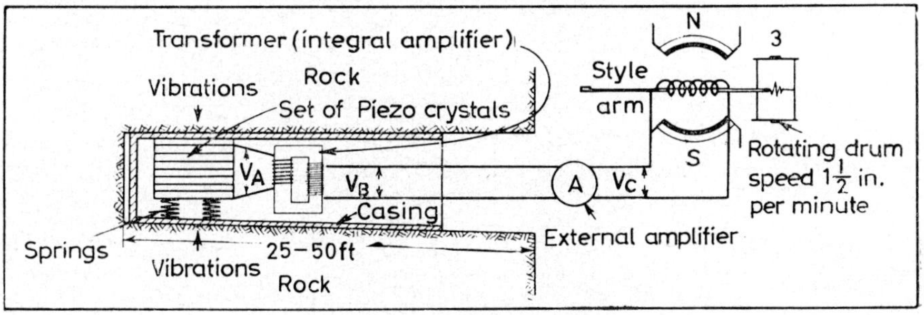

FIG. 13. Geophone for receiving subaudible vibrations.[12]

Microseismic Tests on Rock Specimens

Figure 14 is a record of the microseismic activity in a specimen of coarse-grained granite which was subjected to compressive loads corresponding to various percentages of its ultimate compressive strength. Each record is a chart of the activity over a 2-min period.

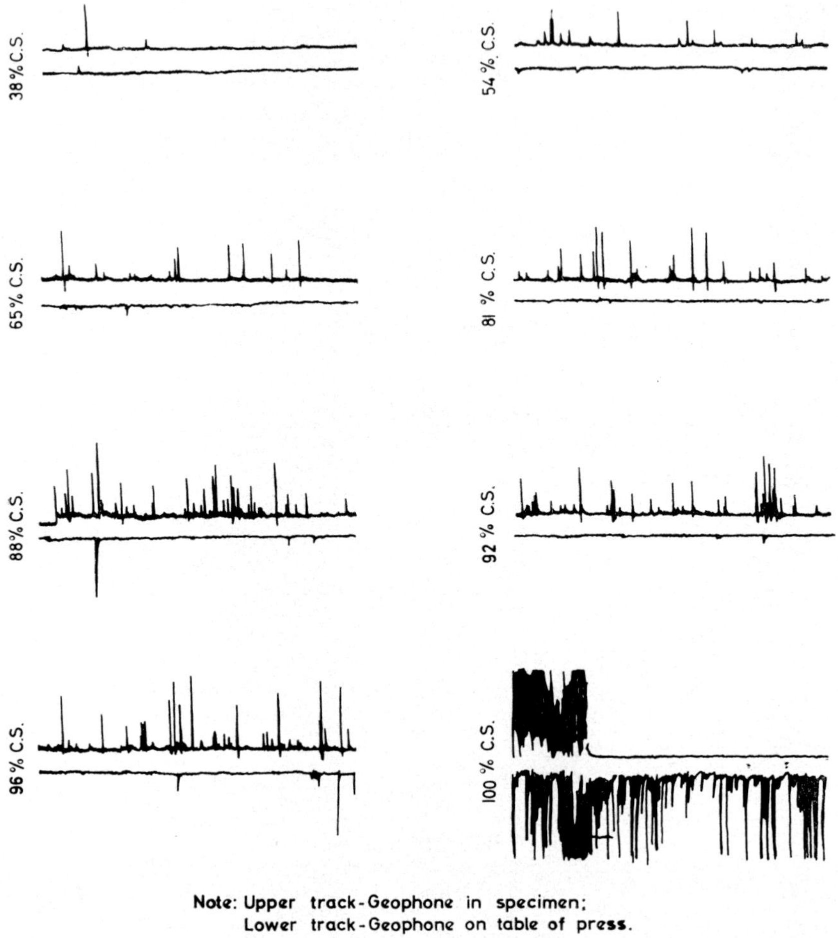

FIG. 14. Sections of microseismic record showing the variation in rate of occurrence of sub-audible noises with the magnitude of the applied stress. (Magnitude of applied stress is indicated as percentage of crushing strength.) Specimen, granite.[36]

Microseismic activity becomes significant at about 54 per cent of crushing strength but tends to die out with time. When loads are increased to values in the range of 65–81 per cent of the crushing strength the microseismic activity does not show any significant decrease during the 2-min test period. At loads between 88–

96 per cent of crushing strength microseisms occur more frequently and the rate of activity indicates that the rock would proceed to failure within a relatively short time.

It appears that microseismic activity may be correlated with "rock creep" or "time strain" and that the same formulas which define the rate of "creep" might apply to the rate of occurrence of microseismic noises in a rock subjected to a given unit stress. The equations for rates of time strain are given in Chapter 4, p. 175.

Conclusions from Laboratory Tests

In laboratory experiments on various types of rock it was found that the total number of microseisms produced during the course of compressing a specimen to failure varied widely.[36] For example, the coarse-grained rocks such as granite produced, on the average, over a thousand microseisms while the fine-grained rocks, such as the greenstones, produced fewer than 200.

It was also demonstrated that microseisms can be produced in homogeneous rock without any visible or audible evidences of failure. It is reasonable to suppose that microseisms do originate by intermovements, or by separations along the grain boundaries such as would occur during the formation of Griffith cracks. (See Chapter 3, p. 142.)

It has been noted that a microseism can be simulated by cracking (not crushing) a single grain of sand on a smooth surface by pressing it with a blunt instrument.[37]

Detection of Incipient Rock Failure in Mines

Oil Shale Mine

Figure 15 is a record of the daily microseismic activity over a 2-month period in a room of the experimental oil shale mine at Rifle, Colorado.[20]

The high microseismic rates between November 13 and 16, 1950 were caused by a 9-ton slab which was found to be partly detached from the rib. This slab was removed on November 16 and the rate immediately dropped to the normal background level.

A similar condition caused by a loose section in the rib in one corner of the room was noted on December 2. On December 5 a high microseismic rate was recorded. Inspection of the testroom showed that a crack about 15 ft long had developed near Station C. On December 20 a high microseismic rate was caused by a loose slab near Station C. After removal of the slab, which weighed approximately 150 lb., the micro-seismic rate dropped to the normal background level.

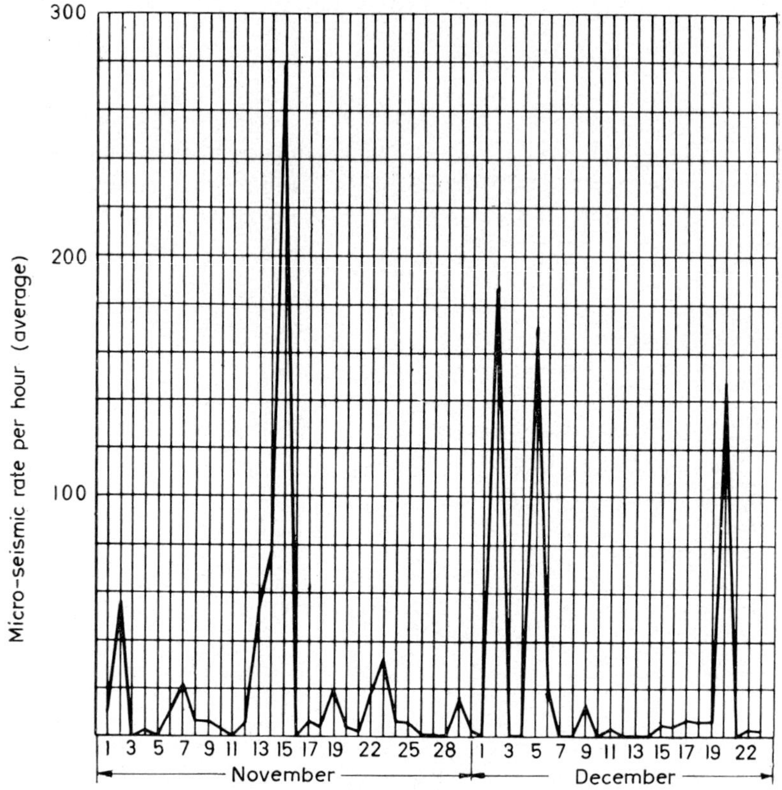

FIG. 15. Average micro-seismic rate from November 1 to December 1, 1950 in a test room of the experimental oil shale mine, Rifle, Colorado.[20]

Baker Mine

A warning device which was used in stopes of the Baker Mine of the St. Joseph Lead Co. during extraction of pillars consisted of a sensitive borehole-type geophone connected to a powerful amplifier. This permitted the subaudible noises associated with ground movements to be heard, and the rate of occurrence of such noises was an indication of the state of structural stability of the roof and pillars.[39]

Conclusions

Important conclusions drawn from laboratory and field research are:[12]

(a) That the geophones are sensitive to subaudible "snaps" occurring within a distance of 50–75 ft.

(b) That there should be the best possible contact between the geophone and the rock so that even small "snaps" may give a fair throw on the stylus.

(c) Light disturbances fail to carry well across discontinuities in the rock mass such as faults or planes of rock layer separation.

(d) With experience in each type of rock and each locality a critical frequency may be determined above which value the rock may be regarded as being in an unstable state of equilibrium and thus unsafe.

TYPES OF STRAIN GAGES

Electrical Resistance Strain Gages

The electrical resistance strain gage shown in Fig. 16 consists of a grid of fine alloy wire cemented to a paper membrane (see far left). The gage is normally covered with a layer of felt both to protect the wire and to reduce the effect of

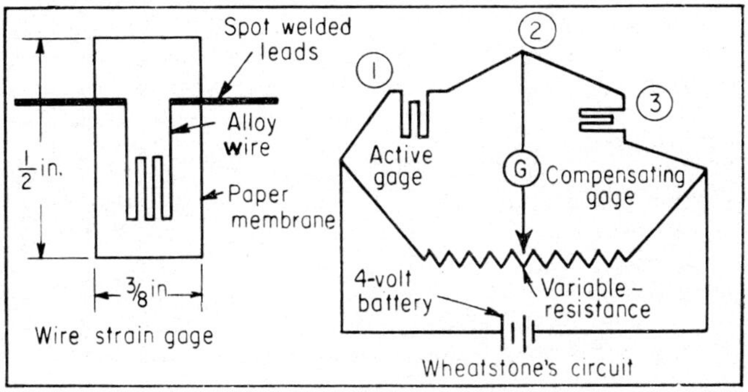

FIG. 16. Electrical resistance strain gage.[12]

draft. The whole may be cemented to the surface of the instrument designed to take the pressure under investigation and any pressure acting parallel to the length of wire will then strain it causing a resultant proportional change in resistance. This latter can be measured accurately using the principle of Wheatstone's bridge (see left) and using a mirror galvanometer to indicate the null-point in the bridge circuit.

Humidity causes breakdown of the insulation and corrosion of gage wire resulting in considerable rise in its resistance. Every precaution should be used to keep the gages dry.

It is customary to include a temperature compensating gage in one branch of the Wheatstone bridge. This compensating gage is mounted so that it is not subjected to strain and compensates for temperature effects on the resistance so that any change in resistance measured by the Wheatstone bridge is entirely due to strain on the active gage.

The electrical resistance strain gages have the advantage that they are much smaller than the vibrating wire type and can be installed within a circle only 2 or 3 in. in diameter. Thus they have been installed in instruments which are inserted into boreholes for measuring pressures at a distance from a free rock face.

Vibrating Wire Strain Gage

A fairly simple, yet accurate, method of measurement for rock strain has been described by St. Q. Isaacson [4]. Two steel bolts, about a foot long, are cemented into holes drilled into the rock on the periphery of a tunnel as shown in Fig. 17. The bolts are drilled to take a length of thin piano wire which is threaded through them and soldered into position under tension. The sonometer itself is then brought into use. It consists of a mild steel rack with a micrometer head. A length of similar piano wire is stretched from one end of the rack to the end of the micrometer screw so that by rotating the micrometer a continuously varying tension may be induced in the wire.

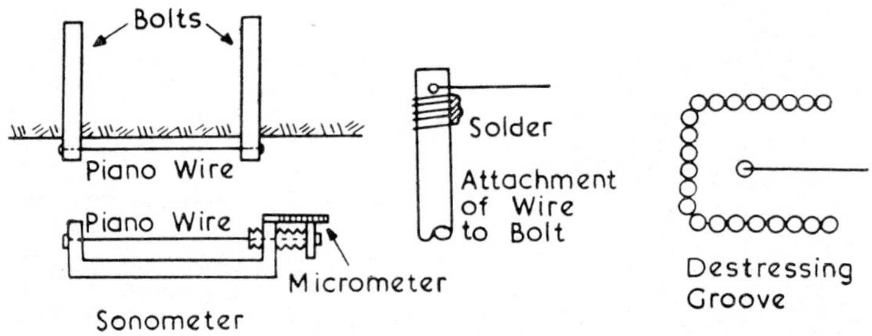

FIG. 17. A simple device for measuring surface strain in rock by the strain relief method.[4]

In use the micrometer is adjusted until the note obtained by plucking the wire is the same as the note emitted by the wire stretched between the two bolts in the rock face. A difference in adjustment of the micrometer by a thousandth of an inch will result in a difference in tone quite appreciable even to the non-musical ear. The micrometer is then read and recorded.

The rock in which the two bolts are situated is then destressed by drilling a groove around it, which consists of a series of ordinary drill holes, located contiguously, which reach to a depth in the rock several inches below the base of the bolts. The rock in which the bolts are located is thus free to expand and the wire stretched between the bolts will therefore be under increased tension. The sonometer wire is then re-tuned to this new note and the change in the micrometer reading is noted. If the original length of the sonometer wire was approximately the same as the length of the wire stretched between the bolts then the strain in the rock will be of similar magnitude. If we have previously determined Young's modulus for the rock we can then compute the original stress acting in the direction of the wire stretched between the two bolts.

A convenient length for the wire to be stretched between the two bolts is about 12 in. and the sonometer wire should be of approximately the same length.

A vibrating wire is very sensitive to changes in tension, that is, to changes in length. For this reason vibrating wires are used as the indicator elements in strain gages where sensitivity is very important.

The vibration frequency of a wire is given by

$$f = \frac{1}{2l}\left(\frac{T}{m}\right)^{1/2}$$

where

l = length of the wire;
T = tensile stress in the wire;
m = specific mass of the wire (mass per unit length).

The vibration frequency of the wire is compared with the vibration frequency of a standard vibrating wire (or a frequency meter) to which we apply a variable stress by means of a micrometer arrangement by means of which we can read the elongation. Wires about 12 in. long have been used in determining rock deformations in French iron mines and in experiments in the Kolar Gold mines of India.

It is also possible to arrange vibrating wire strain gages so that the sensitive elements may be plucked by electrical means and the vibration frequency determined by electrical means. Thus such gages may be installed in remote or inaccessible locations and load readings made at a distant point.

Such vibrating-wire gages have been installed at locations which were later submerged under a 100-ft, head of water and operated successfully by remote control.[5] The vibrating wires, about 4 in. long, were enclosed in a short mild steel cylindrical tube which carried the load axially. The wire was plucked and its free vibration picked up by the same small electromagnet. The frequency was compared against that of a standard vibrating wire with micrometer variable loading.

Determination of the Modulus of Elasticity

Strain relief methods for determining rock stress depend upon the measurement of rock strain and upon the value of the modulus of elasticity (Young's modulus), E. The stress existing in the rock is computed from the formula:

$$\text{unit stress} = (\text{unit strain}) \times (\text{modulus of elasticity})$$

or

$$\sigma = \delta E$$

The rock modulus of elasticity may be determined by subjecting a laboratory specimen to compression and measuring the resultant deformation. However, the laboratory values so obtained are seldom truly representative of the modulus of elasticity of the rock mass *in situ*. Habib[6] conducted field and laboratory experiments in conjunction with investigations for a dam at Tignes. The modulus of elasticity of the quartzite as determined by measurements on laboratory specimens was about 590,000 kg/cm² (about 8,400,000 psi). When measurements were made on rock in the field by means of a hydraulic jack, the *in situ* modulus was found to be about 100,000 kg/cm² (about 1,420,000 psi). (See Chapter 12.)

Similarly, differences were found between the laboratory values and the field-measured values during investigations for the Kemano pressure conduit.[9] Values of E for the quartz diorite as determined in the laboratory averaged about 7 million psi. Values of E as determined from field tests with a large steel sphere to which internal pressure was applied ranged generally from about 2.5 to 4 million psi.

Evidently the lower *in situ* values for rock modulii are due to imperfections in the rock and to numerous small pores, cracks, and fissures in the rock mass which yield by closure when pressure is applied to large masses of rock in the field.

The value of rock modulus of elasticity also tends to decrease with the duration of time for which pressure is applied. This phenomenon is brought about by the plastic flow or "time creep" which occurs in most rocks subjected to long term stresses. Designs which depend upon the modulus of elasticity of rock should be based upon values of the modulus which are anticipated under long-time loading.

Seismic Methods for Determining the Modulus of Elasticity

It was noted previously that the following equation gives the velocity of sound, or seismic waves, in an elastic material:

$$V = \left(\frac{E}{\gamma}\right)^{1/2} \tag{7}$$

where:

V = the speed of travel of longitudinal seismic waves (compressional waves);
E = the modulus of elasticity of the rock;
γ = the density of the rock.

Transposing the equation, we get:

$$E = \gamma V^2 \tag{8}$$

It is possible to determine the dynamic modulus of rock *in situ* by seismic means if its density is known. This may be done by firing shots placed at various depths in drill holes and measuring the travel time of the seismic waves with properly placed geophones. The procedure may be reversed and the geophone may be placed in a drill hole while shots are fired on the surface.

In general, the value of the modulus of elasticity as determined by dynamic means, such as seismic and sound propagation methods, tends to be higher than that determined by compression testing.

MEASUREMENT OF ROCK MOVEMENTS

The relative amount of closure between roof and floor, or between hanging-wall and foot-wall, is an indication of the magnitude of the pressure on the rock above the opening. The amount of strain depends upon several factors, such as the amount of ground which is open, the amount and quality of filling material, characteristics of the country rock, etc.

Measurements of convergence may be found of value in predicting the imminence of failure of roof or floor rock.

Vernier Strain Bar

Probably the simplest device for measuring convergence is the vernier strain bar, such as is illustrated in Fig. 18. This device is convenient for measuring large scale closure of walls such as is common in stopes, or in measuring closure of roof and floor such as is common with mining systems used in bedded deposits.

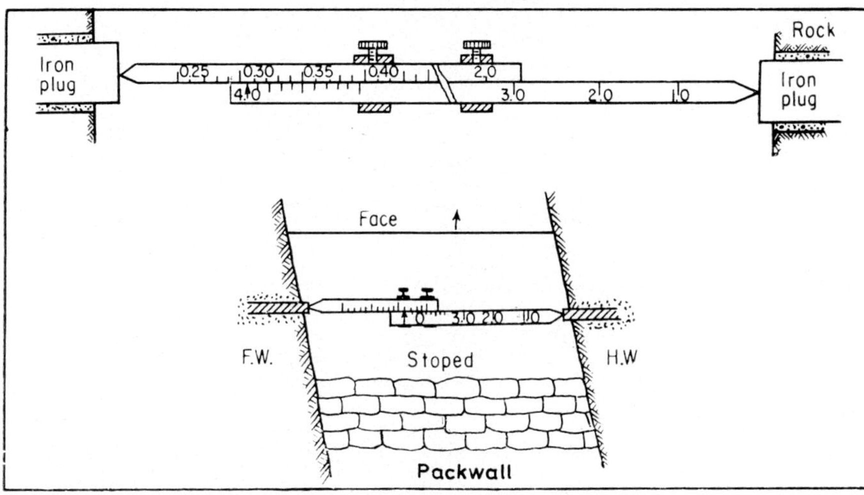

FIG. 18. Vernier strain bar.(12)

The vernier strain bar, as the name indicates, utilizes the principle of the vernier gage. It consists of two steel bars sliding over one another within sleeves. One bar (upper in drawing) is graduated in 0.01-ft divisions up to the 2-ft mark from the taper end. The other bar is graduated only in 1-ft marks from the taper end. The last 0.09 ft from the 4-ft mark is divided in ten equal parts and then gives a least count of 0.001 ft. While in use the bar ends (tapered) are pulled out so as to come into contact with the reference plugs as shown in the drawing. The distance between the two plugs is then equal to 4 ft plus the main scale reading on the vernier. This particular bar has a range of 4–6 ft.

Water Levelling Apparatus

This apparatus, illustrated in Fig. 19, may be used for investigating relative movement of hanging-wall and foot-wall in steeply dipping veins. Operation of water-levelling apparatus is as follows:

As 18-in. long, $\frac{7}{8}$-in. diameter pin with an attached pointer is cemented in the foot-wall hole. A second pin, fitted with a screw bolt, capable of vertical movement, is cemented in the hanging-wall. A water level is established using two tins connected by a rubber tube in such a manner that the two pointers are just touching the water surface in the two tins. In use, the tins are filled with water – care being taken against air locking – and held below the plugs. If necessary, more water is added

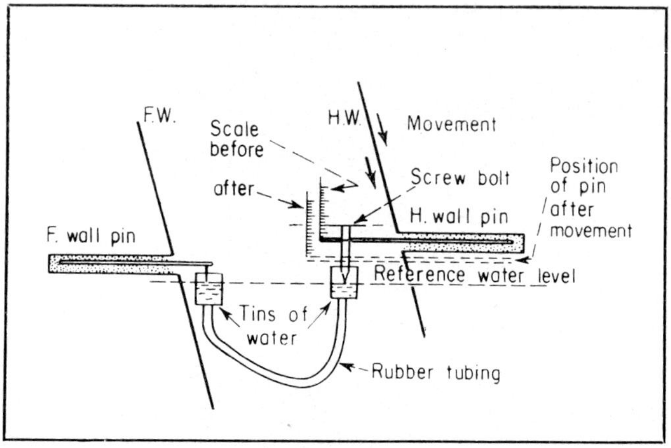

FIG. 19. Water level method for measuring rock movements.[12, 38]

drop by drop to the foot-wall tin until the level just touches the tip of the pointer. The screw bolt is then adjusted such that its tip touches the water level, and the bolt position is then read on a scale. The relative movement of foot-wall and hanging-wall results in changed readings for bolt positions and the difference of successive readings will give the amount of vertical movement.

Micrometer Turnbuckle

For locations where convergence is relatively slight the micrometer turnbuckle may be employed. The apparatus, illustrated in Fig. 20, has the following essential parts:

(a) Swivel rod with cup and bearings and pin-vice head having a blind hole. For clamping the wire in the correct position, the end of the wire must be in contact with the end of the blind hole. Rod connects pin-vice and thimble and frees the wire from tension kinks when connected to the pin-vice head.

(b) Thimble with micrometer screw. The outside of the thimble shell is graduated circumferentially in 50 parts. The screw (20 turns/in.) stud inside advances 0.05 in. in one complete revolution, giving a least count of $0.05/50 = 0.001$ in.

(c) Barrel which is a tube internally threaded (20 turns per in.) from one end to accommodate the advance of the micrometer screw stud and smooth at the other end to accommodate a piston. The outer surface of the barrel, over which the thimble advances, is marked with 0.05-in. divisions. Towards the piston end a view slot is cut in the barrel wall and a reference line is marked on it.

(d) Piston rod with piston, spring, and pin-vice head. The piston fits into the barrel and the spring is fitted over the piston rod between the barrel end and the piston. The piston moves along the axis of the barrel depending on the tension in the wire. There is a reference line on the piston rod to coincide with the reference line on the slot in the barrel, thus indicating a standard tension in the wire. In achieving this tension the spring roughly shortens from $\frac{7}{8}$ to $1\frac{1}{2}$ in.

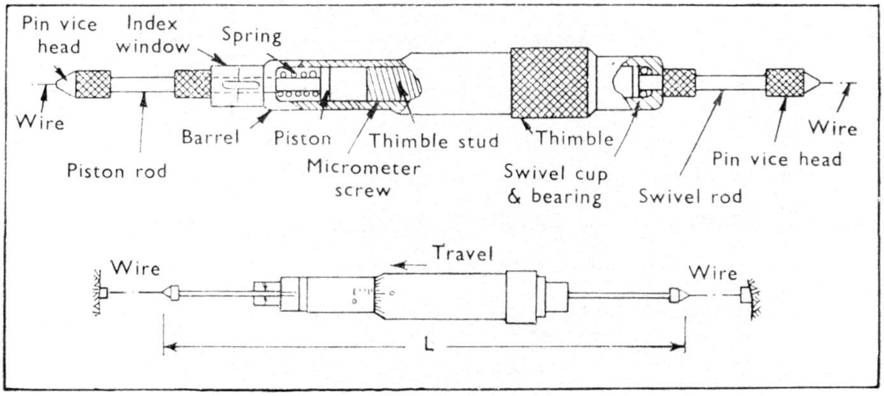

FIG. 20. Micrometer turnbuckle.[12]

The free ends of two wires, the other ends of which are rigidly fixed to plugs in the rock, are attached to the pin-vice ends of the micrometer turnbuckle. When the thimble is advanced, length L shortens and the tension in the wire increases. The increase in tension causes piston rod to move and at one particular tension the mark on piston rod will be in line with the reference mark on barrel. The reading at this position is then recorded. Should the anchor plugs subsequently move nearer or farther from one another, the same movement will have to be met by advancing or retarding the thimble to obtain the required coincidence of marks. The difference of two micrometer readings will then give the extent of the movement.

Roof Sag Measurements

To measure the sag of the immediate roof, steel pins may be set in roof and floor and the distances between pins measured at intervals. In order to measure the differential sag between various roof strata pins may be anchored at various distances into the roof and with their tops protruding from holes. Differential movement between pins then indicates separation of roof strata. An installation of this kind is shown in Fig. 21.[20]

Continuous convergence recorders are sometimes employed in coal mining operations when it is desired to study the behavior of the roof behind an advancing face. Such a recorder consists of two telescopic tubes with a strong spring between them. This allows several inches of movement as the roof and floor converge.[21] Such movement is recorded on a metallized paper chart mounted on a drum which is

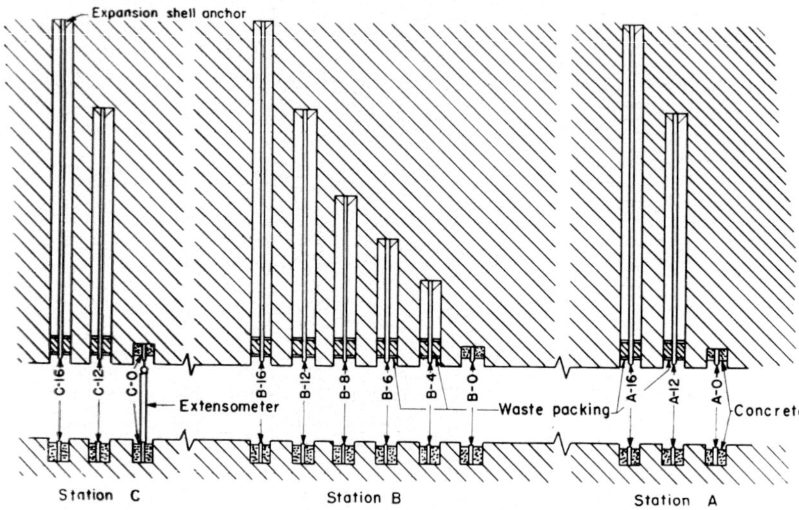

FIG. 21. Section showing installation of sag-measurement pins.[20]

rotated by a clockwork mechanism. Rotation is timed for once in 24 hr or once in 7 days, depending upon the amount and rate of convergence being recorded. Such recorders may be mounted between the roof and floor pins shown in Fig. 21 in order to detect bed separation in roof strata.

Roof Sag Warning Device

Warning devices designed to actuate an alarm when the rate of roof convergence became excessive have been used during pillar recovery operations in the mines of the St. Joseph Lead Co. in Southeast Missouri.[39]

These devices consisted of two spring-loaded telescoping tubes, a micro-switch, and a flashlight. The tubes were adjustable to various stope heights, and the micro-switch could be adjusted so that a convergence of only a few thousandths of an inch was enough to turn on the light. Several of these warning lights were placed around the stopes during noisy activities, such as drilling or loading ore.

Observations in Bore Holes

The stratascope is an optical instrument developed by the Bureau of Mines to investigate $2\frac{1}{4}$-in. diameter boreholes up to a depth of 20 ft. A schematic drawing of the optical and mechanical assembly is shown in Fig. 22.

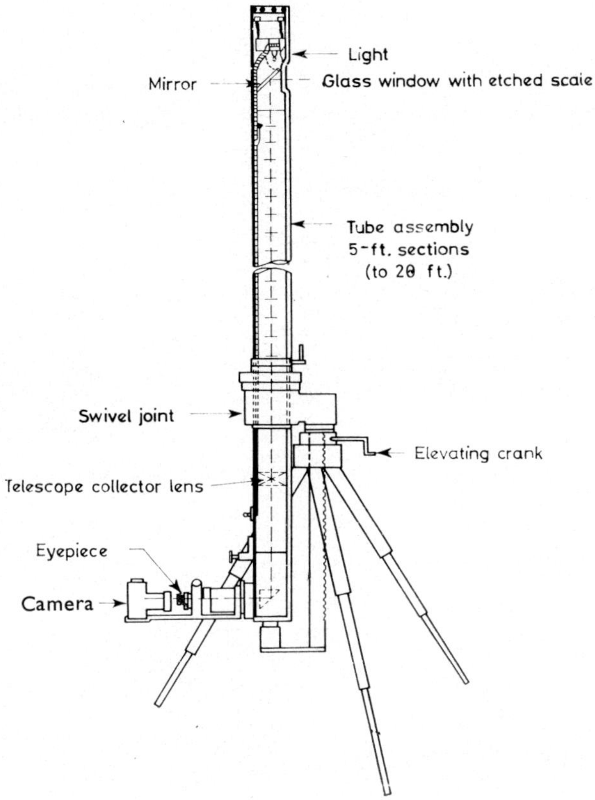

FIG. 22. Schematic drawing of the stratascope.[20]

The instrument is designed to be used in making visual and photographic observations of roof strata to locate separations and fractures and, by periodic re-examination, to note any new defects or changes.[20]

MEASURING PRESSURES ON MINE FILLING MATERIAL

Load Cells

Load cells have been developed for the purpose of measuring the pressure exerted on mine-filling material by a caving and/or subsiding roof behind an advancing long-wall face.

Such cells have been placed in the "packwalls" (stone packs) which are extensively employed in longwall mining operations to partially support the roof and to prevent undue convergence of roof adjacent to access way.

A measuring device installed in a stressed medium usually alters the stress distribution in its immediate vicinity. The difference between the true and the recorded pressure is related to the compressibility of the measuring device related to that of the surrounding material. As a result, even though the true pressure be the same in both instances, the pressure recorded by an instrument placed on the floor of a pack will be different from that recorded by one buried in the floor, since in one case the device is surrounded by broken rock which is readily compressible and in the other case it is surrounded by the intact roof of the floor which is compressible with difficulty.

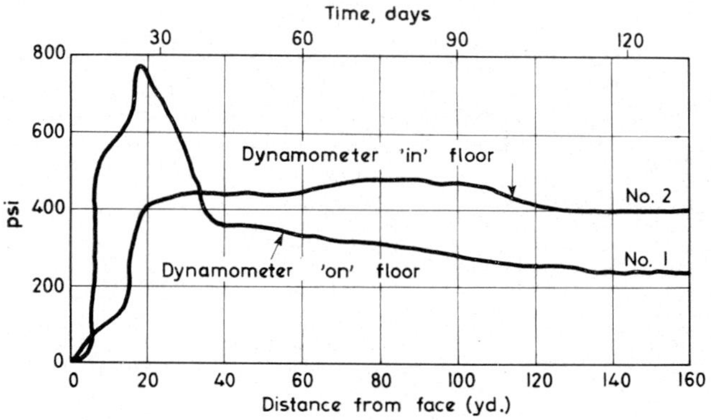

FIG. 23. Pressures recorded by pack dynamometers.[22]

Model experiments have demonstrated that load cells (pack dynamometers) set on pack floors give readings with a peak which is much higher than the actual pressures because the load cells are less compressible than the surrounding pack rock and therefore support more of the load.[22] Subsequent to the model experiments two full size hydraulic pressure cells were installed in a pack 24 in. high in a seam located at a depth of 300 ft. One of these was set on the floor and the rock pack built around it, and the other was set into the floor with its top flush with the floor and the rock pack was built over it.

Figure 23 shows the pressure readings of the two load cells which were installed only 10 ft apart. The experiments indicated that the errors recorded by a load cell in the floor should be less than about 20 per cent.

Figure 24 shows the details of the hydraulic load cell which was used in making these measurements. Oil was used as the hydraulic medium and a small diameter tubing was led from the load cell to a pressure gage which read directly the load on the cell.

Figure 25 shows another type of load cell which is frequently used under similar conditions. This load cell consists essentially of a hollow cylinder which is machined from solid drawn carbon–chromo–molybdenum steel, a high tension steel of 50–60 tons in^2 strength. The top and base are covered by plates of suitable size and the contact surface between cylinder and plates is machined to within 0.001 in. tolerance. In order to insure a perfect contact, and to avoid concentrated

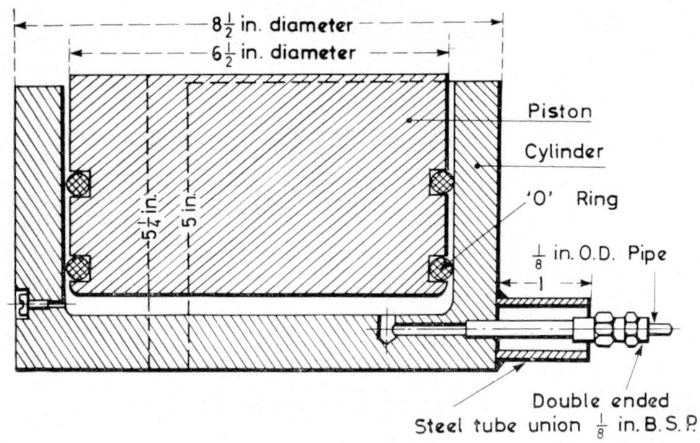

FIG. 24. Constructional details of M.R.E. hydraulic pack dynamometer.[22]

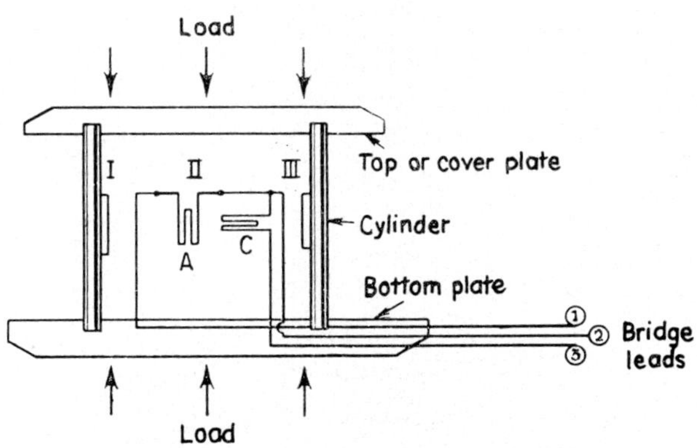

FIG. 25. A typical load cell employing electrical resistance strain gages.[12]

or point loadings on the cylinder, the respective surfaces are rubbed together after a thin layer of liquid carborundum suspension has been placed in between. Three pairs of resistance strain gages, spaced at 120° intervals are attached to the inner surface of the cylinder. Each pair is comprised of an active and a compensating gage, the former having its wire elements parallel to and the latter normal to the direction of the load.

The top cover has provision for countersunk bolts anchoring in the base plate, in order to keep the assembly together. The base plate has provision made for the necessary electrical connections. Mild steel cable glands are provided at the base and the whole is waterproofed with suitable compounds.

Although their use to date appears to have been confined to measuring pressures on pack walls and similar filling in coal mining operations, such load cells could be used for measuring pressures on gob filling in stopes to locate the "pressure abutments".

MEASURING GROUND PRESSURES ON SUPPORTING STRUCTURES

Prop Load Cells (Prop Dynamometers)

Yielding steel friction props and yielding steel hydraulic props are extensively used for roof support in coal mining operations. As an aid to analyzing roof action in a mine it is desirable to measure the loads to which these props are subjected.

When hydraulic props are used, the props themselves serve as load measuring devices. The pressure on the oil in the hydraulic cylinder gives a direct measurement of the load on the prop.

When rigid steel props, or yielding steel friction props are used, the measurement of load on a prop is more difficult. One method for determining the load on a rigid prop uses the body of the prop itself as the compressible element. The steel tube is compressed vertically by an amount proportionate to the load on it, then the amount of such compression is measured by a measuring rod equipped with a micrometer.

Another device consists of a short hydraulic cylinder or "hydraulic jack" which is set either under the foot of the rigid prop or on top of it. The pressure on the hydraulic fluid is a measure of the load.

Still another device utilizes electrical resistance strain gages[(23)]. The load cells are in the shape of cylindrical boxes, made of a short section of steel pipe with steel plates fitted top and bottom. Electrical resistance strain gages are mounted inside this box with wires leading to centrally located recording device, or to a selector switch. This device is quite similar in principle to the load cell which is shown in Fig. 25, except that when used with steel props the body of the load cell may be incorporated as an integral part of the prop, or it may be set between the prop head and the roof bar. Steel props are frequently required to carry loads up to 40 or 50 tons. Load cells may be constructed to obtain maximum sensitivity at the anticipated maximum loads by selecting pipe elements with suitable wall thicknesses.

PRESSURE CELLS FOR MEASURING SOIL PRESSURES

A number of types of pressure cells have been developed for measuring pressures exerted by soils such as clay, sand, and gravel. Such devices can also be used for measuring the pressures of soft rocks or of broken and fragmented rocks on supporting structures.

The characteristic most desirable in this type of instrument is that it should not disturb the actual forces by its presence; in other words, the gage should be as stiff as the supporting structure. The stiffer the soil pressing on the support, the more important it is to meet this condition.

A pressure cell constructed of metal or of some similar rigid material and necessarily constructed to obey Hooke's law in its compressibility can be expected to alter the stress distribution in the soil, producing a concentration of stress in its vicinity in the same manner that a large stone will concentrate stresses in a sand or a plastic soil mass[24]. Theoretically, a very thin compressible (but not flexible) plate will distort the stress pattern in the soil very little; practically, this suggests a cylindrical cell, thin in proportion to its diameter. The probable existence of local stress variations in the soil, due to lack of homogeneity, indicates the desirability for a large pressure responsive area in a pressure cell.

The concentration of stress by a pressure cell would be expected to depend greatly on its compressibility. If cell modulus is stiffer than that of the soil, indicated pressures will be higher than true pressures. Conversely, if the cell modulus of elasticity is less than that of the soil arching action may withhold an appreciable portion of the normal soil stress from the cell.

Most types of earth pressure cells utilize a compressible hydraulic chamber filled with liquid. The cell is fastened to the supporting structure with one side in contact with the earth. Pressure in the cell liquid is measured by various means, such as elastic-wire (vibrating wire) strain meter, by electrical-resistance strain gages, or simply by a tube-connected manometer.

Waterways Experiment Station Earth Pressure Cell

This earth pressure cell, nominally 6 in. in diameter and 1 in. in overall thickness, consists of a circular face plate welded at its perimeter to a thicker base plate and provided with a peripheral slot which forms a flexural joint between the two plates. The thin cavity between the face plate and the base plate is liquid filled, so that pressure on the face plate is averaged and transmitted to an internal diaphragm formed by boring the base plate from the rear to provide a diaphragm section which will respond to loading by increased radial and tangential strains. Electrical strain gages affixed to the rear of this diaphragm undergo resistance changes proportional to the applied pressure, as a result of strains induced in the diaphragm.

W.E.S. earth pressure cells have been constructed with full load capacities ranging from 25 to 500 psi, and there is no foreseeable difficulty in supplying cells with even higher capacity should the occasion arise. Overall accuracy of 0.5 per cent of full load pressure is obtained in the W.E.S. earth pressure cell.

Figure 26 shows a cut-away view of a typical W.E.S. earth pressure cell. This cell was one of the types used in measuring pressures on tunnel support and tunnel lining, at the Garrison Dam Tunnels[25]. Where used for such measurements, the results from many such cells were found to be not dependable because they were subjected to eccentric loading. Such cells must therefore be used with caution in connection with measurements of earth pressures where tangential components may be present.

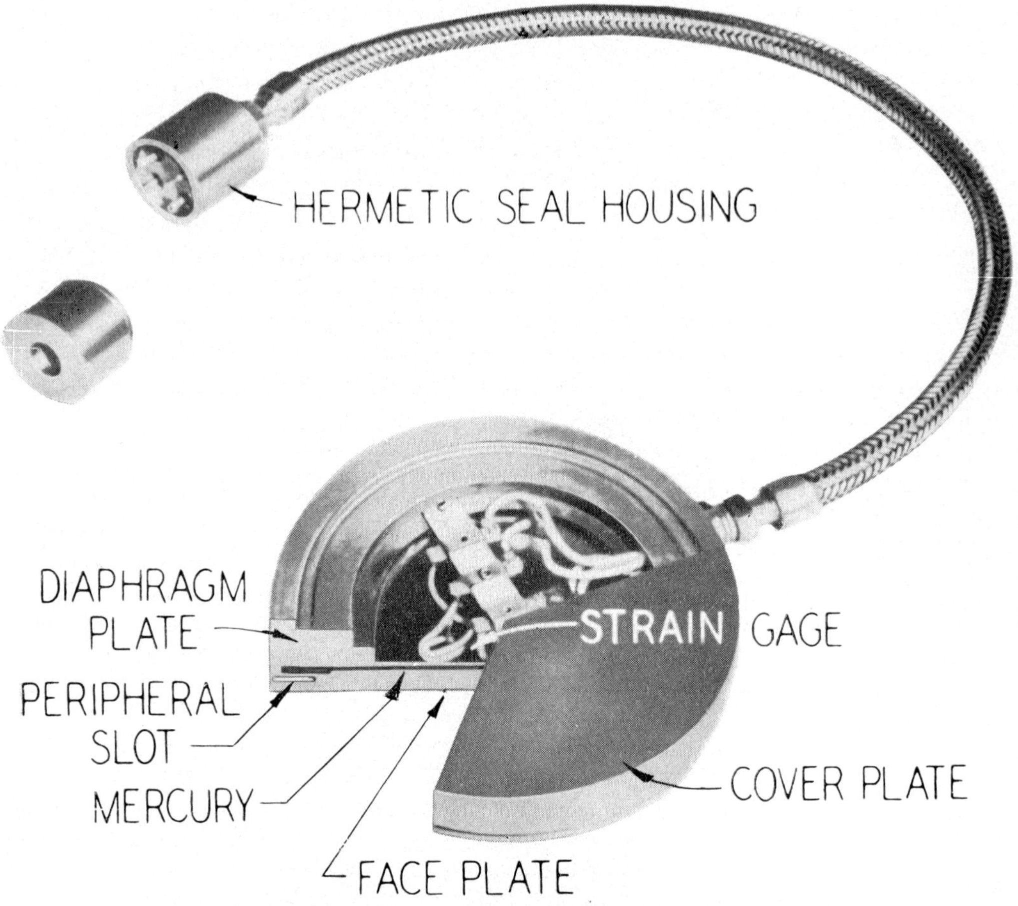

FIG. 26. Waterways Experiment Station earth-pressure cell.[24] (U.S. Army Photo.)

Carlson load cells were also used for these measurements and were found to be more dependable under conditions of eccentric loading.

Two other types of earth pressure cells were especially designed to measure clay pressures on a tunnel lining which was constructed of pre-cast concrete segments in a London water supply tunnel[26]. These were the Hydraulic Radial Pressure Gage and the Vibrating-Wire Radial Pressure Gage.

Hydraulic Radial Pressure Gage

The hydraulic radial pressure gage consisted of a cylinder in which was mounted a piston with its face flush with the outer surface of the concrete lining. The other end of the piston acted on a brass bellows from which copper tubing led to pressure recording gages. Some difficulty was experienced with leakage in the lines in connection with these gages.

Vibrating-wire Radial Pressure Gage

The load sensitive element of this gage is a short mild-steel cylindrical tube which carries the load axially. A thin wire is stretched along the axis of the cylinder and is anchored to circular plates fixed near each end of the cylinder. The wire is both plucked, and its free vibration is picked up, by the same small electro magnet.

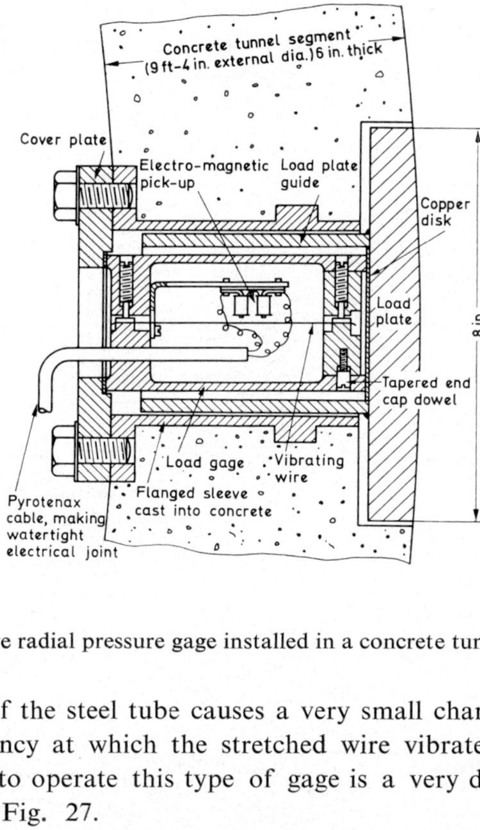

FIG. 27. Vibrating-wire radial pressure gage installed in a concrete tunnel lining segment.[26]

Axial compression of the steel tube causes a very small change in its length and hence in the frequency at which the stretched wire vibrates. The minimum of movement required to operate this type of gage is a very desirable feature. The gage is illustrated in Fig. 27.

The wire vibrations are amplified and fed on to a cathode-ray oscilloscope where they are balanced visually against the frequency of a reference gage which consists of a wire maintained in vibration. The frequency of the reference wire can be varied by spring loading from a screw micrometer. The load gage is calibrated directly in a testing machine against readings of the screw micrometer.

The gages were mounted with the face of the load plate flush with the exterior surface of the concrete lining segment. When properly waterproofed these gages operated successfully under a head of as much as 150 ft of water.

MEASUREMENT OF STRESSES IN SUPPORTING STRUCTURES

Stresses in steel supports and in concrete lining may be determined by measuring the deformations (strains) produced by the supported ground loads. Since stress is proportional to strain within the elastic limit then: Unit Stress = Unit Strain × × Modulus of Elasticity

or
$$\sigma = \delta E$$

The modulus of elasticity of steel is about 30×10^6 psi. It is apparent that unit strains will be very small, even when unit stresses are high. Therefore to be of any value strain measurements must be made with great accuracy.

The modulus of elasticity of concrete varies. A good quality concrete will have an E_c in the range from about 2×10^6 to about 4×10^6 psi. The actual value may be determined for each mix by laboratory tests on specimen cylinders cored from the monolithic concrete. Elastic strains will be very small under the stresses to which concrete is normally subjected.

Portable Strain Gages

Figures 28 and 29 show types of strain gages which may be used to measure strains along a number of gage lines. These are called "portable" strain gages. The Berry gages are available in 2-in. and 8-in. gage lengths.

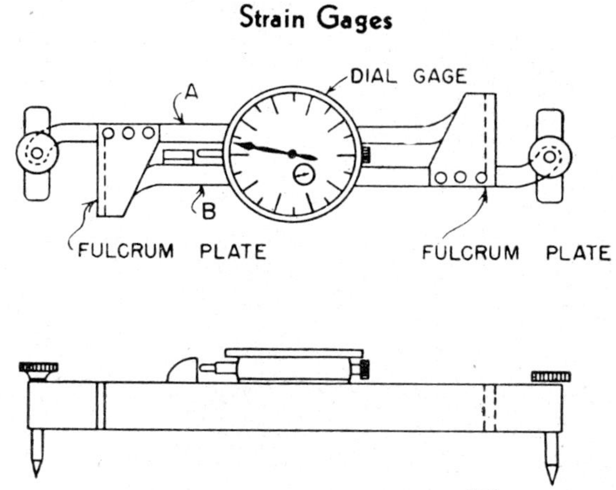

FIG. 28. Whittemore strain gage.[32]

With a 10-in. gage length a strain measurement reliable to 0.0000005 in./in. may be obtained with the Whittemore gage, and similar accuracy can probably be obtained with the Berry gage.[32]

Readings are sometimes made on an auxiliary unstressed bar called a "standard bar" to provide for temperature corrections.

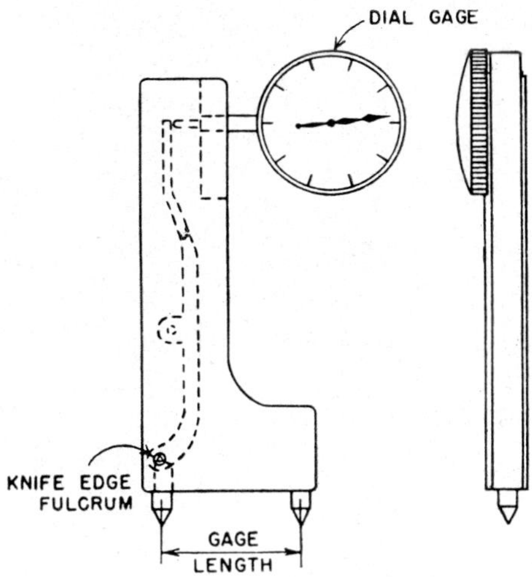

FIG. 29. Berry strain gage.[32]

Stress Measurements in Steel Ribs

In the Niagara Power tunnels deformations of steel ribs were measured in order to determine the stresses in the said ribs. Two striding type deformeters (which measure deformation by mechanical means) were used for determining rib deformation[27]. These 10-in. and 30-in. gages enabled the stresses in the ribs to be determined within accuracies of 300 and 100 psi respectively. Specimen ribs to be tested were prepared for measurement before erection, while still free from load. Holes $\frac{1}{16}$-in. diameter were drilled on both flanges to coincide with nominal gage lengths. An initial set of measurements was made of the exact distance between each pair of holes; the holes were then filled with silicone grease to prevent rusting and were covered with tape to exclude dust.

The first load measurements were made immediately after installation to determine the stresses jacked into the ribs by the timber blocking. Periodic measurements to detect changes in loading were continued until the ribs were covered by the concrete lining.

Results of initial stress measurements indicated that wide variations were prevalent. The stresses induced by the installation procedure varied from 2000 to 12,000 psi, with the average stress being about 5000 psi.

At the Garrison Dam Tunnels[25] strains in steel ribs were measured by means of Whittemore strain gages. These mechanical-type gages were used in the 2-in. and the 10-in. sizes.

In addition SR-4 (electrical resistance strain gages) were installed on some ribs and on some reinforcing steel. These gages were installed on the steel prior to the time that it was embedded in concrete with the intention of measuring stresses after concrete was poured. However, only seven out of sixteen of these gages remained operative after concreting was completed. None of these gages survived the entire observation period and those gages which did operate for part of the time showed extremely inconsistent results. The need for improvement in installation techniques was indicated.

Stress Measurements in Reinforcing Steel

At the Garrison Dam Tunnels[25] measurements of stress were made on several of the transverse reinforcing steel bars in the inner ring of reinforcing steel at the inverts of five tunnels. The concrete was chipped from around the bars and two 10-in. Whittemore strain gage lines were installed on each bar. Where gage lines were installed on stressed bars the true zero stress condition was determined by cutting the bars upon completion of observations.

Measurement of Diametral Deformation of Concrete Lining

The diametral deformation is the simplest strain measurement which can be made in a tunnel and is one which gives a great deal of information on the directions and magnitudes of ground pressures.

For such measurements in the Garrison Dam Tunnels (I.D. was about 26 ft) an extensometer tape was used[25]. This was essentially a Lo-Var nickel-steel tape to which was attached a dial gage reading to 0.001 in. with an extension range of two inches. The tape was punched with holes at intervals of 2.000 in. This allowed the tape to be secured to a frame or head containing the dial gage with a pin and clamp arrangement in precise increments of 2 in. Hooks were provided at both ends of the instrument and corresponding hooks were installed at the ends of the diameters, or chords to be measured. Constant tension was applied to the tape by means of a calibrated adjustable spring device contained in the measuring head. Both diameter and chord measurements were made in order to determine "twist" of the tunnel support as well as diametral deformation.

In the Niagara Power Tunnels (excavated diameter 50 ft) measuring points were set in the rock tunnel ribs and diametral deformations measured with a similar extensometer tape[27]. After the concrete lining had been placed, measuring points were set in the concrete and diametral deformations were measured during the time that pressure grouting was in progress.

For smaller tunnels a "micrometer turnbuckle" such as that shown in Fig. 20 may be used.

Measurement of Surface Strain of Concrete Lining

In the Niagara Power Tunnels[27] surface strain measurements at the inner surface of the concrete lining were commenced as soon as the forms were removed. Stainless steel pins were set in the surface of concrete lining at 30-in. intervals around the circumference and a stridingtype deformeter was used to measure the strain between these points to an accuracy of about 10 micro-in./in.

Measurement of Loads on Rock Bolts

Roof Bolt Compression Pad

One method for determining the amount of tension in a roof bolt is the Roof Bolt Compression Pad[28]. This consists of a layer of tough rubber between two steel plates and with a spring-loaded indicating caliper around the periphery of the rubber. The instrument is placed between the bolt nut and the roof and as the bolt is tensioned the rubber is extruded to expand the caliper. Readings of the caliper scale are translated to tension values by a calibration chart.

Roof Bolt Load Cell (Dynamometer)

The roof bolt load cell has been developed in Great Britain. This consists essentially of a short, hollow steel cylinder with electrical resistance strain gages affixed to its walls. The load cell is slipped over the end of the bolt so that it is in place between the roof plate and the tensioning nut. Bolt tension produces a compressive stress in the annular cylinder which is designed for a maximum load of about 18–20 tons, with a 50 per cent overload factor[13, 28]. Readings are taken by means of a suitable electrical bridge circuit. The principle of operation of this type of load cell is very similar to that shown in Fig. 25.

BIBLIOGRAPHY

1. U. S. Bureau of Reclamation, 1939, Boulder Canyon Project Final Report, Part V (Technical Investigation), Bulletin 4, pp. 265–268.
2. Olsen, O. J., Measurement of residual stress by the strain relief method, *Second Annual Symposium on Rock Mechanics — Behavior of Materials in the Earth's Crust. Quart. Color. School Mines*, Vol. 52, No. 3, July, 1957, p. 186.
3. Terzaghi, Karl and Richart, F. E., Jr., Stresses in rock about cavities, *Geotechnique*, Vol. 3, June, 1952, pp. 57–90. (Same article in Harvard Soil Mechanics Series, No. 41.)
4. St. Q. Isaacson, E. De., *Rock Pressure in Mines*, Mining Publications, Ltd., London, 1958, pp. 171–173.
5. Tattersall, F., Wakeling, T. R. M., and Ward, W. H., Investigations into the design of pressure tunnels in London clay, *Proc. Inst. Civil Engrs., London* (Excerpt) Part I, July, 1955, p. 411.

6. HABIB, M. P., Determination du module d'élascicité des roches en place, *Ann. Inst. Tech. Bâtiment Trav. Publ.*, Sols et Fondations No. 3, September 1950.
7. LAWTON, F. L. and SUTHERLAND, J. G., Kemano pressure conduit engineering investigations, *Proc. Am. Soc. Civil Engrs. J. Power Div.* Vol, 83, No. Po 5, October, 1957, Paper No. 1396.
8. HABIB, P., MARCHAND, R., and TINCELIN, M. E., Measures des pressions de terrains par l'essai de verin plat, *Ann. Inst. Tech. Bâtiment Trav. Publ.*, Serie Sols et Fondations, October 1952.
9. TINCELIN, M. and SINOU, P., Exemple d'application es measures de pressions de terrain enterprises dan les mines de fer de Lorraine, *Rev. Ind. Minerale*, September 1957, pp. 818–844.
10. MOYE, D. G., Rock mechanics in the investigation and construction of T. 1 underground power station, Snowy Mountains, Australia. Paper presented at the Symposium sponsored by the ASCE–GSA Joint Committee on Engineering Geology, St. Louis, Missouri, November 6–8, 1958.
11. HAST, N., Bergtrycksmatningar i gruvor, *Jernkontorets Ann.*, Vol. 141, 1957, Nr. 10, pp. 633–752.
12. GUPTA, K. P., How to measure rock pressures, *Eng. Mining J.*, October 1958, pp. 95–100.
13. POTTS, E. L. J., Underground instrumentation. *Second Annual Symposium on Rock Mechanics — Behaviour of Materials in the Earth's Crust Quart. Color. School Mines*, Vol. 52, No. 3, July, 1957, pp. 135–182.
14. BUCHANAN, J. G., Rock pressure studies in the mines of SPRINGHILL, N. S.; A progress report. Part II. Description of the load cell and analysis of readings, *Trans. Can. Inst. Mining Met.*, Vol. 59, 1956, pp. 242–251.
15. POTTS, E. L. J. and JENKINS, J. D., A study of rock pressures around mining excavations, *International Strata Control Congress, Essen*, October, 1956. Published by *Steinkohlenbergbauverein*, Essen.
16. EVERLING, G., Calculation of stress from measurements made in boreholes in the seam and in the rock, *International Strata Control Congress, Essen*, October, 1956. Published by *Steinkohlenbergbauverein*, Essen.
17. OBERT, L., Measurement of pressures on rock pillars in underground mines. Part II, *U. S. Bur. Mines R. I.* 3444, April, 1939.
18. OBERT, L., Measurement of pressures on rock pillars in underground mines. Part II, *U. S. Bur. Mines R. I.* 3521, June, 1940.
19. DAWANCE, G. and TINCELIN, E., Application de l'ausculatation dynamique a l'étude de development et de la repartition des contraintes autour des galeries, *Ann. Inst. Tech. Bâtiment Trav. Publ.*, Serie: Sols et Fondations (18), March–April 1955.
20. MERRILL, R. H., Design of underground mine openings — Oil-Shale Mine, Rifle, Colorado, *U. S. Bur. Mines R. I.* 5089, December, 1954.
21. ADCOCK, W. J., Measurements of the behaviour of strata and supports in longwall workings, *International Strata Control Congress, Essen*, October, 1956. Papers published by Steinkohlenbergbauverein, Essen.
22. PRICE, N. J. and MAY, N., The use of pack dynamometers, *Colliery Eng.*, September 1960, pp. 379–382; October 1960, pp. 438–442.
23. BRISON, L. and JACQUEMIN, R., Contribution to the study of stresses in roof support at the coal face, *International Conference on Rock Pressures and Support in the Workings, Liège*, 1951, p. 186.
24. Pressure cells for field use, Bulletin No. 40, Waterways Experiment Station, Corps of Engineers U. S. Army, Vicksburg, Mississippi, January 1955.
25. BURKE, H. H., Garrison dam-tunnel test section investigation — a Symposium, *Proc. Am. Soc. Civil Engrs. J. Soil. Mech. Found. Div.*, Vol. 83, No. SM4, November 1957 (Paper 1438).
26. TATTERSALL, F., WAKELING, T. R. M., and WARD, W. H., Investigations into the design of pressure tunnels in London clay, *Proc. Inst. Civil Engrs., London*, July, 1955, pp. 400–471.
27. GIBSON, W. L., Some experiences in investigating stresses and forces in massive structures, *Trans. Can. Inst. Mining Met.* Vol. 59, 1956, pp. 141–153.
28. WRIGHT, A., Investigation into roof bolting technique in Great Britain, *International Strata Control Congress, Essen*, October, 1956.
29. Photoelastic coating technique applied to research in rock mechanics, *Trans. Inst. Mining*

Met. Vol. 71, Part 10, 1961–1962. Also *Bull. Inst. Mining Met.*, July, 1962, No. 668, pp. 581–617. Part 1. Laboratory investigations, A. Roberts, C. L. Emery, and P. K. Chakravarty. Part 2. Field investigations, A. Roberts, C. L. Emery, I. Hawkes, P. K. Chakravarty, and F. T. Williams.

30. Obert, L., Merrill, R. H., and Morgan, T. A., Borehole deformation gage for determining the stress in mine rock, *U. S. Bur. Mines R. I.* 5978, 1962.
31. Merrill, R. H. and Peterson, J. R., Deformation of a borehole in rock, *U. S. Bur. Mines R. I.* 5881, 1961.
32. Marin, Joseph, *Engineering Materials, Their Mechanics, Properties and Applications*, Prentice-Hall, 1952.
33. Terzaghi, Karl, measurement of stresses in rock, *Geotechnique*, June, 1962, No. 2, pp. 105–124.
34. Carlson and Pintz, Design of stressmeters, *J. Am. Concrete Inst.*, November, 1952.
35. Potts, E. L. J., Stress distribution, rock pressures, and support loads, *Colliery Eng.*, August, 1954.
36. Obert, L. and Duvall, W., The microseismic method of predicting rock failure in underground mining. Part II. Laboratory experiments, *U. S. Bur. Mines R. I.* 3803, March, 1945.
37. Obert, L. and Duvall, W., Use of subaudible noises for the prediction of rock bursts. Part II, *U. S. Bur. Mines R. I.* 3654, July, 1942.
38. Report No. 1, Rockburst Research Unit, Kolar Gold Mining Undertaking, Government of Mysore.
39. Reed, J. J., Case history in pillar recovery, *Mining Eng.*, July 1959, pp. 701–705.

CHAPTER 11

PHOTOELASTIC METHODS FOR STRAIN AND STRESS MEASUREMENTS

RECENT developments in the application of photoelastic devices to the field measurements of rock strains, strata pressures, and support loads indicate that these methods have a very promising future in the field of applied rock mechanics.

Because of the exceptional promise of these methods two progress reports which have recently appeared in the literature are reprinted in this chapter.

Quoting the authors of the reports:

> The simplicity of the photoelastic transducers and the fact they are read by visual color comparison, compensation, a straightforward count of fringes or observation of the shape of the fringe pattern, any of which can be done by the ordinary mine personnel, create interest and invite the co-operation of the miners. The relative cheapness of the transducers, compared with other equipment required for a similar purpose, means that many more points of measurement can be established and a significant result achieved. The strata control measurement system may then be incorporated into the daily routine and assimilated into the mining environment.

This chapter consists of three parts as follows:

PART 1. Photoelastic coating technique applied to research in rock mechanics. Part. 1. Laboratory investigations; A. Roberts, C. L. Emery, and P. K. Chakravarty. *Trans. Inst. Mining Met.*, Vol. 71, 1961–62, pp. 581–601.

PART 2. Photoelastic coating technique applied to research in rock mechanics; Part 2. Field investigations; A. Roberts, C. L. Emery, I. Hawkes, P. K. Chakravarty, F. T. Williams, *Trans. Inst. Mining Met.*, Vol. 71, 1961–62, pp. 605–617.

PART 3. The application of photoelastic devices for measuring strata pressures and support loads. A. Roberts, I. Hawkes, *Mine & Quarry Eng.* July, 1963, pp. 298–308.

These papers are reprinted herein by kind permisson of the authors and the publishers.

PART 1. LABORATORY INVESTIGATIONS

by A. Roberts, C. L. Emery, and P. K. Chakraverty

SYNOPSIS

The development of the photoelastic coating technique is outlined, in which a layer of birefringent material is bonded on to the structure the strain on which is to be examined. Details are given of the theory and instrumentation by means of which the technique is applied to the examination of surface strains on rocks.

Part 1 is an account of the application of the technique to various laboratory investigations on the stress-strain phenomena demonstrated by rocks under load, including the observation of intergranular strains and elastic rebound in rocks, the observation and determination of strain relaxation phenomena, shear strain-time phenomena and the effect of inherent strains and granular packing patterns during further loading.

Research on the use of the technique continues, to explore its possibilities and limitations, but sufficient experience has been gained to show that it promises to be a most powerful tool for use in rock mechanics.

Part 2 describes the application of photoelastic transducers to strata control instrumentation. Such devices, being relatively simple compared with other equipment used for similar purposes, can be applied by non-specialist personnel.

The organization of a rock mechanics investigation group is outlined, in which field investigations, supported by systematic laboratory investigations to establish the stress-strain characteristics of the rocks concerned, can provide data for a rational approach to problems in strata control and mine design.

Examples of the application of the technique are included and the results discussed.

INTRODUCTION

Photoelastic methods for the study of strains in engineering components and structures are well known. They depend on the property of birefringence which certain materials display when under strain. When viewed under polarized light with an analyser the resulting interference pattern can be interpreted in terms of strain in the birefringent material.

The use of photoelastic models to predict the stress distribution around mine openings is not new. Phillips used it in England in 1930,[1] it was applied by Duvall in the U.S.A. in 1948,[2] followed by other investigators in many parts of the world, notably Potts in England,[3] and Hoek in South Africa.[4] In general, the optical patterns observed check well with mathematical theory based on the supposition of a continuous elastic homogeneous body. Unfortunately, rocks possess characteristics that are quite different from those of the birefringent materials used in the models, so that the latter can be used only to determine the influence of gross geometry. In problems of mine design, geometry is obviously an important factor, however, and to this extent photoelastic model studies are important and useful.

The photoelastic coating technique, a more recent development, uses a coating of birefringent material bonded on to the structure, the strain in which is to be analysed. The method was suggested by Mesnager in 1930,[5] but owing to lack of suitable materials and bonding cements, successful application of the technique had perforce to await Zandman in 1953[6] and d'Agostino *et al.* in 1955.[7] This work was qualitative, the first quantitative analysis being achieved by Zandman in 1956.[8] Since that time, progress has been rapid, and the technique is proving to be useful in many fields.[9]

Pioneering work on the application of the technique to the examination of stress–strain phenomena in rocks was performed by C. L. Emery in Canada and in England, and has been reported on both sides of the Atlantic.[10, 11] The technique has also been applied by Dantu to record changes of strain under load on concrete specimens.[12] Consequent upon this work, researchers at the Postgraduate School in Mining at the University of Sheffield are exploring the possibilities and limitations of the technique as applied to rocks and have further developed it, both in the laboratory and in underground investigations.

In essence, the technique consists of attaching a coating of some material that is optically birefringent to the rock the surface strain on which is to be observed. If the bond between coating and rock is satisfactory any surface strain in the rock is transferred to the coating and its effects can be observed through a reflection polariscope.

The technique is still in its infancy, but sufficient experience has been obtained to justify the preparation of a progress report. Part 1 deals with the principles of the technique and the instruments by means of which it may be used. Some applications of the technique to laboratory studies in rock mechanics are outlined. Part 2 of the report gives some details of underground investigations performed with the aid of the technique in mines in England and Canada.

DESCRIPTION OF THE TECHNIQUE

Reflection Polariscope

The path of light through a plane reflection polariscope is shown in Fig. 1. Light from the source S is plane-polarized at P and passes through the birefringent coating C to the surface of the work-piece W. From this surface it is reflected back through the coating, from which it emerges as two components polarized in the directions of principal strains. One of the emergent waves is retarded, relative to the other, by an amount δ_n where $\delta_n = d(i_1 - i_2)$, d being the thickess of the birefringent coating, and i_1 and i_2 the indices of refraction for the two components. The two emergent waves combine to form what is known as an elliptically polarized beam in which the two components are vibrating in the directions of principal strains, and they are further resolved into the components A_1 and A_2, parallel to the transmission axis of the analyser A. The observer at O sees the interference pattern of light, which, when the light source S is white, includes coloured lines or isochromatics, which are related to the magnitude of the principal strains.

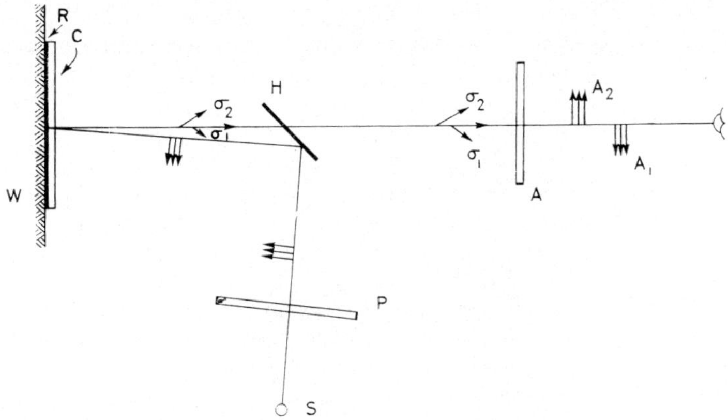

FIG. 1. Principle of action of a plane reflection polariscope.
S, light source. *P*, polarizer. *H*, half-mirror. *C*, birefringent coating. *R*, reflecting surface. *W*, work-piece. *A*, analyzer.

Where the plane of incidence of the polarized light coincides with the direction of one of the principal stresses, the light passes through the coating without interference, and if at the same time the analyser and the polarizer are crossed at right angles, no light will be visible at those points, which then appear as lines of extinction or *isoclinics* in the field of view. At every point on an isoclinic, therefore, the directions of principal stresses coincide with the directions of the crossed polarizer and analyser of the polariscope.

In the plane reflection polariscope, shown diagrammatically in Fig. 1, isochromatics and isoclinics will appear together in the field of view as the analyser is rotated. It is sometimes advantageous to eliminate the isoclinics altogether, so that the isochromatics may be observed more easily. This can be done by inserting two plates, each of which retards the light by one-quarter of a wavelength, one between the birefringent coating and the polarizer and one between the coating and analyser. Light emerging from a combination of polarizer and quarter-wave plates set with their optical axes at 45° to each other is said to be circularly polarized, in that its vibrations are so restricted that they can be represented by a light-vector the extremity of which lies on a circular helix. In such a system isochromatic fringes will appear, irrespective of the orientation of the polarizer-quarter-wave-plate combination, and extinction will not occur.

Materials Used for the Coating

Considerable research has gone into the task of finding suitable materials for the birefringent coating.[12-14] The search for better materials with higher strain-sensitivity factor goes on. As yet two proprietary materials of the epoxy-resin type are most commonly used. Glass is birefringent, but its use as a photoelastic coating is ruled out by reason of its high modulus of elasticity and difficulties in shaping and adhesion. However, Hiramatsu has successfully used glass to measure stresses in concrete, using a reflection polariscope in a manner very similar to that applied in the coating technique, the stresses on the concrete being transferred to glass plugs inserted in shallow bore-holes,[15] and the present authors are also applying this technique in England to the examination of strain in rock.

The Strain-Sensitivity Factor

In a material that exhibits birefringence when stressed, the relative retardation is related to principal stress difference by Newmann's equation.[17] If δ_n is the relative retardation under light incident in a direction normal to the birefringent coating; and since the light passes through the material twice when reflected from the work-piece:

$$\delta_n = (n_1 - n_2)2t = c(\sigma_1 - \sigma_2)2t$$

or $$(\sigma_1 - \sigma_2) = \delta_n/2tc$$

But $$(\sigma_1 - \sigma_2) = (\varepsilon_1 - \varepsilon_2)E/(1+\mu)$$

whence $$(\varepsilon_1 - \varepsilon_o) = \delta_n/2tK \quad (1)$$

where $$K = c\,E/(1+\mu)$$
$$t = \text{thickness of coating}$$
$$\mu = \text{Poisson's ratio}$$

The factor K, which embodies the relevant characteristics of the birefringent material, is termed the strain–sensitivity factor, and when it is intended to apply the technique to field investigations the K factor should be determined periodically for control purposes as near as possible under the conditions in which the coating material is applied in the field. Various methods of doing this are described in a later section.

Application of the Material

The law of direct proportionality between the surface strain and measured strain at a point holds only at low strain gradients. If the strain gradients are high, such that the strain changes greatly over a distance comparable with the coating thickness, then the deformation of the coating will vary appreciably across its thickness. The interference pattern is also affected by changes in curvature of the observed surface under load.

It is desirable in the interests of accuracy to minimize the thickness of coating, but on the other hand such a procedure reduces the sensitivity of the technique. A compromise must therefore be arrived at, and for observing integrated strains it has so far been found that a coating $\frac{1}{8}$ in. thick is generally satisfactory; the plastic is therefore cast into sheets of that thickness. For general application in the field the material is cast or cut into disks of 2 in. diameter and squares of 2 in. side, but for laboratory work it is cut into any size appropriate to the work-piece. To lessen edge effects due to residual strain in the plastic pieces and to obviate annealing the process of cutting must be slowly and carefully performed and heat generation minimized.

If the work-piece has a curved surface, suitably contoured sheet may be prepared. Alternatively, repeated layers of the plastic may be sprayed or painted directly on the work-piece, to polymerize directly on it. The latter procedure creates no complications if strain directions only are to be observed, but the determination of strain magnitude at any point would require also the measurement of coating thickness at that point.

Preparation of the rock surface. The rock surface must be clean, dry and reasonably smooth, the degree of smoothness depending upon the scope of the measurements to be made. If only integrated strains over an area are to be observed the rock surface may not require any treatment, but for the observation of intergranular or crystal-boundary effects the surface would have to be ground and possibly polished also.

Cements. Adhesives that require pressure, heat, or volatilization for setting are not satisfactory. In contrast to electrical or mechanical strain-gauge techniques, where residual strains in the gauge can be compensated to a zero reading, the plastic will show any residual strains and they will be incremental to the measurements subsequently made. The cement must not create residual strains and must, therefore, be a non-shrinking low pressure adhesive. Eastman 910, and epoxy resins of the Araldite type, together with the proprietary materials sold for this technique, have all been used with success, although the beginner must be prepared for initial failures until experience has been gained. Since rock is non-reflective it is necessary either to use a mirrorized plastic or to mix a little reflective medium (aluminium powder) in the cement. The cement should be spread as uniformly as possible, in a thin layer.

The epoxy resins have a variable viscosity, depending on temperature and on the degree of polymerization. When applying the adhesive to rock or concrete its viscosity determines penetration into the pores of the work-piece under the influence of capillary drive. The adhesive may increase the tensile strength of the surface penetrated by it, which is not serious if measurements of overall extension or compression are being made, but can be a source of error when making intergranular measurements. For work of the latter type, therefore, the cement should be applied in as viscous a state as is consistent with the other factors concerned in its application.

INSTRUMENTATION AND MEASUREMENTS

Polariscopes

Large-field Meters

Portable reflection polariscopes capable of analysing the birefringence of comparatively large surface areas in static tests are available. The instruments are employed for quantitative determinations of shear strains.

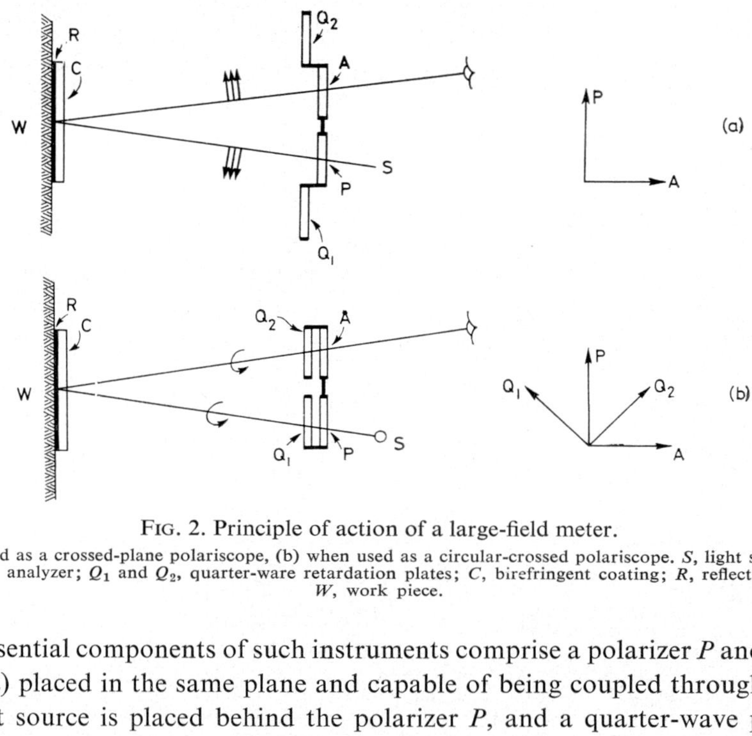

FIG. 2. Principle of action of a large-field meter.

(a) when used as a crossed-plane polariscope, (b) when used as a circular-crossed polariscope. S, light source; P, polarizer; A, analyzer; Q_1 and Q_2, quarter-ware retardation plates; C, birefringent coating; R, reflecting surface, W, work piece.

The essential components of such instruments comprise a polarizer P and analyser A (Fig. 2) placed in the same plane and capable of being coupled through gearing. The light source is placed behind the polarizer P, and a quarter-wave plate Q is placed between polarizer and test object to produce circularly polarized light. The quarter-wave plate Q_2 placed between test-piece and analyser is "crossed" with respect to Q_1. In the absence of any strain in the coating, Q_2 compensates for the relative retardation produced by Q_1, and the resultant plane-polarized light is stopped by the analyser, giving a dark field when the analyser and polarizer axes are crossed at 90°.

If the coating is strained, then under crossed-polarizers a system of fringes will be observed in which $\delta = n\lambda$, where δ is the relative retardation, n the fringe number, and λ the wavelength of the light.

When the polarizing axes are parallel, light passes through the analyser, giving a bright field in the absence of strain. If strain is present, under these conditions the half-order fringes will be visible in which $\delta = (n+1)\lambda/2$.

Figures 2 (a) and (b) show these two arrangements, (b) giving the vector representation of light waves through the crossed circular polariscope. The quarter-wave plates are fixed on hinges so that they can be swung out of the field when required. Using a white light source, two sets of fringes are observed through the analyser, keeping the polarizer and analyser axes crossed and quarter-wave plates swung out of the field. The coloured fringes, or isochromatics, give the values of the difference in principal strains $(\varepsilon_1 - \varepsilon_2)$ and the dark fringes or isoclinics give the direction of principal strains and therefore principal stresses in the test-piece. All isoclinics on the test object may be observed when the polarizer and analyser are coupled together (with their axes crossed) and rotated through 90°.

The instrument is very useful for the measurement of low-order strain, both in the laboratory and in the field. Recording of the isochromatics or isoclinics is done either by photography or by drawing them on the plastic with a chinagraph pencil, to be transferred subsequently on to tracing paper. Integral-order fringes (observed under crossed polariscope) and half-order fringes (observed under parallel polariscope) can be plotted directly on to the plastic. At those other points on the coating that lie between the integral- and half-order fringes the strains may be assessed by colour comparison. Accurate measurement of fractional-order fringes is possible using goniometric compensation by the Tardy method.[16]

Small Hand-viewer

For qualitative work and also for quantitative work where precision is not important, the small hand-viewer is a very simple and useful instrument. It consists of a combined linear polarizer and quarter-wave sheet supported in a circular frame mounted on a handle. It is held between the light source and the work-piece and the work-piece is observed through it. In underground work the miner's lamp can provide the necessary light source.

This simple apparatus, if used with the quarter-wave retardation side towards the observer, acts as a linear parallel polariscope, and those parts of the plastic coating on the work-piece that have their principal strain axes parallel to those of the polariscope will show no interference colour. All other parts will show colours that are functions of the retardation of the polarized light in those regions. The direction of minimum interference colour is the direction of one principal strain in the plastic and hence in the rock surface to which the plastic is attached. The other principal strain direction in the plane of view is normal to the first.

If the hand-viewer is used with the quarter-wave retardation side away from the observer, the instrument becomes a crossed circular polarizer and analyser. The isoclinics are eliminated and an isochromatic colour pattern is seen.

Determination of Strain in Direction. The directions of the principal strains are shown by the isoclinics and can be located by rotating the linear crossed polariscope to the extinction position. At this point directions of the principal strains are the same as the directions of the transmission axes of the polariscope. However, when the small hand-viewer is being used, the principal strain directions are identified as already described, the instrument acting as a linear parallel polariscope. It is customary to use the handle of the viewer as an aid to orientation: a warning is necessary here as not all hand-viewers are so made that the direction of polarization coincides with the direction of the handle. Some hand-viewers are constructed with the direction of polarization at 45° to the line of the handle, and in this case the handle points in one principal strain direction when the hand-viewer is showing maximum interference coloration.

Determination of Stress and Strain in Magnitude. In the following analysis suffix p refers to the birefringent coating and suffix w to the work-piece.

From Newmann's equation

$$(\sigma_1 - \sigma_2) = \delta_n/2tc \tag{2}$$

Since only surface stresses are considered, the third principal stress is zero and

$$\varepsilon_1 = \sigma_1/E - (\sigma_2/E)\mu$$
$$\varepsilon_2 = \sigma_2/E - (\sigma_1/E)\mu$$

Therefore

$$\sigma_1 - \sigma_2 = (\varepsilon_1 - \varepsilon_2)\cdot E/(1+\mu) \tag{3}$$

and

$$(\sigma_1 - \sigma_2)_w = (\varepsilon_1 - \varepsilon_2)_w \cdot E_w/(1+\mu)_w$$
$$(\sigma_1 - \sigma_2)_p = (\varepsilon_1 - \varepsilon_2)_p \cdot E_p/(1+\mu)_p$$

Substituting in equation (1)

$$\delta_n/2tc = (\varepsilon_1 - \varepsilon_2)_p \cdot E_p/(1+\mu)_p \tag{4}$$

Since the coating has exactly the same deformation as the work-piece

$$(\varepsilon_1 - \varepsilon_2)_p = (\varepsilon_1 - \varepsilon_2)_w$$

therefore

$$\begin{aligned}(\sigma_1 - \sigma_2)_w &= (\varepsilon_1 - \varepsilon_2)_p \cdot E_w/(1+\mu)_w \\ &= \delta_n/2tc \cdot E_w/E_p \cdot (1+\mu)_p/(1+\mu)_w \\ &= \delta_n/2t \cdot E_w/K(1+\mu)_w \end{aligned} \tag{5}$$

It is thus possible to determine the difference in the principal stresses in the work-piece if Young's Modulus and Poisson's Ratio for the material of the work-piece are known, together with the strain–sensitivity factor of the coating. Measurement of the birefringence δ_n/t identifies the difference in principal strains (and therefore the relative maximum shear strain) by equation (1), and identifies the relative maximum shear stress by equation (5). The principal stress normal to a free boundary is always zero. Therefore, by measuring the birefringence at normal incidence at a point where a fringe meets a free boundary, the principal stress tangential to the boundary is determined; for, at this point

$$\sigma_1 = \delta/2tK\cdot E/(1+\mu)$$

$$\sigma_2 = 0$$

and

$$\varepsilon_1 = \delta/2tK(1+\mu)$$

$$\varepsilon_2 = -\mu\varepsilon_1$$

Observation and Identification of Colours

The isochromatic colours observed through the crossed circular polarizer and analyzer correspond to a relative retardation

$$\delta = n\lambda$$

It is thus possible to determine the approximate magnitude of the shear strains by the colour of the point concerned. The range of colours representing levels of strain for the particular material and thickness of coating can be recorded at the time of calibration. Thereafter a colour comparator may be constructed in the form of a chart for the operator to carry in his notebook.

Alternatively a colour comparator may be made by stressing a disk of plastic to a central fringe order of three and then freezing in these colours. In this case the natural colour of the plastic is taken into account at all the ordinary fringe levels.

Portable Dynamometer

For the precise correlation of strain levels to colours the authors have constructed a portable dynamometer, by means of which a calibration bar coated with the plastic can be loaded to known levels by the medium of a micrometer. To use this the calibration bar is loaded by turning the micrometer, until the colour seen in the window of the dynamometer corresponds with the colour of the particular portion of the work-piece as seen through the reflection polariscope. The strain is then read on the micrometer.

Compensation

When the strain in the coating is less than one fringe level, or if it is required to determine the stresses at a point between two fringes, a system of compensation may be used by which a "tint of passage" or fringe can be produced at any given point of the field of view in the analyser. In the instruments designed for quantitative work a system of goniometric compensation is employed, the basic principle of which is to add or subtract a known birefringence by rotating the analyser until the tint of passage appears at the point under inspection. The oblique-incidence meter employs a calibrated quartz wedge, in the form of a Babbinet Compensator, for the same purpose.

Oblique-incidence Meter

This instrument makes it possible to measure the individual principal stresses at any point on the test-piece. The instrument, shown diagrammatically in Fig. 3, utilizes Drucker's oblique-incidence method for evaluation of stresses, and its principle of operation is as follows.

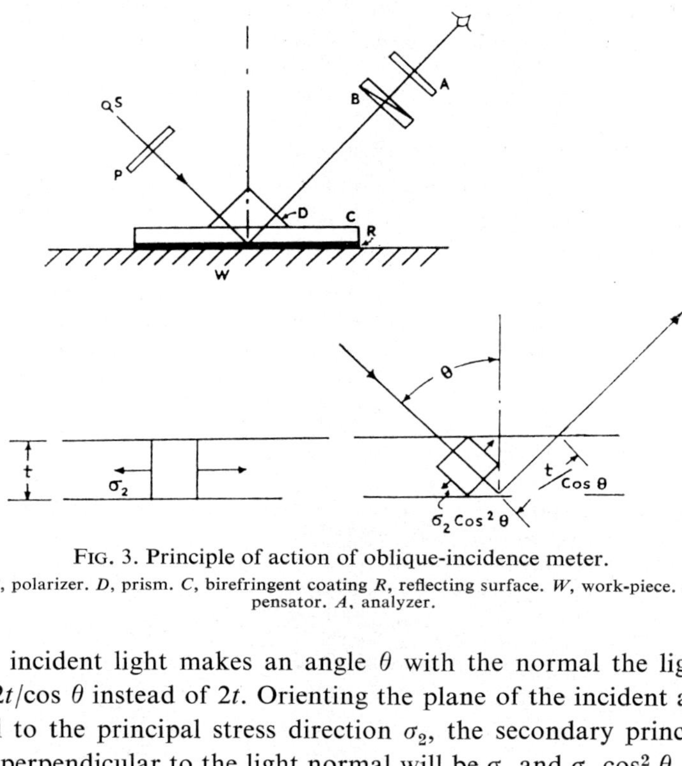

FIG. 3. Principle of action of oblique-incidence meter.
S, light source. *P*, polarizer. *D*, prism. *C*, birefringent coating *R*, reflecting surface. *W*, work-piece. *B*, Babbinet compensator. *A*, analyzer.

When the incident light makes an angle θ with the normal the light travels a distance of $2t/\cos\theta$ instead of $2t$. Orienting the plane of the incident and reflected rays parallel to the principal stress direction σ_2, the secondary principal stresses in the plane perpendicular to the light normal will be σ_1 and $\sigma_2\cos^2\theta$. The relative retardation under this condition will be given by

$$\delta_{0\sigma_2} = c(\sigma_1 - \sigma_2\cos^2\theta)2t/\cos\theta \tag{6}$$

If the plane of the incident and reflected light is oriented parallel to principal stress direction σ_1, the expression for relative retardation becomes

$$\delta_{0\sigma_1} = c(\sigma_2 - \sigma_1\cos^2\theta)2t/\cos\theta \tag{7}$$

Taking $\theta = 45°$ (as in the case of oblique-incidence meter), the values of σ_1 and σ_2 are calculated from equations (6) and (7) to be:

$$\left.\begin{aligned} \sigma_1 &= \frac{\sqrt{2}}{3tc}\left(\delta_{0\sigma_2}+\frac{\delta_{0\sigma_1}}{2}\right) \\ \sigma_2 &= \frac{\sqrt{2}}{3tc}\left(\delta_{0\sigma_1}+\frac{\delta_{0\sigma_2}}{2}\right) \end{aligned}\right\} \quad (8)$$

The principal stresses σ_1 and σ_2 may also be calculated from equations (2) and (6) or equations (2) and (7) by taking one observation under normal incidence and the other under oblique incidence.

Calibration and Limits of Accuracy. The limits of accuracy to which measurements may be made depend on the characteristics of the coating material and the characteristics of the instruments employed. The effect of coating thickness in relation to areas of high strain gradient and to changes in curvature have already been mentioned. If such effects may be neglected the important properties of the photoelastic material are (a) its strain–sensitivity factor K and (b) its stress–strain characteristic.

Determination of Strain-sensitivity Factor, K

There are three methods for determining the strain–sensitivity factor K:

Method 1: A disk of the photoelastic material, of known diameter, is loaded in a diametral direction, with a known load P.

If n_c is the fringe order at the centre of the disk,

$$n_c = 4P/\pi t \cdot F \cdot D.$$

where $F = \lambda/2ct$ and $K = c \cdot E/(1+\mu)$

$$n_c = \frac{4P}{\pi t(\lambda/2ct)D}$$

$$= K(8P/\pi D\lambda)[(1+\mu)/E]$$

Whence $$K = (n_c/P)(\pi D\lambda/8)[E/(1+\mu)]$$

If D, λ, E, and μ are known, K can be found by noting the load P for various fringe orders at the centre of the disk. For example, using a disk of "Stress-X" plastic 1.875 in. in diameter, in sodium light of wavelength 2.27×10^{-5} in. ($E=$ $=4.5\times 10^{-5}$ and $\mu=0.5$):

Load (lb)	Fringe no. (n_c)	K
32.6	1	0.153
65.1	2	0.154
97.6	3	0.154
130.4	4	0.153
162.8	5	0.154

Method 2: The K factor can be measured directly by observing the relative retardation of polarized light traversing the coating under normal incidence, when the coating is attached to a tensile test-piece of a known material.

Then the difference of principal strains

$$(\varepsilon_1 - \varepsilon_2) = P(1+\mu)/A \cdot E$$

where P is the load, and A the cross-sectional area of the metal test-piece; E and μ are the Young's Modulus and Poisson's Ratio of the metal, respectively. But

$$(\varepsilon_1 - \varepsilon_2) = \delta/2tK$$

where δ is the relative retardation and $\delta = n\lambda$

$$K = \delta AE/2tP(1+\mu)$$

Method 3: The coating is bonded on to a tapered cantilever of steel, the shape of which is so made that the ratio of width to loading distance produces a uniform longitudinal strain under the conditions of loading.

Here the stress $= My/I$

where M = moment due to the cantilever load
I = moment of inertia
y = distance to neutral axis

and

$$K = \delta/2t(\varepsilon_1 - \varepsilon_2)$$
$$= \delta/2t \cdot E/\sigma(1+\mu).$$

Method 1 is quick and accurate if E and μ for the coating material are known or can be measured. It is best used as a comparison test against a disk of known K factor. The unknown disk will then have a K factor directly proportional to the fringe order at the disk centre. Method 2 is more time-consuming but lends itself to direct measurement without foreknowledge of E or μ. Method 3 is very useful in that it also facilitates the determination of the stress–strain characteristic of the coating material.

Determination of Stress-strain Characteristic of the Optical Strain Gauge

If in method 3 above, wire resistance strain gauges are mounted on the cantilever to record longitudinal and lateral strains separately, the theoretical stress–strain relationship, calculated from the applied loads and the known value of E for the cantilever, may be compared with the observed relationships, both for the photo-elastic coating and the resistance strain gauges. The results of such an experiment are shown in Fig. 4 from which it can be seen that the stress–strain relationship of the optical strain gauge is linear and it more closely approaches the theoretical relationship than does that of the electrical resistance strain gauges.

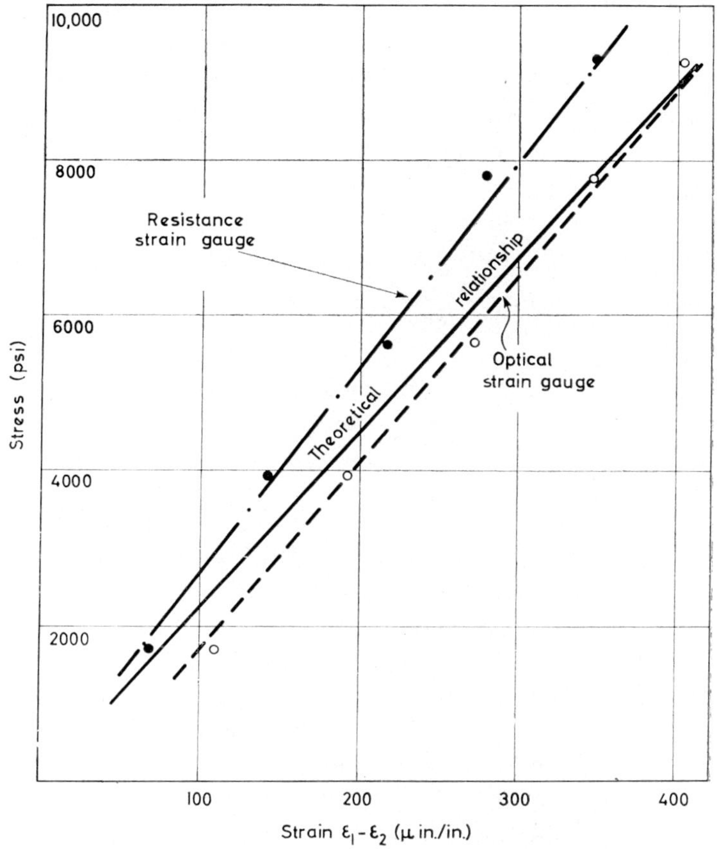

FIG. 4. Calibration of electrical resistance and optical strain gauges.

Calibration of Instruments

If the thickness of the photoelastic coating and characteristics of the light source employed are maintained constant it is possible to calibrate together the coating and the polariscope. A graph of strain versus polariscope reading will then give the value in microstrains for each division of the polariscope and the instrument reads directly in terms of strain.

Limits of Measurement

The accuracy of measurement with the large-field meters depends on how accurately a fringe may be set on the point under investigation. The authors' experience has been that, in using these instruments, there is no difficulty in estimating strain to within $\pm 20\ \mu$ in./in., and the direction of principal stress to within $\pm 3°$.

The quarter-wave plates in the reflection polariscope give an exact quarter-wave retardation only for one particular wavelength, and when using white light the intensity of illumination between the fringes will vary, only those colours being extinguished the relative retardations of which are integral multiples of their wavelengths. The rest of the colours are transmitted with varying intensity. So the field, which is made up of superposition of colours of various wavelengths, will slightly change when the polarizer and analyser are rotated, owing to different errors in relative retardation imparted to them by the quarter-wave plates. In the case of the large-field meters it can be shown that, when the two quarter-wave plates are crossed, the maximum error induced in the fringe pattern will be the algebraic difference of these errors.[(17)] In the small hand-viewer, however, only the one quarter-wave plate is used and this is set to give "parallel" effect. In this case the maximum error will be appreciable, amounting to twice the error of the quarter-wave plate. The isochromatic colour pattern is thus seen to change slightly with changes in orientation of the hand-viewer and it is necessary to adopt a standard procedure for the observation of isochromatics while working with white light. As the nearest approach to bright field is obtained when the polarizing axes are parallel to the principal stress direction, this orientation is adopted as the standard procedure when observing the isochromatics.

Photographic Recording

Photographic techniques have an obvious attraction as a means of recording the phenomena demonstrated by optical strain gauges. Directional phenomena can be faithfully recorded photographically, but for serious quantitative work involving the determination of strains in magnitude direct instrumental measurements are essential. However, photographic records can provide a useful supplement to illustrate the progress of observed data during the course of an experiment.

Reproduction in Colour

When the photoelastic coating technique is applied on materials with a relatively high modulus of elasticity, the order of strain to be measured will be low. In such cases it is an advantage to use white light because then a number of isochromatic colours will exist between the fringes. The colours on a photographic reproduction of a work-piece under such conditions are determined partly by the photoelastic birefringence and partly by the photographic process. Quite apart from the question as to whether, in any particular case, the photograph gives a faithful reproduction of the subject, it follows that no two photographs of an experiment may be used, even on a comparative basis, unless all the factors in the photographic process have been standardized and controlled. This is far from easy but, assuming that it is possible to obtain standard materials, a check on the standardization of exposure and development can be obtained by including on each photograph of the work-piece a photograph of a standard colour scale which has been given exactly the same photographic treatment as the work-piece. Thereafter two photographs illustrating the course of an experiment can really only be compared if the reproduction of the standard colour scale is identical on each.

Reproduction in Monochrome

Panchromatic black and white film or plates may also be used for white light interference fringe photography using the image of a standard colour scale on the same material as a reference. Colours are reproduced in monochromatic greys of varying density, so that a correlation between the reproductions of the standard scale and the work-piece may be made by a densitometer, as in the procedure followed in photographic photometry.[(18)]

However, a complicating factor here is that the photographic emulsion is responsive to actinic brightness, with the result that different colours of the same brightness will be reproduced at the same density. The danger of confusing the colours can be obviated by exposing the scale and the work-piece to red, green, and blue light separately, but such a procedure is lengthy and involved. An alternative solution is to make sketches and notes at the time of observation to identify the colours on the work-piece in relation to the photographic image that will subsequently be produced.

Black and white photography is always used when the work-piece is viewed under monochromatic light, in order to reproduce only the black interference fringes.

SOME APPLICATIONS OF THE TECHNIQUE

Inherent or Primitive Strains in Rocks

Previous research by one of the authors into the time–strain phenomena, or creep, of certain rocks, primarily those of the Elliot Lake area, resulted in the formulation of a stress–relaxation hypothesis, in which the rock is compared with a visco-elastic material represented by the Kelvin model.[(19)] If such a medium is subjected to an initial strain which increases with time, the elastic strain will reach its full value only after infinite time. Conversely, if the strained body is relieved of its load, complete reversion to the unstrained state will take infinite time. If such a material were a rock containing say 25 per cent mineral X and 75 per cent of some other mineral Y, and the rock were subjected to a high load such that mineral Y began to flow but X remained stable, the X particles would be subject to an isotropic stress imposed by mineral Y. If now the load on the rock were reduced to a point such that mineral Y resumed stability, each X particle would remain constrained independently of any external load, and could only be relieved of that constraint by the rupture of mineral *Y* around it. It follows that if a fresh face were created by the fracture of such a rock then a degree of relaxation should follow. There should be some rapid strain release followed by an exponential adjustment with time.

In order to test this hypothesis it was decided to use photoelastic coatings as optical strain gauges, which would register both the directions and magnitudes of the principal strains.

Intergranular Strains

It was hoped to be able to detect and measure intergranular strains, and if so to see how these affect the stress–strain relationship of the material as a whole. In a preliminary experiment to see how two equal-sized crystals of the same material interact under load, a tensile test-piece of aluminium was cast and coated with photoelastic plastic polymerized directly on the metal, to a thickness of 5 mils. The test-piece was then strained in tension.

The results were obviously apparent. In two adjacent crystals the cross-sections of which comprised the whole width of the test-piece, the strain distribution was markedly different. In one crystal the direction of principal strain was along the axis of the test-piece, but in the other the principal strain was aligned at 30° to the same axis. One crystal developed slip lines parallel with the crystal boundary, at an early stage. Slip lines developed in the other crystal later and at an angle of 60° to the boundary. The crystals did not strain uniformly and, in general, the strained condition of the test-piece was quite unlike that which is normally assumed to exist by the theory of elasticity. The significance of dissimilarity in the elastic properties of the several components of a granular material on the stress–strain relationships (a) of the individual components and (b) of the mass as a whole, is discussed elsewhere.[20]

Relaxation Phenomena in Rocks

Several rock samples were examined for evidence of relaxation phenomena. Each sample was marked in the mine, so that its location and orientation were defined and in the laboratory each sample was cut to form a prism with mutually orthogonal faces oriented on the prism to agree with the mine orientation of the original sample. Pieces of photoelastic plastic were attached to each face and after allowing an adequate setting and curing time for the cement the pieces were observed under polarized light. In those rocks which showed isochromatics a point at the highest observable shear strain $(\varepsilon_1 - \varepsilon_2)$ was selected for measurement, on each photoelastic piece.

Three rocks of radically different characteristics were compared. They comprised (*A*) a quartzite from Elliot Lake, (*B*) potash from Saskatchewan, and (*C*) a coal measures shale from a bore-hole in Lincolnshire. In each case the pattern of strain, both in magnitude and direction, appeared to be distinctive. The isoclinics showed that the directions of principal strain varied from point to point through a wide angular range. However, by slowly rotating the polariscope it was possible to estimate the position in which maximum extinction of light was obtained. In this position the axes of polarization gave what may be called "statistical directions" of principal strain.

The pattern of shear magnitude, as shown by the isochromatics, was also distinctive for each rock. The Elliot Lake quartzite tended to develop curved lines of shear, with secondary straight lines of shear. A hammerblow would cause a conchoidal break along the curved shear lines. The coal measures shale developed an irregular patchy shear pattern, the patches being connected by a network of fine shear lines in two directions at 120° to each other. The potash surface showed lines of shear apparently related to the packing pattern of the crystals. In each case the rock could be broken along these shear lines which have been called "preferred shear planes" for the rocks concerned.

Calculation of ε_1 and ε_2 from normal and oblique incidence readings showed that both the principal strains increased with time, but might do so at different rates. Thus if ε_1 increased rapidly at first but more slowly later and if ε_2 increased slowly at first but more rapidly later, $\varepsilon_1 - \varepsilon_2$ would be large at first but would decrease later; ε_1 and ε_2 were also different from point to point. For example, in one case it was noted that the plastic at a point over a quartz crystal developed tensile strains indicating that the crystal was expanding, whereas at a point just outside the crystal compressive strains were observed; ε_1 and ε_2 were quite different at the two points which themselves were close together. It is worth noting that in each case the stresses calculated from ε_1 and ε_2, using values for E and μ determined from conventional laboratory tests, were much higher than could be produced by the weight of overlying rock in the mine.

The rocks appeared to contain residual elastic strain energy which was relieved over a period of time upon the creation of a new surface.

Shear Strain-time Phenomena

In other experiments also the shear strains developed at a rate which differed in various rocks. The general pattern appeared to be one of increasing shear strain with time, up to a maximum, after which $\varepsilon_1 - \varepsilon_2$ would decrease again to a relatively stable value.

A point of speculative interest arising from this phenomenon is that some rock samples taken from areas of known rock-burst potential yielded graphs in which the shear–strain peak in one plane is much higher than the peaks seen in the other two planes, and usually occurs at an early time on the graph. In other rocks the shear strains are generally not radically unequal on the three orthogonal planes.

Figure 5 (a) shows a normal condition, which can be compared with Fig. 5 (b) and (c) the last two being rocks from known bursting regions. Further research into these phenomena is most desirable.

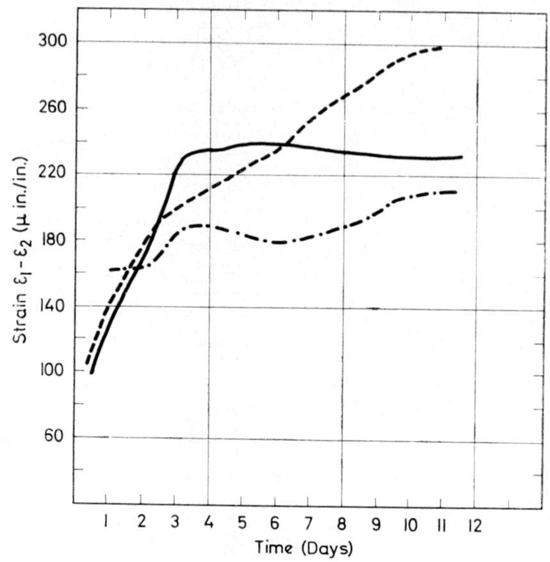

FIG. 5(a). Shear strain–time curves for the three orthogonal faces of a specimen of rock, *A*.

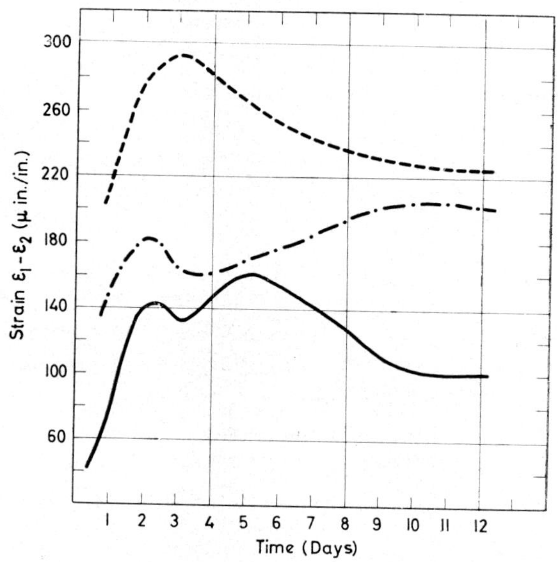

FIG. 5(b). Shear strain–time curves for the three orthogonal faces of a specimen of rock, *B*.

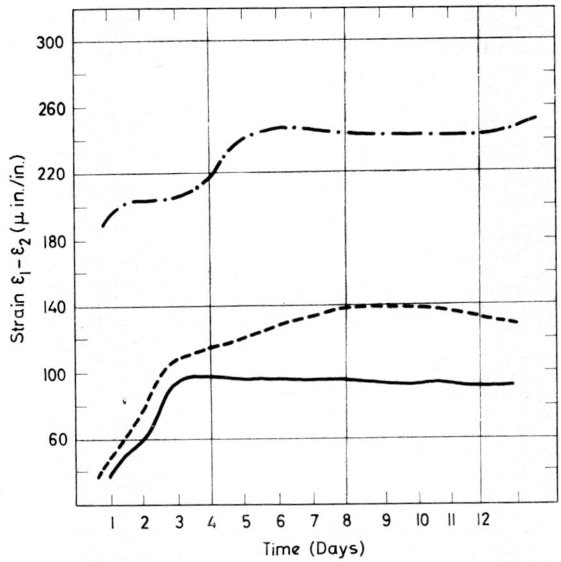

FIG. 5(c). Shear strain–time curves for the three orthogonal faces of a specimen of rock, *C*.

THE EFFECT OF INHERENT STRAINS AND GRANULAR PACKING PATTERNS DURING FURTHER LOADING

A series of experiments to investigate the difference in behaviour between a granular rock (a coal measures sandstone) and an isotropic elastic material under load was conducted in the following manner:

(1) 2-in. diameter disks, each 0.5 in. thick, were prepared from the rock, the faces of the disks being normal to the bedding plane. A similar disk was cut from $\frac{1}{2}$-in. thick Perspex sheet.
(2) The face of each disk was coated with photoelastic plastic, 0.12 in. thick.
(3) The disks were loaded diametrically in compression, to an amount well below that required to cause rupture. They were then unloaded, rotated 90°, and then reloaded.
(4) For each disk the isochromatic pattern was observed and the isoclinics recorded at intervals of 15° on the analyser.

On examination, all the sandstone disks exhibited isochromatics which were taken to be evidence of the existence of inherent or primitive strains. Upon loading, the Perspex disk showed the isoclinic pattern characteristic of an elastic disk under such conditions. The sandstone disks, however, had quite a different appearance from that of the elastic disk under load.

When the sandstone disks were loaded in a direction normal to the bedding there was no similarity in the distribution of shears, or principal strains, to that shown by the Perspex disk at any isoclinic angle. This may be due to the fact that the packing pattern of the components in the sandstone distributed the strain in a way that could not be related to the Perspex disk.

However, when the sandstone disks were loaded in a direction parallel to bedding, a recognizable isoclinic pattern could be discerned, but this was not at the angle corresponding to that shown by the Perspex. For example, the pattern of the zero isoclinic in the Perspex appeared at 15° in the sandstone. It is suggested that the existence of inherent strain in the sandstone had the effect of adding a component to that produced by the external load, to give a resultant 15° off that expected from the external load itself. If this is so it becomes evident that the existence of inherent strain may be an important factor in mine design in that it influences the reaction of the rocks to the loads imposed by mining.

EXAMINATION OF ELASTIC REBOUND IN ROCK

The phenomenon of "elastic rebound" is usually observed by attaching a strain gauge rosette or bore-hole plug to rock *in situ* and then relieving the portion of rock to which the strain gauge or plug is attached, by sawing or cutting a slot around it. The strain which results from the expansion of the de-stressed plug is a function of the stress–strain characteristics of the rock and the original confining stress to which it was subjected. The phenomenon was demonstrated very simply and effectively in the laboratory in the following manner.

A 2-in. diameter cylinder of Darley Dale sandstone 2 in. long was subjected to uniform radial confining stress by shrinking a length of steel pipe of wall $\frac{1}{8}$ in. thick around its circumferential perimeter. A disk of photoelastic plastic was attached to one end (not touching the steel pipe), and under the initial equilibrium of rock cylinder in its confining ring the plastic showed no initial strain.

A core of rock 1 in. in diameter, with the plastic attached, was then drilled and extracted from the cylinder, leaving a cylindrical annulus of rock still confined within the section of steel pipe. The plastic attached to the rock annulus immediately showed a strong isochromatic colour pattern, up to the third fringe order, this representing the shear strain resulting from the confining stress. The detached rock cylinder also exhibited an isochromatic change, of half a fringe, which was the shear strain resulting from the expansion of the rock core upon its relief from confinement. At the ends of both annulus and core, a radial principal stress pattern was evidenced by the isoclinics.

A STUDY OF PLATEN EFFECTS IN COMPRESSION TESTING

It has long been known that the conventional uniaxial compression test on rock gives extremely variable results. Values of compressive strength obtained from such tests depend on several factors the effects of which are, as yet, not fully understood.

One of these factors is the end effect, probably due to frictional restraint on the test-piece, imposed by the platens of the testing machine.

A full discussion of these matters is worthy of a separate paper, but it is appropriate to mention here that, as part of a general investigation in which the phenomena associated with compression tests on rocks are being examined, the photoelastic coating technique is being used to determine surface strains throughout the loading of a specimen, and thereby shed light on the possible cause of failure.

In a preliminary experiment the strain patterns on an aluminium test-piece were examined under two different end conditions. In the first case the ends of the test-piece were in direct contact with the platens, and in the second a layer of polytetrafluorethylene $\frac{1}{100}$ in. thick was interposed between the test-piece and the platens. A marked change in the strain pattern was observed on a photoelastic coating bonded to one side of the test-piece.

The strain pattern when the ends were in contact with the platens agreed substantially with that calculated theoretically by Lisowski for similar end conditions.[21] From the isochromatics it was noticed that, although the greater part of each end face is restricted by friction, a portion near each edge is sufficiently free to permit a degree of lateral extension.

A similar experiment was conducted on a rock. The specimen tested was a 2-in. cube of Darley Dale sandstone, on one face of which a 2-in. square of plastic, $\frac{1}{8}$ in. thick, was bonded. The specimen was then loaded up to 10,000 psi, this being about 60 per cent of the breaking load under such conditions. The specimen was in direct end-contact with the platens of the testing machine. The isochromatics were seen to take the form shown in Fig. 6 and the distribution of shear stress on the surface of the specimen was determined as indicated on the figure.

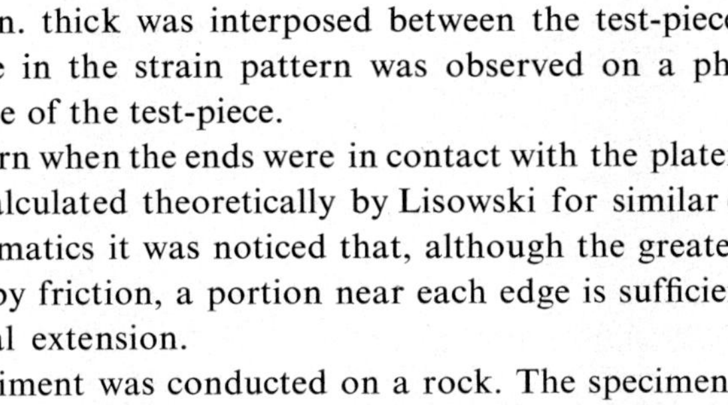

FIG. 6. Isochromatic pattern on a birefringent coating attached to one vertical face of a rock cube under load.

The strain pattern on the surface of the sandstone test-piece resembles to some extent that of the metal cube. However, the shear strains show a lack of symmetry which may be due to heterogeneity in the rock and possibly to some eccentricity in loading.

The research continues. The results to date are extremely interesting in that they indicate the character and importance of the plastic deformation that occurs before the initiation of cracks around the circumferential edges of the test-piece adjacent to the platens. It is expected that the research will shed light on the problem of choosing the most suitable size and shape of rock specimens for compression tests.

CONCLUSIONS

The principles by which photoelastic coatings may be used as optical gauges to determine strain both in magnitude and direction have been outlined, with details of the instrumentation developed for the application of the technique in the field of rock mechanics. Research on the use of the technique is continuing, to explore its possibilities and limitations, but sufficient experience has been gained to show that it promises to be a most useful tool in rock mechanics not only in the laboratory but also in the mine. Some details of laboratory experiments made by its aid have been given.

Acknowledgements. This part of the paper describes work performed at the Postgraduate School in Mining of the University of Sheffield, England, and at the Mining Department of Queen's University, Kingston, Canada. It is based, in part, on research undertaken by C. L. Emery and P. K. Chakravarty, under the supervision of A. Roberts, Director of the Postgraduate School in Mining, at Sheffield. Grateful acknowledgement is made to Messrs. Dowty Mining Equipment, Ltd., for a grant of funds in aid of the research by Mr. Chakravarty, and to Professor A. V. Corlett and the executive officers of Queen's University, Ontario, who, by granting leave of absence, made it possible for C. L. Emery to undertake his research in England.

PART 2. FIELD INVESTIGATIONS

by A. Roberts, C. L. Emery, I. Hawkes, P. K. Chakravarty, and F. T. Williams

It is the task of the mining engineer to construct, maintain, and operate functional structures in rock which is subject to loads that are partly the result of natural inherent and tectonic forces, and partly the result of forces induced by mining. The movement of rock that is strained to the point of fracture must be controlled. In some parts of the mine fracture must be minimized and the ground supported. In other parts it may be necessary to induce fracture and to control the movement of the resulting broken and caved ground. Rocks are the materials that the mining engineer must understand and use.

For his task he requires information. Much of this he may obtain by measurements and the resulting data must be interpreted in the light of experience. Based on his interpretations mining operations can be planned and designed. During the operations further observations and measurements will serve either to confirm the original interpretations or to indicate that events do not conform to predictions. In the latter case the mining engineer must decide to what extent he will change his original interpretations and what steps he will take to control the situation. Observation, measurement, and interpretation are three mutually interdependent components of an effective strata control system.

As yet the science of measurement in strata control is in its infancy. There is considerable difficulty, for example, in deciding what to measure and where to take measurements, and the situation is not improved by theories based on assumptions that are contrary to observation. However, some measurements can be informative. For example, measurements of changes in shape, or volume, of a borehole during mining operations give positive evidence of the nature of the force fields that are active the rock. Useful results have been forthcoming also from the measurement of reactions such as prop loads, roof-bolt loads, floor penetration resistance, and pressures in caved ground. Progress is also being made in the use of strain gauges and the application of geophysical devices to investigate the state of strain in rocks.

Unfortunately, the more refined techniques have a limited applicability in mines because of their complex and highly technical nature. Much of the equipment is elaborate and costly. Some of it cannot be used at all in gassy mines, and in general, the equipment can only be applied and understood by highly specialized personnel. Not infrequently, much tedious labour has produced quite simple, almost obvious, results, and the mine staff have remained unimpressed. The most outstanding merits of the photoelastic transducers are that they require no elaborate ancillary apparatus and can be used with complete safety in any mine. They are cheap and easily applied. Their recordings are visual and the normal mine personnel can be trained to use them without difficulty.

ORGANIZATION OF WORKING PARTIES

The instruments used in rock mechanics investigations will measure and record deformations in the strata and in the mine supports. From the strains so measured the operative stresses can be deduced, provided that the stress–strain characteristics of the deformed materials are known. Therefore, if the fullest information is to be obtained each investigation must be supported by a complementary laboratory investigation, in the form of a detailed study of the physical properties of the particular rocks concerned.

SUPPORTING LABORATORY INVESTIGATIONS

Samples of rock are sent from the mine to the laboratory, suitably marked so that their orientation relative to the geometry of their surroundings is known. Depending upon the objectives of the field programme a selection of physical tests will be made. Such tests may include an examination of component grains size and structure, determination of hardness, porosity, specific gravity, and thermal conductivity. The strength of the rocks can be determined in various ways, in flexure, compression, and tension, by direct and indirect loading to determine the elastic moduli. Particular attention may be paid to the strength of the rocks under confinement, in triaxial loading, or to time–strain or creep characteristics. The strength of the rocks under dynamic loading may be investigated by impact, abrasion and vibration tests, and the dynamic moduli and damping factor determined. Tests using photoelastic coatings on three orthogonal faces of rock specimens may be made, to determine whether the rocks contain inherent or residual stress and, if so, what are the directions and magnitudes of the principal stresses and the shear patterns to which the rock is subject.

The object of the laboratory work is to obtain as much information as possible about the rock as a material and thus to aid the mining engineer in his interpretation of the phenomena observed in the mine.

FIELD INVESTIGATIONS

In addition to the use of strain gauges, extensometers, rock bolt dynamometers, load cells, bore-hole plugs, and various forms of deformation meter, the field investigators may use auxiliary techniques which, although of a simple nature, will be found most useful—such devices as, for example, a short survey line between two plugs, with marked points at intervals between; any movement of an intermediate point with respect to an end point can be detected without instruments if it is in excess of 0.1 in. Additional holes drilled into the strata and left empty during rock bolting will serve as useful observation holes. Such holes often show lateral displacements along their length and thus give useful evidence of the forces which exist in the strata.

CO-ORDINATION OF LABORATORY AND FIELD WORK

Laboratory and field work should be co-ordinated by an experienced mining engineer who has made the study of rock mechanics his special field. Data can be fed to him from the laboratory and from the mine, in each of which a supervisor has the responsibility of organizing the measurement systems and collecting the results. These data should then be reported on to the responsible mining engineers, with an assessment as to its probable significance in relation to mine design, layout or mining techniques, or the special objective sought in the particular operations concerned.

The rock mechanics group does not take any responsibility from the shoulders of the mining engineer in charge of production, nor in many circumstances can it supply him with a solution of his strata control and support problems. But it can and does give him facts and data on matters which he might otherwise overlook or, at best, conjecture. By presenting him with rational explanations of observed phenomena, and by making suggestions as to their significance, they enable him, in a difficult situation, to depend more upon knowledge and less upon intuition.

EXAMPLES

The following examples may serve to illustrate the application of the technique in the field.

Mine A

The first example was a uranium mine, operating a room-and-pillar system, with rooms 45 ft wide by 250 ft long, separated by rib pillars along full dip 15 ft wide running along the full length of the rooms. The formation dipped 15° and consisted of a pebble conglomerate. A stratum 10 ft thick was mined, under a roof of fine-grained quartzite, in which a distinct parting occurred about 5 ft above the opening. The roof was bolted with $\frac{3}{4}$-in. diameter by 22-in. high-tensile steel rock bolts. Falls of roof in the stopes were frequent and, although minor in extent, constituted a serious hazard to safety in that there was a continual violent spalling of slabs a few inches thick from the roof stratum.

In consequence the mine management brought in the strata control research group to examine and report. Photoelastic strain gauges were attached to the strata in the stope, and specimens of roof strata, marked to identify the direction of orientation in relation to stope geometry, were removed to the laboratory for examination. The strain gauges in the stopes gave evidence of high lateral compressive stresses in the lower surface of the roof structure, of a minimum magnitude of 2000 psi in a direction normal to full dip, i.e. across the stope, while at the sides of the stope, near the abutments formed by the rib pillars, there was a horizontal stress of about 1200 psi. Photoelastic gauges attached to the laboratory specimens of roof rock showed maximum shear strains $(\varepsilon_1 - \varepsilon_2)$ developing about three days after the creation of a free face, and then a subsequent diminution of strain. A graphical plot of strain against time was of the type illustrated in Fig. 5. The phenomenon is that of some form of relaxation of the constituent particles of the rock, which is apparently brought about by the relief of constraint.

One of the present authors having formulated and published a theory of roof control in mines based on the design principles of pre-stressed reinforced concrete now widely used in civil engineering,[19] it was decided to test this theory in the mine. Considering unit length of the stope, the immediate roof stratum below the parting was imagined as analogous to a beam 1 ft wide and 5 ft deep, spanning the stope between the rib pillars. It was argued that the high compressive strains and therefore stresses that were measured on the lower surface of this roof beam could be counteracted by increasing the span of the beam, which would thereby impose a horizontal tensional stress on the same surface due to bending. Calculations based on the evidence produced by the measurements of the stress–strain phenomena of the roof stratum indicated that the required horizontal stress due to bending would be produced at the lower surface of the centre of the beam if the span were increased to 160 ft.

The engineering staff at the mine decided to experiment by removing one pillar completely, to give a total room span of 105 ft. Equipment was installed to measure convergence between roof and floor. Survey lines were established to detect relative lateral movements on dip and on strike. Roof-bolt tensions were measured and inspection holes were drilled in the roof to a depth of 8 ft.

The first rib pillar was drilled and blasted out completely in one operation, and the newly exposed roof was bolted in the usual manner. Observations and measurements over a period of several months showed no appreciable roof movements and the roof appeared to be stable and free from cracking and spalling. Thereafter the management changed their mine design to include the driving of stopes 100 ft wide in areas where the measured primitive force fields were similar to those measured in this experiment.

Mine B

In an attempt to correlate photoelastic coating techniques with actual roof movement an experiment was conducted on the roof of a drift which was advancing in an easterly direction, approximately on strike. The drift required the complete removal of the ore which was about 8 ft thick and dipped 30° to the south. The roof was a pebble conglomerate with small pebbles in a quartzite matrix. Fig. 7 illustrates the situation.

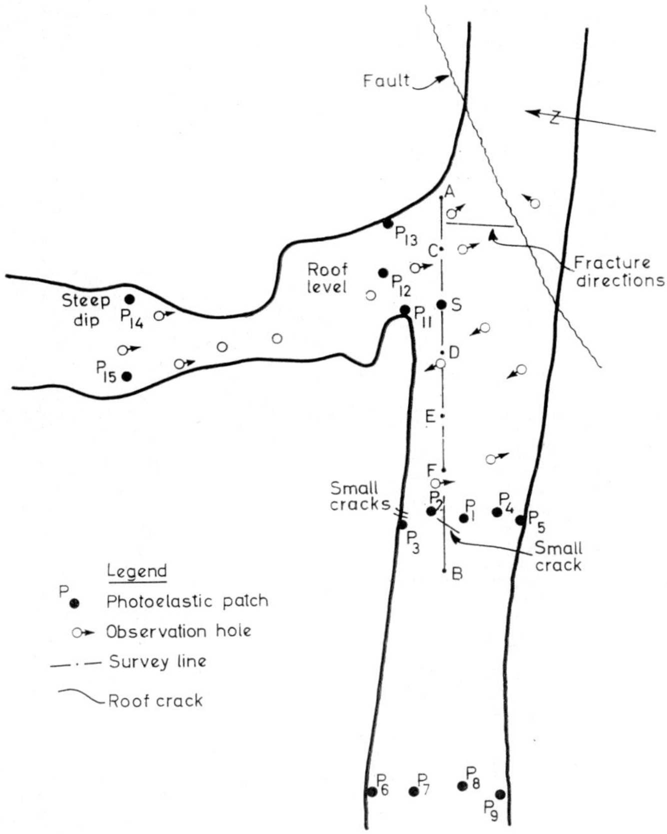

FIG. 7. Plan of layout of experiment in Mine B.

The experimental procedure was to install patches of photoelastic plastic, 0.12 in. thick, at random on the roof immediately following its formation in the sequence of mining. Observation holes were drilled in rows in the roof at the time of bolting.

Within three days of the creation of a new roof as the drift progressed, buckling and spalling would occur in the roof, usually resulting in a fall of conglomerate 2–3 ft thick. The plane of fracture appeared to be a newly-sheared surface with breaks running across the pebbles of quartz. The roof would work and buckle, during which period the plastic patches recorded a general down-dip compression, exceeding 2000 μ in./in. in some cases. A strain of this order in the plastic represents a rock surface stress of over 16,000 psi and it appeared therefore that the heading was under a very high compressive force in a down-dip direction. In any event the plastic patches indicated that large shear strains were building up and observation showed this to occur at about the same rate as the deterioration of the roof. The principal strain directions were the same in the plastic as those observed in the subsequent displacement of the rock. From the point of view of the miner the patches predicted roof failure.

Mine C

An opportunity to study the actual strata movements in a stope advancing against the dip through faulted ground was afforded at this mine, where the strata were similar to that in Mine A.

The primary object of the research was to seek a useful system of strata control measurement which could form part of a normal mining operation. The stope was in active production with only a limited time available for measurement. The whole process had to be within the scope of the operating crew and the measurement techniques had perforce to be simple and yield useful results.

The conditions of the experiment are shown in Fig. 8. Plastic patches (P_1 to P_6) were attached to the roof, observation holes were drilled to a depth of 4 ft and plugs were set across the width of the stope to establish survey lines Nos. 1 and 2. The pedestal pillar was apparently under heavy pressure as evidenced by thin spalls of rock flaking from it. There was also some spalling on the west rib pillar. The top slice of the stope was to be completed prior to benching. It was obvious that the pedestal pillar would be unstable at the finished stope height, but it was feared that the load on it was due to roof weight which would produce a collapse if the pillar were removed.

Laboratory tests showed that the rock contained high primitive strains and therefore it was probable that in so short a span as 55 ft large shear strains would exist in the roof strata at its lower surface. Within the first two days after installation, patch P_1 exhibited shear strain of 350 μ in./in., the principal strain direction being towards the dip. The other patches showed lesser shears and various strain directions, as indicated in Fig. 8. It appeared that there was some general roof movement down the dip but that there was also some tendency for the roof to rotate in an anticlockwise direction about the pedestal pillar.

After two weeks the directions of principal strain indicated by the plastic patches remained unchanged, but all the shear magnitudes had increased. Patch P_1 showed a strain of 1000 μ in./in., P_4 and P_6 400 μ in./in. P_5 had increased to 500 μ in./in. The rock to which P_3 was attached slabbed-off. Survey line No. 1 showed a slight general movement in a direction down the dip, while line No. 2 showed a slight movement up-hill at C_2, no measurable change at D_2, and a slight movement down-hill at E_2 and F_2.

There was now no doubt about the kind of movement actually in progress. The strains were obviously time-dependent and were being accommodated, at least partly, by flow rather than by fracture. Where fracture did occur it was in the direction predicted by the "preferred shear" planes demonstrated by the rock

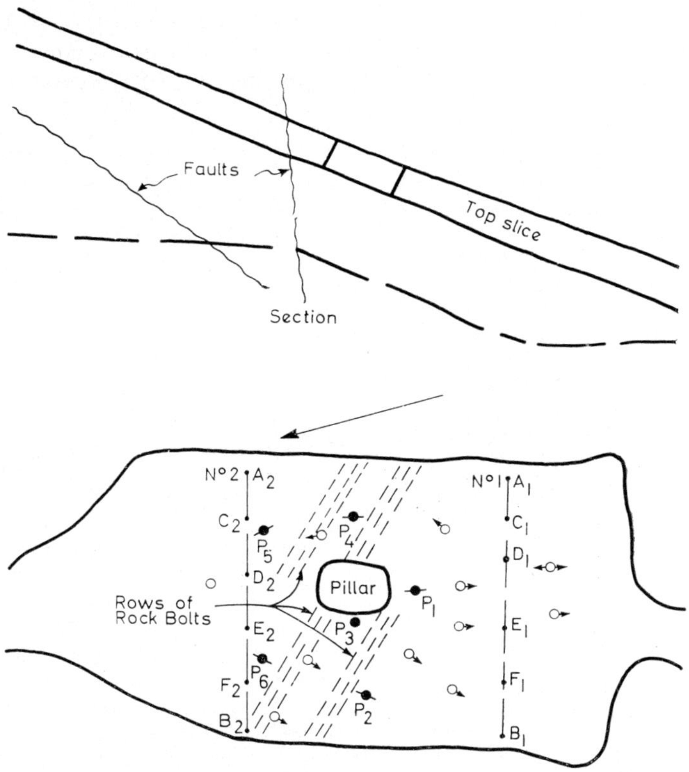

FIG. 8. Plan of stope in Mine C.

specimens in the laboratory. While a pedestal pillar in the particular situation concerned was an obvious point of stress concentration of vertical or "weight" stress if such existed, the presence of high primitive shear strain in the roof strata meant that lateral stresses were high and possibly were much more important than weight stress. The pedestal pillar probably was therefore not offering support to the roof, but was being loaded by forces that tended to rotate it. The west rib-side pillar was shearing probably because of the eccentricity of this rotation around the pedestal, which caused a resultant thrust against the rib-side.

It was decided to blast out the pillar, having first inserted a number of rock bolts set in parallel lines at an angle across the stope and designed to reinforce the rock in the expected direction of crack formation. The function of the rock bolts was to offer resistance to lateral rock movements, and to constrain and increase the structural stability of the rock mass.

After the removal of the pillar there was an immediate reduction of shear strain demonstrated by P_1, and to a lesser extent in the other patches. No roof failure occurred and the shearing on the west rib pillar ceased. Both survey lines exhibited a slight movement down-hill at their centres.

The directions of principal strain in the plastic patches reverted to a general down-hill direction; no further displacement was noted in the observation holes, nor were any measurable changes noted during the next two weeks.

Mine D

Figure 9 shows the progressive development of shear strains in three plastic patches which were placed at different times along the longitudinal axis of a stope, as the face advanced.

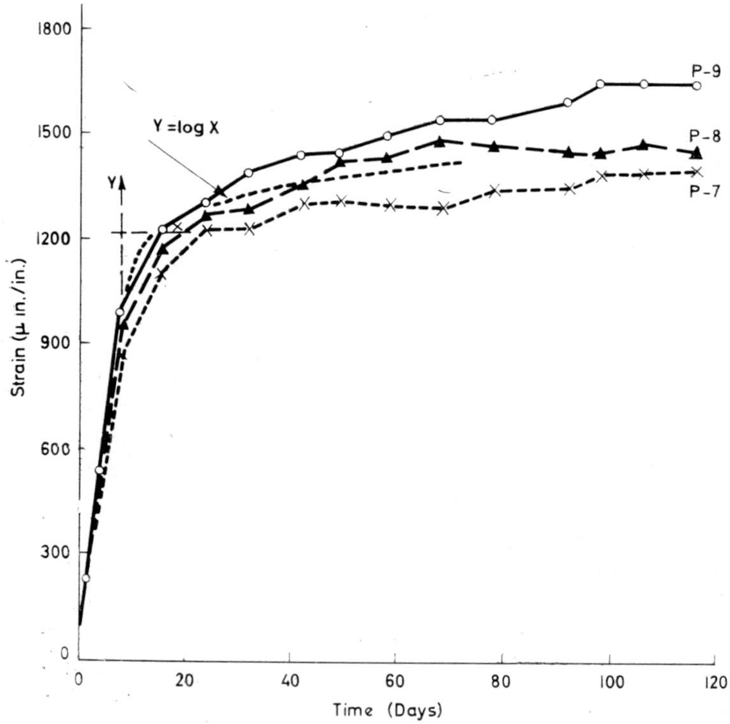

FIG. 9. Shear strain–time curves.

The shear strains increased approximately linearly with time for the first eight days and then followed very closely a logarithmic curve. The rate of increase of shear strain was greatest within the first few days. The movement then slowed and the strain–time curve approached asymptotically a maximum shear strain. The logarithmic portion of the curve is characteristic of the second stage of creep usually recorded in conventional creep tests and is that sometimes attributed to Kelvin creep, or more properly, the aquisition of full strain with time as shown by the equations for a Kelvin solid. During the early part of the observations it appears that there was approximately a steady strain rate. Such a strain rate can be produced by pseudo-viscous flow which often precedes or accompanies Kelvin movement.

Mine E

In Mine E coal was being extracted on a 200-yard advancing longwall face in a seam 2 ft 11 in. thick, under a shale roof. Gradient was 4–5 per cent, the line of the face being approximately "on end", or at right angles to the main direction of cleat. The face was advanced continuously over three shifts in steps of 2 ft 3 in. by a drum shearer, giving a rate of advance of 80 ft per week.

The roof was supported by hydraulic self-advancing supports set at 3-ft centres, each unit having a cantilever extension, 3 ft 6 in. long, to support the roof over the face conveyor and the machine track. The goaf was caved between the roadside packs. Photoelastic patches 2 in. square were bonded to the roof 6 in. from the coal face. This was done over the weekend, to allow some 40 hr for the cement to harden. The patches were examined after each passage of the face machine.

At the first transit of the shearing machine strain differences of the order of 700 μ in./in. were recorded at several points along the face. Near the ends adjacent to the gate roads, strain of around 900 μ in./in. were noted. Some decrease in strain of from 60 to 100 μ in./in. was observed in the interval that elapsed before the second transit of the shearer. This decrease was gradual and was apparently not connected with the advance of the hydraulic support units. The second transit of the shearer resulted in an increase in the observed strain, to some 900 μ in./in. over most of the face, and up to 2000 μ in./in. near the roadside packs. Again, the advance of the support units produced no noticeable changes in roof strains, but a slow decrease was observed.

By the time of the third transit of the shearer the forward extremities of the main roof members of the support units had approached the observation points, and no increase in strain was observed. As further mining proceeded and the supports were advanced, a gradual reduction in strain was recorded until strain differences, or shears, decreased to approximately 400 μ in./in. before the plastic patches were lost in the caving goaf.

The isoclinics indicated by the plastic squares were examined regularly and were found to be quite consistent. Along the caved length of the face the directions of the principal strains were at right angles to and also parallel to the face line. However, in the vicinity of the roadside packs near each end of the face the directions of principal strain lay at approximately 45° to the face line.

From the observed strain in this experiment it was deduced that the removal of coal from the face, and the exposure of roof at each transit of the machine, produced horizontal compressive stresses in the roof ahead of the support units. Within the intensively supported zone on the goaf side of the conveyor and machine track, the observation that the movement of the support units produced no sudden changes in roof strain confirmed the adequacy of the support system.

Mine F

At coal mine F photoelastic techniques were used to investigate the strains in the strata at the side of a roadway which formed the return airway of two panels; that on one side had been extracted, while at the other side a longwall coal face was advancing at some distance behind. The road therefore had goaf on one side and a solid coal rib on the other.

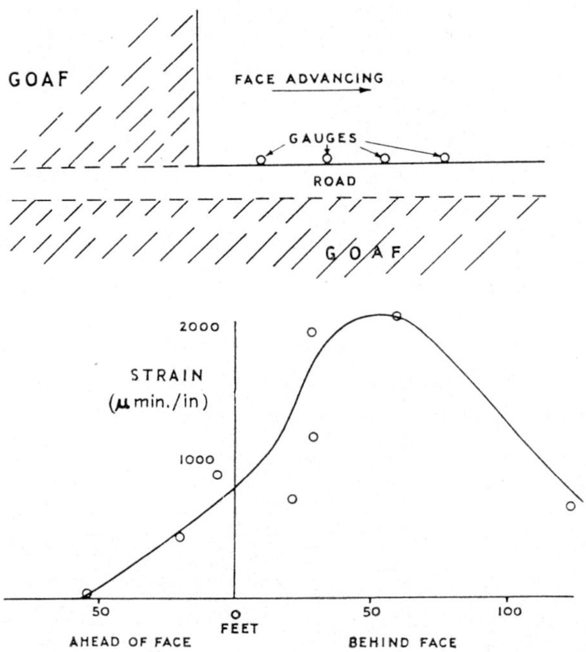

FIG. 10. Change of shear strain in roof strata adjacent to a longwall coal face.

Two types of transducer were used—photoelastic plastic patches $\frac{1}{8}$ in. thick bonded on to the rock and photoelastic hollow glass plugs $1\frac{1}{4}$ in. long, $1\frac{1}{2}$ in. external diameter and $\frac{1}{2}$ in. internal diameter. The plugs were bonded inside holes drilled into the rock, in the roof strata above the coal rib. The plastic patches had a sensitivity of 1500 μ in./in. fringe.

Six observation points were selected at intervals of approximately 50 ft, the farthest being 300 ft in advance of the face. Measurements of $(\varepsilon_1 - \varepsilon_2)$ and of the direction of principal stress were made at regular intervals. There was wide variation in the direction of principal stress but in general this inclined from the vertical towards inbye when the gauge was ahead of the face and inclined towards the outbye side of vertical when the gauge was behind the face. A curve of strain developed in the plastic patches which is shown graphically in Fig. 10. The ultimate loss in strain was due to the rock on which the gauge was affixed becoming detached from the roadway wall.

DISCUSSION OF RESULTS

The interpretation of results in strata control systems and measurements is liable to be influenced greatly by the conditions that exist in the particular experiments concerned and by the philosophy of the observer. However, certain generalizations may be made.

In the first place the results of this work, both in the laboratory and in the mine, lead inevitably to the conclusion that the classical concepts of strength of materials based on the theory of elasticity do not fit entirely into the pattern of the phenomena demonstrated by rocks under load. Rocks are, in fact, crystalline and granular heterogeneous media and upon experiencing a change of load they react and adjust by intergranular movements and by crystal deformation. To a limited extent, depending on the grain size in relation to the thickness of the coating and the magnitude of the strains, the photoelastic coating technique can be used to examine such movements, as was described in Part 1 of this paper.

In several instances the rocks observed underground have demonstrated plastic deformation or "creep", and this has been observed, as in Mine D, to comprise two stages. In the first stage the shear strains increased linearly with time, and in the second stage, usually within a few days after exposure of the rock face, the rate of deformation decreased, to approach a maximum shear strain asymptotically. There is a close resemblance between the time strains measured underground and creep measured in the laboratory. While, as S. A. F. Murrell has pointed out,[(23)] the so-called "viscosity coefficients" of crystalline solids depend very markedly on stress, and are therefore not constants, yet it may be helpful from an operational point of view to picture the observed creep phenomena as including a pseudo-viscous element. Using such an analogy one of the present authors (C. L. Emery) suggests that the linear time-dependent strains are primarily due to deformation of the cementing matrix, giving pseudo-viscous flow of a Maxwellian character, while the subsequent deformation following a logarithmic law occurs when the constituent grains have substantially established contact, resulting in a Kelvin-type adjustment.

One feature common to all the underground experiments is the evidence of lateral stresses, sometimes, as in Mines A, B and C, obviously outweighing vertical stress in importance. Even in the coal mines E and F, where observations were made in zones that were very largely de-stressed and subject only to minor local loads, lateral strains were very much in evidence. Lateral movements of the roof relative to the floor have been observed throughout mining history. When individual non-yielding-type props are used as supports, these lateral movements are seldom of great importance, but with the modern type of composite hydraulic yielding support unit such movements can be ruinous. In spite of the fact that their existence has been known for so long a time, very little is known about lateral strata movements in mines.

The simple assumption of vertical "weight stress" directly proportional to depth has for so long dominated the thinking of mining engineers, and the conventional treatment of rock as an elastic material has had such a strong influence on theories of rock behavior, that to talk of strata movement in terms of the flow of a viscous mass may seem fanciful in the extreme. Nevertheless it has long been known that a rock material, which in conventional uniaxial compression tests appears to be brittle, may, under suitable conditions of constraint, demonstrate phenomena of a plastic character. The researcher may well argue that the mechanism which results in the observed phenomena of strata flow cannot truly include the property of viscosity. Yet, to the mining engineer faced with the problem of strata control underground, at the present state of knowledge this may appear to be simply playing with words. To him the important facts are that he may be faced with difficulties in strata control and support due to flow, as was demonstrated in Mine C, and the photoelastic strain gauges make it possible for him to detect and measure such movements more easily than has hitherto been possible or can otherwise be done.

Mining engineers have also been long aware of the existence of inherent or primitive stress in rocks, but the magnitude of the primitive strains and the phenomena of relaxation had never, so far as the authors are aware, been demonstrated in detail before the application of the photoelastic coating technique to rocks. To those whose experience has been confined to weaker sediments, where one of the main preoccupations is to determine how wide an excavation may be made without danger of roof collapse, the idea of deliberately widening a stope as described in Mine A, in order to reduce lateral compressive stresses in the roof, may appear to be contrary to logic Nevertheless, experience in the Elliot Lake quartzites, where high primitive strains were measured, has fully justified the application of the "pre-stress" theory, whereby a calculated tensile stress due to bending is caused to oppose and neutralize the high primitive compressive stress.

The simplicity of the photoelastic transducers and the fact that they are read by visual colour comparison, compensation, a straightforward count of fringes, or observation of the shape of the fringe pattern, any of which can be done by the ordinary mine personnel, create interest and invite the co-operation of the miners. The relative cheapness of the transducers, compared with other equipment required for a similar purpose, means that many more points of measurement can be established and a significant result achieved. The strata control measurement system may then be incorporated into the daily routine and assimilated into the mining environment.

By using this equipment it may be soon possible to attach a transducer to every unit of the support system on a face or in a stope, and to determine how the loads are distributed in space and time. However, if the transducer takes the form of a photoelastic coating directly applied to a steel support, any observed changes in the plastic are not likely to be very pronounced, so long as the steel remains within its elastic range. The extensometer, however, is eminently suited for this purpose.

The coating technique also has a limited application in coal mines, or in any mine where the strata immediately accessible is broken and fissured and, as a result, substantially de-stressed. In such circumstances it is necessary to penetrate to some depth in the roof, floor and sides, and at the face, in order to determine the state of stress around the mine workings, and the technique is being developed towards this end. Nevertheless, even in the de-stressed zone the fractured rocks are subject to local stresses and strains, although it is probable that these will be of a low order. The observations at Mine F were of this nature.

Acknowledgements. The work described in this paper was done in Britain and Canada, and forms part of the rock mechanics research programme being undertaken at the University of Sheffield under the supervision of Dr. A. Roberts, Director of the Postgraduate School in Mining, and Professor F. S. Atkinson, Head of the Department of Mining. It is based, in part, on a thesis submitted in the University of Sheffield, by C. L. Emery, for the degree of Doctor of Philosophy. Grateful acknowledgement is made to Messrs. Dowty Mining Equipment, Ltd., for a grant of funds in aid of the research by P. K. Chakravarty. Such photoelastic transducers and equipment described in both Parts 1 and 2 of this report, as have not been designed by the authors, are the proprietary materials and equipment of: Baldwin-Lima-Hamilton, Waltham, Massachusetts, U.S.A.; The Budd Company, Phoenixville, Pa., U.S.A.; C-I-M Ltd., Toronto, Canada; Budd, S. A., Paris, France; and Horstman Ltd., Bath, England.

PART 3. THE APPLICATION OF PHOTOELASTIC DEVICES FOR MEASURING STRATA PRESSURES AND SUPPORT LOADS

by A. Roberts and I. Hawkes

DURING recent years considerable advances have been made in development of instruments for the measurement of stress and strain in the strata surrounding mining excavations and for the determination of the loads on mine supports. The same instruments are applicable to the measurement of deformation and load in buildings and other structures in mining and civil engineering generally. Broadly speaking, this equipment has in the past been such that the signals originated from mechanical, electrical, or hydraulic components, but a more recent introduction is a range of devices that makes use of the birefringent optical properties that certain materials display when under strain. These devices have been designed by research workers at the University of Sheffield Postgraduate School in Mining as part of an extension of an investigation into the possibilities and limitations of the photoelastic coating technique. The theoretical basis of the technique and its mode of application to rock mechanics research both in the laboratory and in the mine have been described elsewhere.[24, 25]

The basic principle of the system is that when a birefringent material is subjected to strain or load and the material is viewed by polarized light an optical pattern is produced due to the interference of light within the birefringent material. To make use of this phenomenon as a practical tool of measurement a photoelastic component is incorporated into various gauges, constructed in such a way that th, gauge is subjected to strain in a known manner. The resulting optical pattern ie identified either by a direct count of fringes, or by recognition of a specific patterns which is related by calibration to load, stress, or strain, depending upon the type of gauge.

PHOTOELASTIC TRANSDUCERS

The optical fringe pattern or signal is characteristic of the type of gauge being used and the loading system to which the gauge is subjected. In general, the signals may be designated as uniaxial, biaxial, or transient.

Figure 11 shows the photoelastic pattern obtained when a birefringent disk is loaded externally across a diameter. The fringes appear bow-shaped at the periphery of the disk at either end of the axis of loading y, and move inwards along this axis to merge at the disk centre as the load increases. With further increase of load this central fringe divides into two separate bands which move towards the peripheral ends of the x-axis. A continuous increase of load builds up a succession of fringes one by one in a similar manner and the number of fringes, counted along the x-axis, is directly proportional to the axial load on the disk. In certain types of

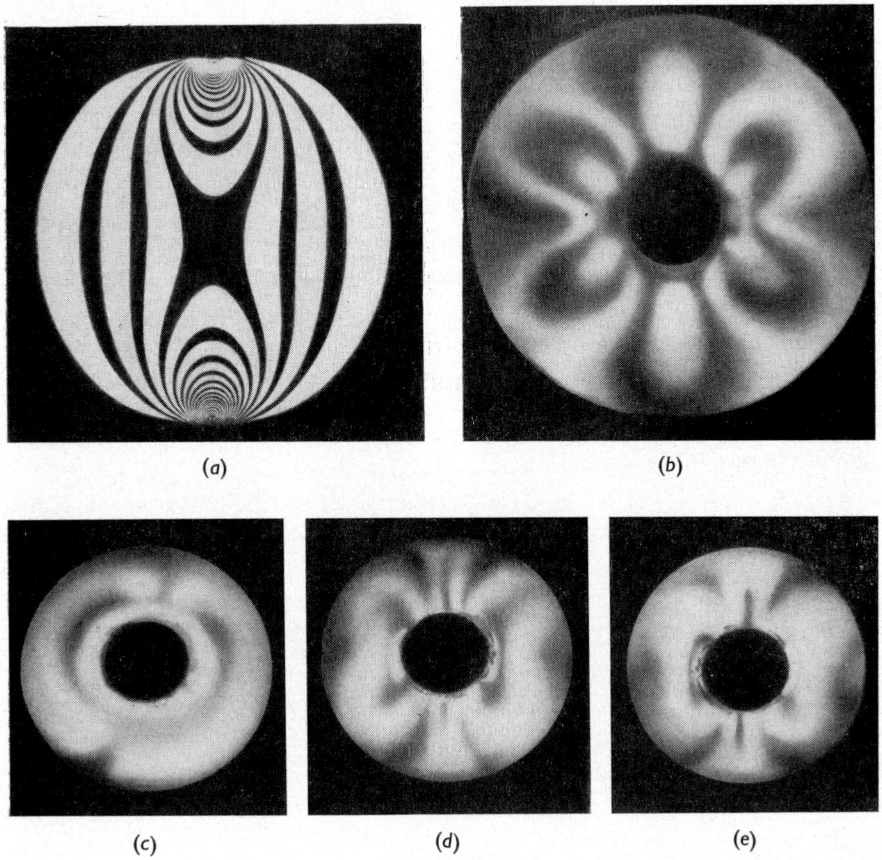

FIG. 11. Standard fringe patterns. (a) disk loaded diametrically. (b) biaxial gage loaded uniaxially.
(c) biaxial gage in stress field, ratio of principal stress 1: 1. (d) biaxial gage, ratio of principal stresses $1:\frac{1}{2}$. (e) biaxial gage, ratio of principal stresses $1:\frac{1}{3}$.

gauge a transducer may be used in which a transient fringe pattern occurs. That is, a succession of fringes follows one after the other with change of load, but these fringes do not remain visible on the gauge. As with the type shown in Fig. 11(a) the gauge is calibrated so that a count of fringes gives the magnitude of the imposed load.

A photoelastic member in the shape of a circular disk or cylinder, with an axial hole, forms a biaxial strain gauge. Such a gauge is used either in the form of a disk cemented around its periphery to the surface of the object under examination, or in the form of a disk or cylinder bonded around its circumferential surface inside a hole drilled into the body of the object under examination. When the gauge is subject to strain as a result of a uniaxial stress field normal to the direction of viewing, the pattern takes a form similar to that shown in Fig. 11b. In this the major axis of symmetry of the pattern identifies the direction of the imposed load and the fringe order gives its magnitude. The shape of the pattern on a biaxial gauge identifies the ratio between the principal strains and hence, stresses, in the plane of measurement. The limiting conditions range from uniaxial loading when one of the principal strains is zero, to hydrostatic loading, when both the principal strains are equal, in which case the fringes form concentric circles. Intermediate patterns showing other loading conditions are illustrated in Fig. 11c.

A number of devices have been constructed, employing such photoelastic transducers. These include dynamometers, load cells, extensometers, stressmeters, and associated viewing equipment.

ROCK BOLT DYNAMOMETER OR BOLT TENSION METER

This device (Fig. 12a) employs the transducer of type 1a. It consists of a cylindrical block which is placed over the rock bolt, between the end nut and the bearing plate, in such a manner that when the bolt is tightened, the block is compressed by the tensional load in the bolt.The compression of the block is transferred to a photoelastic disk on which a fringe count can be made when it is observed through an appropriate optical viewer. The device is designed so that each fringe denotes a uniform increment of load, and it reads directly to half a fringe. It is found in practice that up to six fringes can be recognized with ease, and by a system of pre-loading the disc, so that the first fringe appears at up to 6 tons, the range of the device can be extended from 0–12 tons, the ultimate load being 20 tons.

For accurate readings under eccentric loading conditions, spherical seating washers are used, up to 15°, with angled anchor plates when the degree of eccentricity is greater than 15°.

Early models of bolt tension meter used a birefringent plastic transducer, in combination with a compression block of nylon. While such materials will give satisfactory service when used within the limitations imposed by their physical characteristics, circumstances may exist that call for long-term measurements. The Mk. 1 type bolt tension meter, as illustrated, is primarily intended for the measurement of relatively load loads (up to 8 tons reading directly to $\frac{1}{2}$ ton) in conditions of low humidity and ambient temperature (below 80° F). A Mk. II bolt tension meter uses aluminium for the compression block and a glass disc transducer, for measurements in more arduous site conditions and over very long periods of time.

(*a*)

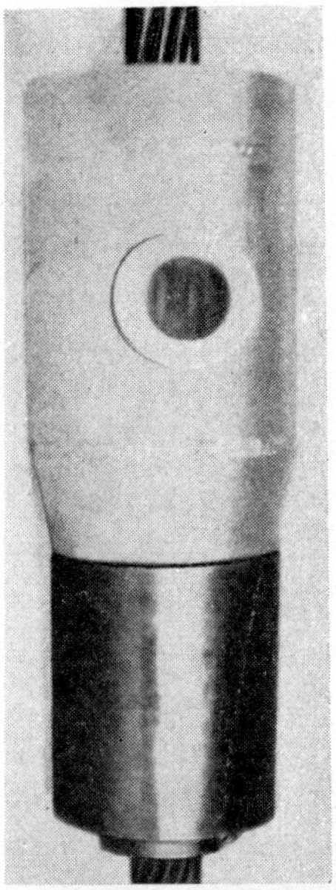

(*b*)

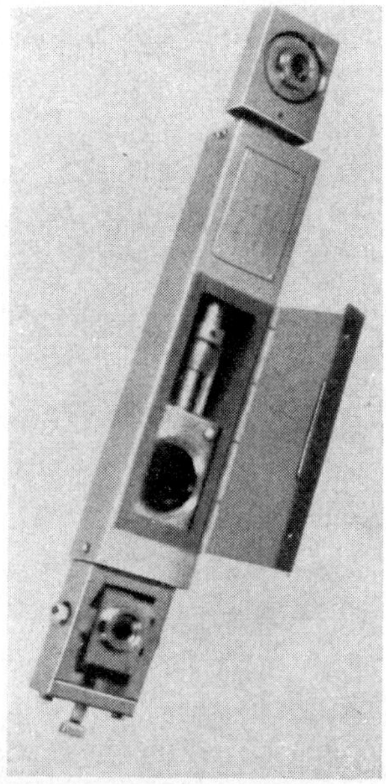

(*c*)

FIG. 12. Instruments incorporating photoelastic transducers.

(a) bolt tension meter. (b) load cell (attached to prestressing rope). (c) extensometer.

PRE-STRESS LOAD CELL

The stability of the glass transducer is also taken advantage of, in order to produce a device which is employed to measure the tension in the pre-stressed strands used in reinforced concrete construction. The prestress load-cell is intended to be used with $1\frac{1}{8}$ in. diameter prestressing strand which has a breaking load of 185,000 1b. In essence it consists of a steel body with a transverse hole into which the optical transducer, a disc of glass, is inserted. The purpose of the hole is to produce a concentration of stress in the steel, the strain in which is communicated to the glass disc insertion and measured as an optical fringe pattern when the glass is observed under polarized light. In a transducer of the type used in the load cell the maximum number of fringes that can easily be recognised is five, and the load represented by one fringe is determined by the geometry of the system and the dimensions of the photoelastic material. The response of the load cell is not linear over the whole of its five-fringe range, but the deviation from linearity over any one imited portion of the range is quite negligible. It is therefore possible to obtain a very sensitive indication of a load of any magnitude over a limited range.

The method of using the instrument is to pre-set it to register any required load over its range (25–65 tons in the model illustrated) which it will do with great accuracy. This can be done quite easily, in a matter of minutes, on a compression testing machine, and in effect it enables the accuracy of the testing machine to be used *in situ* even under the most adverse conditions. For ease of application the instrument is designed and calibrated so that the condition to be identified is the tint of passage between red and green in the formation of the second fringe, as this has been found by experience to be the clearest pattern and the one least likely to be confused.

In its basic form the instrument was found to be susceptible to the degree of eccentric loading that is likely to result from the use of the standard form of cone anchor that is commercially available for strand. However, this problem was overcome by applying the load to the instrument through a pair of ball-bearings at each end. The instrument is produced to the same diameter as a standard anchor, i.e. $3\frac{3}{4}$ in., and when the compensating plates for eccentric loading are added it is $6\frac{1}{2}$ in. long (Fig. 12b).

The esssential requirement of stability is met by using a steel body in which stresses are kept very low and by using as the photoelastic material glass, which is completely free from drift or creep and the effects of such changes in temperature and humidity as are likely to be experienced on site. When the instrument is used solely to rekord a change in the load it is not essential to compensate for eccentricity. The initial setting may be made at the time of stressing, using behind the jack another instrument against which the load cell to be left in place is matched. Subsequently the load will probably diminish, due to relaxation and creep in the prestressed stucture. This will be observed on the load cell as a fall in the fringe order, the extent of which can be measured.

EXTENSOMETER

The type la transducer is also used as the controlling medium of the extensometer (Fig. 12c). This consists of a differential micrometer held in a rigid harness which is set on two plugs attached to the surface on which measurements are to be made. The optical transducer is used to set the micrometer to the standard tension at which readings are observed. The datum setting of the instrument consists of turning the micrometer until the optical gauge shows an easily recognizable fringe pattern. Subsequent observations are made by setting the extensometer over the measurement pins and turning the micrometer until the optical gauge shows the standard fringe pattern, then reading the micrometer.

Rigorous tests have been conducted to determine the limits of accuracy to which readings can be made. These have included (i) taking off measurement from the measurement pins, then re-setting and repeating the measurement, and (ii) stripping the extensometer to its component parts and reassembling it between readings. It has been found that changes of length of 0.0001 in. between the anchor points can easily be detected and measured.

DETERMINATION OF *IN-SITU* LOADS USING PHOTOELASTIC STRESSMETERS

A problem which constantly presents itself to strata control engineers in mining, and indeed to all engineers who have to deal with massive structures of rock, concrete, or masonry, is that of determining the load within the structure. Various designs of borehole plug are applied for this purpose. Most of these devices are deformation meters which respond to changes in load within the medium in which they are set, although by application of the "stress relief technique" certain of them can also be applied to determine the absolute load. A more recent introduction to this range of equipment is the M.R.E. borehole plug which is a high modulus inclusion or stressmeter.(27) Photoelastic gauges are also applicable, both as high modulus borehole plugs or stressmeters and low modulus deformation meters.

An authoritative description of the stress relief technique is given by Slobadov.(28) Similar techniques are also described by Hast(29) and Obert(30). Slobadov's method consisted of drilling a hole in the rock, grinding the back of this hole flat, and then bonding two electrical resistance strain gauges at right angles to each other on the end face, by special tools. The strain gauges and the rock to which they were bonded were subsequently trepanned out in the form of a core so that the rock element to which the gauges were attached could expand elastically and provide information regarding the degree of constraint.

P. K. Chakravarty has applied a similar technique, using the photoelastic gauge instead of electrical resistance strain gauges.(31) The hole was drilled to a depth of 5 ft by a $5\frac{1}{2}$ in. diameter diamond core drill, any core left at the end of the hole being wedged out by a tapered rod. A grinding tool was then attached to the drill rod and the back of the hole made flat. A 2-in. diameter disk of photoelastic material was then bonded to the back of the hole and subsequently trepanned out. A study of the stress relaxation pattern in the disc by direct and oblique incident polarised light enabled the initial state of stress in the rocks to be deduced. Figure 13 shows typical cores extracted in this manner with the photoelastic disc attached.

FIG. 13. Stress relieved cores of rock and detached rock fragments with photoelastic gages attached.

The stress relief technique is also applicable for the observation of surface strains, by bonding the plastic gauge on to the rock and subsequently scaling off the rock element containing the cemented gauge. In such an investigation on the roof stratum of a mine at a depth of 625 ft shear-strains of 175 micro-in/in. were observed by the reflection polariscope, which, for the rock in question ($E = 6 \times 10^6$, $\mu = 0.26$) means that the maximum shear stress in the rock is approximately 700 psi. Observation under oblique incidence showed the principal stresses to be 4,000 psi and 3,240 psi, both compressive. In the same mine the principal stresses at a depth of 5 ft at the backs of boreholes ranged from 2,380–3,100 psi vertical stress, and 2,160–2,880 psi horizontal stress. The overburden stress in this mine, calculated on the basis of an average specific gravity of 2.4 for the overburden strata, would be no more than 1,600 psi.

PHOTOELASTIC BOREHOLE PLUG STRESSMETER

The elastic moduli of rock vary considerably with stress, which makes it extremely difficult to correlate strain and load when the strain is measured on the surface or inside the body of the rock. It is preferable to correlate directly the two quantities load and strain by calibration in a measurement system which is not effected by variations in the values of the moduli.

The theoretical basis of a suitable technique has been described by Sezawa and Nishimura,[(32)] and by Coutinho,[(33)] whereby the biaxial stress field around a high modulus solid inclusion within a material of lower modulus is considered. It can be shown that if the modulus of elasticity of the solid inclusion is approximately two or more times that of the surrounding material, the effect of variation of the elastic modulus of the material is negligible.

A photoelastic stressmeter has been constructed of glass ($E = 10 \times 10^6$ psi) in the form of a cylindrical plug, through the axis of which a hole is drilled, so that it can function as a biaxial stress gauge. This simple device in itself serves as a load

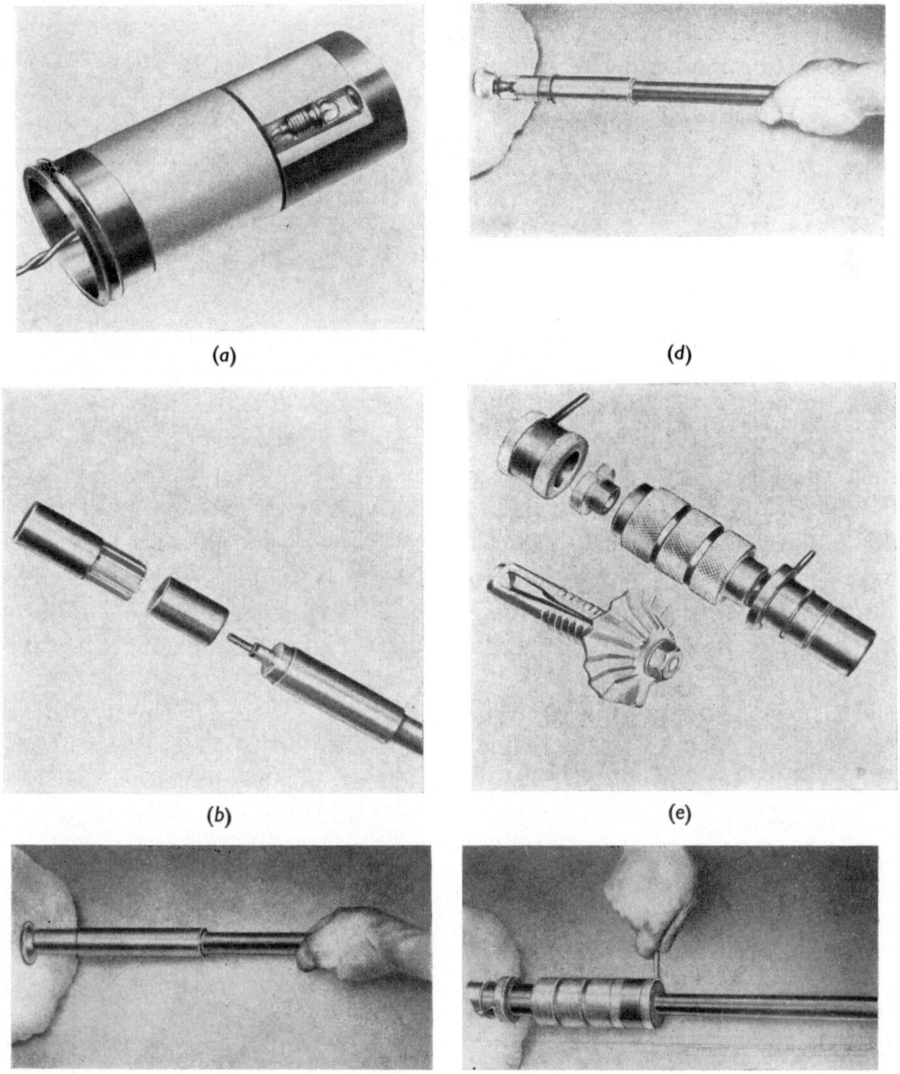

FIG. 14. Application of the photoelastic borehole stressmeter.

(a) the stressmeter unit. (b) method of assembling the cement box to the end of the drill rod. (c) insertion of the cement box into the borehole. (d) insertion of borehole plug. (e) assembly at the mouth of the borehole, for forcing the drill rods holding the stressmeter a measured distance during the setting procedure (plug and cap for sealing the mouth of the hole as shown). (f) operating the setting gear at the mouth of the drill hole.

gauge and when inserted at shallow depth can be observed by reflection polariscopes. Observations at depths beyond the range of reflection polariscopes may conveniently be made if the plug is viewed by transmitted light, and the stressmeter constructed by the authors adopts this principle. The stressmeter therefore consists of a glass plug behind which is fitted a light source, carried in a transparent capsule of acrylic resin, the light from which is polarised by a combined quarter wave and linear polarizing film. Leads to supply current for the light source are brought through the axial hole of the stressmeter.

The plug is inserted and cemented at the back of a borehole, using a simple setting apparatus attached to the drill-rod (Fig. 14). Observation of the stressmeter is effected by supplying current to illuminate the plug which is then either viewed by telescope or photographed. The optical fringe pattern observed is related directly to stress by calibrating the plug in a cube of rock subjected to known loads.

LOAD CELLS

The determination of the loads on mine supports is customarily performed either by the aid of hydraulic dynamometers or by load cells inserted in the support system. The loads so measured are likely to have no readily identifiable relationship to the strata pressure, since they are a function of the elasticity of the materials in contact with the cell, or with the mine support. This is not important if our concern is to determine the load at which the support will yield or fracture, but it does lead to difficulties when we wish to determine the loads carried by packs or constructions of brick, masonry, or concrete.

In such circumstances the body of the load cell itself tends to act as a high modulus inclusion and measures the stress concentration produced by the redistribution of the load. This would not matter if the load cell can be calibrated under the exact conditions of loading, but this is not always possible. Ideally the body of the load cell should consist of the same material as will surround it in use. The photoelastic stressmeter can very easily be applied in this manner, since it can be inserted in a brick or in a masonry block, or cast in a cement block, to be incorporated later as an integral part of the structure concerned.

The determination of the load on a pack is a particularly difficult strata control problem and it is not suggested that the insertion of one photoelastic dynamometer will, in itself, yield a useful reading. Nevertheless the photoelastic devices can be used for this purpose and, since them are relatively cheap to produce, more of them can be used and a result assessed statistically from the results of a number of observations.

Load cells that are to be used for the direct measurement of point loads, for example on props and in structural steel work, can be constructed by setting a stressmeter in a designed steel body which is directly calibrated in terms of applied load.

THE STRESS CONCENTRATION TECHNIQUE

If a photoelastic transducer in the form of a birefringent coating is attached over the whole of its surface to a material of much higher modulus, such as steel, the birefringence that will result from elastic strain in the steel will be measurable, but will not be obviously apparent as an isochromatic pattern of high fringe order. To observe it accurately would require the use of optical viewing equipment and techniques that, while presenting little difficulty in the laboratory, may possibly

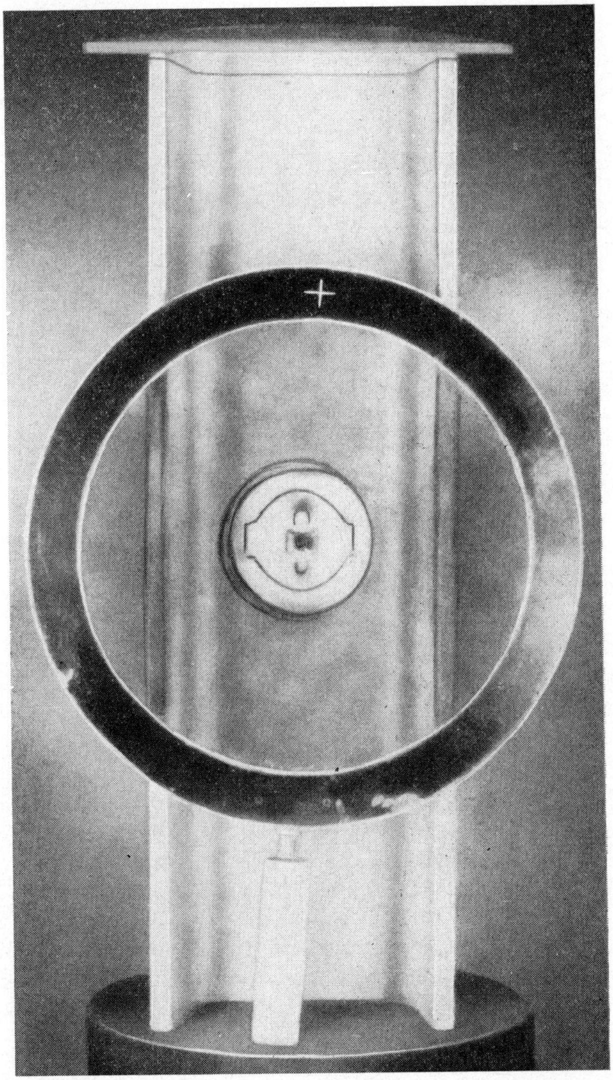

FIG. 15. Steel joist under load, with stress concentration disk attached, and seen through the simple hand-viewer.

be inconvenient in the field. There are other occasions also when a photoelastic coating may show relatively small birefringence although the stresses in the material being observed may be high. This is because the optical pattern seen in the coating when viewed under light incident at right angles to it is determined by the shear strain, which in turn is proportional to the difference between the principal stresses, and not to their individual magnitudes. In other words, if the strain is uniform, there may not be much to see. In a gauge which is cemented over the whole of its surface to the work piece under observation, the fringes may also be transient, i.e. they might not remain visible on the disk no matter what the fringe order, but appear and later disappear as higher or lower fringe orders develop. As a result, the value of the fringe order on such a gauge may be difficult to identify, unless a count can be made from a known condition of strain.

In such circumstances it is advantageous to use a technique that results in an intensification of the strain in the gauge so that high fringe orders are produced is such a way that they remain visible on the gauge and can be easily identified.

It is known that the existence of a hole in an elastic material under load causes an intensification of stress in that material around the hole, and this fact can be made use of to facilitate observation of the elastic strains in steel, using a birefringent plastic. The technique consists of attaching a plastic disk over a hole in the material being studied. The disk records the stress concentration produced by the hole. If at the same time a hole exists in the plastic, as is the case in a biaxial gauge, the stress in the gauge itself is correspondingly multiplied and the plastic shows the characteristic four-winged pattern with its major axis of symmetry indicating the direction of load. Using this system of measurement it is necessary to calibrate the fringe pattern in terms of applied load. Fig. 15 shows a section of 4 in. $\times$ 4 in. steel joist in which the load is being observed by the stress concentration technique. The range of the device attached to this joist is from 5 to 50 tons, measurable to within $\pm\frac{1}{2}$ ton over the whole of its range.

It should be remembered that when this technique is applied to a joist under load and the gauge is attached before the stress concentration hole is drilled, the gauge will respond to the total stress in the joist, including such residual stresses which may be locked into the steel as a result of the material fabrication processes.

VIEWING INSTRUMENTS

An essential part of the system lies in the various instruments used for viewing the gauges. A simple hand-viewer is used to read photoelastic gauges within arm's length, to one half-fringe of accuracy. It consists of a combined linear polarizing filter and quarter-wave retardation filter, set between glass and mounted in a frame of a size suitable to fit into the pocket. A monochromatic filter may also be included for viewing gauges that require a count of fringes.

When looking at photoelastic coatings or transducers at very close range the light source, which for convenience may be a cap lamp, must be directed through the viewer which thus serves as a reflection polariscope. When this is not practicable or convenient two separate polarizing filters are used, one for the light source and one for the photoelastic transducer. The polarizer for the light source contains similar components to those in the hand-viewer, held in a retaining cup which clips over the headpiece of the cap lamp. The axes of polarization on the simple hand-viewer and on the light polarizer are marked by directional arrows, to facilitate their proper orientation.

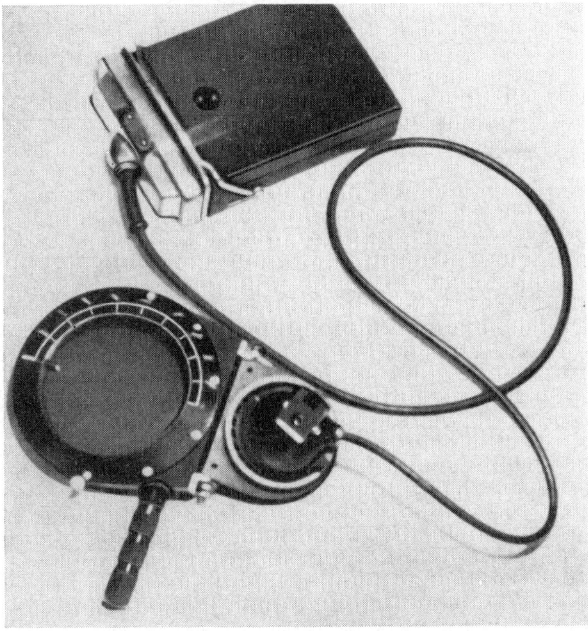

FIG. 16. The precision hand-viewer.

A precision hand-viewer is used to read the gauges to a high degree of accuracy by resolving the line pattern in the gauge to an exact fringe order or to a standard recognizable fringe pattern. By this technique, known as "goniometric" fringe compensation, which simply involves the rotation of a linear polarizing filter in the viewer, the gauges can be read to within 5 per cent of a fringe.

The viewer is illustrated in Fig. 16. Its optical system consists of a light polarizer which includes a linear polarizer and quarter-wave filter the respective optical axes of which are set with a 45° phase difference. The polarizing filter faces the light source (cap lamp) and its axis of polarization is in line with the viewer handle. A separate viewing system is provided which includes a quarter-wave filter fixed to the main body and a linear polarizer which can be rotated through an angle of 180° in the viewing plane. When the viewer is in its zero position as read by an index and scale its axis of polarization is crossed relative to the light polarizer. The quarter wave plate in the viewing system is fixed at 90° to that in the light polarizer.

VIEWERS FOR DISTANT OBSERVATION

When the gauge is at more than arm's length from the observer the hand-viewers are not suitable. Viewers for such application include a periscope viewer which is intended to read gauges up to 15 ft above ground level, and was originally designed for use with the bolt tension meter. The essential components of the precision hand-viewer are here divided between a lampholder that can be hooked on to the gauge and the viewing eyepiece, between which runs an extensible tube containing the necessary lenses for optical magnification. As with the precision handviewer, provision is made for goniometric compensation.

For reading gauges at distances up to 50 ft from the observer, telescope viewers may be used, and a prototype viewing telescope was constructed by adding a fixed quarter wave filter and a movable analyser to a surveyor's level of the reversible type. This was a suitable instrument to adapt, as the rotation of the telescope about its optical axis could be measured on a scale. To use this instrument it was first levelled and aligned on the direction of the borehole. The eye-piece was then rotated until the crosshair coincided with the axis of symmetry of the fringe pattern seen on the borehole plug. This marked the directions of principal stress and also formed the datum position for the rotation of the analyser in the process of goniometric compensation to measure the fringe order.

Viewers used to read simple glass plugs by reflected light must project a beam of light along the axis of the borehole, which must also be the line of sight. A suitable instrument is described by Hiramatsu, Niwa, and Oka[35], and a simple modification of this, using a cap lamp as the light source, has been designed by the present authors.

Gauges that are inaccessible to distant view by the periscope or telescope can be monitored by remote control. Where the transducer gives out a fringe count, photoelectric monitoring to transpose the optical signal to an electrical digital count presents little difficulty. The application of photoelectric readout devices for precision measurements in photoelastic stress analysis is now widely practised. Using such devices measurements can be made with extreme accuracy, to one-thousandth parth of one fringe order.

Transducers of the type that give out a signal pattern can be monitored photographically. Borehole television camera equipment is available if the project is of sufficient scale and importance to justify the cost of this. Otherwise, film cameras can be applied on a more modest budget.

SYSTEMS OF MEASUREMENT

One of the primary objects of this work is to provide a simple means by which information about the strata pressures and support loads can be obtained by ordinary mine personnel, during normal working conditions. That is, the aim is to integrate the strata control observations into the mining routine. If this objective be is to achieved the gauges must be simple to set and operate and easy to read.

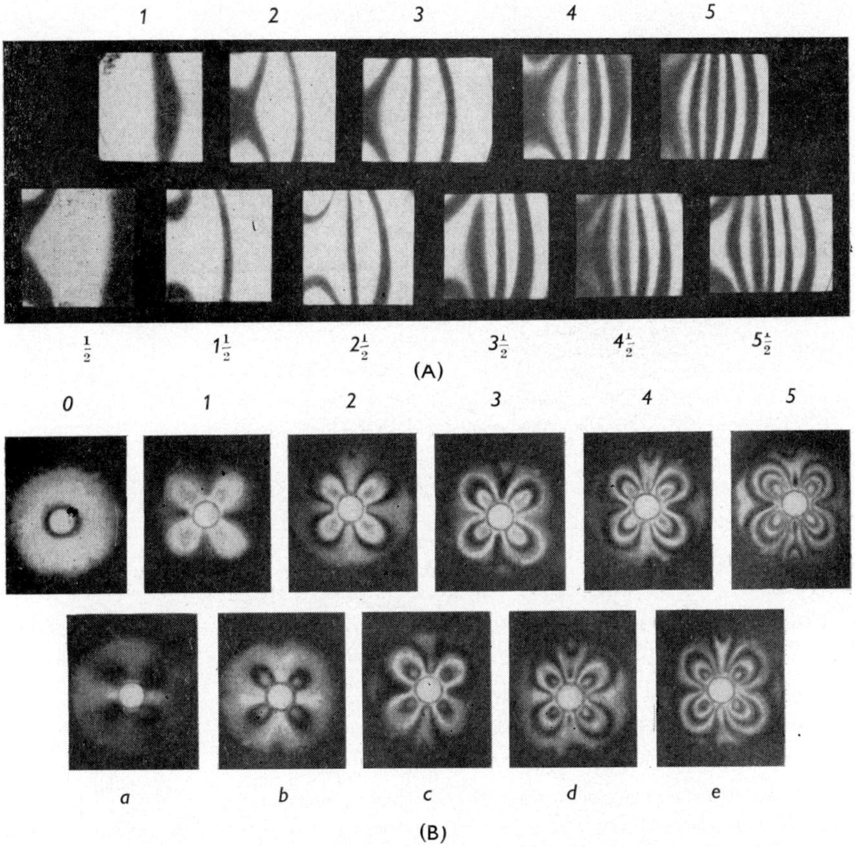

FIG. 17. Optical scales of measurement

A. A fringe count scale.
B. A typical scale operated by pattern recognition.

There are two basic methods by which the gauges are read, one involving a count of fringes, the other depending upon the recognition of specific patterns.

In Fig. 17A is shown a simple count of fringes, such as is associated with transducers of type la. The half fringe orders are shown and there is no difficulty in reading to half a fringe, directly, through the simple hand-viewer. Precise measurements, to within 5 per cent of the unit interval, are possible with the aid of viewers fitted for goniometric compensation.

PATTERN RECOGNITION

Figure 17B shows another series of fringe patterns which form a scale such that each pattern is characteristic of its place on the scale. Patterns would not always be exactly of this shape. Each system of loading would produce its own characteristic pattern depending upon the geometry of the system and the material properties of the various components. With this type of gauge it is best to calibrate, not on the basis of a fringe count, but on the basis that each step on the scale is recognizably different from its neighbours on either side. The upper series of patterns, reading from left to right, represent unit steps in progression up the scale from 0-5.The lower series of patterns represent intermediate positions between the unit steps on the scale. These may not be immediately recognizable as such to the casual observer, although those accustomed to the technique and familiar with the manner in which the fringe pattern grows with increase of load will be able, not only to recognize them as intermediate stages, but also to estimate their position relative to the two numerically designated patterns above and below.

However, the precise location of the intermediate patterns on the scale is possible by goniometric compensation.

GONIOMETRIC COMPENSATION

The theory of the system of goniometric compensation is that described by Tardy.(34) In this technique it is applied by rotating the analyser of the viewing instrument, which must be of a precision type, while observing the fringe pattern. The effect of such rotation is to change the apparent birefringence to the extent that 180° rotation of the analyser changes the retardation of light by one fringe order. The extent of this rotation is therefore measured on a scale marked in percentages, over the angle of 180°. The change in birefringence of the observed pattern with rotation of the analyser is such that readings can be made to an accuracy within 5 per cent of one fringe, i.e. to ± 0.025 of each scale.

The operation of making a reading consists therefore of observing the fringe pattern through the viewing instrument with the index of the analyser at zero. The viewer is then rotated bodily, if necessary, until the zero – 180 axis of the instrument is perpendicular to the major axis of symmetry of the pattern, i.e. perpendicular to the axis of loading on the gauge. Holding the viewer firmly in this position the analyser is then rotated until one of the standard scale patterns 0–5 is seen. The reading is then that scale number plus the percentage indicated by the index of the analyser. The observer need make no attempt to identify the pattern as it is first seen on the gauge, but only to recognize the standard patterns. The value of this will be readily apparent by reference to Fig.17b where it can be seen that the intermediate patterns *c* and *d* are very much alike and could easily be confused, one for the other. However, pattern *c* compensates back to pattern No. 2

while pattern *d* compensates back to pattern No. 3—two patterns which are identifiably different.

A similar system of goniometric compensation can be used with gauges of type 1a, relating to fringe counts. In this case the intermediate fringe orders i.e. the lower row of photographs in Fig. 17a, are goniometrically compensated to the exact unit fringe counts or upper row of photographs in Fig. 17b, and the magnitude of the intermediate order read off the scale of the viewer.

CALIBRATION

Gauges which consist of a layer of birefringent material attached to the surface of the object under investigation are calibrated in the normal manner associated with the photoelastic coating technique. Details are given in the bibliography[25, 26].

Gauges that contain an optical transducer in a particular type of housing or enclosure must be calibrated under controlled conditions, in which the factor, the effect of which is to be observed, be it strain, stress, or load, is brought to bear on the transducer in the same manner as that which is expected in the field. In some of the instruments this presents no difficulty, as the devices are so constructed that the transducer is always subjected to strain or load in the same manner, no matter where it is used, but the same may not apply to load cells or stressmeters. The photoelectric instruments are not unique in this requirement, which is often overlooked when applying dynamometers and load cells in field investigations. The requirement of similarity in the manner of loading means, for example, that a load cell must be set on the testing machine and subjected to known loads with the same materials above and below it that it will encounter in its field setting, and with high friction surfaces of contact between the cell and its neighbouring materials, both in the laboratory and in the field.

The stressmeter is a special case in that it need not be calibrated for every type of material in which it is to be used.

The modulus of elasticity of the photoelastic stressmeter is 10×10^6 psi, which is more than twice that of many of the sedimentary rocks likely to be encountered in British mines. Recent research in the Postgraduate School at Sheffield[31] has shown that, in uniaxial compression at stresses ranging from 1 to 10×10^3 psi. Youngs Modulus for Darley Dale sandstone, which is a moderately hard Carboniferous rock, varies from $1–3.5 \times 10^6$ psi. In triaxial loading, under confining pressures from zero to 5000 psi the Youngs Modulus of the same rock has been observed[35] to range from 2.77 to 4.51×10^6 psi.

Coutinho's theory for a high modulus inclusion in an elastic material is applicable to the photoelastic stressmeter when inserted in materials of less than $E = 5 \times \times 10^6$ psi. In such cases variations of Young's Modulus in the material in which the plug is inserted have no appreciable influence upon the response of the stressmeter.

The method of calibration is to insert the stressmeter in a cube of material which is subsequently subjected to known loads. Theoretically it can be shown[37] that the minimum size of cube required for a $1\frac{1}{4}$-in. diameter inclusion is about $6\frac{1}{4}$-in. side, if the inclusion is to have no effect on the stresses at the edge of the cube. This means that a calibration made under these conditions would be representative of the stresses in an infinitely large body subjected to a similar intensity of loading.

A linear calibration, giving a constant ratio of stress/fringe order is obtained in all materials having E less than 5×10^6 psi. The calibration constant and the range of the stressmeter in such materials is determined by the optical characteristics of the glass and the geometry of the plug. As at present designed, optical stressmeters can be constructed to measure a range of stress increase from 0 to 3000 psi with a sensitivity of ±12 psi to 0–400 psi with a sensitivity of ±21 psi.

It is important that care should be exercised if the stressmeter is to be used in rocks or other materials with a high elastic modulus approaching that of glass, for in that case a linear calibration may not apply. If it is to be used in materials where E is greater than 5×10^6 psi the meter should be calibrated in that particular material.

ADVANTAGES OF PHOTOELASTIC GAUGES

There are several features of the photoelastic gauges which render them excellently suited for field applications in industry, and in mining and engineering generally. The gauges are robust and self-contained. Simple in construction and in use, they need no associated costly or delicate mechanical or electrical equipment. The gauges that are observed by reflection polariscope need no ancillary equipment at all, other than the viewer and hand torch or lamp, the operation of which has been brought to a very simple procedure. Indeed it is no exaggeration to say that any person who is able to count can install the meters and observe their readings. They can therefore be used as control tools, by which the men on the job can observe stress, strain, and load, in mines and earthworks, in structures and in support systems, as the operations proceed. At the same time, the gauges, when used in conjunction with the more refined measurement techniques and instrumentation, lend themselves very usefully as research tools for the purpose of stress and strain analysis.

The photoelastic stressmeter is unique in that it produces an optical pattern which gives directly, without mathematical computation, the following information:

(1) The magnitude of the increase in stress subsequent to the time of setting the meter in place, and the magnitude of the two principal stresses perpendicular to the axis of the borehole in which the meter is set.

(2) The direction of the two principal stress perpendicular to the axis of the borehole.

(3) The relationship between the two principal stresses, i.e. it defines the system of loading as to whether it is hydrostatic, uniaxial, or biaxial, in the plane of measurement.

The biaxial birefringent plastic gauge, when applied together with the over-coring technique, can give in addition to all the above information:

(4) The magnitude of the load in that part of the body on which the gauge is set, at the time the gauge is set. That is, the gauge measures absolute load as well as change of load.

(5) The sense of the load in the body, that is, whether the area to which the gauge is attached is in tension or compression.

It is doubtful whether a simpler method could possibly be devised for finding the state of stress in a rock body at depth than the photoelastic disc used in conjunction with the over-coring stress relief technique. The whole operation requires no complicated equipment or any involved mathematical assumptions or calculations. When the core, with the gauge attached to it, is examined in the laboratory, the determination of each of the principal stresses can be reduced to a simple routine of instrument reading and the application of appropriate multiplication factors.

The application of the surface gauges to the rocks around mining excavations and engineering earthworks is limited by the fact that the rock surface may contain fractures so that the strains observed can be interpreted only as being applicable to the place at which they are observed. In this respect, however, the technique is no different from other techniques of measuring surface strain and it is true to say that the photoelastic coating records the strain on the surface to which it is attached in a manner that is inherently more comprehensive than are other methods, within the limitations imposed by the physical properties of the materials used.

These properties are such that one can have every confidence in the technique for short-term investigations, and for long-term investigations where the ambient temperatures lie below 80° F. The technique is still new and confidence will grow with experience in using it intelligently, and applying it within its limitations.

The stressmeter and gauges employing glass as the photoelastic material are entirely free from the objections frequently levelled at the plastics, i.e. they are perfectly stable and unaffected by moisture, humidity and temperature.

The optical gauges are also unique in that the validity of the signal given out, in relation to the calibration of the gauge, is self-evident. If the gauge is functioning properly, the shape of the pattern indicates that this is so. Conversely a pattern shape that does not conform to the standards for that type of gauge indicates that the gauge is malfunctioning and should not be read. These properties are not possessed by electrical or mechanical gauges, from whose signals it is often impossible to tell whether or not the gauge is in order and properly set. When one also takes into consideration the simplicity and robustness of construction of the photoelastic gauges, there can be no doubt that they promise to have a wide application in many fields of engineering, and particularly in mining.

Acknowledgments: The authors have been assisted in this work by their colleagues on the staff and by the research students of the Mining Department at Sheffield University, and particular mention should be made of Mr. F. T. Williams, Mr. P. K. Chakravarty and Mr. J. W. Strafford. To all of these the authors express their appreciation and thanks.

The equipment described in this article is patented and is manufactured, under licence, by Horstman Limited, Locksbrook Road, Bath, England.

BIBLIOGRAPHY

1. PHILLIPS, D. W. and HUDSPETH, H. M., The forces induced by the extraction of coal and some of their effects on coal-measure strata, *Trans. Inst. Mining Engrs., London* Vol. 85, 1932-33, Pt. 1, April, 1933, pp. 37–56.
2. DUVALL, W. I., Stress analysis applied to underground mining problems. *U.S. Bur. Mines Rep. In.* 4192, March, 1948.
3. POTTS, E. L. J.,, Stress distribution rock pressures and support loads, *J. Leeds Univ. Min. Soc.*, Vol. 30, 1954, pp. 107–121.
4. HOEK, E., The application of photoelastic models as a means of solving problems of stress and fracture around excavations at depth, C.S.I.R. (South Africa) Publication RN. 114, 1960.
5. MESNAGER, M., Sur la détermination optique des tensions intérieures dans les solides à trois dimensions, *Compt. Rend. Ass. franc. Avanc. Sci.*, Vol. 190, June, 1930, pp. 1249–1250.
6. ZANDMAN, F., Analyse des contraintes par vernis photoélastiques. *Groupement pour l'Avancement des Méthodes d'Analyse des Contraintes*, Vol. 26, March, 1953, pp. 1–12.
7. D'AGOSTINO, J., DRUCKEV, D. C., LIU, C. K., and MYLONAS, C. Epoxy adhesives and casting resins as photoelastic plastics, *Proc. Soc. Exptl. Stress. Anal.*, Vol. 12, 1955, pp. 123–128.
8. ZANDMAN, F. and WOOD, M. R., Photostress—a new technique for photoelastic stress analysis, *Prod. Eng.*, Sept. 1956, pp. 167–178.
9. ZANDMAN, F., Stress analysis of a guided missile tail section, *Proc. Soc. Exptl. Stress. Anal.*, Vol. 17, 1959, pp. 135–150.
10. EMERY, C. L., The strain in rocks in relation to mine openings, *Trans. Inst. Mining Engrs., London*, Vol. 120, 1960–61, Oct., 1960, pp. 54–61.
11. EMERY, C. L., The measurement of strains in mine rocks, *International Symposium on Mining Research*, Pergamon Press, Oxford 1962, pp. 541–553.
12. DANTU, M. P. Étude des contraintes dans les milieux hétérogènes. Application au beton, *Ann. Inst. Tech. Bâtiment et Trav. Publ.* No. 121, Jan., 1958, pp. 55–77.
13. SPOONER, H. and MCCONNEL, L. D., An ethoxylene resin for photoelastic work, *Brit. J. Appl. Phys.*, Vol. 4, 1953, pp. 181-184.
14. KAWATA, K., Analysis of elasto-plastic behaviour of metals by means of photoelastic coating method, *J. Sci. Res. Inst., Tokyo*, Vol. 52, 1958, pp. 17–40.
15. HIRAMATSU, Y., OKA, Y. and OGINO, S., Investigations on the stress in circular shaft linings, *Mem. Fac. Eng. Kyoto*, Vol. 23, Jan. 1961, pp. 90–110.
16. TARDY, H. L., Méthode practique d'examen et de la mesure de la birefringence des verres d'optique, *Rev. Opt.* Vol. 8, 1929, pp. 59–69.
17. JESSOP, H. T. and HARRIS, F. C., *Photoelasticity*, Cleaver-Hume, London, 1949, p. 70.
18. ROBERTS, A. Photographic photometry applied to the study of mine lighting, *Trans. Inst. Mining Engrs., London*, Vol. 112, 1952–53, Sept. 1953, pp. 1029–1046.
19. CORLETT, A. V. and EMERY, C. L., Prestress and stress redistribution in rocks around a mine opening, *Trans. Can. Inst. Mining Met.*, Vol. 62, 1959, pp. 186–98. *(Can. Mining Met. Bull.*, No. 566, June 1959, pp. 372–84.)
20. EMERY, C. L., Testing rock in compression, *Mine & Quarry Eng.*, Vol. 26, 1960, pp. 164–168, 196–202.

21. LISOWSKI, A., Failure types of rock cubic specimens in the light of the theory of elasticity, *Bull. Acad. Polon. Sci.*, Vol. 7, No. 5, 1959, pp. 341–351.
22. OPPEL, G. U. Photoelastic strain gauges. *Proc. Soc. Exptl. Stress. Anal.*, Vol. 18, 1961, pp. 65–73.
23. MURRELL, S. A. F. and MISRA, A. K., Time-dependent strain or "creep" in rocks and similar non-metallic materials, *Trans. Inst. Mining Met., London*, Vol. 71, 1961–62 (*Bull. Inst. Mining Met., London* No. 665, April 1962), pp. 353-378.
24. EMERY, C. L., The strain in rocks in relation to mine openings. *Trans. Inst. Mining Engrs., London*, Vol. 120, 1960, p. 54.
25. ROBERTS, A., EMERY, C. L., HAWKES, I., WILLIAMS, F. T., and CHAKRAVARTY, P. K., The photoelastic coating technique applied to research in rock mechanics, *Trans. Inst. Mining Met.*, Vol. 71, 1962, p. 581.
26. HOLLISTER, G. S., The photoelastic method applied to materials research, *Appl. Mat. Res.*, Vol. 1, 1962, p. 149.
27. WILSON, A. H., The measurement of rock stress, *Colliery Guardian*, 25 January, 1962, pp. 118–127.
28. SLOBODOV, M. A., The method of relief from load in studying stresses in the depth of rock mass, *Ugol*, Vol. 33, No. 7, 1958. pp. 30–35.
29. HAST, N., *The Measurement of Rock Pressure in Mines*, Arsbok 52, Novstedt and Sonts, Stockholm, 1958.
30. OBERT, L., *In situ* determination of stress in rock, *Mining Eng.* August, 1962, pp. 51–58.
31. CHAKRAVARTY, P. K. The application of the photoelastic technique to the problems of rock mechanics, University of Sheffield, Ph. D. Thesis.
32. SEZAWA, K. and NISHIMURA, G., Stresses under tension in a plate with heterogeneous insertion, *Rep. Aero. Res. Inst. Tokyo*, Vol. 6. 1931, pp. 25–43.
33. COUTINHO, A., Theory of an experimental method for determining stresses, not requiring an accurate knowledge of the elasticity modulus, *International Assoc. of Bridge and Structural Engineers Congress*, No. 9 (1949), pp. 83–103. (N. C. B. Translation A 632/AB.)
34. TARDY, H. L., Methode practique d'examen et de la mesure de la birefringence des venes d'optique, *Rev. Opt.*, Vol. 8, 1929, pp. 59–69.
35. MURRELL, S. A. F. The effect of triaxial stress systems on the strength of rocks, University of Sheffield, Ph. D. thesis.
36. JENKINS, J. D. and STOREY, I., Support loads at the coal face, *Trans. Inst. Mining Engrs., London*, Vol. 119, 1960, p. 699.
37. SAVIN, G. N., *Stress Concentration around Holes*, Pergamon Press, 1961.
38. HIRAMATSU, Y., NIWA, Y. and OKA, Y., Measurement of stress in the field by application of photoelasticity, Technical Report Engineering Research Institute, Kyoto University. VII 3, No. 37 (1957).

CHAPTER 12

LABORATORY METHODS FOR DETERMINING PHYSICAL CHARACTERISTICS OF ROCKS

THE following are the most important of the structural characteristics of a rock which may be determined by laboratory examination and testing.

Petrographic description
Hardness
Apparent specific gravity
Apparent porosity
Compressive strength
Compression fracture characteristics
  Fracture angle
  Fracture classification
  Strain energy at failure
Flexural strength
Tensile strength
Shear strength
Coefficient of internal friction
Modulus of elasticity
Poisson's ratio

The foregoing are the most important of the physical characteristics which determine the structural behavior of a rock. Other properties which may be determined in the laboratory are:

Scleroscope hardness
Impact toughness
Abrasive hardness

These properties are indices of the amenability of a rock to drilling, grinding, and blasting.

The following elastic properties may also be measured. These values may be used in connection with seismic prospecting or other geophysical exploration methods:

Modulus of rigidity
Specific damping capacity

Petrographic Description

Each rock sample which is to be tested in the laboratory should be throughly examined and the following characteristics should be recorded:

Rock type
Major and minor constituents
Grain size
Cementing material
Percentage of grains to cementing material
Type and degree of structure present
Moh's hardness of the specimen

Apparent Specific Gravity

The apparent specific gravity may be calculated from the following equation:

$$\text{apparent specific gravity} = \frac{M_1}{M_w - M_s} \tag{1}$$

where

M_1 = weight of dry specimen
M_w = weight of wet specimen after water immersion
M_s = weight of specimen suspended in water

When a specimen has a regular geometric configuration its volume may be calculated from measurements and its specific gravity calculated by dividing its dry weight by that of an equal volume of water.

Specimens may be oven-dried, dried in a dessicator, or air-dried at room temperature for two weeks. The latter procedure is probably the most common.

Apparent Porosity

The apparent porosity represents the ratio of the volume of open pore space to the exterior volume of the specimen. This can be calculated from the following equation:

$$P = \frac{M_w - M_o}{d_w V} \times 100 \tag{2}$$

where

P = apparent porosity, per cent
M_w = weight of water-saturated specimen
M_o = weight of oven-dried specimen
V = external volume
d_w = density of water

SPECIMEN PREPARATION

Rock specimens for laboratory tests are commonly taken from cores from exploratory diamond drill holes and are therefore cylindrically shaped. In cases where run-of-mine rock is to be tested, cylindrical specimens are usually cored from a large rock block in the laboratory with a diamond bit. Diamond tipped cutting wheels are customarily used for sawing the cylindrical specimens to proper length.

The diamond saw requires considerable quantities of fluid (usually water) to be applied to the cut to prevent the build-up of excessive heat which would damage the saw. However, some of the softer sedimentary rocks contain clay minerals which soften when exposed to water and therefore do not provide satisfactory test specimens. If an attempt is made to cut "dry" with a diamond wheel, certain types of rocks tend to "fire" as a result of excessive heat at the bottom of the cut and the resulting glaze effectively stops further cutting. Hackett[(1)] has found that certain types of sedimentary rocks can be successfully cut and shaped by using carborundum-grit cut-off wheels. A shellac bonded wheel with a grit size between 36 and 46 proved to be best. Wheels with smaller grit sizes clogged rapidly while wheels with larger grit sizes were inefficient cutters and wore rapidly.

It was also found that ordinary metal-working machine tools could be adapted successfully to the working of rock specimens by "machining" of irregularly shaped specimens of run-of-mine rock to produce the required geometrical shapes for testing.

A 36-in. bed surface grinder was used to produce flat surfaces on the rough specimens. The depth of cut which could be taken with safety varied considerably with the nature of the rock. With tough siltstones and shales it appeared that cuts of $\frac{1}{16}$ in. or more could be taken until the final pass when the depth of the cut had to be reduced to a few thousandths of an inch for best results.

After two parallel flat surfaces were formed on a rough specimen the specimen was clamped between the rubber-faced spring-loaded end plates of the cylinder and ground to a smooth cylindrical shape.

Because all grinding was done dry it was necessary to install a dust collector system. A high-volume low-pressure collector system was found to be most efficient.

COMPRESSIVE STRENGTH TESTING

Compression tests are normally made on specimens which are cut from diamond drill cores. The largest size diamond bit in common mine use is the NX which cuts a core $2\frac{1}{8}$ in. in diameter. The hydraulic press provided for the laboratory should therefore have a capacity sufficient to break rock cylinders with a diameter of up to $2\frac{1}{8}$ in. Since the stronger rocks have compressive strengths of 30,000 psi or more a hydraulic press with a capacity of about 120,000 lb would be the minimum desirable size.

In softer rocks $5\frac{5}{8}$ in. diameter cores are frequently taken for laboratory testing purposes. If it is intended to test specimens of this size it would be desirable to have available a press with a capacity of at least 250,000 lb. The larger hydraulic press will also permit triaxial tests to be performed.

Procedures

After a specimen is cut to the required length on the diamond saw the ends are usually lapped (ground smooth) with No. 200 silicon carbide to obtain load-bearing faces which are as nearly true and parallel as practicable.

The degree of overall flatness can be checked by observing the contact with the surface of a plate which is coated with Prussian blue.

Care should be taken to center the specimen when placing it in the testing machine, and the initial load should be applied at a low rate to permit adjustment of the bearing plates so that when the load is applied the surface of the plate will be parallel to the end of the specimen.

Bearing plates should be equipped with spherical bearings to permit alignment of the face of each plate with the bearing surface on the specimen.

The U.S. Bureau of Mines standard procedure[(2)] calls for no capping to be used on the ends of the specimens.

The compressive strength is calculated from the applied load and the cross-sectional area as follows:

$$\text{compressive strength} = \frac{\text{total load}}{\text{cross-sectional area of specimen}}$$

Short Period Uniaxial Compressive Strength Tests

The compressive strength of a rock specimen varies with the rate at which the compressive load is applied.

The breaking strength of a rock under a dynamically applied compressive load (for example an impact load which lasts only a few milliseconds) may be several times the breaking strength as determined when a load is applied to the rock slowly as in the usual laboratory test. If the same rock is subjected to a continuously applied compressive stress of constant magnitude for a period of weeks or months it will be found that many rocks progress to fracture at loads much less that those applied in ordinary laboratory tests.

Because the duration of the load application affects the strength, it is customary, in laboratory compression testing of rock specimens, to apply the load at a rate of 100–200 psi/sec. The U.S. Bureau of Mines standardized procedure[(2)] calls for the application of stress at the rate of 100 psi/sec.

Compressive strength tests discussed in this section are "short period uniaxial tests" – with the term "short period" being used to distinguish these tests from test where impact loading or tests where long-term loading is used.

The term "uniaxial" indicates that the specimen is subjected to compressive stress along one axis only and is not confined or restrained on other axes as are specimens which are subjected to triaxial compression testing.

The Effect of Specimen Length/Width Ratio on Compressive Strength

It has been found[2, 3] that the compressive strength of rock is a function of both the diameter and the length of the test specimen.

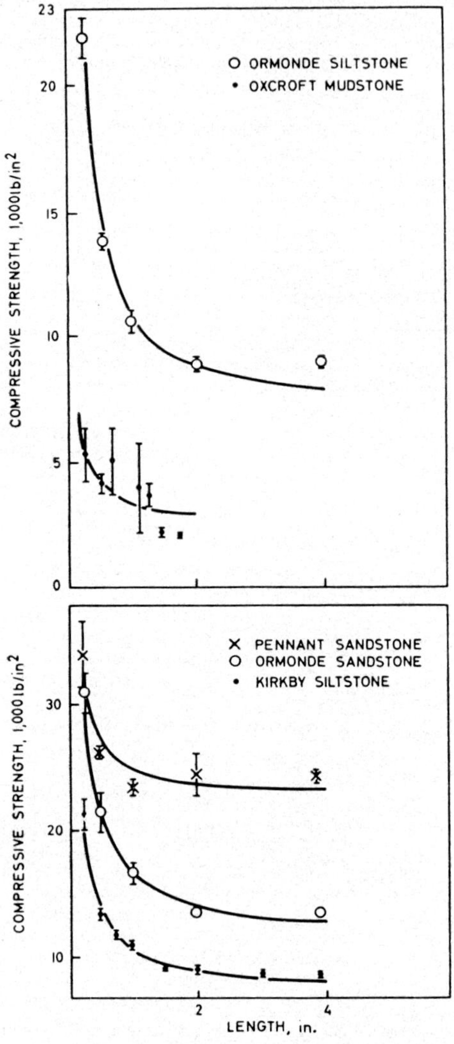

FIG. 1. Relationship between mean uniaxial compressive strength and length of 1-in. diameter rock specimens.[3]

Generally it is considered desirable[(4)] that the length (or height) of the specimen should be at least twice its lateral dimension so that there will be no interference by the platens with the development of shear planes.

Hobbs[(3)] conducted experiments on specimens of siltstones, sandstones and mudstones to determine the effect of cylinder length on compressive strength. The test cylinders were 1 in. in diameter and ranged up to 4 in. in length. The results of the tests are shown in Fig. 1 and the graphs indicate that when the length-to-width ratio is 2 or more the true compressive strength of the rock is determined directly from the test readings.

Hobbs also found that the strength relationship between length and width of samples tested was closely expressed by the equation:

$$\frac{C_h}{C_2} = 0.848 + \frac{0.304}{h} \tag{3}$$

where

C_2 = the compressive strength of a 2-in. length cylinder
h = specimen length
C_h = is the compressive strength of a specimen with a length-to-diameter ratio other than 2.

The diameter of all specimens tested by Hobbs was 1 in. and the constants derived for equation (3) are therefore based on the use of cores 1 in. in diameter.

Previous work by Obert, Windes, and Duvall[(2)] had found that the following equation gave good results for correcting strength tests to a basis of length/diameter = 1. The equation was found to be applicable to specimen tests for a wide variety of rocks.

$$C_1 = \frac{C_h}{0.778 + \dfrac{0.222D}{h}} \tag{4}$$

where

C_1 = compressive strength of a specimen with a length/diameter ratio of 1
C_h = is the compressive strength of a specimen with a length/diameter ratio greater than 1
D = specimen diameter
h = specimen length

Alternative Techniques for Determining Compressive Strength of Rock

Because it is sometimes difficult to obtain finished specimens with the proper dimensions for laboratory tests, Hobbs[(3)] suggests the following alternative procedures for determining the compressive strength of rock:

1. The "lump" test in which irregularly-shaped lumps of rock are crushed between parallel steel platens.

2. The "impact strength index" test in which rock fragments are broken inside a hollow cylinder by a succession of drop hammer blows.

The Irregular Lump Test

The crushing of irregularly-shaped lumps of rock provides a method of estimating the compressive strength of cylindrical rock specimens having a 1-in. diameter and a 1-in. length. Either the fracture load or the average applied stress at fracture may be used as a measure of the strength of an irregular lump of rock. (The average applied stress at fracture is the fracture load divided by the mean area of contact of the two surfaces of the irregular lump in contact with the loading platens.)

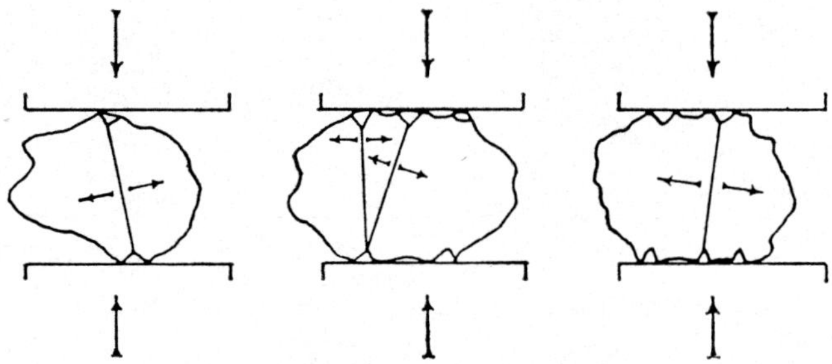

FIG. 2. Mode of fracture of an irregular lump of rock.[3]

The technique is particularly valuable for determining the compressive strength of weaker rocks, from which regularly-shaped specimens are extremely difficult to prepare. The method is also applicable to the stronger rocks where the preparation of regularly-shaped test pieces requires special apparatus and is both time consuming and laborious.

The procedures for assessing the compressive strength of rock by crushing irregularly-shaped lumps is as follows:

The test pieces are obtained by breaking blocks of rock with a succession of hammer blows. Fragments which have a diameter of less than 4 in. are broken no further.

Sufficient rock is broken to enable 30 or more pieces, lying in the weight range 30–34 g, to be selected from the fragments of rock. Each test piece is crushed parallel to its shortest dimension between parallel steel platens of a press of at least 5000 lb capacity.

As the load is increased, localized fracture generally occurs at the points of contact of the test piece and the loading platens. Fracture is assumed to have occurred when the test piece breaks into two or more pieces in a manner illustrated in Fig. 2. The load at which failure occurs is noted for each test piece.

Area of Contact. To measure the area of contact of the irregular specimens with the loading platens, a piece of graph paper and a piece of carbon paper are placed between irregular specimen and each loading platen. When the specimen is loaded, a number of well-marked blackened imprints are formed on the graph paper. The total area of these imprints at fracture is obtained by counting the blackened squares, and the mean of the two total areas thus found is taken as the area of contact of the platens with the specimen.

Selection of Specimens. When specimens are selected for the irregular lump test it is important that specimens are not disregarded because of peculiarities in their shape, for although the average applied stress at fracture is not related to specimen shape (see Fig. 3), the fracture load may increase with increasing specimen height.

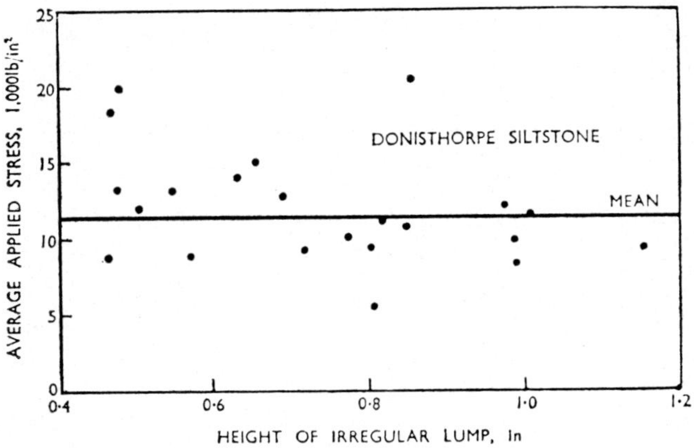

FIG. 3. Variation in average applied stress at fracture with irregular lump height.(3)

Thus, higher mean fracture loads can be obtained if specimens of short height are disregarded. In addition it is important that the irregular lumps selected for testing should lie in the mass range 30–34 g. This can be seen from Fig. 4 which shows the mean fracture load plotted against the mean specimen mass for six rock types. On a log–log scale the observations fall on a straight line. Thus a correction can be made if the mean specimen mass of the irregular lumps tested is not 32 g. The corrected load is given by:

$$\frac{6.96 I_L}{M^{0.56}} \tag{5}$$

where M is the mean specimen mass and I_L is the mean fracture load.

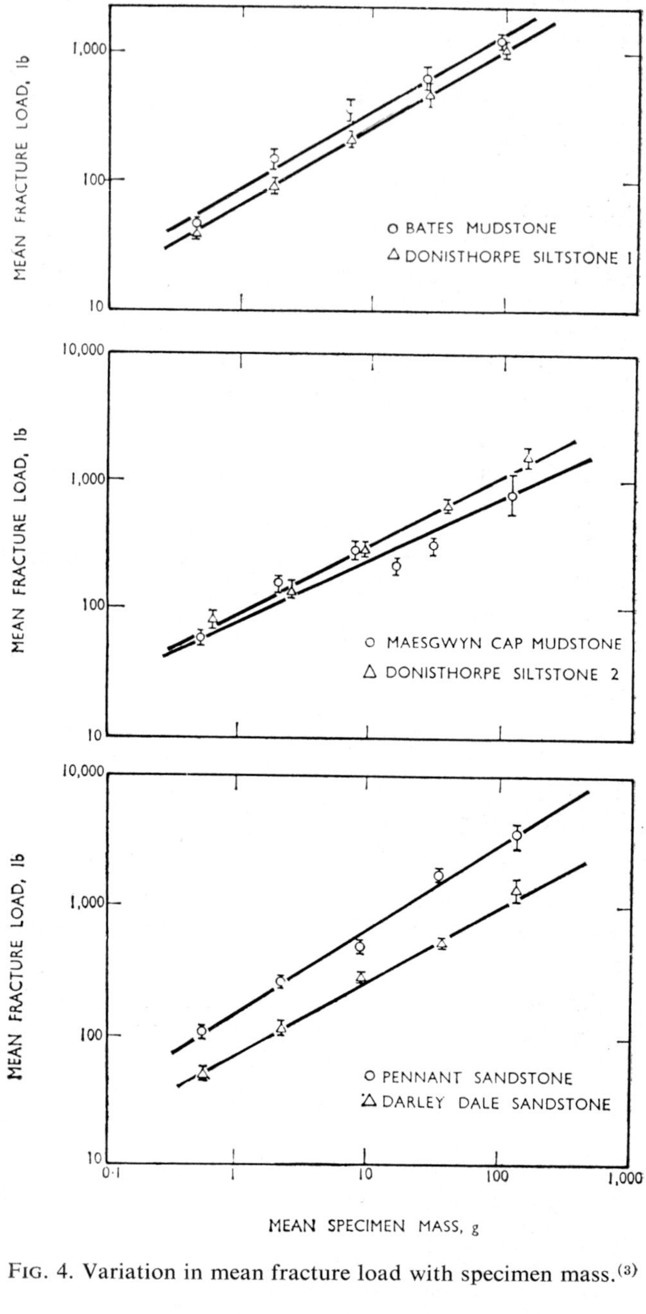

FIG. 4. Variation in mean fracture load with specimen mass.(3)

Impact Strength Index

It was shown by Pomeroy[5, 6] that the crushing of coal fragments by a succession of hammer blows provides a method for estimating the compressive strength of coal. Because of this a number of experiments were carried out to determine whether the Impact Strength Index is suitable as a measure for the compressive strength of rock types other than coal. The technique adopted by Pomeroy is, briefly, as follows:

A 100 g sample of rock in the size range $\frac{1}{8}/\frac{3}{8}$ is placed inside a cylinder of $1\frac{11}{16}$ in. diameter and a 4-lb weight is dropped 20 times from a height of 12 in. on to the rock sample. The test is repeated six times. The amount of rock remaining in the initial size range after the test is termed the Impact Strength Index.

Correlation of Foregoing Tests with Compressive Strength

In Figs. 5, 6 and 7 the mean fracture load, the mean average applied stress at fracture, and the mean Impact Strength Index are shown plotted against the compressive strength of cylinders of 1-in. length and 1-in. diameter for a variety of rock types. The straight lines shown were obtained by the method of least squares. The

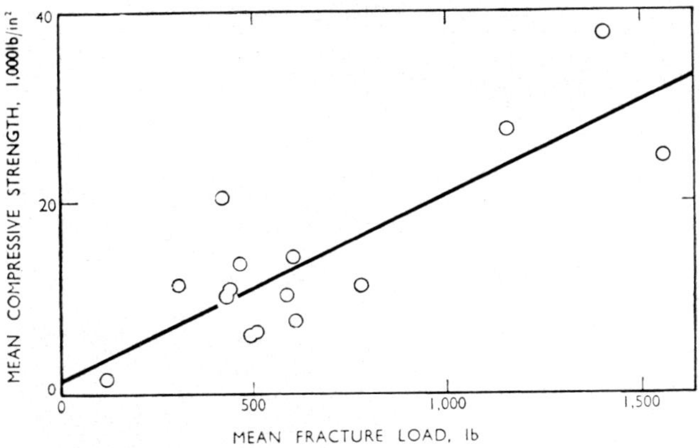

FIG. 5. Relationship between compressive strength and fracture load.[3]

cylinders were prepared from the same parent blocks of rock as the irregular lumps and fragments for the Impact Strength Index test. The rocks tested included siltstones, sandstones, mudstones, limestones and a siderite. Two coals were also included in the irregular lump tests.

A high degree of correlation exists for each comparison with the compressive strength. The closest correlation with the compressive strength is shown by the average applied stress at fracture. However, it does not necessarily follow that the average applied stress at fracture is the best measure of rock compressive strength, since an estimate of the area of contact requires a degree of skill and patience. For this reason the measure of fracture load may be equally suitable as a measure of rock compressive strength. Thus if tests are carried out on irregularly-shaped lumps of rock using the procedure outlined above, then an estimate of the compressive strength of the rock can be obtained from Figs. 5 or 6. If a measurement is made

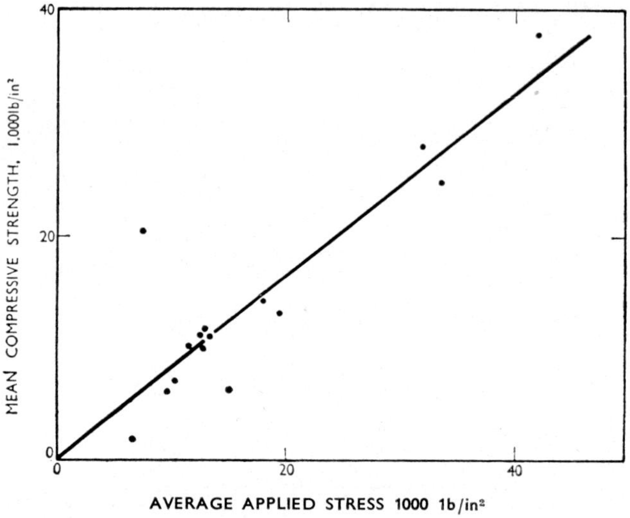

FIG. 6. Relationship between compressive strength and average applied stress at fracture.(3)

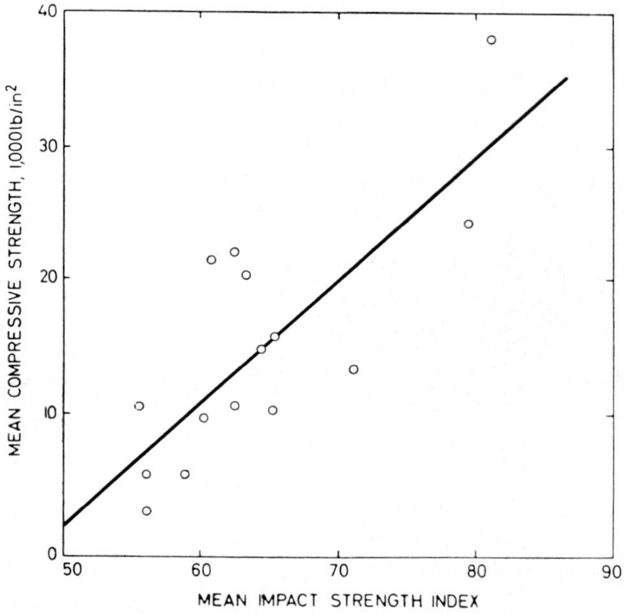

FIG. 7. Relationship between compressive strength and Impact Strength Index.(3)

of the Impact Strength Index, then an estimate of the compressive strength can be obtained from Fig. 7.

It should be noted that the compressive strength obtained from Figs. 5 to 7 is an estimate of the strength of a cylindrical specimen of rock which would have a diameter of 1 in. and a length of 1 in. To estimate the compressive strength of a specimen with a diameter of 1 in. and a length of 2 in. the following equations can be used[3]

$$C_2 = 16.75 I_L + 1210 \quad (6)$$

$$C_2 = 0.697 I_A + 270 \quad (7)$$

$$C_2 = 769 I_I - 36{,}360 \quad (8)$$

where

C_2 is the estimated compressive strength of a specimen 1 in. in diameter and 2 in. long (psi)
I_L is the fracture load (lb)
I_A is the average applied stress at fracture (psi)
I_I is the Impact Strength Index

The Effect of Moisture on Compressive Strength

The effect of moisture content on the compressive strengths of some rock types has been noted by Obert, Windes, and Duvall.[2]

Specimens in three different states of dryness were tested:

1. Air-dried two weeks at room temperature.
2. Oven-dried 7 days at 100 °C.
3. Saturated 7 days in water.

In general for all rocks the compressive strength is decreased by wetting and increased by drying. The average increase in strength for oven-drying was about 6 per cent and the average decrease for the saturated state was about 12 per cent.

It appears to be a safe assumption that the harder igneous and metamorphic rocks will be the least affected by saturation and that the rocks, such as shales and mudstones, which contain a large proportion of clay or silt will be most affected by variations in moisture content.

In general it appears that consistent test results may be obtained by simply air-drying the test specimens for two weeks at room temperature prior to testing.

Fracture Characteristics to be Noted

The Ultimate Compressive Stress. This may be defined as the stress necessary to induce failure in a specimen, with failure being considered to take place when a sudden drop of applied load is observed and no further load can be supported.

In order to correlate results of tests it is necessary to reduce test results to a common length/diameter ratio for all specimens. The U.S. Bureau of Mines[(2)] has accepted The American Society for Testing Materials procedures which require that all compressive strength test results be reduced to values based on a specimen with a length/diameter ratio of 1. A similar standard procedure is followed by the Canadian Department of Mines and Technical Surveys.[(7)]

It has previously been noted herein that a more realistic value of the true compressive strength of a specimen is obtained if the length/diameter ratio is not less than 2.

The Fracture Angle. It was noted in Chapter 4 that the inclination of shear planes produced when a brittle material fractures under compressive stress is a function of the internal friction of the material.

From theory the angle, β, which the shear plane makes with the direction of the major principal stress is related to the angle of internal friction, ϕ, by the following equation:

$$\beta = 45^\circ - \frac{\phi}{2} \tag{9}$$

Thus the angle of fracture is an indication of the coefficient of internal friction of the rock.

Degree of Violence of Fracture. Hardy[(7)] classifies fracture by degree of violence and by the size and shape of the resulting fragments.

Stored Strain Energy. The violence of fracture of a rock specimen depends upon its capacity to store strain energy. The capacity to store large quantities of strain energy is characteristic of those rocks which are susceptible to "bursting".

The specimen subjected to compressive stress is shortened in the manner of a compressed spring and, if its strength is exceeded, it may fracture suddenly, releasing the stored elastic energy in the projection of fragments and the production of sound, or seismic, waves.

The strain energy in rock specimen at fracture is computed from the equation

$$U = \frac{1}{2} \cdot \frac{Q^2}{E} \tag{10}$$

where U is the energy in in.-lb or ft-lb, Q is the ultimate compressive stress at fracture, and E is the modulus of elasticity of the rock.

HARDNESS

The concept of hardness does not lend itself to exact definition in terms of the customary physical units. In the study of metals and other homogeneous materials, the hardness is expressed in arbitrary units that depend upon the design and application of the measuring instrument. In both instances the success of the method depends largely on having homogeneous specimens. Obviously, most rocks do not provide such homogeneity, hence it is necessary to perform a type of integration in order that a hardness figure may be obtained that is representative of the rock, as distinguished from that of the minerals or particles composing it. The integration can be performed mechanically as when a comparatively large area is subjected to an abrasion test, or it can be performed statistically by taking a series of measurements at random points (as with a sclerometer) and averaging the readings.

The latter method has the advantage that, by examining the spread of the data about the average, information pertaining the composition or mineral content of the rock can be obtained.

The Shore Scleroscope

The Shore scleroscope is a hardness-measuring device which indicates hardness (on a scale of 0–120 divisions) by the height of the rebound of a diamond-pointed hammer dropped vertically on the test surface.

Although it is designed primarily for use on metals or plastics, measurements can be made on rocks with no change in instrument or procedure. Hence the manufacturers instructions for use of the instrument are applicable.

In using the scleroscope for testing rock, it should be noted that as the diamond point is small (approximately 0.03-in. diameter) it may strike different mineral components within a given types of rock. Therefore an average hardness value for the rock type must be obtained by averaging a sufficient number of readings.

Table 1 lists scleroscope hardnesses as determined for some quarried rocks.

TABLE 1. SCLEROSCOPE HARDNESS OF QUARRIED ROCKS[2]

Rock type	Scleroscope hardness	Std. deviation, %	No. of readings*
Marble	56	9	1200
Limestone	27	22	1200
Granite	95	7	1200
Sandstone 1	31	23	1200
Sandstone 2	20	45	1200
Slate	56	9	1200
Greenstone	81	10	240

* 10 readings per specimen.

Correlation of Strength Properties with Rebound Hardness

An investigation into the relationship between strength properties and rebound hardness for some coal-measure rocks was carried out by Greenland.[8] The specimens examined were divided into four main rock types. These were:

1. Medium-grained sandstones, with a dominant grain size of 0.25 mm to 0.5 mm.
2. Coarse-grained siltstones, with a dominant grain size of 0.05 mm to 0.1 mm.
3. Carbonaceous mudstone.
4. Laminated shales.

This grouping was purely for convenience in testing and recording, and there were obvious variations in the physical properties between samples in any one group.

Rock cubes measuring 1 in. on each side were prepared and crushed in a compression testing machine. The maximum load supported was taken as the crushing strength of the rock. During compression the rock specimen was in direct contact with the platens of the machine. No lubricant was used between the platens and the rock specimen surfaces. Spherical seating of the upper platen ensured axial loading even with slight non-parallelism of the specimen surface.

Rebound Hardness

The rebound hardness of the rock was determined with a Shore scleroscope. It was measured once on each face of the cube, and the mean value was obtained. Rebound hardness values obtained on the rock samples prior to cutting the cubes

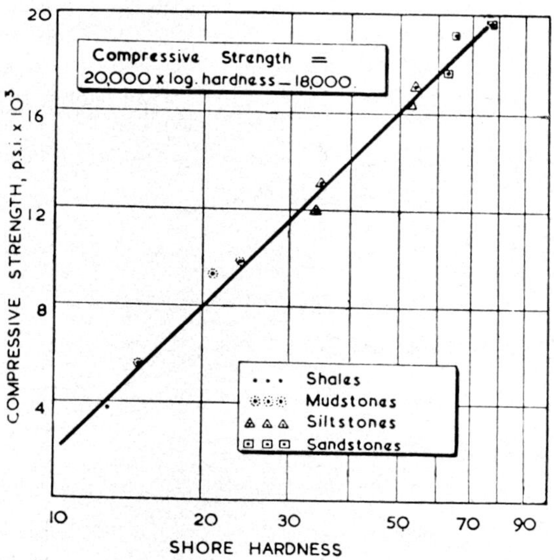

FIG. 8. Showing the logarithmic relationship between compressive strength and Shore hardness.[8]

were between 5 per cent and 20 per cent lower than the values obtained as above. This can be attributed to irregularities of the rock surface prior to cutting and or to workhardening of the rock surface by the cutting saw.

For the rock types examined, Fig. 8 indicates that a substantially linear relationship exists between strength and the logarithm of rebound hardness. The relationship is of the form:

$$\text{compressive strength (psi)} = 20{,}000 \times \log \text{rebound hardness} - 18{,}000.$$

It will be realized that the content refers only the scleroscope used in this investigation. Also in the case of the laminated shales, the compressive strength measured parallel to the bedding planes did not follow the above relationship.

MODULUS OF RUPTURE

The method of flexural testing used by the U.S. Bureau of Mines for determining the modulus of rupture of diamond drill core specimens, as reported in Ref. 2, differed from the A.S.T.M. procedures in two respects: (1) the self-adjusting knife-

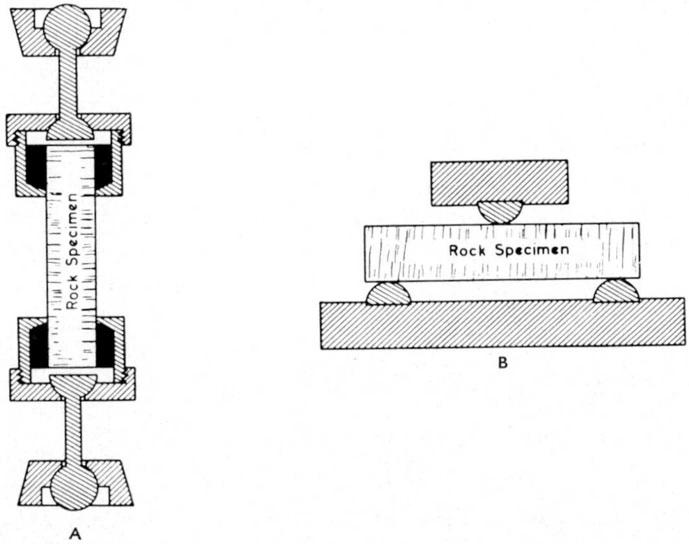

FIG. 9. A, Tension grips for tensile strength test.[2] B, Bearing and load application device for modulus of rupture test.[2]

edge bearing and load application device was not employed because the circular cross-section specimens (from diamond drill cores) did not require the self-adjusting feature; and (2) owing to the small cross-sectional area of the diamond drill core the length of the specimen was reduced to 6 in., and correspondingly, the separation between the centers of the knife edges was reduced from 10 to 5 in. A drawing of the bearing and load application device used in the tests is shown in Fig. 9B.

Loads were applied to test specimens at the rate of 100 lb/min.

The modulus of rupture was calculated by the following equation:

$$R = \frac{8Wl}{\pi D^3} \tag{11}$$

where

R = modulus of rupture
W = applied load
l = span between centers of knife edges
D = diameter of specimen

Effect of Specimen Diameter

There was some question[2] as to whether or not small-diameter cores would give reproducible results and whether the values of the modulus of rupture would be the same for cores of different diameters. A test was performed to check this point in which four (or more) 6-in. specimens, cut perpendicular to the bedding, for each of the four diameters ($\frac{7}{8}$, $1\frac{1}{8}$, $1\frac{5}{8}$, and $2\frac{1}{8}$-in.) were prepared and tested in the air-dried state (two weeks air-dried at room temperature). The results of these tests showed that the modulus of rupture was independent of specimen diameter, in the sizes tested.

TENSILE STRENGTH[2]

Experimental difficulties encountered in tensile testing of rock specimens include difficulties in preparing suitable test specimens, difficulties in mounting the specimens in tension grips without damaging the surface of the specimen, and difficulties in applying the load parallel to the axis of the specimen.

Regarding the latter point it is obvious that if the load cannot be applied parallel to the axis of the specimen within close limits there will be a tendency to produce bending, which, in turn, will produce abnormal stress concentrations.

In the U.S. Bureau of Mines experiments[2] an attempt was made to overcome these difficulties by designing a pair of tension grips, as shown in Fig. 9A, in which the core specimen is held in place with Leadite compound. The spherically seated joint (in the mounting cup) was placed as close to the end of the specimen as possible and in line with the axis of the specimen. The spherically seated joints in both the stationary and the movable heads of the press permitted the coupling-link to align itself.

This grip was tested on a group of quarried rock specimens. Four specimens from the $\frac{7}{8}$, $1\frac{1}{8}$, $1\frac{5}{8}$-in. diameter cores were prepared from each of the quarried rocks. The specimens were tested after two weeks of air-drying. The rate of loading was maintained at 100 lb/min. The results of the tests are given in Table 2.

TABLE 2. TENSILE STRENGTH OF QUARRIED ROCKS[2]

Rock type	No. of specimens	Diameter core size, in.	Average tensile strength, psi	Standard deviation, %
Marble	4	$\frac{7}{8}$	700	20
Marble	4	$1\frac{1}{8}$	900	6
Marble	4	$1\frac{5}{8}$	990	7
Limestone	4	$\frac{7}{8}$	520	1
Limestone	4	$1\frac{1}{8}$	570	14
Limestone	3	$1\frac{5}{8}$	490	15
Granite	4	$\frac{7}{8}$	930	6
Granite	4	$1\frac{1}{8}$	1100	7
Granite	4	$1\frac{5}{8}$	1070	7
Sandstone 1	1	$\frac{7}{8}$	105	—
Sandstone 1	4	$1\frac{1}{8}$	119	6
Sandstone 1	4	$1\frac{5}{8}$	135	5
Sandstone 2	4	$\frac{7}{8}$	149	14
Sandstone 2	4	$1\frac{1}{8}$	147	59
Sandstone 2	4	$1\frac{5}{8}$	206	58

It was found that the Leadite compound was not strong enough to hold specimens having a tensile strength greater than 1200 psi; and, as a result, the greenstone specimens could not be tested.

Although the percent standard deviations for the quarried rock specimens, given in Table 2, are not much greater than the percent standard deviations characteristic of the other test procedures reported herein, it was found in subsequent tests that for rock types of less uniform nature the deviations were so great that the test was not considered satisfactory and therefore no standard procedure is specified. Occasionally the test has been employed to test the strength of recemented fractures, joints, or seams to determine whether or not the joint is weaker or stronger than the rock.

An indication of the tensile strength of a rock may be obtained by subjecting a disc of rock to compressive stresses applied by two diametrically opposite edges. The pressure required to split the specimen is a function of its tensile strength.

Splitting Strength

In the splitting test, commonly known as the "Brazilian" test,[9] a cylinder or disc of rock is subjected to a compressive line load acting along a diameter. The resulting stresses comprise a uniform horizontal tensile stress over most of the loaded diameter and a variable vertical compressive stress.

Near the loaded edge, both stresses are compressive and rise to a maximum value. Initial failure is assumed to result from the tensile stress at the center of the disc, although this might be influenced by the high compressive stresses at the loading points.

As indicated in Table 3 the splitting strength decreases as the specimen size increases.

TABLE 3. TABULATION OF STRENGTH PROPERTIES OF AN ANDESITE[9]

(a) Compressive Strength

Confining Pressure	No. of Tests	Extreme Values		Mean	Coeff. of Variation
		High	Low		
ksi		ksi	ksi	ksi	%
0	27	51.2	14.1	33.5	28
0.1	20	43.2	15.0	29.3	28
2.0	28	63.4	25.5	42.7	24
5.25	35	117.2	36.7	66.8	26

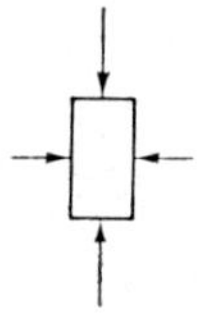

(b) Direct Shear Strength

End Pressure	No. of Tests	Extreme Values		Mean	Coeff. of Variation
		High	Low		
ksi		ksi	ksi	ksi	%
0	8	8.39	7.33	7.89	6
0.69	9	10.1	7.22	8.44	11
5.51	8	20.3	14.4	16.2	12
11.0	8	25.6	16.5	20.3	14

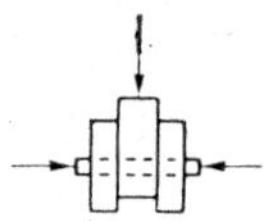

(c) Modulus of Rupture

Distance Between Load. Prs.	No. of Tests	Extreme Values		Mean	Coeff. of Variation
		High	Low		
in.		ksi	ksi	ksi	%
0	14	7.81	4.97	6.40	13
1	14	7.11	5.12	6.26	9
2	18	7.50	3.55	5.38	23
3	19	6.48	3.21	5.21	17

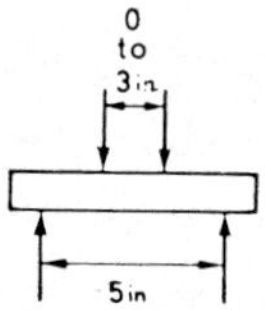

(d) Splitting Strength

Dia. × Length	No. of Tests	Extreme Values		Mean	Coeff. of Variation
		High	Low		
in.		ksi	ksi	ksi	%
0.83×0.47	7	4.04	3.15	3.73	9
1.24×0.36	6	3.72	2.30	2.93	18

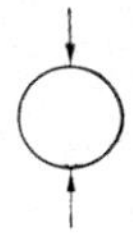

(Note: ksi = kilo-pounds per sq. in.)

Relationships between Modulus of Rupture, Splitting Strength, and Tensile Strength

The relationship between the modulus of rupture, the splitting strength, and the tensile strength has been evaluated for specimens made of plaster and for specimens made of concrete.

For the plaster specimens the tensile strength was about one-half of the modulus of rupture and about 0.8 of the splitting strength. For the concrete specimens the tensile strength was about one-half of the modulus of rupture and about 1.5 times the splitting strength.

In Table 3 one-half of the mean of the four modulus of rupture values is 2.9 ksi (kilo-pounds per sq. in.) and the mean of the splitting strength values is 3.3 ksi.

A review of the results of tests of physical properties of rocks as reported in the literature indicates that the value of the modulus of rupture for a rock is generally between 3 and 6 times its tensile strength and that most often the value of modulus of rupture is in the range of 4 to 5 times its tensile strength.

The "Brazilian" test appears to offer a convenient method for obtaining an approximate value for the tensile strength of a rock without the inconvenience involved in preparing specimens and conducting tensile (extension) tests.

DIRECT SHEAR STRENGTH TESTS

The direct shear block consists of three plates through which a hole has been drilled to accommodate the specimen. By loading the center plate a portion of the specimen may be sheared out along two failure planes.

By mounting the shear block in a loading frame an axial load may be applied to the specimen with a small hydraulic ram at the same time that the shearing load is applied to the edge of the plate. The resultant readings will enable the coefficient of internal friction for the specimen to be determined.

The cylindrical specimens may be ground to the proper diameter with a toolpost grinder mounted on a lathe.

Direct shear strength (shown in Table 3) increased from 7.89 to 20.3 ksi as the normal stress on the failure plane was increased from 0 to 11.0 ksi.

DETERMINATION OF THE MODULUS OF ELASTICITY

The modulus of elasticity or elastic modulus (or Young's modulus) is normally determined by measuring the axial deformation of a rock specimen which is subjected to axial loading.

The elastic modulus is computed as:

$$E = \frac{\sigma_1}{\delta} \tag{12}$$

where

E = modulus of elasticity
σ_1 = unit stress
δ = unit strain

Unit strain may be determined by measuring the total shortening of the specimen and dividing by its length, or it may be determined by mounting resistance strain gages on the surface of the specimen, parallel to its axis, to record unit strain.

If it is desired to determine Poisson's ratio during the same compression test then strain gages will be mounted on the specimen to measure circumferential strain and there will be no space to mount longitudinal gages. In this case the total axial strain can be determined by employing one of the types of mechanical strain gages to measure the decrease of distance between the compression plates.

The Stress-Strain Curve

A typical stress–strain curve for a brittle material is shown in Fig. 10. It is taken from a concrete compression test. In such case it is difficult to find a straight line portion. Three separate modulii of elasticity can be determined from this curve. They are:

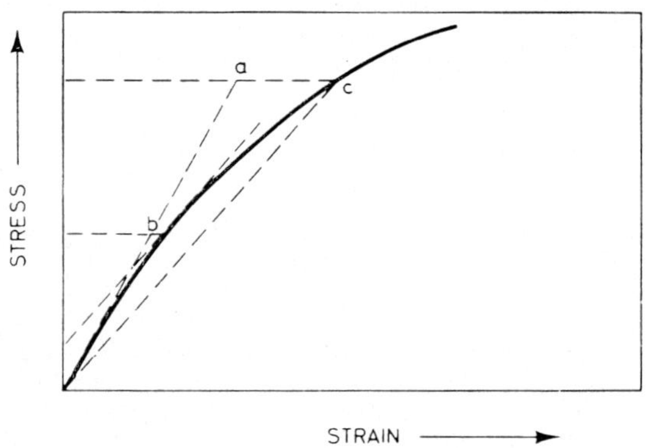

FIG. 10. Tangent and secant modulii for a brittle material.[8]

1. The initial tangent modulus is the line, *a*, drawn as a tangent to the stress–strain curve through the point of origin in Fig. 10.
2. The tangent modulus for a given stress, *b*, in Fig. 10.
3. The secant modulus at a given stress, usually the proportional limit stress, if this point can be found on the curve, are at any assumed stress, such as *c* in Fig. 10.

The modulus of elasticity, as determined by sonic methods, is always the initial tangent modulus, that is the elastic modulus as determined at no load.

In order to establish a uniform basis for comparisons the Canadian Department of Mines and Technical Surveys[7] considers a compressive stress of 6000 psi to be the standard and the value of E to be the slope of the tangent line drawn to the stress–strain curve at an applied stress of 6000 psi.

Modulus of Elasticity of Stress Relief Cores Determined by a Biaxial Device

The U.S. Bureau of Mines has developed a device[10] which permits the modulus of elasticity of a rock to be determined by hydraulically loading the external surface of the stress relief core and measuring the corresponding deformation of the central borehole.

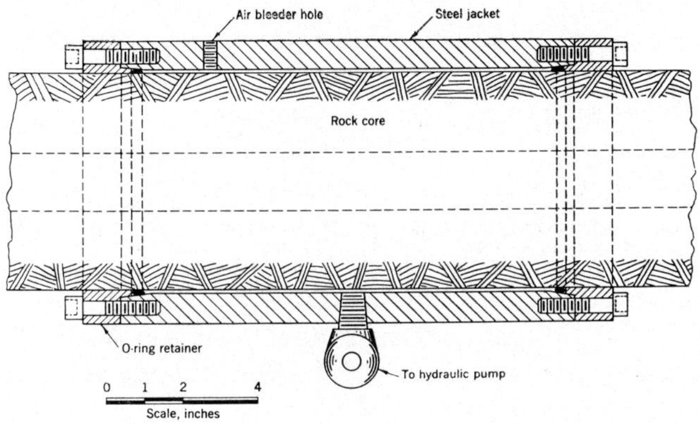

FIG. 11. Cross-section of hydraulic loading device.[10]

Figure 11 shows a cross-section of the loading device which consists essentially of a steel sleeve which fits around the stress relief core and which can be sealed, by means of rubber "O" rings, at each end.

A borehole deformation gage (see Appendix to Chapter 2, and Chapter 10) is installed inside the central borehole in the core and hydraulic pressure is applied to the exterior of the core. The diametral deformation at various pressures is noted and E is calculated from the equation:

$$E = \frac{4ab^2 p_0}{(b^2 - a^2)U} \tag{13}$$

where

a = radius of central borehole
b = radius of exterior of the core
p_0 = unit pressure applied to exterior of the core
U = diametral deformation of the borehole

The steel jacket is designed for a pressure of 10,000 psi and is designed to accommodate stress-relief cores $5\frac{5}{8}$ in. in diameter. An ordinary handoperated jack pump is used to create the hydraulic pressure for the apparatus.

The biaxial device provides a convenient means for determining the modulus of elasticity of a stress-relief core under a state of stress approximating that exist in the *in situ* rock. Also it is portable and measurements can be made in the field. However, the method has one disadvantage. Because the ends of the specimen are unconstrained the radially applied load creates a tensile stress acting in the axial direction. If the tensile strength of the rock is low the core will fail in tension at a comparatively low biaxial load.

To overcome this difficulty the device for triaxial testing, described elsewhere in this chapter, was developed.

Effect of Moisture on Elastic Properties

In general the effect of variations in moisture content of samples varies with the rock types. From some rock types[(2)] the modulus of elasticity increased with increasing moisture content (marble and granite, for example) while in others (limestone, for example) it decreased. The modulus of rigidity showed similar changes which agreed in direction with those of the modulus of elasticity.

The percentage change in going from an air-dried state to a saturated state in a sandstone was a 38 per cent decrease in the modulus of elasticity. It is a reasonably safe assumption that the softer sedimentary rocks will be adversely affected by saturation with water and that the elastic moduli will be greatly reduced.

The values given in Table 4 illustrate the effect of moisture on the elastic properties of various types of rock.

TABLE 4. EFFECT OF MOISTURE CONTENT ON YOUNG'S MODULUS AND THE MODULUS OF RIGIDITY

Rock type and bedding direction[1]	Moisture content, % by volume[2]			Change in E, %[2]		Change in G, %[2]	
	Oven-dried	Desiccator dried	Wet	Desiccator-dried	Wet	Desiccator-dried	Wet
Marble ⊥	−0.13	−0.10	0.36	0	19	0	4
Limestone ⊥	− .20	− .17	15.9	1	−21	0	−20
Granite ⊥	− .11	− .09	.74	−1	30	−2	4
Sandstone 1 ⊥	− .22	− .19	15.4	0	−38	1	−42
Sandstone 2 ⊥	− .52	− .46	15.3	0	−30	0	−38
Slate ‖	− .22	− .02	.72	0	− 6	−2	−23
Greenstone	− .12	− .02	.59	0	− 2	0	− 7

1) The symbols ⊥ and ‖ are used to indicate that the axis of the core was perpendicular to or parallel to the bedding.

2) Referred to the air-dried state.

POISSON'S RATIO

The usual definition of Poisson's ratio is as follows:[7]

If a uniform stress be applied to a specimen which is allowed to expand freely in all directions at right angles to the direction of applied stress, the ratio of the linear lateral (transverse) strain to the linear longitudinal strain within the elastic limit of the material is defined as Poisson's ratio.

Note: Poisson's *number*, designated by m, is preferred for use by the European writers and is used frequently also in this book. Poisson's *number* is the reciprocal of Poisson's ratio; that is:

$$\text{Poisson's number} = m = \frac{1}{\text{Poisson's ratio}}$$

The transverse or diametric strain in a cylinder is proportional to the circumferential strain[7]; therefore, when cylindrical specimens are subjected to axial compression (unconfined), the transverse strain may be determined by measuring the circumferential strain by means of resistance strain gages which are cemented to the surfaces of the specimen.

For materials which exhibit non-linear stress–strain curves, restrictions must be placed on the previous definition. It is necessary, as in the case of longitudinal compression modulus, to state Poisson's ratio at a particular load since the slope of the transverse, as well as the longitudinal stress–strain curves, varies with stress.

Rock Deformation under Long-Term Loads

Tests to determine rock deformation under long-time stresses are usually carried out on loaded rock "beams" rather than on compression specimens.

Hardy[11] has described systems being developed by the Canadian Department of Mines and Technical Surveys. One system involves a pneumatic–hydraulic servo-loading system which will enable accurate constant-strain and constant-stress experiments to be carried on over periods of time up to a year.

Another system (continuous electro-hydraulic servo-loading) will allow constant-strain rate and constant-stress rate experiments to be carried on over periods of time up to 24 hours.

A high-pressure confining system has also been designed for carrying out long-term deformation experiments under triaxial conditions. Oil pressures up to 20,000 psi can be generated in this system, which relies on a 10/1 pressure intensifier driven by nitrogen gas, rather than relying on motor-driven pumps to maintain the required pressures. This arrangement eliminates the possibility that a power failure will interrupt an experiment.

TRIAXIAL COMPRESSION TESTS

In triaxial tests rock is subjected to restraining stresses on all sides. The compressive strength, shear strength, and the angle of internal friction can be determined by means of such tests, and, as explained in Chapter 4, these characteristics can be graphically represented by Mohr's stress diagram which allows the behavior of the rock *in situ* to be predicted.

In addition to the characteristics mentioned above, the modulus of elasticity and Poisson's ratio for stress-relief cores can be determined at stress levels comparable to those existing in underground mines.(12)

Triaxial Apparatus

Obert(13) has designed an inexpensive apparatus which can be fabricated at nominal cost and which is designed to test $2\frac{1}{8}$, 4-, and $5\frac{5}{8}$-in. diameter cylindrical specimens which have a length-to-diameter ratio of about 2.

These cylindrical specimens can be prepared from conventional NX ($2\frac{1}{8}$ in. diameter) diamond drill cores or from cores taken in the laboratory or in the field with 4- or a $5\frac{5}{8}$ in. i.d. thin-walled coring bits. The maximum axial stresses which can be applied to the $2\frac{1}{8}$-, 4-, and $5\frac{5}{8}$-in. diameter specimens are 70,000 psi, 20,000 psi, and 10,000 psi respectively with the 250,000 lb hydraulic press. The maximum radial stress is the same for each specimen, namely 10,000 psi.

Compression Testing Machine

The hydraulic press used with this equipment is a type normally used for testing concrete cylinders or concrete blocks. The maximum clearance between the swivel head and the bed of the machine is $14\frac{1}{2}$ in., which is the length required for testing

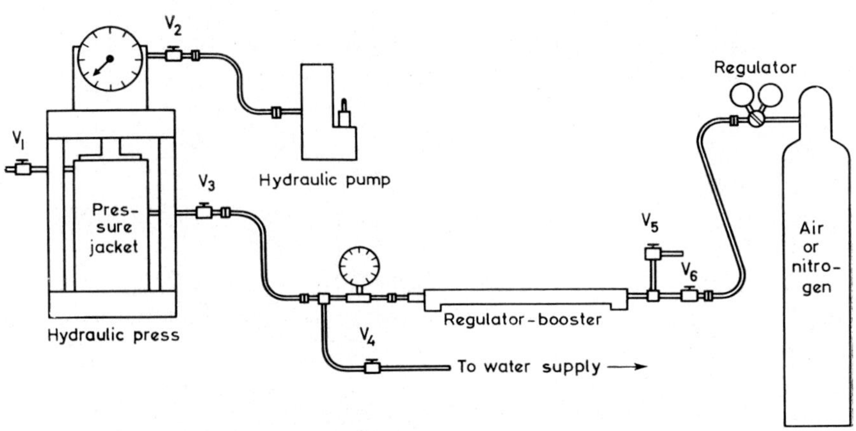

FIG. 12. Triaxial equipment — schematic diagram of hydraulic system.(13)

the largest size triaxial specimen. The capacity of the machine is 250,000 lb and the manufacturer's specified accuracy is 1 per cent.

The hydraulic ram on the machine is sealed with "O"-rings and the machine will hold an applied load almost indefinitely. Fig. 12 shows a schematic diagram of the hydraulic system for the apparatus. Fig. 13 shows a half section of the pressure

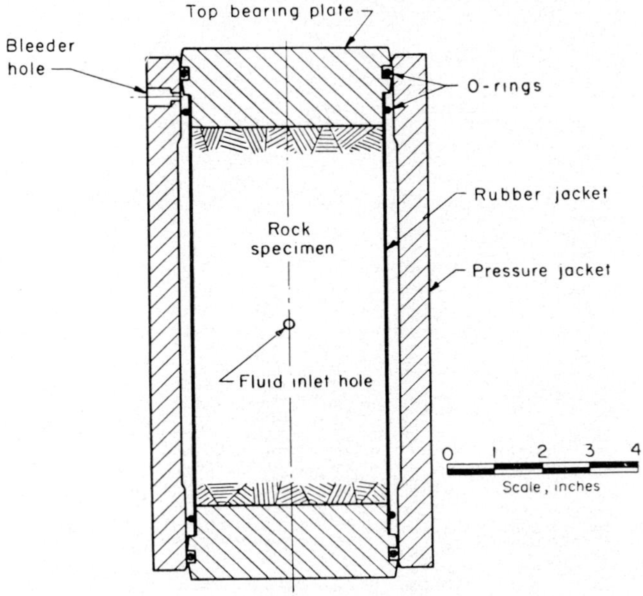

FIG. 13. Half section of rock specimen, pressure jacket, and bearing plate assembly.[13]

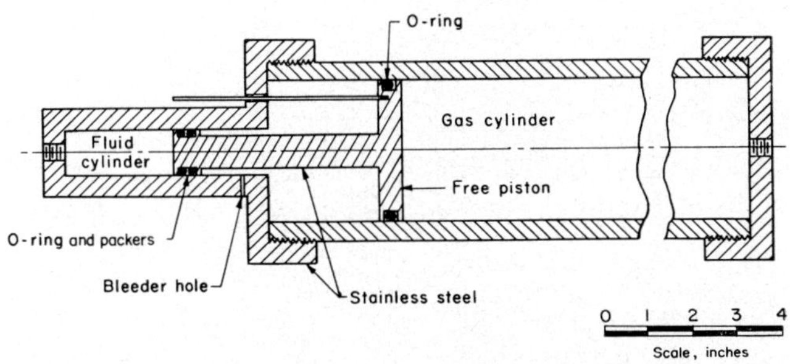

FIG. 14. Half section of the regulator-booster for the hydraulic system.[13]

cylinder assembled with rock specimen while Fig. 14 shows a half section of the regulator-booster. This device makes it possible to use compressed gas (nitrogen or air) at a maximum pressure of 1000 psi to develop a fluid (water) jacket pressure of 10,000 psi.

The total cost of this triaxial equipment, including the hydraulic press, but not including labor costs, was approximately $1000.

Procedures for Determining Elastic Constants by Triaxial Tests[12]

For determination of elastic constants it is necessary to measure the axial stress, hydraulic pressure, axial strain, and borehole deformation.

The following equations may be used:

$$\frac{\sigma_1}{E} = \frac{2\nu b^2 p_0}{E(b^2-a^2)} + \varepsilon_1 \tag{14}$$

$$U = \frac{4(1-\nu^2)ab^2 p_0}{E(b^2-a^2)} - 2\nu\varepsilon_1 a \tag{15}$$

$$p_0 = -\frac{E(b^2-a^2)U}{4ab^2} - \frac{\nu(b^2-a^2)\sigma_1}{2b^2} \tag{16}$$

where

σ_1 = axial stress (psi) (tensile stress or strain is taken as positive in these equations)

ε_1 = axial strain (in./in.)

p_0 = pressure applied to exterior surface of specimen (psi)

a = radius of hole in specimen (in.)

b = outside radius of specimen (in.)

E = modulus of elasticity (psi)

ν = Poisson's ratio

U = borehole deformation (in.) (change in diameter of the borehole)

SONIC TESTS APPLIED TO ROCK SPECIMENS[2]

Longitudinal Velocity of Sound, Torsional Velocity of Sound, Modulus of Elasticity, Modulus of Rigidity, Poisson's Ratio and Specific Damping Capacity

A method of sonic testing of rock specimens to determine their elastic properties has been described in Ref. 2. This method involves the following procedures:

A cylindrical specimen is supported at the center, and vibrated, first, in the fundamental longitudinal mode, which gives the fundamental longitudinal frequency; and, second, in the fundamental torsional mode which gives the fundamental torsional frequency. From these frequences, together with the dimensions of the specimens, it is possible to calculate the longitudinal and torsional velocities of sound by the following equations:

$$v_l = 2f_l L, \tag{17}$$

$$v_t = 2f_t L, \tag{18}$$

where

v_l = longitudinal velocity of sound,
v_t = torsional velocity of sound,
f_l = fundamental longitudinal frequency,
f_t = fundamental torsional frequency,
L = length of the specimen.

From the longitudinal and torsional velocities, together with the density, it is possible to calculate Young's modulus, the modulus of rigidity, and Poisson's ratio by the following equations:

$$E = v_l^2 d, \tag{19}$$

$$G = v_t^2 d, \tag{20}$$

$$\nu = \frac{E}{2G} - 1, \tag{21}$$

or

$$\nu = \frac{f_l^2}{2f_t^2} - 1, \tag{22}$$

where

E = Young's modulus
G = modulus of rigidity,
ν = Poisson's ratio,
d = density.

The specific damping capacity, which is the ratio of the energy dissipated per cycle of vibration to the total vibrational energy, is determined by the following equation:

$$b = \frac{2\pi \Delta f}{f} = \frac{\Delta U}{U}, \tag{23}$$

where

b = specific damping capacity,
Δf = width of resonance curve in cycles per second measured at an amplitude equal to $\frac{1}{\sqrt{2}} A_{\max}$ ($A_{\max}$ = maximum amplitude),
f = resonance frequency,
U = total vibrational energy,
ΔU = energy dissipated per cycle.

In determining Young's modulus and the modulus of rigidity by the sonic method, the value obtained is the tangent modulus at zero pressure in contradistinction to methods of determining these quantities where the specimen is under an applied pressure and where the value is either the secant modulus or the tangent modulus for a specific pressure.

In the theoretical development of equation (17) it has been assumed that the ratio of the radius to the length of the specimen is small. If this condition is not satisfied a correction factor must be applied which is given by the following equation:

$$v_l = v_{lm}\left(1-\frac{\pi^2 \nu^2 r^2}{L^2}\right); \tag{24}$$

where

v_l = true longitudinal velocity,
v_{lm} = longitudinal velocity calculated by equation (17)
r = radius of specimen.

For the specimens used in these tests (described in Ref. 2) the radius: length ratio was, in all cases, less than 1 : 10; for this ratio the correction factor produces a change less than 0.25 per cent and therefore was neglected.

Apparatus

The apparatus used in the sonic tests described in Ref. 2 had two functions (1) to cause the specimen to vibrate in the fundamental longitudinal or torsional mode; and (2) to provide a means of measuring the frequency and amplitude of the vibration so produced. A schematic diagram of the apparatus is shown in Fig. 15. The

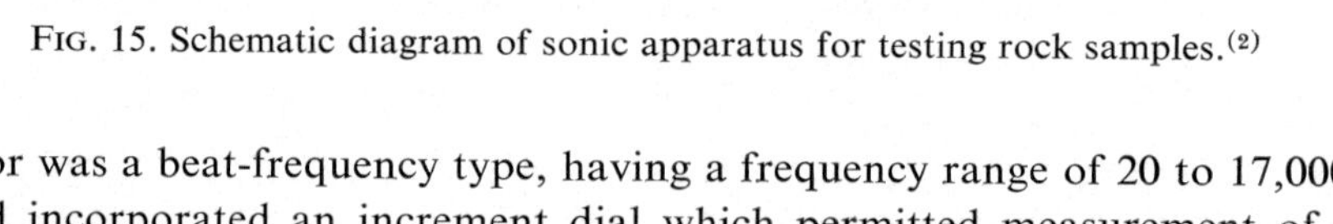

FIG. 15. Schematic diagram of sonic apparatus for testing rock samples.[2]

oscillator was a beat-frequency type, having a frequency range of 20 to 17,000 cps, and had incorporated an increment dial which permitted measurement of small changes in the frequency. The oscillator was mounted in a thermal insulated box and the temperature maintained within a maximum variation of +1 °C. The dial on the oscillator was readable to an accuracy of −1 per cent. The oscillator was connected to a 15-watt linear power amplifier, the output of which was supplied to an electromechanical driver. This driver produced vibration in the specimen through electromagnetic interaction with the soft iron pole piece, which was cemented on the end of the specimen. The type of vibration produced — that is, longitudinal or torsional – depended upon the type of driver employed. A drawing of the longitudinal and torsional drivers is given in Fig. 16. A magnetic pick-up, identical in construction to the driver, was placed symmetrically at the other end of the specimen. The voltage from the pick-up was amplified by a linear high-gain amplifier, the output of which was read on a high-impedance voltmeter. The voltage generated

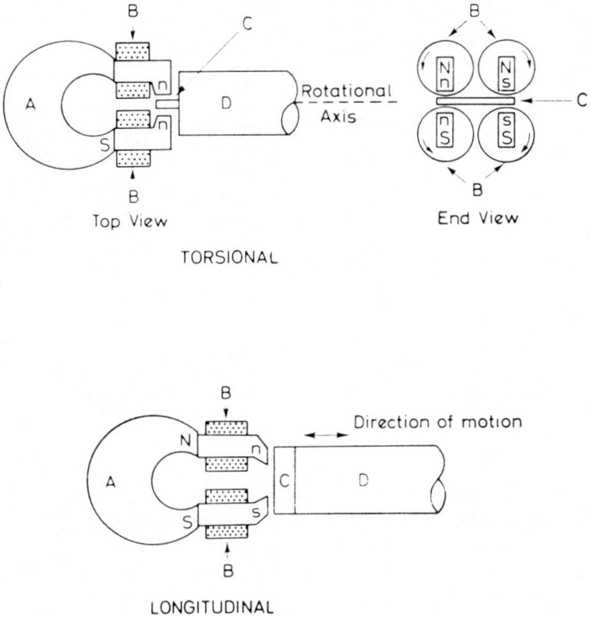

FIG. 16. Torsional and longitudian driver and pick-up. A, Alnico magnet; B, coils; C, pole piece; D, Specimen; N — S, permanent magnetic poles; n—s, alternating magnetic poles; →, coil-winding direction.[2]

by this type of pick-up is proportional to the velocity of the vibratory motion rather than the amplitude. In determining the frequency of the vibration this makes no difference, but in determining Δf the amplitude resonance curve is required by theory. However if Δf is very small compared with f, either the amplitude or the velocity resonance curve gives approximately the same results.

Some difficulty was encountered with stray radiation between the driver and the pick-up. This was partly reduced by the use of shielding, and the remainder was eliminated by introducing an out-of-phase component from the power amplifier into the high-gain amplifier. The "zero control" in the schematic diagram (Fig. 15) was used to perform this function.

The type of mounting clamp used to hold the test specimens was known from previous investigations to have no significant influence on either the longitudinal or torsional frequency but tends to increase the specific damping capacity owing to the loss of vibrational energy through the mounting system. It was found that this loss could be reduced to less than 2 per cent by the proper design of clamp. For the f_l and Δf measurements the specimen was held in the mounting without the aid of the clamping screw. For the f_t measurements it was necessary to clamp the specimen.

Test Procedure(2)

Briefly, the test procedure was as follows: Iron pole pieces were cemented to the ends of the specimen, which was then allowed to stand 24 hours to permit the cement to harden. The specimen was then mounted in the apparatus and (1) using the torsional driver and pick-up, the oscillator was tuned until f_t was found (indicated by a maximum reading on the voltmeter); and (2) using the longitudinal driver and pick-up, the oscillator was tuned until f_l was found (also indicated by a maximum reading on the voltmeter). Then, with the main tuning dial of the oscillator set at the longitudinal frequency, the frequency was decreased with the increment dial until the voltmeter reading was reduced to $\frac{1}{\sqrt{2}} X_{max}$ where X_{max} is the maximum voltmeter reading corresponding to f_l. Next, the frequency was increased until the voltmeter reading was again reduced to $\frac{1}{\sqrt{2}} X_{max}$. The total change in frequency required to produce these amplitude reductions is Δf.

Correction for Pole Pieces

The soft-iron pole pieces cemented to the ends of the specimens reduce both the longitudinal and torsional frequency, thereby producing an error. This error can be corrected by using the following equations:

$$f_l = f_l'\left(1-\frac{M'}{M}\right), \tag{25}$$

$$f_t = f_t'\left(1-\frac{k^2M'}{M}\right), \tag{26}$$

where

f_l = longitudinal frequency (without pole pieces),
f_l' = longitudinal frequency (with pole pieces),
f_t = torsional frequency (without pole pieces),
f_t' = torsional frequency (with pole pieces),
M' = weight of both pole pieces,
M = weight of specimen,
k = ratio of the radius of gyration of the pole pieces to the radius of gyration of the specimen.

The validity of equation (25) was confirmed by measuring the longitudinal frequency of three annealed steel rods, both with and without the pole pieces.

It was possible to vibrate the steel specimens without pole pieces because the driver and pick-up interacted directly with the ends of the rod. The results of these tests showed that the calculated and measured change in the frequency produced by the pole pieces agreed within the limit of error of measurement. Although the correction for the pole pieces on the steel rods amounted to less than 1 per cent, the correction for rock specimens (owing to their lighter weight) amounts to as much as 4 per cent; hence, this correction is applied to all longitudinal frequency measurements. Similarly, equation (26) is used to correct all torsional frequency measurements.

Effect of Size of Specimen

Except for the small correction indicated by equation (24), the longitudinal and torsional velocities are not affected by the size of the specimen. To show this, 12 duraluminium rods of different lengths and diameters were tested. The results showed that, within the limits of experimental error, the velocities are independent of the dimensions of the specimen. These duraluminium specimens were preserved and are used periodically to check the accuracy of the apparatus.

Accuracy

The estimated experimental error in measuring f_l or f_l was ± 1 per cent, and for the metal rods the estimated error in measuring the lengths, diameters, and weight was negligible. Thus, for the metal rods, the estimated error in determining v_l or v_t was ± 1 per cent and E or $G+2$ per cent. The values of v_l and v_t, and the values of E and G showed that for the metal specimens the reproducibility was within the limits of error and that the value of E and G for both duraluminum and annealed steel compares favorably with handbook values. For rock specimens the error in the length and weight measurements was negligible, but the error in determining the diameters was estimated to be ± 1.5 per cent, and the estimated error in determining E or G was ± 5 per cent. The error in determining Poisson's ratio depends upon the value obtained and was substantially larger than the error in either f_t^2 or f_l^2 (see equation 22) because it depends upon the determination of a small difference, namely, that between f_l^2/f_t^2 and 1. The error in measuring Δf and, hence, the error in determining the specific damping capacity depends upon the value of Δf. For values within the range of most rocks ($b = 0.0016$ to 0.016), the error is estimated at ± 5 per cent.

STATISTICAL TREATMENT OF TEST DATA[2]

General

Most physical tests involve tabulation of a series of readings, with computation of an average said to be representative of the whole. The problem arises at once as to just how representative this average is as the measure of the characteristic under test. Three important factors introduce uncertainties in the result; these are:

1. Instrumental errors.
2. Variations in the sample being tested.
3. Variations between the sample and the other samples that might be drawn from the same source.

If there were available for test a number of identical specimens, or if the test were nondestructive and could be repeated a number of times on the same specimen, determination of the instrumental errors (or errors of observation) would be comparatively simple because the variations could be observed independently of the specimen. Inasmuch as perfect specimens are not available, the most nearly perfect specimens available must be substituted. The variations encountered in testing these selected specimens are then described as "instrumental errors", although variations inherent in the specimens as well as those caused by the instrument or test apparatus are included.

Inasmuch as most of the tests used in determining the physical properties of rock are destructive (completely, or in part) it is difficult to distinguish between instrumental variations and those caused by changes in composition or structure of the rock because the test cannot be repeated with the identical specimen (or part of specimen). As specimens can be selected with a minimum of variation in the characteristic that constitutes the test variable, it is advantageous to establish certain selected specimens as standards. This affords a means of periodically checking the apparatus or equipment to make certain that the precision of measurement is maintained. The standards should be similar to the specimens being tested, so that the mechanism of the test is essentially duplicated.

The usual method of recording the highest, lowest, and average readings fails to give an adequate conception of the physical property under study when a material, such as rock, is tested. For example, in determining Scleroscope hardness, one sample may be essentially uniform with occasional stray inclusions or impurities that account for the low and high readings, whereas another, with the same average hardness, may be a heterogeneous mixture of constituents, in which the hardness ranges gradually from the same low to high value. It is important to know not only the size of the deviations from the average but also their frequency of occurrence because of the physical picture provided. The same information is useful for various statistical reasons.

Determination of Average and Standard Deviation

A useful and well-known method of determining the best value for a series of readings made under presumably identical conditions is to compute the arithmetic average as follows:

$$a = \frac{X_1+X_2+X_3+\ldots+X_n}{n};$$

where

X_n = nth reading,
n = number of readings.

One measure of the spread of the readings is the standard deviation, σ, determined from:

$$\sigma = \sqrt{\frac{(X_1-a)^2+(X_2-a)^2+(X_3-a)^2+\ldots+(X_n-a)^2}{n}};$$

or

$$\sigma = \sqrt{\frac{\Sigma d_a^2}{n}};$$

where

d_a = deviation of each reading from the average.

The per cent standard deviation is:

$$\sigma\text{, per cent} = \frac{\sigma}{a}\times 100.$$

In general, the average gives the best value of the characteristic being measured; the standard deviation shows the spread in data caused by random variations and by variations in the material or objects being tested; and the number of readings indicates the completeness and accuracy of the test when considered in conjunction with the standard deviation.

As these factors can be determined quickly and easily, yet yield considerable information, their computation and presentation have been made a part of the standard procedure for investigations of this type.

Determination of Significant Differences

When a series of readings is taken, occasional values are encountered that appear to have an abnormally large deviation from the mean. Accordingly, they may be suspected of containing a gross error or of having been made under some condition differing from that of the remainder of the test. One criterion that determines whether or not a suspected reading may be rejected is based on showing that the reading is not likely to be a random variation. The degree of likelihood, however, depends on the distribution of the data. Readings that deviate by more than three times the standard deviation may ordinarily be rejected, because such data will not exceed 11 per cent of the total data regardless of the distribution, and in the instance of a normal frequency distribution there is less than one chance in a hundred that a reading would fall outside the same range due to random variation.

A problem akin to the rejection of data is the question of presenting data. The question has been dealt with by the ASTM(14) and is quoted herewith because of its general applicability:

> Presentation of data presumes some intended use either by others or by the author as supporting evidence for his conclusions. The objective is to present that portion of the total information given by the original data that is believed to be essential for the intended use. ESSENTIAL INFORMATION will be described as follows: [quoting from Shewhart].(16) "We take data to answer specific questions. One can usually say that a set of statistics (functions) contains the essential information given by the data when, through the use of these statistics, we can answer the questions in such a way that further analysis of the data will not modify our answers to any practical extent."

When a series of readings is made on a group of specimens, an average can be computed. If a similar series of readings is made on a second group of specimens a second average will be obtained which may differ from the first. The question immediately arises as to whether or not the difference can be considered significant.

Although criteria for determining significant differences are well known(15) particularly in statistical work, they are not generally applicable to data in which the number of readings is small and the distribution (of readings about the average) unknown or irregular. Knowledge of the standard deviation does provide, nevertheless, a means of determining significant differences which is independent of the distribution of the data and which is useful for tests with a limited number of readings. The method is based on Tchebycheff's(16) theorem, which delimits the range within which a predetermined minimum percentage of the data will fall; that is, taking advantage of Tchebycheff's theorem, *at least* $100\left(1-\frac{1}{Z^2}\right)$ per cent of the observations will fall in the range:

$$a \pm Z\sigma,$$

where

$$Z > 1.$$

For example, let $Z = 2$, then *at least* $100\left(1-\frac{1}{Z^2}\right)$ per cent or 75 per cent of data will fall in the range,

$$a \pm 2\sigma,$$

or let $Z = 3$, then *at least* $100\left(1-\frac{1}{3^2}\right)$ per cent or 89 per cent of the data will fall in the range

$$a \pm 3\sigma,$$

regardless of the distribution of the data. This condition is shown graphically in Fig. 17 where the range about average (a) is shown.

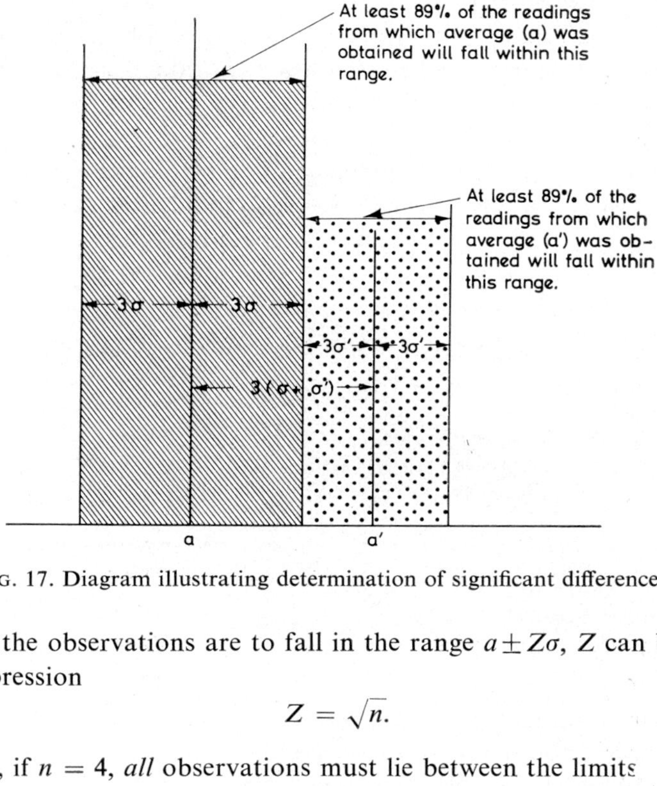

FIG. 17. Diagram illustrating determination of significant differences.[2]

Also if *all* the observations are to fall in the range $a \pm Z\sigma$, Z can be determined from the expression

$$Z = \sqrt{n}.$$

For example, if $n = 4$, *all* observations must lie between the limits

$$a \pm \sqrt{4}\sigma \qquad \text{or} \qquad a \pm 2\sigma.$$

To illustrate how this theorem can be applied to determine significant differences, assume two sets of data with averages of (a) and (a') and standard deviations σ and σ'. From Tchebycheff's theorem, of the readings which resulted in average (a) at least 89 per cent fall within the range $a \pm 3\sigma$, or not more than 11 per cent fall outside the range $a \pm 3\sigma$; similarly, of the readings from which (a') was derived not more than 11 per cent fall outside the range $a' \pm 3\sigma'$. Graphically this relationship is shown in Fig. 17.

If (a) and (a') are separated by an amount $3(\sigma+\sigma')$, as shown in the figure, *not more than* 11 per cent of the data contributing to average (a) can extend into the range including at least 89 per cent of the data associated with (a') and vice versa. Such a separation of data means that the difference approaches a certainty.

If (a) and (a') are separated by the amount of $2(\sigma+\sigma')$, not more than 25 per cent of the data contributing to average (a) can extend into the range including at least 75 per cent of the data associated with (a') and vice versa. This separation of data, while not as certain as with $3(\sigma+\sigma')$, still represents a strong probability of a real difference. For this investigation[(2)], $2(\sigma+\sigma')$ was adopted as the criterion for determination of significant differences. Increasing the magnitude of the constant by which the expression $(\sigma+\sigma')$ is multiplied increases the probability that the difference is significant. It should be noted also that as the number of readings decreases the worth of any statistical analysis decreases, and the necessity for experience and interpretation increases.

If the above method for determining a significant difference yields a result that is not significant, it is possible to utilize more precise methods of analysis which, if properly applied, may reveal significant differences below the limits set up on the basis of Tchebycheff's theorem. This might be desirable, for example, when a correlation in data is noted. Generally speaking, however, if the number of readings is small (say 10 to 20), application of Tchebycheff's theorem provides a quick, reliable method of determining significant differences.

MISCELLANEOUS FIELD AND LABORATORY METHODS
EVALUATION OF IN SITU STRESS DETERMINATIONS

In connection with the construction of an underground power house at Oroville Dam the California Department of Water Resources,[(17)] with the cooperation of the U.S. Bureau of Mines,[(18)] conducted a series of flatjack (flat hydraulic pressure cells) and borehole deformation tests to determine the natural (primitive) stresses existing in the rock at the site. (Procedures for standard flatjack stress determinations are described in Chapter 10 and procedures for borehole deformation stress determinations are described in Chapters 4 and 10.)

Flatjacks and borehole deformation gages were installed at sites near the end of an exploratory tunnel (with a minimum cross-section 5×7 ft) which was driven in the area of the proposed underground power house. Cover over the test sites was about 350 ft.

The rock at the site was a hard, dense, generally fine grained amphibolite with a specific gravity of about 2.94, a modulus of elasticity averaging 12.9 million psi and a compressive strength of about 40,000 psi. Tensile and cohesive (shear) strengths were determined to be about 3300 psi and 5400 psi, respectively.

Results obtained from flatjack installations were checked by measurements made by strain relief methods in which borehole deformation gages were installed at adjacent sites and rock strain was determined by the overcoring method.

Stresses as determined by the two methods checked within limits considered adequate for design purposes. The results, when corrected for stress concentration factors, indicated that the primitive vertical and horizontal stresses were approximately equal and that their magnitudes were about 500 psi.

Evaluation of Flatjack Performance

Where multiple reference pins are used for strain measurements for flatjack stress determinations it has generally been found to be impossible to obtain simultaneous cancellations of all strains. Therefore an average value must be used.

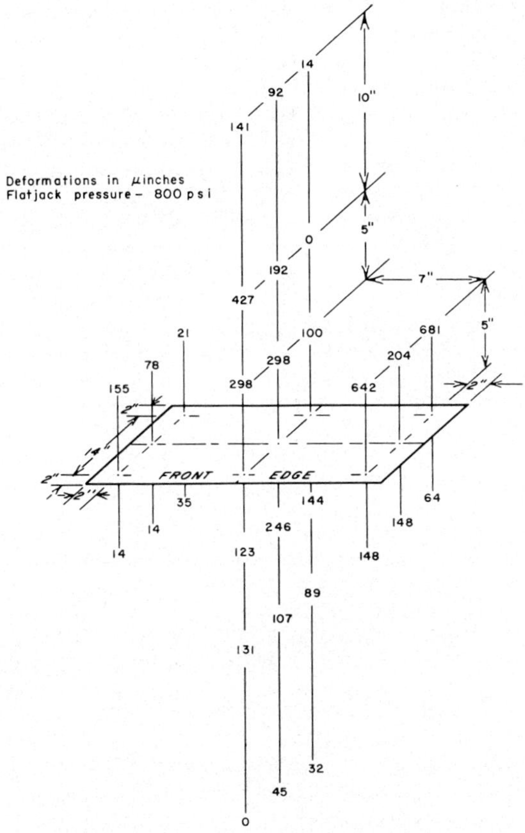

FIG. 18. Borehole deformation produced by flatjack pressure, flatjack 22.[17, 18]

As part of the test program an investigation was made of the effect of slotting the rock and then repressuring the slotted area with a flatjack as is normal procedure for flatjack stress determinations. To determine the effects of normal flatjack procedure holes were bored and borehole deformation gages were installed in patterns above and below the flatjack installations.

The borehole deformation measurements indicated a lack of stress symmetry around the flatjack. The deformations compared on opposite sides were in frequent disagreement and deformations were not necessarily largest in the holes nearest the flatjack. At some points, deformations were largest near the front of the flatjack and at other points they were largest near the center.

Figure 18 is an isometric drawing showing borehole deformation recorded at 30 points within the zone of influence of a horizontal flatjack when the pressure in the flatjack was 800 psi.

Although variations in the elastic modulus of the rock and proximity of the boreholes may account for some of this non-uniformity, the schistocity of the rock and irregularities in the surface of the opening are believed to be the principal reasons for the non-symmetrical stress distribution.

The accuracy of the flatjack method of stress measurement is not necessarily influenced by the non-symmetrical distribution of stress if the pin displacements created by the rock stress when the slot is cut are equal and opposite to those created by an equivalent pressure in the flatjack.

The use of a multiple reference pin pattern is considered to be a useful safeguard in averaging out the effects of non-symmetry of stress.

Laboratory Tests for Stress Concentrations

The foregoing tests were made in openings with rough interior surfaces and irregular configurations. Stress concentration factors for openings of regular geometric cross-section can be computed from equations (see Chapter 1, Part B); however, the determination of stress concentration factors at the boundary of an irregular opening requires special laboratory techniques.

Such determinations are commonly made by photoelastic methods, or by other methods which relate strain to optical interference patterns.

Figure 19 shows stress concentrations around the irregular opening at Test Site No. 1 (the location of the flatjack installation which is shown in Fig. 18), as determined by the Moiré method.[19]

The Moiré effect is an optical phenomenon produced when two similar arrays of dots or lines are superimposed, resulting in the formation of light and dark areas representing interference bands.

In this study a rectangular template was prepared from a sheet of transparent elastic material with smooth mirrored surfaces. A typical cross-section of the opening at the flatjack location was cut from the center of the template. The image of a grid of alternate black and white lines, as reflected by the template, was photographed by a camera mounted in the center of the grid. Double exposures of this image were made with the template both unstressed and with a uniaxially applied stress simulating rock load on the tunnel. The patterns thus produced and properly interpreted provide solutions of Airy's stress function.

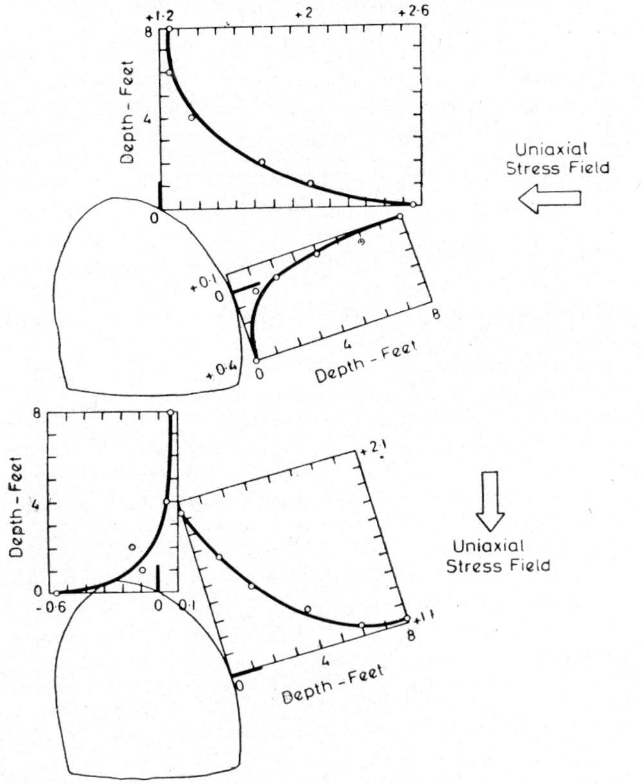

Stress Concentrations Determined by Moire Method
Negative Value Implies That Concentrated Stress Is Opposite in Sign to the Applied Load

Stress Concentrations Normal to Flatjack Planes Versus Depth from Edge of Opening — Site 1

FIG. 19. Stress concentrations as determined in models by the Moiré method. Negative value implies that the concentrated stress is opposite in sign to the applied load. Stress concentrations normal to the flatjack planes vs. depth from edge of opening — Site 1.[17]

A STRESS MONITORING STATION FOR LONGTIME SERVICE

Devices for measuring strain and deformation with magnitudes in the range of micro-inches commonly incorporate electric resistance strain gages. Such gages are satisfactory for short-term tests but are, by their nature, extremely sensitive to moisture and, no matter how well sealed or insulated, the failure rate is high when they are subjected to the moist conditions common to mine workings.

In tests conducted by the U.S. Bureau of Mines with such gages installed in a mine[20] it was found that the probability is that only 50 per cent of strain gages will survive for $13\frac{1}{2}$ months or longer and none of those installed remained reliable for more than 1140 days (see Fig. 20).

Experiments were conducted for the purposes of developing a flat hydraulic pressure cell (flatjack) which could be installed in a slot in typical mine rock and which would be sufficiently strong and yet sensitive enough to measure micro-inches of strain in the rock as changes in pressure which could be read on a Bourdon pressure gage.

Figure 21 shows cross-sections of three types of cells which were tried. The first cell (Fig. 21A) was found to have insufficient strength and was replaced by that of

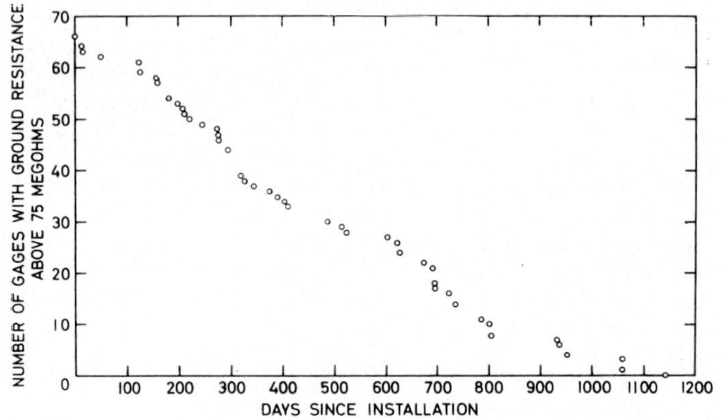

FIG. 20. Mortality of strain gages.[20]

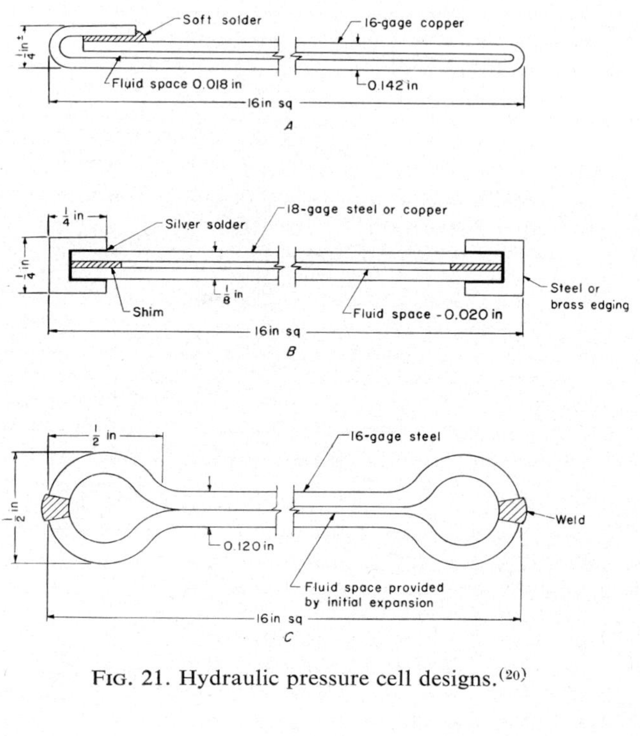

FIG. 21. Hydraulic pressure cell designs.[20]

Fig. 21B. However, it was found that this latter cell, when installed underground, ruptured at pressures which ranged from 480 to 4400 psi. In order to meet the requirement that such a cell should, when installed in the rock, be capable of resisting an internal pressure of 10,000 psi the cell shown in Fig. 21C was designed.

In a strength test in which this latter cell was cemented into a slot in diabase at a depth of about 1800 ft it was pressured to 8500 psi before failure. The strength of this cell is considered to be adequate for pressures up to 10,000 psi if it is properly installed in the rock.

The fluid thickness of these hydraulic cells was made about 1/100 of the width of the cells.

TABLE 5. REPEATED MEASUREMENT OF INDICATED VERTICAL COMPRESSIVE ROCK STRESS (PSI)

Borehole cell	Installed 3/60	Recycled 10/62	Recycled 6/63
1	2680	2250	2380
2	1780	1720	1890
3	1550	1210	1240

The stability of an array in which flat hydraulic pressure cells acted as the sensing devices was investigated by repeated determinations of rock stresses over a period of 3 years. Table 5 shows that the reproducibility of results relative to the known precision was acceptable.

Design for A Monitoring Station

Figure 22 shows an arrangement of elements for an all-hydraulic-pressure cell array which can be used a a permanent station for monitoring ground stress. The essential criteria for design of the array are (1) the pressure transmitter must occupy about 90 per cent of the slot width, and (2) the sensing device must be located in the zone of 100 per cent relative response, that is H/V must be greater than 1, where H is the horizontal distance, normal to the rib, from the edge of the sensor to the edge of the transmitter, and V is the vertical distance from the transmitter to the extremity of the sensor.

The array may be installed at any convenient distance from the rock surface of an underground opening.

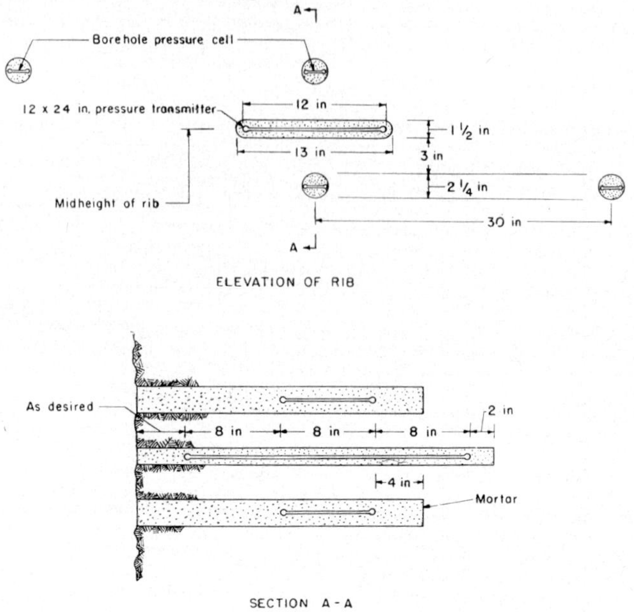

FIG. 22. Permanent ground-stress monitoring array. Borehole cells are installed in Lumnite or portland cement mortar. Pressure transmitter is installed in portland cement mortar.[20]

Installation Procedure

The installation procedure is as follows: First, the borehole cells are installed, preferably in lumnite mortar. (*Note:* Portland cement mortar is also satisfactory, but gypsum based mortars were found to be subject to excessive creep.) After the mortar has cured, the pressure in each cell is cycled two or three times to the estimated existing ground stress (or somewhat higher), and the valve is closed. When the pressures in the borehole cells have become stable, the flat slot is drilled out and the large hydraulic cell is installed in mortar.

After the mortar has cured, the pressure in the slot is cycled, to obtain the response graphs for the borehole pressure cells, from which the average indicated ground stress normal to the slot is determined. The slot pressure is raised approximately to the existing ground stress and the valve is closed to seal off the pressure cell.

Once the components of the array have been installed and stabilized in this manner, then, irrespective of subsequent changes in ground stress, the pressure change in the two central borehole cells (assuming for simplicity that they have identical responses) will always be equal to the average pressure change in the two outer borehole cells, *provided* the effective modulus of elasticity of the combined hydraulic pressure cell (with valve closed, acting as a sensor) and mortar in the flat slot is equal to the modulus of elasticity of the rock; the fluid pressure acting normal to the plane of the flat slot is then equal to the existing rock stress component acting normal to the slot.

On the other hand, even if the effective modulus of the hydraulic pressure cell and mortar does not match that of the rock, the hydraulic pressure cell in the slot can be brought to the correct pressure simply by pumping in or bleeding off fluid until the central borehole cells show the same pressure change as the average pressure change of the outer borehole cells. The slot pressure is then the same as if the effective modulus of the hydraulic cell and mortar were equal to the modulus of the rock; the slot pressure should therefore be equal to the existing ground stress. As the two central borehole cells will not, in general, have identical responses, the best estimate of existing ground stress is the average of the two values of indicated ground stress corresponding to the two central borehole cells.

Actually the fluid pressure in the flat slot need not be altered, because the change of slot pressure required to raise or lower the pressure in either central borehole cell by a given amount can be determined from the graph of borehole cell response to the pressure transmitter, which is obtained from the pressure cycling performed at the time of installation. That is, given the pressure readings of the four borehole cells and of the hydraulic cell in the slot, the new ground stress normal to the slot can be determined from the response graphs of the two central borehole cells.

The above procedures are independent of the effective moduli of the pressure cell and mortar in the slot and of the pressure cells and mortar in the boreholes.

If the ratio

$$C = \frac{\text{change of borehole cell pressure}}{\text{change of rock stress}}$$

is known for the rock in which the borehole cells are installed, then a separate estimate of ground stress values subsequent to the time of installation can be obtained by regarding the outer borehole pressure cells as devices to measure change of rock stress as follows:

New ground stress = Initial ground stress + Change of ground stress.
Change of ground stress = $(1/C) \times$ (Average pressure change in outer borehole cells since installation).

MODULUS OF ELASTICITY OF ROCK AS MEASURED IN THE LABORATORY AS COMPARED WITH MODULUS MEASURED IN THE FIELD

It has long been recognized that there are extensive differences between the physical properties of a rock mass as determined from small samples and from large-scale tests. In the design of pressure tunnels, shafts, and dams it it very important to be able to predict the modulus of elasticity of the rock mass against which the structure will bear.

As an example the underground penstocks for the Kemano Power plant[21] were designed with steel penstock liner thick enough to carry only part of the stress due to interior water pressure. The concrete and the rock surrounding the steel liner were depended upon to furnish resistance to expansion of the liner and thereby to carry the larger portion of water pressure.

Laboratory measurements of the compressive strength and modulus of elasticity of the rock (a fresh, sound, medium-grained quartz diorite) were made on samples cut from $2\frac{1}{8}$-in. diamond drill cores and the average modulus of elasticity for samples subjected to compressive stress was found to be about 7 million psi. Poisson's ratio at minimum stress was found to be 0.115 and at maximum stress 0.275. Compressive strength ranged around 24,000 psi, although one of the four samples failed at about 17,000 psi.

Methods were then considered for determining the modulus of elasticity of the rock mass *in situ*. This has been done frequently in European practice by constructing short sections of test tunnel which are lined with steel plate liner, backed by concrete, and are then subjected to internal water pressure and deformations noted.

At Kemano, in lieu of a test tunnel, a hollow sphere of $\frac{1}{2}$-in. steel plate, 10 ft in diameter, was concreted into an excavation extended from an exploratory tunnel. Extensive instrumentation was installed to measure deformations of the steel sphere and the surrounding rock mass.

The installation was tested with internal pressures up to 3600 psi without rupturing, and the modulus of elasticity of the rock mass was determined to be 2 million psi, a value which was less than $\frac{1}{3}$ of that measured on drill core specimens.

It was concluded, after making allowances for creep, that the value of the rock modulus would probably decrease to a minimum of about 1.94 million psi after 30 years of continuous load application of 3600 psi. (*Note:* the maximum test pressure, 3600 psi, was about 3 times the maximum operating pressure in the penstocks.)

The difference between elastic moduli of rock measured on small specimens and that measured *in situ* is generally ascribed to the fact that rock *in situ* contains minute cracks and fissures, as well as occasional streaks of softer rock, and the deformation measured in compression tests includes the movement necessary ot close the voids and to compress the softer rock.

Modulus of Elasticity Investigations for Dam Foundations

In designing arch dams which will exert high compressive stresses on the rock at the abutments, it is customary to drive exploration galleries into the rock and to carry on deformation tests in which a circular plate is jacked against the rock surface and the resultant rock deformations measured. The modulus of elasticity of the rock mass can be determined by using this data in appropriate equations.

Results of tests conducted for the foundation of the Karadj Dam (Iran)[22] will be described as an example.

The dam is founded on diorite which is fundamentally strong and has an elastic modulus of about 5 million psi as determined by tests on laboratory specimens. Field modulus tests were performed at 7 locations in the abutment rock over a vertical range of 110 ft. The range of the elastic modulus in sound rock varied from 2.05 million psi to about 2.61 million psi. A value of 2.33 million psi was used for dam design purposes.

The rock was cut by three well marked joint sets and some hydrothermal alteration had taken place along certain joint planes. It appeared that the difference between the laboratory and the field values for the elastic modulus was primarily due to closure of joints under applied compressive stress as well as to the lower moduli of some of the hydrothermally altered zones.

LABORATORY TESTS APPLYING TO THE LOAD-BEARING CAPACITY OF CYLINDERS OF GRANULAR MATERIAL

In Chapter 5, the general equation (equation (7)) for stresses in a horizontal plane around a circular opening in granular cohesionless material is given as:

$$p_i = \sigma_{r_R} \left(\frac{a}{R} \right)^{j-1} \tag{7}$$

where

p_i is the unit pressure which must be applied to the interior surface of the opening to maintain stability (e.g. support pressure)

σ_{r_R} is the radial exterior stress at distance R from the center of opening

a is the radius of the opening

R is the distance from the center of the opening at which radial exterior stress is applied.

$$j = \frac{1 + \sin \Phi}{1 - \sin \Phi}$$

Laboratory experiments were carried out on cylinders of sand in the apparatus illustrated in Fig. 23, in which a cylinder of sand, confined between two thin rubber membranes, was subjected to differential internal and external gas pressures, with the external pressure being increased until shear failure occurred in the sand cylinders.[23]

The results of these experiments indicated that the theoretical relationship expressed by equation (7) Chap. 5 correctly described the observed phenomena, but that the effective angle of internal friction, ϕ, of the sand was considerably higher than that observed from triaxial tests on the same soil at the same initial density.

Figure 24 illustrates the values of ϕ for various values of the void ratio as determined from triaxial tests for the sand and also shows the values of ϕ as calculated from the ratio p_0/p_i where $p_0 = \sigma_{r_R}$ in equation (7) Chap. 5.

It will be noted that the values as determined from cylinder tests are generally about 6 degrees higher than the values as determined from triaxial tests.

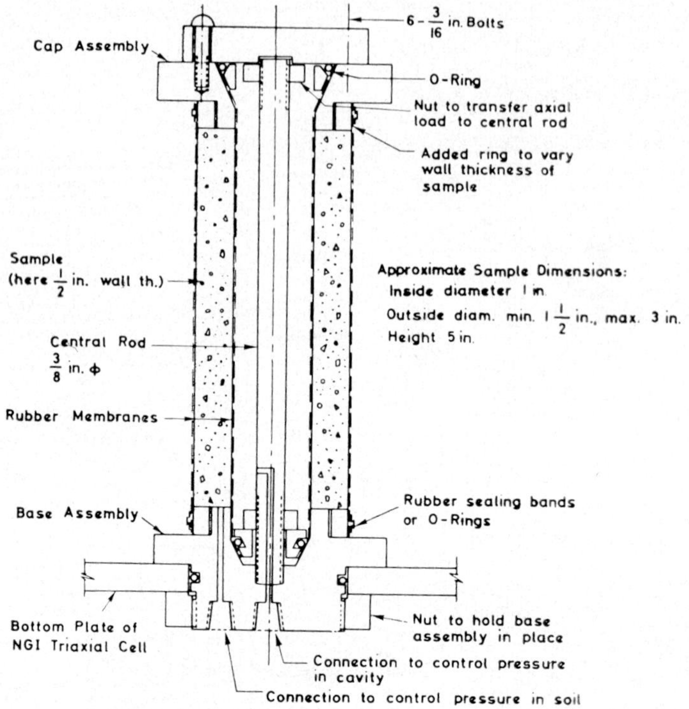

FIG. 23. Apparatus for soil cylinder tests.[23]

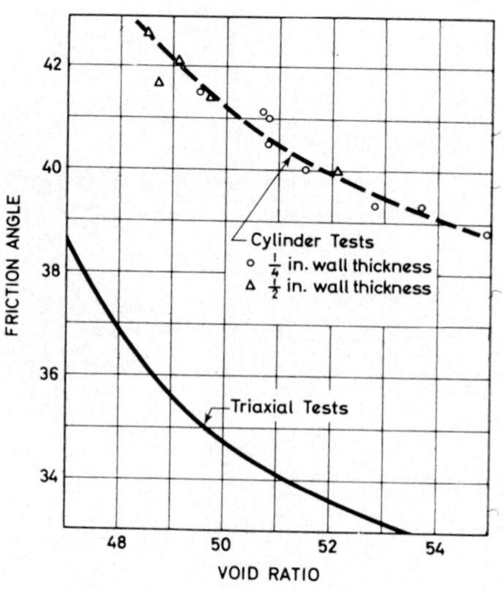

FIG. 24. Angles of internal friction plotted against void ratios for triaxial tests and for cylinder tests.[23]

The increased angle of internal friction, if employed in equation (7), Chap. 5 for calculating pressures on supports for openings in granular materials would give lesser values for support pressures. The use of the higher values of the angle of internal friction, ϕ, as determined by sand cylinder tests would give values for pressures on supports which would be 35 to 40 per cent less than if the values of ϕ determined by triaxial tests are used.

BIBLIOGRAPHY

1. HACKETT, P., Specimen preparation for rock mechanics research, *Mine and Quarry Engineering*, October 1963, pp. 438–441.
2. OBERT, L., WINDES, S. L., and DUVALL, W. I., Standardized tests for determining the physical properties of mine rock, *U.S. Bur. Mines R.I. 3891*, August 1946.
3. HOBBS, D. W., Rock compressive strength, *Colliery Engineering*, July 1964, pp. 287–292.
4. GROSVENOR, N. E., Specimen proportion — key to better compressive strength tests, *Mining Engineering*, Vol. 15, 1963, pp. 31–33.
5. POMEROY, C. D., A simple method for the assessment of coal strength, *Jour. Inst. Fuel*, Vol. 34, 1957, pp. 50–54.
6. POMEROY, C. D., Routine strength tests on coal, *Steel and Coal*, Vol. 186, 1963, pp. 640–645, 698–704.
7. HARDY, H. R., Jr., Standardized procedures for the determination of the physical properties of mine rock under short period uniaxial compression, *Canadian Department of Mines and Technical Surveys, Mines Branch Technical Bulletin TB 8*, December 1959.
8. GREENLAND, B. J., Rock mechanics, *Colliery Guardian*, September 21, 1961, pp. 346–348.
9. ZAHARY, G., A study of the strength characteristics of an andesite, *The Canadian Mining and Metallurgical Bulletin*, December, 1963 *(Trans. C.I.M.M.*, Vol. LXVI, 1963, pp. 380–382.)
10. FITZPATRICK, J., Biaxial device for determining the modulus of elasticity of stress-relief cores, *U.S. Bur. Mines R.I. 6128*, 1962.
11. HARDY, H. R., Jr., Concepts and measurements of the mechanical properties of rock, *Paper 64FM 40* presented at the annual meeting of the *A.I.M.E.*, New York, February 16–20, 1964.
12. OBERT, L., Triaxial method for determining the elastic constants of stress-relief core, *U.S. Bur. Mines R.I. 6490*, 1964.
13. OBERT, L., An inexpensive triaxial apparatus for testing mine, rock, *U.S. Bur. Mines R.I. 6332*, 1963.
14. Manual on Presentation of Data, including Supplements A and B; Sponsored by Committee E-1 on Methods of Testing: Reprinted April 1943, 73 pp. *American Society for Testing Materials* (ASTM).
15. CRUMPLER, T. B., and YOE, J. H., *Chemical Computations and Errors*, John Wiley, New York, 1940, p 184.
16. SHEWHART, W. A., *Economic Control of Quality Manufactured Products*, D. Van Nostrand, New York, 1931, p 95.
17. Static stress determinations at Oroville underground powerhouse, *California Department of Water Resources Rock Mechanics Report No. RM-4*, December 1963.
18. MERRILL, R. H., WILLIAMSON, J. V., ROPCHAN, D. M., and KRUSE, G. H., Stress determinations by flatjack and borehole-deformation methods, *U.S. Bur. Mines R.I. 6400*, 1964.
19. MORSE, S., DURELLI, A. J., and SCIAMMARELLA, C. A., Geometry of Moire stress fringes in strain analysis, *Trans. A.S.C.E.*, Vol. 126, Part 1, 1961.
20. PANEK, L. A., and STOCK, J. A., Development of a rock stress monitoring station based on the flat-slot method of measuring existing rock stress, *U.S. Bur. Mines R.I. 6537*, 1964.
21. LAWTON, F. L., and SUTHERLAND, J. G., Kemano pressure conduit engineering investigations, *Proc. A.S.C.E.*, Vol. 83, No. PO 5, October 1957, (Paper 1396)
22. WALDORF, W. A., VELTROP, J. A., and CURTIS, J. J., Foundation modulus tests for Karadj arch dam, *Proc. A.S.C.E.*, Vol. 89, No. SM 4, July 1963 (Paper 3576).
23. WHITMAN, R. V., and LUSCHER, U., Basic experiment into soil-structure interaction, *Proc. A.S.C.E.*, Vol. 88, No. SM 6, December 1962 (Paper 3366).

INDEX

MADE IN GREAT BRITAIN